ADVANCES IN COMMUNICATIONS

Advances in Communications

Volume I of a selection of papers from INFO II, the Second International Conference on Information Sciences and Systems, University of Patras, Greece, July 9-14, 1979

Edited by

D. G. LAINIOTIS
Dept. of Electrical Engineering,
State University of New York at Buffalo, Amherst, N.Y., U.S.A.

and

N. S. TZANNES
University of Patras, Chair of Information Theory,
School of Engineering, Patras, Greece

D. Reidel Publishing Company

Dordrecht : Holland / Boston : U.S.A. / London : England

Library of Congress Cataloging in Publication Data

International Conference on Information Sciences and Systems, 2d, University of Patras, 1979.

A selection of papers from INFO II, the Second International Conference on Information Sciences and Systems, University of Patras, Greece, July 9–14, 1979.

Includes indexes.

Contents: v. 1. Advances in communications. v. 2. Advances in control. 3. Applications of information and control systems.

1. Electronic data processing—Congresses. 2. Telecommunication—Congresses. 3. Automatic control—Congresses. 4. Pattern recognition systems—Congresses. I. Lainiotis, Demetrios G. II. Tzannes, Nicolaos, 1937– III. Title.

QA75.5.I58 1979 001.64 80–18203

ISBN 90–277–1140–2 (v. 1)
ISBN 90–277–1129–1 (v. 2)
ISBN 90–277–1143–7 (v. 3)

Published by D. Reidel Publishing Company
P.O. Box 17, 3300 AA Dordrecht, Holland

Sold and distributed in the U.S.A. and Canada
by Kluwer Boston Inc., Lincoln Building,
160 Old Derby Street, Hingham, MA 02043, U.S.A.

In all other countries, sold and distributed
by Kluwer Academic Publishers Group,
P.O. Box 322, 3300 AH Dordrecht, Holland

D. Reidel Publishing Company is a member of the Kluwer Group

Printed in The Netherlands

CONFERENCE COCHAIRMEN

D.G. Lainiotis   N.S. Tzannes

SPONSORED BY

The Ministry of Culture & Sciences of Greece
The University of Patras
The IEEE Automatic Control Society
The IEEE Information Theory Group
The IEEE Systems, Man and Cybernetics Society
The World Organization of General Systems and Cybernetics
The IEEE Greece
The Greek Society for Cybernetics and Systems

STEERING COMMITTEE

D.G. Lainiotis,	U.S.A.
N.S. Tzannes,	Greece
J. Cruz,	U.S.A.
G.F. Ferrate-Pascual,	Spain
K. Kerner,	Austria
J.L. Lions,	France
Y. Sunahara,	Japan
C. Tapiero,	Israel
C.P. Tsokos,	U.S.A.
M. Younis,	Egypt

TECHNICAL PROGRAM COMMITTEE

D.G. Lainiotis,	U.S.A.
J. Aguilar-Martin,	France
E. Baptista,	Brazil
A. Bensoussan,	France
E.J. Davison,	Canada
A. Ephremides,	U.S.A.
M. Gevers,	Belgium
H. Kwakernaak,	Holland
E. Mosca,	Italy
K. Najim,	Morocco
G. Saridis,	U.S.A.
P.D. Scott,	U.S.A.
C. Tapiero,	Israel
C.P. Tsokos,	U.S.A.
N.S. Tzannes,	Greece
M.I. Younis,	Egypt

LOCAL ARRANGEMENTS COMMITTEES

GREECE

A. Sissouras, Chairman
G. Megalokonomos
P. Panagiotopoulos

U.S.A.

C. Nikias, Chairman
E. Cosmidis
P. Groumpos
C. Holtz
S. Thomopoulos

CONFERENCE SECRETARIES

Mrs. Niki Sarantoglou, Greece
Mrs. Helen Alexandridis, Greece
Miss Ismini Kariofili, U.S.A.

THE CONFERENCE HAS BEEN AIDED FINANCIALLY BY

The Ministry of Culture & Sciences of Greece
The University of Patras
A.B.E.E.A. Co.
A.B.E.X. Co.
ACHAIA-CLAUSS Co.
Astir Hotel
BIOSOL Co.
DINAMIKI Co.
ELLENIT Co.
KAZA Co.
Kalogerakis Co.
Koutsadonis N.
Piraiki-Patraiki Co.
Pirelli Hellas S.A.
Titan Cement Co.

TABLE OF CONTENTS

PREFACE

This volume contains a selection of papers presented at INFO II, the Second International Conference on Information Sciences and Systems, held at the University of Patras, Greece, from July 9th to 14th, 1979. Additional selections of papers from this Conference are published in separate volumes, also by D. Reidel and are entitled "Advances in Control" and "Applications of Information and Control Systems", respectively.

INFO II was hosted by the University of Patras and sponsored by the Ministry of Culture & Sciences, IEEE Automatic Control Society, IEEE Information Theory Group, IEEE Systems, Man and Cybernetics Society, World Organization of General Systems and Cybernetics, IEEE Greece, and the Greek Society for Cybernetics and Systems. It was the second such Conference held at the University of Patras, the first one having taken place from August 19th to 24th, 1976, with its papers published by the Hemisphere Publishing Corporation in a three volume set, entitled "Applications and Research in Information Sciences and Systems".

INFO II, then, was the second high level Conference on these important topics in the Eastern Mediterranean Region. The reason that these conferences have been established is because we feel strongly that the topics covered have reached a high degree of maturity in this region, both from the point of view of theory and application. The applications areas are especially important to the Mediterranean region in view of the Common Market connections of such countries as France, Italy and Greece and the anticipated admission of Spain. In view of this, a major objective of the conferences is to focus attention on these important technical subjects and to facilitate the exchange of information and cross-fertilization of ideas between researchers of the Mediterranean region and those of the rest of the world, thereby promoting research and development by the region's engineers, researchers and academicians.

The Conference, keynoted by Prof. J.L. Lions, France, Prof. L.A. Zadeh, U.S.A., and Prof. M. Dertouzos, U.S.A., was highly successful as evidenced by the attendance of over 250 scientists and engineers from approximately 30 countries, and by the presentation of over 200 technical papers by well-known scientists and engineers from around the world.

D. G. Lainiotis and N. S. Tzannes (eds.), Advances in Communications, xiii–xiv.

The technical program of the Conference consisted of technical sessions that covered all-important aspects of control, information processing/communications, and computers. In particular, there were numerous technical sessions covering all-important aspects of control, such as control theory, control applications, estimation, identification, adaptive systems, linear systems, stability, cybernetics, computational methods, and simulation. The program also included many sessions on information processing/communications, such as information theory, coding, signal analysis, signal processing, communication theory, satellite communications, pattern recognition and image processing. Moreover, there were several sessions on computers, in particular on computer systems, computer communication networks, and automata. Finally, the technical program included numerous sessions on important applications of systems technology, such as power, energy modeling and planning, earth resources, transportation, economics and management, and physiological systems.

In view of the broad scope of the Conference technical program and the extensive coverage of many important aspects of systems theory and applications by internationally known researchers, we hope that this collection of papers will be a useful supplement to the published literature and textbooks used for research and teaching.

For the success of the Conference we are indebted to a great many people and institutions, primarily to the authors of the papers, without whom the conference would not have existed, and this book would not have materialized. We are particularly indebted to the great majority of them who paid some or all of their own expenses. We are also indebted to the institutions throughout the world that gave permission to the authors to attend and covered their expenses.

We are greatly indebted to the University of Patras for hosting the Conference, and for its financial support, as well as to the various other institutions and companies, whose financial support was essential for the success of the Conference. We would also like to thank the above mentioned scientific societies for adding their names as conference co-sponsors.

Finally, in a personal vein, we would like to thank the session chairmen, and the members of various conference committees, for their enthusiasm and dedication with which they worked in making INFO II a reality.

Nicolaos S. Tzannes　　　　Demetrios G. Lainiotis

SECTION I

INFORMATION THEORY,

CODING

AND

DIGITAL SIGNAL PROCESSING

GENERALIZED INFORMATION AND NATURAL (FUZZY) LANGUAGE

John F. Cyranski

Theoretical and Physical Chemistry Institute
The National Hellenic Research Foundation
48 Vassileos Constantinou Ave., Athens, GREECE

ABSTRACT: Can an information theory of arbitrary natural languages be constructed? I suggest how this may be done via construction of a ("fuzzy") logic cum semantics and the use of generalized informations (based on order). Included is a complete model for a fully interpreted scientific theory.

I. INTRODUCTION

Information theory is best known as cornerstone of statistical communication theory, based on the work of Shannon (1). More recently, information has been defined in terms of complexity (2). Neither of these approaches, however, is suitable for discussing the amount of information contained in a proposition, for example. For both approaches are concerned with structural (syntactical) aspects of communications and not their content. Bar-Hillel and Carnap (3) have attempted an approach to semantic information theory, based on Carnap's "degree of belief" model for probability (4). This restriction unfortunately limits the generality of the approach.

In recent years a new explication of "information" has been based on the pioneering study of Kampê de Fériet and Forte (5). Defining information most generally in terms of the ordering of statements (6) permits elimination of all reference to probability. Moreover, as logic consists of statements ordered by implication, we gain the possibility of defining semantic information on logics, and inevitably on languages.

D. G. Lainiotis and N. S. Tzannes (eds.), Advances in Communications, 3–10.

In this presentation I will show how one can generate the "logic" of a fuzzily defined object, how an abstract theory is developed, and the role of "semantics" in a "fully interpreted theory" for the objects. This will involve a discussion of the general notions of information, evaluations, and evidences. One can extend the description to the (finite) universe of observable (and conceivable) "objects" and properties thereof. I argue how such a description should be imbedded in a "language". I also cite a number of open problems and a few suggestions as to their solution.

II. THE "LOGIC" OF AN OBJECT AND INFORMATION

How do we go about defining an object? In essence, we determine by observation a set of properties relevant to the object. In physics, for example, an electron is defined by its rest mass, spin, charge, and the various values of position and velocity it may have. Properties are observed by experiments which classify the objects according to the various achievable manifestations: In effect, properties define relations among the objects, mappings $R:A^{n_R}\rightarrow\{0,I\}$ (classically) where A is the class of all the objects. Such binary-valued relations are essentially characteristic functions. From the viewpoint of logic, $R(x_1,\ldots x_{n_R})$ has truth value I if and only if $(x_1,\ldots x_{n_R}) \in A^{n}R$. The natural extension based on the work of Zadeh (7) and Goguen (8) is to define $R:A^{n}R\rightarrow Q$, where Q is a quoset (a set with a quasi-order $\prec$: $q \prec q$ and $q \prec r$, $r \prec s$ implies $q \prec s$). If $F=\{R:A^{n}R\rightarrow Q\}$ is the class of all relevant empirical (n_R finite) relations on A, then Q induces a quasi-order $\Rightarrow$ on F in a very natural way (9). Because $R\Rightarrow S$ and $S\Rightarrow R$ does not mean $R=S$, we define P to be the poset (partially ordered set) obtained from F by identifying such equivalent relations. We take P as the abstract empirical definition of the object. The original construct $\langle A,F,Q\rangle$ is an empirical relational system which will be associated with the semantics of P.

Of course, what I have just described is an impractical prescription. In practice one chooses P and then tests whether it adequately describes what is observed. Generally we do not observe all the manifestations of all the properties of an object. Rather, we sample various derived properties. We then compare the observations with what our model of the object predicts, and modify the model as needed. In order to obtain sufficient mathematical power, we must extend the poset P--which corresponds, eg., to the rational numbers--to the complete lattice L(P)--which corresponds to the reals. A natural imbedding exists that corresponds precisely to this Dedekind completion of the rationals "by cuts" to the reals. L(P) has particular physical significance (9) which lends weight to its interpretation as the idealized definition of the objects. The complete lattice L(P) is an ordered set ($\subseteq$) in which every set $X\subseteq L(P)$ has a greatest lower bound ($\bigcap_{x\in X} x$) and least upper bound

($\bigvee_{x\epsilon X} x$) with respect to the ordering. Information enters the story as follows:

Let us consider, for the moment, a logic of propositions ordered by implication. Let 0 be the "absurd" proposition which is always false, and let I be the "trivial" tautology--a statement whose truth is implied by all propositions in the logic. These statements are thus universal bounds to the poset of propositions. What should an information measure on such a structure satisfy? Certainly, if $A \lneq B$, a meaningful information J should assign greater value to A than to B, since evidence defining the "truth" of A automatically fixes the truth of B. Thus, if we assume $J:P\rightarrow[0,\infty]$, (i) $A \leq B$ implies $J(A) \geq J(B)$. Since I is always "true", there is no information involved with I, so we assume (ii) $J(I)=0$. Finally, since 0 is "absurd", its information is undefined: (iii)$J(0)=\infty$. These are the basic properties that define the most general concept of information. They were first proposed by Sallantin in 1972 (6).

Of concern to us is the structure of Z(P), the family of all informations defined on a poset (lattice) P with universal bounds. We proved (10) that Z(P) is a complete, convex, Brouwerian lattice and is metrizable. Every complete automorphism on Z(P) is a homeomorphism in the order topology. Each $\mu \in Aut(P)$ induces a $\gamma \in Aut(Z(P))$ via $\gamma(J(A))=J(\mu(A))$, $A\epsilon P$. Detailed connections between probability theory and this generalized information theory are given in ref. 9. We point out only that if p is a probability on a Boolean algebra, eg., then $J(A)=-\log p(A)$ defines an information.

III. FULLY INTERPRETED THEORIES (FITS)

Once we have defined our object abstractly, the obvious question is: Does our object definition fix only the original objects? The answer is the same as in formal logic: Many "interpretations" of P are possible, and the class $M(P)=\{\langle A,F,Q\rangle : P$ is order isomorphic to $\bar{P}$ and $\bar{P}$ is obtained from $\langle A,F,Q\rangle\}$ can be called the semantic space or space of "meanings" (interpretations) of P. Obviously, when we perform experiments on the objects defined by P, we are automatically selecting a particular interpretation, $\mathfrak{a}=\langle A,F,Q\rangle$.

We next ask: What do we typically observe in a scientific measurement on an object? Numbers! I.e., measurements concerning P are in terms of numerical functions X on A. (Such a coding is always possible and--by common prejudice--essential for a scientific experiment.) Let $B_X = \{\Delta:R^1\rightarrow Q'(X)\}$ define the "fuzzy" relations associated with the "real-life" observation of X. If Q'(X) can be imbedded in Q, then for all $\Delta \in B_X$, $\Delta X \in F$ and we say that X is a scale. Every scale X imbeds B_X in L(P) via $\hat{x}(\Delta)=\vartheta(\Delta X)$, where ϑ

imbeds P in L(P) under completion. *Observables* are scales satisfying a further technicality related to how $\hat{x}(\Delta)$ transforms under transformations of R^1 (9). Let $\Gamma(\mathcal{A})$ be the family of all observables associated with the model $\mathcal{A}$. Observables thus depend on the interpretation.

The next question we ask is: How is evidence obtained from experiment and used to calculate and predict in the theory? Indeed, what *is* evidence? From our viewpoint, *evidence* is any data which either serves to define a class of appropriate information functions or else serves to restrict this class.

The first evidence we require, B_o, is the definition P of the object class we are studying. This fixes L(P) and thus Z(L(P)). All other evidence serves to restrict informations in Z(L(P)). In particular, we may have a priori reasons (B_1) for considering only a sub-class $Z_o \subset Z(L(P))$--such as those informations associated with probabilities (when applicable!)

Let us suppose, for a moment, that we have sufficient (posterior) evidence e to single out some $J_e \varepsilon Z_o$. That is, J_e represents the information content of e. The concern of science is to evaluate any observable X that is well-defined for the model of interest. The evaluation of X will depend on $\hat{x}$ (in L(P)) and on J_e. I.e., the evaluation is given by $e(X)=G(\hat{x},J_e)$, where the *evaluation procedure* G is to be specified as prior evidence B_2. (See ref. 9 for examples.) In effect, once P and $\hat{x}$ are specified, only the information content $J_e(\hat{x})$ about X given e is relevant to the evaluation.

Once we have fixed the evaluation algorithm, it remains to discover how empirical evidence defines J_e. In fact, the laboratory determines not a single information, but rather a subset $Q \subseteq Z_o$ in general. (It is well-known, eg., that "evidence" in statistical physics consists of the set of probabilities over micro-states consistent with observed expectation values of a few functions like energy (11).) Note that the evidences depend on the model, $\mathcal{A}$. We thus have $K(\mathcal{A})=\{Q \subseteq Z_o\}$ as the set of possible evidences available from the lab. It is assumed that $K(\mathcal{A})$ is in 1-1 correspondence with experiments on A. We require *some* map $\alpha: K(\mathcal{A}) \rightarrow Z_o$ which fixes an unique information $\alpha(Q)$ depending on the evidence $Q \varepsilon K(\mathcal{A})$. (The Maximum Entropy Principle (11) is such a procedure, in a probabilistic context.) It is desirable that the inference map α not depend on $\mathcal{A}$, so we include its specification as part of the *abstract theory*, $T=(P,G,\alpha)$. Thus, given evidence $Q \varepsilon K(\mathcal{A})$, we may unambiguously evaluate any observable $X \varepsilon \Gamma(\mathcal{A})$ via the formula $e(X)=G(\hat{x},\alpha(Q))$. We define a *fully interpreted theory* (FIT) to be the class $\langle (P,G,\alpha);\ \Gamma(\mathcal{A}),\ K(\mathcal{A}),\ \mathcal{A} \varepsilon M(P) \rangle$.

IV. OPEN PROBLEMS FOR OPEN MINDS

In our discussion we have exploited what is probably the most important feature of logic--the ordering of propositions according to "implication". This feature is admirably suited for the discussion of information (in the general sense). Our approach is "bottoms up". We start with a particular model, generate the "logic" P and then return to semantics. The problem that confounds this natural approach is that P (and L(P)) does not have all the formal properties of a logic. First, statements as "R implies S" are not generally elements of P or L(P). This ailment is cured in ref. 12 when P (L(P)) is an orthomodular ortholattice (eg., Brouwerian or Boolean): $(R \rightarrow S) = R^{\perp} \vee (R \wedge S)$. One must generalize the conditions contained in this paper. Second, quantifiers do not seem to exist in P (L(P)). The remedy has been suggested by Halmos (13): Quantifiers and substitution are described as *mappings* in an algebraic model of Boolean logic, so that the information associated with quantification and substitution can be studied in terms of the mappings induced in L(P) and Z(L(P)). It is necessary to extend this description to non-Boolean lattices.

Another problem that arises is the choice of ordering the evidences $K(\mathfrak{a})$ and Q in each $K(\mathfrak{a})$. As Q are subsets of Z(L(P)), perhaps the partial order of set inclusion is most appropriate. This question is (or ought to be) mixed up with the problem of assigning $K(\mathfrak{a})$ to $\mathfrak{a} \varepsilon M(P)$ and Q to $K(\mathfrak{a})$ corresponding to experiments. As each problem has several alternative solutions, one must investigate these various possibilities with the goal of developing a general viable theory. This is--from our viewpoint--a central question in the study of semantics. For alternatives, see refs. 17,18.

Finally--as we have noted in refs. 10,14--our inference procedures differ from Bayesian concepts in that the hypotheses and evidences are distinguished explicitly and cannot be mixed in the sense of "H&e"--conjunction in some "language". We have investigated the idea of relating evidence and hypotheses via Z(L(PxK))--assuming a partial order can be defined on K. Surprisingly, most of Bayesian theory remains (in the probability case, of course) with the basic change that "H&e" be treated as the ordered pair (H,e) in L(PxK)= L(P)xL(K) (10). Nevertheless, "prior evidences" defining these structures are not in L(PxK), so further extension, say to $L(PxK)x\hat{L}$ is required. But, evidences defining *this* structure are metastatements not in this structure. And so on...

Nevertheless, language appears to evade these (infinite regress) difficulties. This suggests that the formal statements in P and $K(\mathfrak{a})$ must be somehow expressed in a common language after all.

V. LOGIC VS. LANGUAGE

In this section I would like to speculate about how logic relates to language. My ideas seem quite distinct from previous attempts (15-18).

To begin with, we can extend the description of one object class to any (finite) number of object classes. If two objects are non-interacting, a poset product PxP' defines a poset for the "universe" consisting of these objects. More generally, two objects may combine to form a new object. Eg., the composite system called a "family", composed of interacting male(P_m) and female(P_f) adults, requires additional descriptive relations, namely those defining subsystems called children. Thus, $P_m \otimes P_f \subseteq P_{fam}$. The "tensor product" $\otimes$ is defined so that the component systems preserve their identity in the interaction (9).

Thus, in principle, at least, one can describe the observable universe in terms of some poset P consisting of all component objects, P_n. Likewise, one can consider M(P) and evidence classes $K(\mathfrak{a})$, $\mathfrak{a} \epsilon M(P)$--i.e., a FIT applicable to the universe. Needless to add, such a description is not available, and without doubt, many would believe that it is impossible for any number of reasons. It is not our point, however, to discuss feasible theories. Rather, our goal is to better understand the inter-connections among logic, observation (i.e. "real-life") and language.

Now language is defined most generally as a set of sentences (finite length strings composed of symbols from an alphabet) over an alphabet (15). Languages are structured by grammars, which serve to identify allowable sentences. Formal linguistics is divorced from the semantics of the symbols. Formal logic is, in a sense, an example of a language. Logicians are interested in logical truth relationships ("intensional semantics"(16)) as well as formal syntax. However, it is only in model theory--the interpretation of a logic by specifying a relational system $\langle A,F,Q \rangle$ -- that semantics comes into being.

Let me now conjecture as to how all these ideas relate. First, our universe P of objects must be describable in the language Λ. Thus, P is somehow imbedded in Λ. Moreover, M(P) must also be imbedded in Λ, as the models correspond to specific evidence statements. The imbedding must preserve the ordering of relationships!! I.e., the language must reflect the logical structure it contains. Obviously, not all statements in a language are logical. In fact, questions, commands, requests, directions, doubts, wishes, etc. are not logical statements: They have no intrinsic truth value. Rather, they are transformations (defined by the grammar) of declarative statements. Note that declarations are, by nature, logical propositions.

We therefore are led to conjecture that our logic cum semantics corresponds to $\Lambda_o \subset \Lambda$, where Λ_o is the class of declarative statements in Λ; Λ_o inherits the logical order of P and M(P) and is thus at least quasi-ordered. (As the ordering in M(P) is still uncertain, we cannot be more specific.)

One can thus define information theory on the complete lattice generated by Λ_o--L_o. Let Λ_i, i=1,2,...N, denote language structures like commands, questions, wishes, etc. and let $t_i:\Lambda_o \rightarrow \Lambda_i$ denote the grammatical transformation that turns a declarative statement into a question, command, etc. (Eg., λ_o="I go to the party.", $t_1(\lambda_o)$="Do I go to the party?", $t_2(\lambda_o)$="Let me go to the party.", etc.) Apparently each Λ_i is in 1-1 correspondence with Λ_o. Order can be induced on Λ_i by assuming that t_i preserves the order of Λ_o. A possible model for L, the complete lattice generated by Λ, is the product $\times_{i=0}^{N} L_i$, where L_i is generated by Λ_i. Informations in $Z(\times_{i=0}^{N} L_i)$ can be related to the informations in $Z(L_i)$ in an obvious way, and each $Z(L_i)$ can be defined in terms of $Z(L_o)$. Thus an "information theory" can be assigned to L once it is known for L_o.

VII. CONCLUSIONS

In this all-too-brief survey, I have tried to describe a new approach to semantical information theory that takes into account "real life" as well. Needless to add, the discussion has been rather speculative at times. To some extent I have obtained relevant theorems which tend to substantiate this approach, but space does not permit their inclusion. I hope that I have conveyed, at least, the potential of my approach and perhaps enthused some of you about the interesting problems that can be formulated--maybe even solved--from this viewpoint.

ACKNOWLEDGEMENTS

This work was done under a grant from the National Hellenic Research Foundation. I wish also to acknowledge the hospitality of the Theoretical and Physical Chemistry Institute (director Prof. C.A. Nicolaides).

REFERENCES

1. C.E. Shannon, Bell Syst. Tech. J. 27 (1948), p. 379.
2. G.J. Chaitin, J. Assoc. Comp. Mach. 22 (1975), p. 329: and refs. therein.
3. Y. Bar-Hillel and R. Carnap, Chap. 15 in Y. Bar-Hillel, "Language and Information" (1964), Addison-Wesley.
4. R. Carnap, "Logical Foundations of Probability" (1950), Univ. Chicago Press.
5. J. Kampé de Fériet and B. Forte, C.R. Acad. Sci. Paris A 265 (1967), pp. 110, 142, 350.
6. J. Sallantin, C.R. Acad. Sci. Paris A 274 (1972), p. 986.
7. A. Kaufmann, "Introduction to the Theory of Fuzzy Subsets" (1975), Academic.
8. J.A. Goguen, J. Math. Anal. Appl. 18 (1967), p. 145.
9. J.F. Cyranski, Found. Phys. 9 (1979), p. 641.
10. J.F. Cyranski, Inform. Contr. 41 (1979), p. 275.
11. E.T. Jaynes, Phys. Rev. 106 (1957), p. 620; 108 (1957), p. 171.
12. G.M. Hardegree, Z. Naturforsch. 30a (1975), p. 1347.
13. P. Halmos, "Algebraic Logic" (1962), Chelsea.
14. J.F. Cyranski, in "Théorie de l'Information", ed. C.F. Picard (1978), p. 431, Paris.
15. J.E. Hop and J.D. Ullman, "Formal Languages and their Relation to Automata" (1969), Addison-Wesley.
16. Y. Bar-Hillel, "Language and Information" (1964), chap. 2, Addison-Wesley.
17. H.C. Longuet-Higgins, Proc. Roy. Soc. Lond. B 182 (1972), p. 255.
18. G. Jumarie, Ann. Telecommunic. 33 (1978), p. 13.

DISTORTION-RATE THEORY AND FILTERING[1]

R. S. Bucy

University of Southern California,
Los Angeles, California 90007

ABSTRACT

The constraint of realizability is introduced and with it the Distortion Rate Function is redefined. For the Orenstein Uhlenbeck process, the constrained Distortion Rate function is shown to satisfy a dynamic programming equation. Further, a filtering problem is shown to be equivalent to determination of the constrained distortion rate function.

Introduction

The theories of Information and Filtering have in the past developed along independent paths motivated by different problems and imposing different assumptions. Recently, in [1], a method of bridging the gap between the two disciplines has been suggested. In [2], Rate Distortion theory was used to provide quite general lower bounds on the filtering error based on the Shannon Rate Distortion theory. These bounds have been shown to be rather pessimistic in the context of a simple problem, i.e., that of the cubic sensor; see [3]. In order to improve these bounds, the definition of the distortion rate function can be modified to include the condition of realiz-

[1]This research was supported in part by the United States Air Force, AFC under Grant #AF-AFOSB-76-3100.

D. G. Lainiotis and N. S. Tzannes (eds.), Advances in Communications, 11–15.

ability, i.e., the Constrained Distortion Rate Function is the minimum mean square error when the estimate is dependent on the future of the process and the joint information is fixed.

Not only does the theory provide better nonlinear filtering bounds but also the estimate which achieves the minimum Distortion Rate is given by the optimal filter for a noise-corrupted signal. In fact, the filter is the infinite lag filter in the unconstrained case and the ordinary filter in the realizability constrained case. Further channels which realize the Distortion Rate are easily described. Here, we restrict our attention to a single process, the Orenstein-Uhlenbeck process.

1. THE CONSTRAINED DISTORTION RATE FUNCTION

We consider the Orenstein-Uhlenbeck process

$$X_n = a\, x_{n-1} + q^{\frac{1}{2}}\, u_{n-1} \tag{1.0}$$

with u_{n-1} a unit level Gaussian white noise process, x_o independent of $u_{\cdot}$ and Gaussian of zero mean and variance ℓ. At the same time consider an observation process

$$z_n = x_n + \nu_n$$

with ν_n a zero mean Gaussian white noise process of variance $1/\ _n$, independent of x_o and u_i. The following variational problem is of interest; given I_o choose $\sigma_1, \ldots, \sigma_n$, so that A_n:

$$I(x_1, \ldots, x_n ; z_1, z_2, \ldots, z_n) = I_o$$

and

$$D_n(I_o, \ell) = \sum_{i=1}^{n} E[x_i - Ex_i \,|\, z_i\ z_{i-1} \ldots z_1)]^2 = \text{minimum} \tag{1.1}$$

It is shown in [5] that this A_n problem is the same as the seemingly more general variational problem; given I_o choose $y_i, \ldots, y_n$ such that

B_n: 1) y_i is realizable (i.e., is independent of $x_k - Ex_k | x_{k-1} \ldots x_1$) for all $k \geq j$.

2) $I(x_1, \ldots x_n; y_1, \ldots, y_n) = I_o$

3) $\sum_{i=1}^{n} E(x_i - y_i)^2 =$ minimum

It is convenient to repose A_n in order to make progress as

$$I(x_1, x_2, \ldots, x_n; z_1, z_2, \ldots, z_n) = \frac{1}{2} \ln \prod_{i=1}^{n} (1+p_i \sigma_i) \qquad (1.2)$$

with

$$p_{i+1} = a^2 \sigma_i + q$$

$$S_i = \frac{1}{p_i^{-1} + \sigma_i} \qquad (1.3)$$

$$p_i = \ell$$

$$D_n = \sum_{i=1}^{n} S_i \qquad (1.4)$$

In the reformulation of problem A_n, we see that its solution involves the solution of a filtering problem where the noise level is varied, the σ_i are to be chosen. In particular, the minimum distortion is achieved by choosing the y_i as the outputs of an Optimal Filter and p_i are the optimal, one-step prediction error, while S_i is the optimal filter estimate. Notice that in the unconstrained problem the filter in question is the infinite **lag** filter, while in the constrained case the y_i are the filter based on the past and present.

In fact, it is known - see [6] -

$$D_1(I, \ell) = \begin{cases} \ell e^{-2I} & I \geq 0 \\ \ell & I < 0 \end{cases}$$

and using Dynamic Programming

$$D_n(I, \ell) = \min_{\sigma_i} \left\{ \frac{1}{1/\ell + \sigma_1} + D_{n-1}\left(I - \frac{1}{2} \ln(1 + \ell \sigma_1), \frac{a^2}{1/\ell + \sigma_1} + q\right) \right\} \qquad (1.5)$$

One can iterate (1.5) to obtain $D_2(I, \ell)$ as

$$D_2(I,\ell) = \begin{cases} \ell\, a^2\, e^{-2I} + 2\sqrt{\ell q}\;\; e^{-I} & \text{for } I \geq \frac{1}{2}\, \ell n\, \frac{\ell}{q} \\ q + (1+a^2)\ell e^{-2I} & \text{otherwise} \end{cases} \tag{1.6}$$

In fact, (1.6) can be compared with the unconstrained Distortion Rate Theory; see [5]. Finding D_3 involves the solution of a cubic and does not seem to be analytically tractable.

The most interesting aspect of the Distortion Rate Theory is the description of the channel which achieves the minimum; see [5], as an optimal filter.

2. CONCLUSIONS

In this paper, we have reposed one of the fundamental problems of information theory which not only reveals a connection with filtering theory but also is useful in deriving more precise bounds for the optimal filtering problem. Although we have considered only the Orenstein-Uhlenbeck process here, it is clear that our methods generalize. The dynamic programming equation (1.5) is central to the complete understanding of the connections between filtering and information theories.

References

[1] A. Glados, "Information and Distortion in Filtering Theory", Thesis, E.E. Dept., M.I.T., January 1975.

[2] M. Zakai and J. Ziv, "Lower and Upper Bounds on the Optimal Filtering Error of Certain Diffusion Processes," I.E.E.E. Trans. Inf. Theory, IT-18, 1972.

[3] R. S. Bucy, J. Pages, "A Priori Bounds for the Cubic Sensor Problem," AC-23, 1, 1978, 88-91.

[4] R. S. Bucy, "Filtering and Information", Information Sciences, 17, 1979, 179-187.

[5] R. S. Bucy, "Rate Distortion Theory and Filtering", submitted to Information and Control.

[6] R. G. Gallager, "Information Theory and Reliable Communications", Wiley, New York, 1968.

ASYMPTOTICALLY MEAN STATIONARY SOURCES AND CHANNELS[1]

Robert M. Gray, John C. Kieffer, and Robert J. Fontana

Stanford University, University of Missouri, and
Carnegie-Mellon University

ABSTRACT

Several properties of information sources and channels that are asymptotically mean stationary (a.m.s.) are described and discussed.

INTRODUCTION

In this paper we summarize without proof several recent results for asymptotically mean stationary sources and channels. These are sources and channels which may be neither stationary nor block stationary, but which converge to a stationary source or channel in an arithmetic or Cèsaro mean sense. In particular, such models have ergodic theorems and Shannon-McMillan and Shannon-McMillan-Brieman Theorems and their properties can always be determined from an appropriately defined stationary mean. The mathematical details and proofs of these results along with several examples may be found in Gray and Kieffer (1980) and Fontana, Gray and Kieffer (1980).

SOURCES

In information theory the mathematical model of a source is usually a one-sided random process $\{X_n\}_{n=0}^{\infty}$ or a two-sided random process $\{X_n\}_{n=-\infty}^{\infty}$ defined on some probability space $(\Omega,\mathfrak{F},\mu)$. Hence when treating processes we will focus on the appropriate

[1]This research was supported by the U.S. National Science Foundation.

D. G. Lainiotis and N. S. Tzannes (eds.), Advances in Communications, 16–21.

sequence space with the distribution rather than the underlying probability space and T will be the shift. The basic coding theorems of information theory are applications of the ergodic and Shannon-McMillan Theorems and it is customary in the information literature to assume that the process distribution μ is stationary ($\mu(T^{-1}F) = \mu(F)$, all $F \in \mathcal{F}$) in order to invoke these results. This typically involves placing restrictions on the allowed sources, coders, and channels in order to ensure that all sources arising within a communications system are stationary. Several important models arising in information theory are not stationary. Examples are the outputs of codes mapping blocks of a source into encoded blocks of fixed or variable length, the output of a finite state machine driven by a stationary one-sided source, and a stationary source when given side information in the form of conditioning.

More general conditions under which the ergodic and Shannon-McMillan Theorems hold under additional assumptions on T have appeared in the literature. Dowker (1951) showed that if T is invertible and nonsingular (T is nonsingular if $\mu T^{-1} \ll \mu$) then the ergodic theorem holds if and only if

$$\lim_{n \to \infty} n^{-1} \sum_{i=0}^{n-1} \mu(T^{-i}F) \quad \text{exists, all} \quad F \in \mathcal{F}, \tag{1}$$

in which case it follows from the Vitali-Hahn-Saks Theorem that

$$\bar{\mu}(F) = \lim_{n \to \infty} n^{-1} \sum_{i=0}^{n-1} \mu(T^{-1}F), \quad F \in \mathcal{F}, \tag{2}$$

is a probability measure. Since $\bar{\mu}$ is obviously stationary, (2) can be interpreted as saying that the measures μT^{-n} converge to the stationary measure $\bar{\mu}$ in an arithmetic mean or Cèsaro mean sense. Hence we define a probability measure μ to be <u>asymptotically mean stationary</u> with respect to T if (1) holds and we call $\bar{\mu}$ stationary mean of μ. If T is clear from context we merely say that μ is a.m.s. A process is said to be a.m.s. if its distribution is a.m.s. If T is invertible, it follows easily that $\bar{\mu} \gg \mu$ and that μ is a.m.s. iff there exists a stationary measure η such that $\eta \gg \mu$.

Rechard (1956) generalized these ideas to nonsingular transformations that need not be invertible (such as the one-sided shift). It follows from his results that the ergodic theorem is satisfied iff μ is a.m.s. It need no longer be true, however, that $\bar{\mu} \gg \mu$. Rechard replaced this condition by an asymptotic form; a measure η on $\mathfrak{F}$ is said to asymptotically dominate a measure μ on $\mathfrak{F}$ (with respect to T) if $F \in \mathfrak{F}$ and

$\eta(F) = 0$ implies that $\lim_{n\to\infty} \mu(T^{-n}F) = 0$.

The Shannon-McMillan Theorem has not been shown to hold under conditions as general as those for the ergodic theorem. Jacobs (1959) proved that if η satisfies the Shannon-McMillan Theorem for T and $\eta \gg \mu$, then μ also satisfies the theorem. This immediately implies that if μ is a.m.s. and T is invertible, then the Shannon-McMillan Theorem holds for T since $\mu \ll \bar{\mu}$ and $\bar{\mu}$ is stationary. If T is not invertible, however, then $\bar{\mu}$ only asymptotically dominates μ and the Shannon-McMillan Theorem has not previously been proved for this case.

THE ERGODIC THEOREM

Let $(\Omega, \mathfrak{F})$ be a measurable space and $T: \Omega \to \Omega$ a measurable transformation. If μ is a.m.s. with stationary mean, then it is clear from (2) that $\bar{\mu}(F) = \mu(F)$ for every T-invariant set $F \in \mathfrak{F}$. This observation yields

Theorem 1: Let μ be a probability measure on $\mathfrak{F}$. Then μ is a.m.s. if and only if for every bounded measurable $f: \Omega \to (-\infty, \infty)$ $\{n^{-1} \sum_{i=0}^{n-1} fT^i\}$ converges a.e. $[\mu]$ as $n \to \infty$.

It is sometimes difficult to determine if a measure μ on $\mathfrak{F}$ is a.m.s. by using the definition. The following theorems provide some alternate characterizations.

Theorem 2: If μ, η are probability measures on $\mathfrak{F}$, where η is stationary and asymptotically dominates μ, then μ is a.m.s.

Define $\mathfrak{F}_\infty = \bigcap_{n \geq 0} T^{-n}\mathfrak{F}$. Note that if T is invertible, then $\mathfrak{F}_\infty = \mathfrak{F}$ and several of the following results are trivial. Note also that if Ω is a one-sided sequence space, then $\mathfrak{F}_\infty$ is the tail σ-field $\bigcap_{n \geq 0} \sigma(X_n, X_{n+1}, \ldots)$, where $\{X_i\}_{i=0}^{\infty}$ are the coordinate mappings.

Theorem 3: Let μ, η be probability measures on $\mathfrak{F}$, where η is stationary. The following are equivalent:

(a) η asymptotically dominates μ;

(b) If $F \in \mathfrak{F}$ is T-invariant and $\eta(F) = 0$, then $\mu(F) = 0$;

(c) If $F \in \mathfrak{F}_\infty$ and $\eta(F) = 0$, then $\mu(F) = 0$.

Remark: Note that the equivalence of (a) and (b) implies that if μ is a.m.s., then $\bar{\mu}$ asymptotically dominates μ.

Theorem 4: Let μ,η be probability measures on $\mathfrak{F}$, where η is invariant.

(a) $\mu \ll \eta$ implies η asymptotically dominates μ;

(b) If T is invertible and η asymptotically dominates μ, then $\mu \ll \eta$.

Thus asymptotic dominance and ordinary dominance are equivalent in the invertible case and $\bar{\mu} \gg \mu$.

SOME ASYMPTOTIC CONVERGENCE THEOREMS

Fix in this section $(\Omega,\mathfrak{F}),T$, and probability measures μ,η on $\mathfrak{F}$. If λ is a probability measure on $\mathfrak{F}$, let λ_a denote the part of λ absolutely continuous with respect to η.

If μ is a.m.s. and T is invertible, then $\mu \ll \bar{\mu}$ and hence from the Radon-Nikodym Theorem $\mu(F) = \int_F (d\mu/d\bar{\mu})\,d\bar{\mu}$. The Radon-Nikodym Theorem is useful in studying a.m.s. sources with invertible shifts. The following theorem provides an asymptotic form of the Radon-Nikodym Theorem for asymptotically dominated measures.

Theorem 5: Define

$$f_n = \frac{d(\mu T^{-n})_a}{d\eta}, \quad n = 0,1,\ldots$$

If η asymptotically dominates μ, then

$$\lim_{n\to\infty} \left\{ \sup_{F\in\mathfrak{F}} \left| \mu(T^{-n}F) - \int_F f_n\, d\eta \right| \right\} = 0 .$$

Henceforth if λ is a probability measure on $\mathfrak{F}$, let λ_∞ denote the restriction of λ to $\mathfrak{F}_\infty$.

Corollary 1: If η is stationary and asymptotically dominates μ, then $\mu_\infty \ll \eta_\infty$.

Theorem 6: Let η be stationary and asymptotically dominate μ. Then

$$f_n T^n \underset{n\to\infty}{\rightarrow} \frac{d\mu_\infty}{d\eta_\infty} \quad \text{a.e. } [\eta] \text{ and } L^1(\eta) .$$

Corollary 2: If η is stationary and asymptotically dominates μ, then

$$\lim_{n \to \infty} \sup_{F \in \mathfrak{F}} \left| \mu(T^{-n}F) - \int_{T^{-n}F} f_\infty d\eta \right| = 0 ,$$

where $f_\infty = d\mu_\infty / d\eta_\infty$.

THE SHANNON-McMILLAN-BREIMAN THEOREM

The preceeding asymptotic convergence results yield the following.

Theorem 7: Let η be stationary and asymptotically dominate μ, then

$$n^{-1} \sum_{i=0}^{n-1} f_i \underset{n\to\infty}{\to} \frac{d\bar{\mu}}{d\eta} \quad \text{in } L^1(\eta) \text{ norm.}$$

Thus if $\eta = \bar{\mu}$,

$$n^{-1} \sum_{i=0}^{n-1} f_i \underset{n\to\infty}{\to} 1 \quad \text{in } L^1(\bar{\mu}) \text{ norm.}$$

Theorem 8: Let η be stationary and asymptotically dominate μ. Let h be any tail function (i.e., measurable with respect to $\mathfrak{F}_\infty$) such that

$$\frac{\log \eta(X_1^n)}{n} \to h \quad \text{a.e. } [\eta] ,$$

then also

$$\frac{\log \mu(X_1^n)}{n} \to h \quad \text{a.e. } [\mu] \text{ and in } L^1(\mu) .$$

Corollary 4: The Shannon-McMillan-Breiman and Shannon-McMillan Theorems hold for any one-sided a.m.s. processes with finite-state space.

CHANNELS

Let ν denote a channel, i.e., for each input sequence x $\nu_x(\cdot)$ is a measure on the output space and for each output event F, $\nu_x(F)$ is a measurable function of x. A channel is stationary if $\nu_x(T^{-1}F) = \nu_{Tx}(F)$ for all x,F. For a source μ the input-output process $\mu\nu$ is defined by $\mu\nu(F) = \int \nu_x(F_x)d\mu(x)$ where F_x is the section of F at x. We define a channel to be a.m.s. if μ a.m.s. implies $\mu\nu$ is a.m.s. The following provides a characterization of two-sided a.m.s. channels in terms of stationary channels analagous to Theorems 2 and 3 and equation (1) for sources.

Theorem 9: Given a two-sided channel ν the following are equivalent:

(a) ν is a.m.s.

(b) There exists a stationary channel $\bar{\nu}$ such that for any stationary measure μ and say $F \in \mathcal{S}_{A \times B}$

$$\lim_{n \to \infty} n^{-1} \sum_{i=0}^{n-1} \nu_{T^{-i}x}(T^{-i}F_x) = \bar{\nu}_x(F_x), \ \mu\text{-a.e.}$$

(c) There exists a stationary channel $\bar{\nu}$ such that for any stationary measure μ

$$\nu_x \ll \bar{\nu}_x , \quad \mu\text{-a.e.}$$

REFERENCES

Dowker, Y.N. (1951), "Finite and σ-finite Invariant Measures," Ann. Math., 54, pp. 595-608.

Fontana, R.J., Gray, R.M. and Kieffer, J.C. (1980), "Asymptotically Mean Stationary Channels," submitted for publication.

Gray, R.M. and Kieffer, J.C. (1980), "Asymptotically Mean Stationary Measures," Ann. Prob. to appear.

Jacobs, K. (1959), Die ubertragung diskreter informationen durch periodishe und fastperiodische kavale. Math. Annalen, 137, pp. 125-135.

Rechard, O.W. (1956), "Invariant Measures for Many-One Transformations," Duke J. Math., 23, pp. 447-488.

AN INFORMATION THEORY APPROACH TO THE APPROXIMATION OF FUNCTIONS

N.S. Tzannes and T.G. Avgeris*

Chair of Information Theory, Univ. of Patras, Greece
*Hellenic Communications Organization, Athens, Greece

ABSTRACT

The problem of the approximation of unknown functions on the basis of prior measurements is examined from an information theory point of view. A new method is proposed, which is based on the MEP (Maximum Entropy Principle) when the prior data is errorless, and on the MIP (Mutual Information Principle) when the data is obtained with errors. Examples are given to illustrate the method, and its relation to the method of the least squares is discussed.

1. INTRODUCTION

The problem of approximating an unknown deterministic function from a given set of its values is of fundamental importance in many branches of science and engineering. If these values are given exactly, we usually try to find a polynomial or set of polynomials which equals the unknown function at the given points. In the more practical situation where the given data points include some measurement error, we pick that function (usually a polynomial) which best approximates the unknown function in some sense, such as the least squares criterion.

In the present work a new method of approximation is proposed, which does not arbitrarily assume the type of the interpolating function and which takes into account any information given a priori about the unknown function, such as range, area, duration. The method is applicable to positive integrable functions, to which an entropy can be assigned by a suitable extension of the consept of entropy of a signal source. The function, which best approximates

D. G. Lainiotis and N. S. Tzannes (eds.), Advances in Communications, 22–29.

the unknown function is the one that maximizes the above entropy, while all the given pieces of information are used as constraints.

If the given data contain some error of known average value, then the approximating function is found by minimizing a mutual information functional which relates the unknown function to its observed version in a manner analogous to the mutual information relating the p.d.f.'s of transmitted and received signals through a noisy channel.

Both of the above parts of the proposed method take into account all of the available a priori information about the unknown function without making any arbitrary assumptions. The method is applicable to a large class of practical functions as it is shown by examples. It is believed that the introduction of information theory concepts in the classical approximating theory will bring to light some hitherto unknown links between the two theories, and will stimulate further research in this new direction.

2. APPROXIMATION BASED ON EXACT DATA

Consider the class of positive integrable functions x(t), i.e. those functions satisfying the following conditions:

$$\int_a^b x(t)dt = k \qquad \text{and} \qquad x(t) > 0 \qquad \text{for all } t \tag{1}$$

Since each member of the above class has the properties of a p.d.f. we may define the entropy

$$H(t) = -\int_a^b x(t) \log x(t)\, dt \tag{2}$$

where the variable t now plays the role of a "random variable" with p.d.f. x(t).

If we have some partial a priori information about x(t), such as the values of x(t) at some points $t_1, t_2, \ldots\ldots, t_n$, one way to approximate the entire x(t) is to assume that it has a certain functional form (usually polynomial) and try to find the "best" such function which passes through the given points. If the a priori information is the integral form, i.e.

$$\begin{aligned} \int_a^b x(t) f_i(t)\, dt &= M_1 \\ \vdots \quad & \quad \vdots \end{aligned} \tag{3}$$

$$\int_a^b x(t) f_n(t)\, dt = M_n \tag{3}$$

we propose as an "optimum" function approximation to the unknown $x(t)$ the function which maximizes the functional $H(t)$ of eq. (2) above under the constraints of eqs. (3). Using the Maximum Entropy Principle, or MEP [1] we find the optimum $x(t)$ as,

$$x(t) = \exp\left[-\lambda_0 - \lambda_1 f_1(t) - \ldots - \lambda_n f_n(t)\right] \tag{4}$$

where λ_j are the Lagrangian multipliers determined from eqs. (3) and (1).

If we are given exact values of $x(t)$ at some points t_i, the constraints (3) become:

$$\begin{aligned} \int_a^b x(t)\delta(t-t_1)dt &= x_1 \\ &\vdots \\ \int_a^b x(t)\delta(t-t_n)dt &= x_n \end{aligned} \tag{5}$$

and $x(t)$ takes the form of Fig. 1.

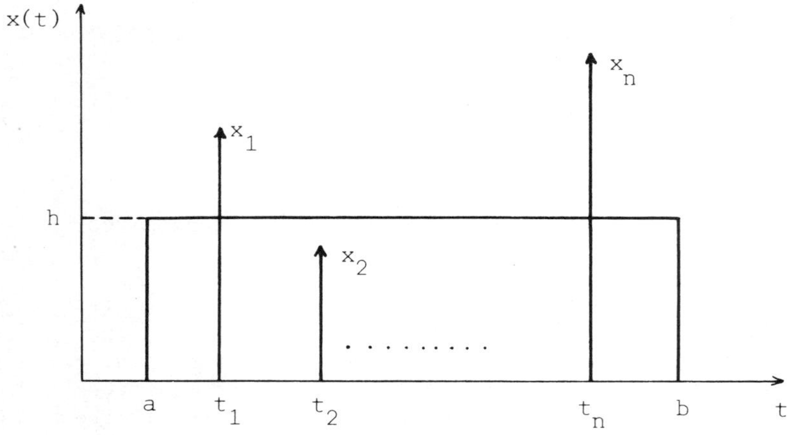

Fig. 1. MEP approximation with given exact values.

where the height of the rectangle is:

$$h = \frac{k - \sum_{i=1}^{n} x_i}{b - a} \tag{6}$$

In this case, our method leads to a constant value h and discrete values $x_1, x_2, \ldots, x_n$, with no arbitrary variation between them.

If, in addition to points x_i, we are given moments:

$$\begin{aligned} \int_a^b x(t)\,dt &= M_0 \\ &\vdots \\ \int_a^b x(t)t^m dt &= M_m \end{aligned} \tag{7}$$

we subtract from these moments, the moments of each x_i:

$$\begin{aligned} \int_a^b x(t)\,dt &= M_0 - \sum_{i=1}^{n} x(t_i) = M_0' \\ &\vdots \\ \int_a^b t^m x(t)dt &= M_m - \sum_{i=1}^{n} t_i^m\, x(t_i) = M_m' \end{aligned} \tag{8}$$

where we have introduced the new moments $M_o', \ldots, M_m'$, as defined by the above equations. Our method gives the MEP estimate of x(t) in its unknown regions, except at the points $x_1, x_2, \ldots, x_n$. For example, the zero and first order moments lead to a double exponential at:

$$t_0 = M_1'/M_0' \tag{9}$$

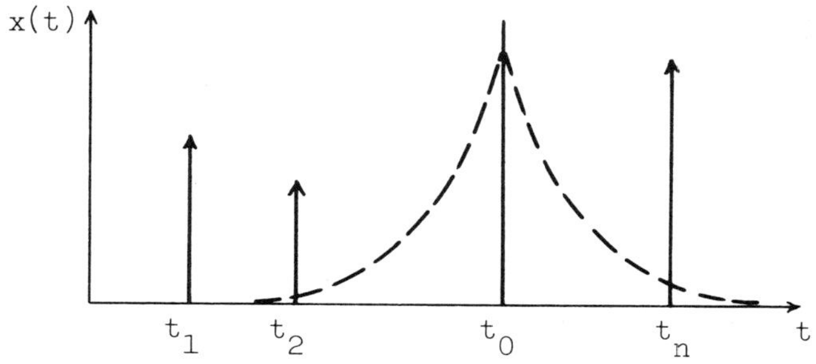

Fig. 2. MEP approximation with given moments.

3. APPROXIMATION BASED ON DATA WITH ERRORS

In most practical problems our a priori information is not exactly related to $x(t)$ but to its distorted version $y(\tau)$ at the output of our observation system, as in Fig. 3, where $g(\tau|t)$ is a conditional function description of the observation system.

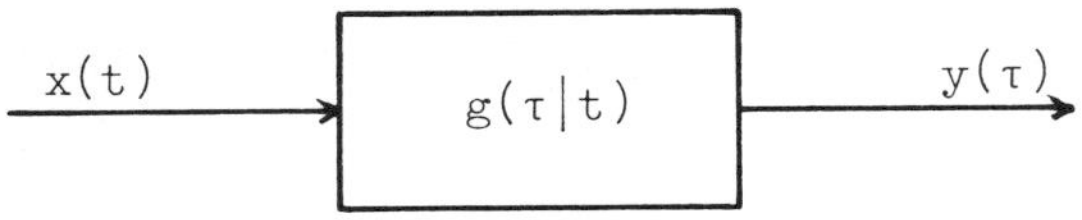

Fig. 3. Observation system relating $x(t)$ and $y(\tau)$.

Assuming that both $x(t)$ and $y(\tau)$ are positive and normalized, i.e.

$$\int_a^b x(t)\,dt = 1 \quad , \quad \int_c^d y(\tau)\,d\tau = 1 \tag{10}$$

and considering $x(t)$ and $y(\tau)$ to be p.d.f.'s of the "random variables" t and τ we formulate a mutual information functional:

$$I(t;\tau) = \int_a^b \int_c^d x(t) g(\tau|t) \log \frac{g(\tau|t)}{y(\tau)}\, dt\, d\tau \tag{11}$$

As the MEP approach was used in the previous section to approximate $x(t)$ using exact a priori information, the Mutual Information Principle or M.I.P. [2] can be used in the case of the distorted information given by $y(\tau)$ about $x(t)$. Namely, we consider as the best approximation to the unknown $x(t)$ that function $x(t)$ which minimizes the mutual information $I(t;\tau)$ under the constraints (10) and:

$$\int_c^d g(\tau|t)\,d\tau = 1 \qquad \text{for all } t \tag{12}$$

as well as the "fidelity criterion", or average distortion:

$$\int_a^b \int_c^d x(t) g(\tau|t) \rho(t,\tau)\,dt\,d\tau \leq D \tag{13}$$

where $\rho(t,\tau)$ is an error function relating t and τ.

In the above MIP function the observed function $y(\tau)$ is assumed to be known or it may be approximated through MEP or any other method of approximation based on exact data. The conditional function $g(\tau|t)$ is a new concept that emerges from the above discussion, as a description of the observation system. It expresses the value of the output at time τ due to a unit impulse input at time t. In the MIP framework the functions $x(t)$ and $y(\tau)$ are related by the equation:

$$\int_a^b g(\tau|t)x(t)\,dt = y(\tau) \tag{14}$$

Thus, $g(\tau|t)$ is a MIP (or maximum uncertainty) description of the observation system, whose only known characteristics are: its output $y(\tau)$ and its average distortion D given by eq. (13). It is a new type of "transfer function" relating input and output of a system, about which we have very limited information.

The classical method of approximation of $x(t)$, when $y(t)$ is given, is to minimize the average square error function:

$$E = \int_a^b \left[x(t) - y(t)\right]^2 dt \tag{15}$$

assuming a certain functional form for $x(t)$, such as polynomial, and that the errors $\varepsilon(t) = x(t) - y(t)$ are uniformly distributed, i.e.

$$p_\varepsilon(t) = \text{const. for all } t \tag{16}$$

Since most measuring systems do not introduce the same error throughout their measuring scale, a weighting function, assumed to be known a priori, is often included in the integrand of eq. (15), i.e.

$$E_w = \int_a^b \left[x(t) - y(t)\right]^2 w(t)\,dt \tag{17}$$

In order to relate our method to the above method of least squares consider the error function:

$$\rho(t,\tau) = \left[x(t) - y(t)\right]^2 = \varepsilon^2(t) \tag{18}$$

and eq. (13) becomes:

$$\int_c^d g(\tau|t)\, d\tau \int_a^b x(t)\ x(t) - y(t)\ ^2 dt \leq D \tag{19}$$

or, taking into account the constraint (12):

$$\int_a^b x(t)\left[x(t) - y(t)\right]^2 dt \leq D \tag{20}$$

Thus our method leads, in this particular case, to an average square error with a weighting function equal to x(t). This is not without some intuitive justification, as often the devoce errors are proportional to the measured quantity.

As an example of the application of the MIP method, we take up the following common case.

Let us assume that measurements have led to a collection of specific points for $y(\tau)$ i.e. we know $y(\tau_k)$ for $k = 1, 2, 3, \ldots, n$, and they are all positive. Let us further assume that,

$$\rho(t,\tau) = d(t - \tau_k) \tag{21}$$

i.e. it is a difference error function. With the above constraints a solution of the MIP is [4],

$$x(t) = A \sum_k y(\tau_k)\ e^{-s(t-\tau_k)} \tag{22}$$

where

$$A = \sqrt{\frac{\pi}{s}} \tag{23}$$

and

$$a = \frac{1}{2D} \tag{24}$$

More specifically, if

$$d(t - \tau) = (t - \tau)^2 \tag{25}$$

the final solution is,

$$x(t) = A \sum_k y(\tau_k)\, e^{-s(t-\tau_k)^2} \qquad (26)$$

i.e. a sum of Gaussian functions, whose width (variance) depends on the average error. It is obvious from the above that the solution is a sum of exponentials whose form depends on the error function $\rho(t,\tau)$.

Before closing this paper, we make the following important remark. The method, theoretically, is applicable to everywhere positive functions. In practical applications, though, the $y(\tau_k)$ can be shifted upward by an amount equal to the maximum negative value, which is then corrected at the end (in $x(t)$) with some minor error. Nevertheless, this aspect of the work requires further research.

4. CONCLUSION

A new method of approximating an unknown function was proposed using measurements of the function taken either directly or through an observation system introducing a known average distortion. In the first case the approximation was based on the Maximum Entropy Principle and, in the second case, on the Mutual Information Principle. It is believed that the extension of Information Theory concepts from p.d.f.'s to positive integrable functions in general, will prove useful to many applications. In particular, the MIP method may provide a means of estimating the input of a system with given output, while identifying the system itself in the MIP sense, i.e. taking into account any limited information we may have about the system.

REFERENCES

1. Jaynes, E.T., "Prior Probabilities", IEEE Trans. on System Science and Cybernetics, Vol. 4, pp. 227-241, 1968.

2. Tzannes, N.S., Noonan, J.P., "The Mutual Information Principle and Applications", Information and Control, Vol. 22, p.12, Feb. 1973.

3. Avgeris, T.G., Tzannes, N.S., "On Reconstituting a Process from its Quantized Form in the Presence of Noise", Information Sciences, July 1978, pp. 127-141.

4. Cyranski, J.F., Tzannes, N.S., "An Alternate Mathematical Formulation of the Mutual Information Principle", submitted for publication to Information and Control.

UNCERTAINTY AND ENTROPY-QUANTUM INFORMATION SYSTEMS

C. Syros

Laboratory of Nuclear Technology
University of Patras, Patras, Greece.

Abstract: Information is generated by microscopic systems. Fluctuations in the characteristics of macroscopic systems are frequently sources of information about the system. The uncertainties in microscopic systems may give rise to macroscopic fluctuations and, therefore, to the generation of information. The entropy of a given system of particles is calculated on the basis of the Uncertainty Principle. The dimensionality of the physical space is determined by the position of the minimum of the entropy as function of the degrees of freedom per particle. If the total number of degrees of freedom of the system is very small, then the entropy is not additive.

1. INTRODUCTION

If the chaos is defined as a state of a given composite system in which all states of each one of the system constituents are simultaneously present in the system, and if information is defined as a set of data characterizing a sub-set of the states of the system constituents, then the two sets of data associated on the one hand with the chaos and on the other hand with the subset of states for which the information is available have a non-empty intersection. Of course, the above situation is strongly idealized and in physical reality to which belong all practical information systems, the states and the probabilities correspondingly may not differ by continuously changing quantities. These facts determining the properties of the information systems fix to a great extent the properties of the data defining

D. G. Lainiotis and N. S. Tzannes (eds.), Advances in Communications, 30–38.

the information. Since the appearance of Information Theory important developments have been achieved in various aspects of this theory. An area which will probably be very important in future is the set of problems on the quantum level related to the production of information data and the influences during handling and transmission. The idea of quantizing a source is not new (1) but the application of the quantum principles throughout the area of Information Theory where this is feasible is still an open problem. For definiteness let us mention the following examples:
(i) The visual receptors can be activated by a single quantum and thus they act as single quantum counters. At night time a receptor might catch the light quanta at the rate of about one per hour and when playing on the sunny beach of Patras the photon registration rate might reach hundreds of counts per second (2). (ii) A photomultiplier produces an output if it is excited by a single photon. (iii) A Geiger-Müller tube detects single γ-quanta or charged particles entering it. (iv) The Josephson contacts constitute an important example where quantum effects determine the output of the source, etc. All these devices constitute sources which give rise to a more or less well defined output which may be of interest to Information Theory. It is, therefore, clear that Information Theory on the long range will have to adapt its mathematical tools to the conditions imposed by the quantum behaviour of the information sources. In view of the above the equation

$$I(x) = -\sum_n P_n(x) \ln P_n(x) \qquad (1.1)$$

will acquire new aspects when $P_n(x)$ is subject to the Quantum principles. Moreover, since $P_n(x)$ is related to the entropy of the system there will be needed clarification as to what should be the meaning of entropy of a system of a few particles.

2. UNCERTAINTY PRINCIPLE AND ENTROPY

Let us consider a system producing information by means of the F degrees of freedom of s particles,

$$F = 2 \nu N s, \qquad (2.1)$$

where ν is the dimension of the space, and 2N the number of degrees of freedom per dimension. In applying the Uncertainty Principle for the calculation of the entropy we have to specify our understanding of the uncertainties

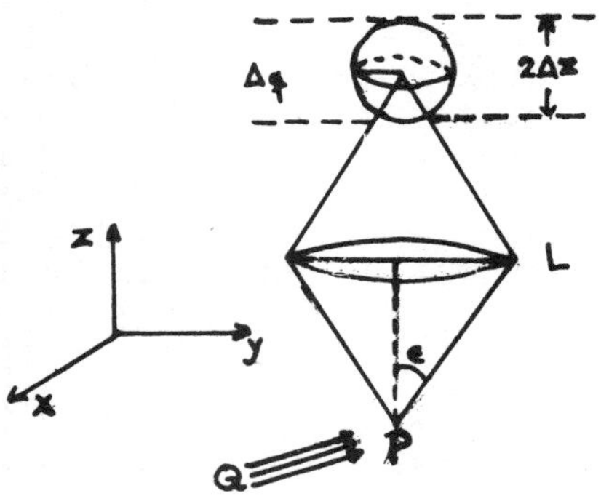

Fig. 2.1
The resolving power of the lens, L, is Δq and equals the radius of the uncertainty region. The shape of this region depends among other parameters on the resolving power of L along the z-axis.

of the relevant quantities.
First it is assumed that the Uncertainty Principle applies to directly observable quantities. This becomes clear by considering the gedankenexperiment of fig. 2.1.
Assuming that the resolving power of the lens is the same in all directions, the observation point of the scattered photon at P will be found inside the sphere of radius $\Delta q \sim \lambda/\sin\varepsilon$, where λ is the wave length of the photons scattered (3).
On the other hand the uncertainty of the momentum as calculated from the relation $p=h/\lambda$, where the wave length λ, corresponds to the total momentum but not to a particular component. The result is, of course,

$$\Delta p \cdot \Delta q > h. \qquad (2.2)$$

The hypothesis of the statistical fluctuations leading to the formation of dissipative structures as those proposed by Prigogine (4) is used in connection with the Uncertainty Principle to calculate the entropy of a quantum mechanical system of particles. The first assumption made here is that the physical mechanism according to which ordered motion energy is generated in a particle system consists of the cooperative addition of the statistical fluctuations in position and momentum coordinates of a sufficiently large number of particles.
The second assumption is that the uncertainties Δp, Δq in the coordinates of the various degrees of freedom of a particle result from the collisions as a consequence of the Uncertainty Principle independently of whether the system is being observed or not. This is in accordance with the fact that to a given eigenvalue of, e.g., the momentum operator is associated a root mean square deviation which is calculated with the help of the density matrix of the system. The uncertainties are independent of $\{p, q\}$. The third assumption is that the uncertainties,

as mentioned above, appearing in the relation $\Delta p \cdot \Delta q > h$ correspond to the absolute value of the vectors to which they refer. This expresses the fact that the uncertainties are isotropically distributed around the point of collision.
Let the system consist of s particles. The phase space, R^f, has the dimension $f=2\nu s$, where ν is the number of dimensions of the physical space. If p_i, q_i are the momentum and the position vectors of the i-th particle, then the statistical weight $\Delta\Gamma_f$ is given in the classical case by

$$\Delta\Gamma_f = \prod_{i=1}^{s} \prod_{j=1}^{\nu} (\Delta p_{ij} \cdot \Delta q_{ij} / h). \tag{2.3}$$

The quantity $\Delta\Gamma_f$ to be calculated on the basis of the Uncertainty Principle is usually called the statistical weight of the system of the f degrees of freedom.
Since the visual receptors interact with single quanta in the same way as in the case of any other macroscopic systems of degrees of freedom, the present situation may be considered as a macroscopic system. This fact justifies using $\Delta\Gamma_f$ as the macroscopic statistical weight of the corresponding state of the f degrees of freedom, although Δp, Δq obey the Uncertainty Principle.
In the quantum mechanical case the space R^Q has the dimension $Q=2\nu sN$, where 2N is the number of quantum mechanical degrees of freedom corresponding to each dimension of the space (for classical point particles 2, for quantum mechanical particles 2N). The statistical weight, $\Delta\Gamma_Q$, in the present case will be defined in polar coordinates by

$$\Delta\Gamma_Q = \prod_{i=1}^{sN} \left\{ \left(\frac{p_i^{\nu-1} \cdot q^{\nu-1}}{h} \right) \cdot \prod_{j=1}^{\nu-1} (\sin\vartheta_j \sin\varphi_j)^{\nu-j-1} \Delta\vartheta_j \Delta\varphi_j \cdot \Delta p_i \Delta q_i / h \right\}. \tag{2.4}$$

The space R^Q is generated via the prescription for the definition of the Wigner functions. To proceed, the distribution function of the particle system in the space of the uncertainties is needed. In contrast to the classical case, in the quantum case the distribution function, $\rho_s(\vec{p}_1, \ldots \vec{q}_{sN}; \tilde{a}_1, \ldots \tilde{a}_{sN})$ will depend not only on $\{\vec{p}_i, \vec{q}_i\}$ **but** also on the fluctuations $\{\Delta p_i \cdot \Delta q_i = h + \tilde{a}_i\}$. This fact emerges naturally from the relation

$$f_s^W(\vec{p}_1, \vec{q}_1 \ldots \vec{p}_s, \vec{q}_s) = (8\pi^3)^{-s} \int dk_1 \ldots dk_s \cdot \exp\left[i \sum_{n=1}^{s} \vec{k}_n \vec{p}_n \right]$$
$$\mathrm{Sp}\left[\hat{\rho}\ \psi^+(\vec{q}_1 - \tfrac{h}{2}\vec{k}_1) \ldots \psi^+(\vec{q}_s - \tfrac{h}{2}\vec{k}_s) \cdot \psi(\vec{q}_1 + \tfrac{h}{2}\vec{k}_1) \ldots \psi(\vec{q}_s + \tfrac{h}{2}\vec{k}_s) \right],$$

where $\{\frac{h}{2}\vec{k}_n\}$ **can** be visualized as the uncertainties in question, if by a slight change of the above definition the k-integration is eliminated.
The distribution function will in general be defined on

$R^Q \subset \tilde{R}^{Q/2\nu}$, where $\tilde{R}^{Q/2\nu}$ is the space of the action uncertainties, $\tilde{a}_i = \Delta p_i \Delta q_i - h$. Obviously, $\tilde{a}_i > 0$ for all i. For the present only the $\tilde{a}_i$-dependent part of the distribution function is of interest to us.
Let $g(\vec{p}_1 \ldots \vec{p}_s)$ represent the probability for the event

that from a sub-set $\tilde{A} \subset \tilde{R}^{Q/2\nu}$ of the fluctuations a macroscopically observable cooperative phenomenon emerges.
Since the fluctuations are stochastic and $\rho_s(\vec{p}_1, \ldots \vec{q}_{sN}; \tilde{a}_1, \ldots \tilde{a}_{sN})$ is independent of the polar angles, $\tilde{g}$ will be assumed constant on A.
Next the relation $\rho_s \cdot g\Delta\Gamma_Q \simeq 1$ is used from which it follows that (5)

$$\rho_s = \frac{\left[g \cdot \prod_{i=1}^{sN} (p_i^{\nu-1} q_i^{\nu-1} / h^{\nu-1} \cdot d\Omega_i \right]^{-1}}{\prod_{i=1}^{s} (\Delta p_i \cdot \Delta q_i / h)} \tag{2.5}$$

The factor g accounts for the fact that a much larger element, $\Delta\Gamma_Q$, of the space R^Q is required in order that the cooperative fluctuations be included in this space element.
According to the definition of the Uncertainty, $\tilde{a}_i$, the equality

$$\Delta p_i \cdot \Delta q_i = h + \tilde{a}_i \tag{2.6}$$

results from the Uncertainty Principle.

3. SOME PROPERTIES OF THE ENTROPY

Taking the product of all the equalities with respect to i the expression is obtained

$$\prod_{i=1}^{sN} (\Delta p_i \cdot \Delta q_i / h) = \prod_{i=1}^{sN} (1 + a_i); \quad a_i = \tilde{a}_i / h. \tag{3.1}$$

Therefore, $\{a_i\}$ is for the present physically more relevant than $\{\Delta p_i, \Delta q_i\}$ and ρ_s is normalized to the unity on $R^{Q/2\nu} = \tilde{R}^{Q/2\nu}/h$ where h is Planck's constant. From (2.5) and (3.1) it follows for $a_i < A_i$ that

$$\rho_s = \{\prod_{i=1}^{sN} [\ln(1+A_i)](1+a_i)\}^{-1}. \tag{3.2}$$

Since ρ_s is approximately equal to the distribution function of the uncertainties $\{a_i\}$ according to (2.5), the density distribution, $S(a_1, \ldots a_{sN})$, of the entropy in $A^{Q/2\nu}$ is

$$\begin{aligned} S(a_1, \ldots a_{sN}) &= -\rho_s \ln \rho_s \\ &= -\sum_{i=1}^{sN} \ln[(1+a_i)\ln(1+A_i)]\{\prod_{i=1}^{sN} (1+a_i)\ln(1+A_i)\}^{-1}. \end{aligned} \tag{3.3}$$

$S(a_1, \ldots a_{sN})$ has an extremum because

$$\frac{\partial S}{\partial a_l} = -(\frac{\partial W}{\partial a_l} + \frac{\partial W}{\partial a_l} \ln W) = 0, \qquad \frac{\partial^2 S}{\partial a_l^2} < 0, \tag{3.4}$$

$$\prod_{i=1}^{sN} (1+\bar{a}_i)\ln(1+A_i) = e. \tag{3.5}$$

The position in $A^{Q/2\nu}$ is $\{\bar{a}_i | i=1,sN\}$. This is indeed a maximum, $\partial^2 S/\partial a_l^2$, $(\frac{\partial W}{\partial a_l})^2 > 0$ for all l. The entropy $\bar{S}$ is

$$\bar{S} = \nu \int_0^{A_1} da_1 \ldots \int_0^{A_{sN}} da_{sN}\, S(a_1, \ldots a_{sN}). \tag{3.6}$$

Assuming that $A_i = A$ for all $i=1,sN$, we get the expression

$$\bar{S} = \frac{Q}{2}[\frac{2\nu}{Q} + \frac{1}{2} \frac{\exp(Q/2\nu}{1+\bar{a}} - \ln(1+\bar{a})]. \tag{3.7}$$

$\bar{S}$ is not an additive quantity, if $2\nu Q^{-1} \neq 0$. The entropy satisfies

$$(\partial\bar{S}/\partial Q) = 0, \quad (\partial^2\bar{S}/\partial Q^2) > 0; \quad \bar{a} = 0. \tag{3.8}$$

For one particle

$$Q = 2\nu; \quad (s = 1) \tag{3.9}$$

and consequently, N=1 for $\bar{a}=0$.
The maximum order is observable for $Q=2\nu$, because at $Q=2\nu$ the entropy attains its minimum value.
For a classical point mass particle

$$Q = 2 \cdot 3 = 6. \tag{3.10}$$

In relativistic quantum mechanics new degrees of freedom appear, e.g., (q^0, q^1, q^2, q^3) and (p^0, p^1, p^2, p^3) and the number of degrees of freedom increases, e.g., spin, magn. quantum number, space parity, helicity and isospin, charge, charge conjugation, strangeness. This gives in total 16 degrees of freedom per particle but there may be even more.

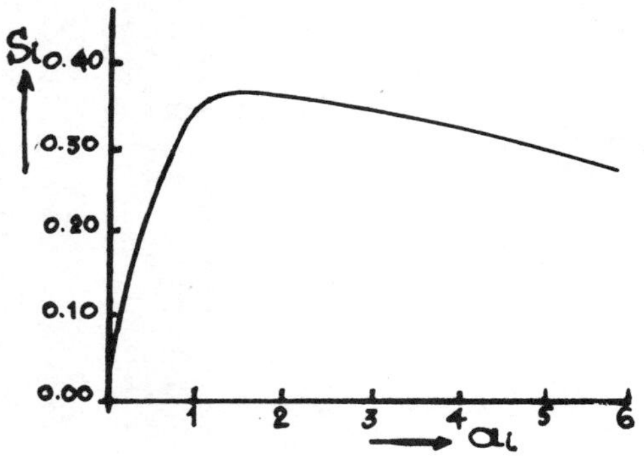

Fig. 2.1
A plane cut of the entropy distribution density in the action uncertainty space, $A^{Q/2\nu}$, as function of a_i. The flat maximum implies high probability for fluctuations. The action a_i is measured in h units (Planck's constant). The larger the uncertainty $\bar{a}$ determining the maximum position the larger the expectation value of the fluctuating physical quantity.

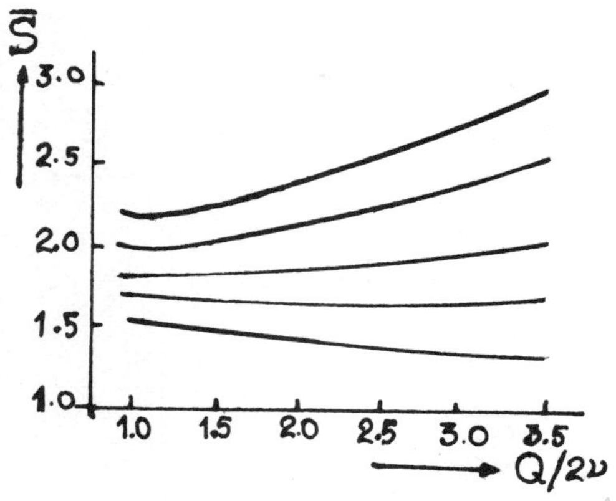

Fig. 2.2
The entropy for $\bar{a}=0$ exhibits an absolute minimum if the number of degrees of freedom equals 2ν. For larger uncertainties, $\bar{a}$, the entropy minimum diminishes and its position shifts toward larger numbers of degrees of freedom. ν is the dimension of the physical space.

For the first 8 degrees of freedom the dimensions are

$$Q = 2\cdot 4 = 8. \qquad (3.11)$$

The value $\bar{a}=0$ has been considered because the probability density distribution introduced by Einstein (6)

$$P(a_1,\ldots a_{sN}) \simeq \exp[S(a_1,\ldots a_{sN})] \qquad (3.12)$$

exhibits the sharpest maximum at $\bar{a}_i=0$ for all $i<sN$. For $Q\rightarrow$ large number, the entropy becomes additive,

$$\bar{S} = \frac{1}{2}\,\nu sN, \qquad (3.13)$$

where νsN is the total number of quantum mechanical degrees of freedom, and

$$\Delta E = \frac{1}{2}\,\nu\ s\ N\ k\ T. \qquad (3.14)$$

In another treatment the entropy appears as the imaginary part of the expectation value of a non-hermitian Hamiltonian (7).

4. CONCLUSIONS AND DISCUSSION

We have analyzed the concept of the entropy in the case of quantum mechanical information systems. This is in fact physically meaningful because degrees of freedom with zero uncertainty in the variable of action have very small probability and the entropy becomes equal to zero. Similar is the probability behaviour also in very large uncertainty values, where the entropy vanishes again. This result is very important from the information theoretical point of view, because it shows that information theory can yield reliable conclusions even in the case of information systems of very small number of degrees of freedom obeying Quantum Mechanics. On the other hand, it is possible to give an expression of the entropy on the basis of the density matrix obeying the von Neumann equation.
For uncertainty values near 1 (in Planck's constant units) the system shows fluctuations which diminish for increasing uncertainty. For smaller uncertainty the fluctuations disappear.
The entropy considered as a function of $\bar{a}_i$, the position of the maximum, shows a minimum. The highest value of the minimum corresponds to $\bar{a}_i=0$ in the situation for which we have 2 degrees of freedom per space dimension. Since the minimum number of degrees of freedom of a relativistic

particle is 8, it follows that the dimensionality of the space is 4.
Finally, the important result is pointed out that only for large numbers of degrees of freedom the entropy becomes an additive quantity.

REFERENCES

1. Gallager, R.G.:"Information Theory and Reliable Communication", 1968, John Wiley and Sons, Inc. N.Y.
2. Bouman, M.A.:"Theoretical Physics and Biology", 1969, Ed. M. Marois, North-Holland Publ. Co., pp. 246-249.
3. Schiff, L.I.:"Quantum Mechanics", 1955, McGraw-Hill Publ. Co., London, p. 81.
4. Prigogine, I.:"Theoretical Physics and Biology", 1969, Ed. E. Marois, North-Holland Publ. Co., pp. 23-52.
5. Landau, L.D., and Lifschitz, E.M.:"Statistical Physics", 1970, Pergamon Press, pp. 22-28.
6. For the description of the Einstein method see for example ref. 5, p. 341.
7. Syros, C.: 1977, Phys. Lett. 64 A, pp. 17-18.

THE REDUCTION OF THE WAVE-FUNCTION AND THE REDUCTION OF ENTROPY

H. Ioannidou

Chair of General Mathematics,
University of Patras
Patras, Greece.

The wave-function ψ represents in quantum mechanics the state of a physical system and is supposed to describe it completely, i.e. to furnish all information conserning the system at a given time instant t. ψ is developed as a superposition of the eigenfunctions y_n of a hermitean operator $\hat{A}$ which corresponds to the measured physical quantity A. During measurement, ψ is reduced according to the von Neumann "projection postulate"; namely ψ undergoes a non-causal change transforming it from a superposition of eigenstates to a particular eigenstate $y_{n'}$. This "reduction" is similar to the reduction of the initial entropy when information is received. Nevertheless, this analogy may not reflect quite the same situation.

1. INTRODUCTION

Quantum theorists have been recently interested in Information Theory (1-4). This is due to the fact that both theories have something in common, namely the probabilities they both involve. The question is, what kind of probabilities are these, since, as it is known, there exist several theories of probability introducing different definitions (5), each assigning a different meaning to the word. Let us now see briefly how probabilities enter the quantum theory.

The state of a quantum physical system is represented by a function ψ, vector of Hilbert space H. The time evolution of ψ is given by the causal Schrödinger differential equation

$$\hat{H}\psi = i\hbar \frac{\partial\psi}{\partial t} \qquad (1.1)$$

D. G. Lainiotis and N. S. Tzannes (eds.), Advances in Communications, 39–44.

with $\hat{H}$ the Hamilton operator, corresponding to the total energy of the physical system.
A physical quantity, often called "observable", is represented by a hermitean operator $\hat{A}$, with eigenvalue equation

$$\hat{A}y_n = \alpha_n y_n \tag{1.2}$$

where y_n eigenfunctions and α_n eigenvalues of $\hat{A}$; $\{y_n\}$ is a basis of H, and ψ can be expanded as

$$\psi = \sum_n c_n y_n \tag{1.3}$$

The coefficients $c_n = \langle y_n|\psi\rangle$ are such that

$$|c_n|^2 = p(\alpha_n) \tag{1.4}$$

i.e. their square modulus give the probabilities of finding the value α_n when we measure the physical quantity A of a physical system found in state ψ.
The mean value of A is given in terms of ψ as

$$\langle A \rangle = \langle\psi|A|\psi\rangle \tag{1.5}$$

From the above we see that knowledge of the quantum state ψ provides us with probabilities, in spite of the fact that ψ obeys the causal equation (1.1). This leads to the foundamental problem of what is the nature of the quantum state, which is intimately connected with the nature of the quantum probabilities.

2. THE QUANTUM - MECHANICAL MEASUREMENT PROBLEM

This problem, introduced in 1936 (6) is considered as still unsolved, at least in the frame of quantum theory. In measuring a physical quantity A on a physical system (ψ), we consider before measurement the initial state of (ψ) as a sum

$$\psi_i = \sum_n c_n y_n \tag{2.1}$$

Immediately after measurement, which takes time $\Delta t=\tau$, we read the result $\alpha_{n'}$. This means that after measurement, i.e. for $t \geqslant \tau$, the final state of (ψ) is an eigenfunction $y_{n'}$,and not the sum (1.3). In other words we have to accept that the state of our physical system undergoes a change of the form

$$\psi_i = \sum_n c_n y_n \xrightarrow{\text{measurement}} \psi_f = y_{n'} \tag{2.2}$$

which is discontinuous and unpredictable. But theory predicts a causal change of ψ in time τ. In fact, according to eqs. (1.1), the final state of the system will be

$$\psi_f = e^{(i/\hbar)\hat{H}\tau} \psi_i \tag{2.3}$$

The unpredictable process (2.2) is called "the reduction of the wave-function". Research in quantum-mechanical measurement consists mainly on efforts to make the reduction to enter in a consistent and persuasive way into the theory. The main problem is <u>how</u> the reduction is done and <u>by whom</u>, and also what is the role of the measuring apparatus and especially of the human observer in the process of measurement.

3. THE SUBJECTIVE INTERPRETATION OF ψ AND ITS CONNECTION WITH INFORMATION THEORY

If we consider the so-called subjective interpretation of ψ, then the difficulties mentioned in the previous paragraphs seem to be resolved(7). Heisenberg was the first to consider the quantum-mechanical probabilities as "potentialities" (8), according to the Aristotelean notion of "potentia" (in greek δύναμις). Namely $p(\alpha_n)$ means that one could find the value α_n of a physical quantity A measured on a physical system (ψ), with probability (potentiality) $p(\alpha_n)$. These potentialities are assigned to the system by the human observer, so that the state ψ represents not the physical system itself , but the observer's knowledge (or rather ignorance) about it. Then, as Landé (9) put it, "..The state is a well-ordered list of betting odds, for the diverse outcomes of specific tests of a microscopic object with a macroscopic instrument.." .These "betting odds" are the probabilities of the several eigenstates, assigned by the human observer; in other words they are inductive or subjective probabilities[1] .

The above statement about the meaning of the quantum state ψ, as a sum of potential outcomes of measurements, reminds one strongly the information-theoretical concept of the a-priori uncertainty or "entropy" of a physical system (ψ), which is given as a function of the probabilities, i.e.

$$S(\psi) = -\sum_n p(\alpha_n) \log p(\alpha_n) \tag{3.1}$$

The $p(\alpha_n)$'s in (3.1) are inductive probabilities assigned by the observer to the several outcomes of measurement. Therefore we see that the state ψ of a quantum system can be envisaged as the initial uncertainty of the observer about the physical system in question.

After measurement, the uncertainty becomes zero, since the final probabilities are

$$p_f(\alpha_n) = \begin{cases} 0 \text{ for } n \neq n' \\ 1 \text{ for } n = n' \end{cases} \quad (3.2)$$

This leads to the reduction of entrory, i.e.

$$\begin{aligned} S_i(\psi) &= -\sum_n p(\alpha_n) \log p(\alpha_n) \geqslant 0 \xrightarrow{\text{measurement}} \\ S_f(\psi) &= -\, p(\alpha_{n'}) \log p(\alpha_{n'}) = 0 \end{aligned} \quad (3.3)$$

which is essentially the same process with the reduction of the wave-function (2.2).

4. THE REDUCTION OF ψ ACCORDING TO THE SUBJECTIVE INTERPRETATION

Once we have accepted that ψ is a state in the observer's mind conserning a physical system and not an objective representation of a system, then it is clear that the reduction (2.2) is performed in the observer's mind as a "gedankliche Vorgang" (10), by the "abstract ego of the observer" (11), at the time of the reading of the result $\alpha_{n'}$. This is also suggested by the information-theoretical picture (3.3), where it seems that the initial and final entropies, althought refering to the object (ψ), are spanned by the observer, viewed as a second system which communicates with (ψ).

Now in order to explain how the reduction is done, in a structural and not simply operational way, we have to consider the (rather non-physical) problem of human perception. In other words, we have to plunge into physiology of the human brain, or even psychology. This is not impossible for mathematical physicists, especially for ones interested in biological problems. It is always realizable to construct one or several mathematical models for this purpose; actually such models of physiological processes do exist already. One has only to modify and apply such a model to the special case of quantum-mechanical measurement, or to quantum mechanics in general. Such an approach naturally would be extremely interesting, and the idea is stimulating, but it would involve several implications.

5. DISCUSSION

In accepting the subjective interpretation of ψ, we have to confront the fact that ψ is the solution of the Schrödinger differential equation (1.1), which is linear and with coefficients involving physical constants. Surely it is not reasonable to accept that a function representing the observer's uncertainty or knowledge about something would obey such an equation. In order to include the reduction of the wave-function in a consistent way into the theory, we have to modify quantum mechanics on the whole, perhaps as Wigner (12) proposes, by adding one or more non-linear terms to the Schrödinger equation. But this would lead to a revision of the quantum-mechanical formalism and consequently to severe difficulties. One must take into account the fact that the present formalism is quite satisfactory and in agreement with experiment; one then has to consider seriously whether it is worth the labour to affect a successful mathematical formalism in order to satisfy a deficient "philosophy".

Another objection to subjective considerations is related to the Einstein "element of reality" (13). According to Einstein, a physical theory must be complete, and in order to be complete, every element of the theory must correspond to an element of physical reality. In other words, Einstein believed in the objective existence of the external world and also that the mission of theoretical physicists is to construct models or theories making this world understandable.

Presumably Einstein was right, and physics has preserved till today its "element of reality". Introduction of subjective concepts such as the acceptance at large of "the observer's influence" in physical theories, is sure that would destroy it. For these reasons, quite a few physicists today accept the Statistical Interpretation of the quantum state, (14), according to which ψ referes to a statistical ensemble of similarly prepared physical systems, so that the quantum-mechanical probabilities are the relative frequencies of the several eigenstates in the ensemble.

Note:
[1] Landé however believed that the quantum-mechanical probabilities can be given an objective physical meaning, since they are usually based on statistical evidence.

REFERENCES

1. V. Majerník: *The Entropic Model of Quantum Measurement,* 1970, Acta Phys. Austr. 31, pp. 271-299.
2. M.D. Srinivas: *Quantum Generalization of Kolmogoroff Entropy,* 1978, J.Math. Phys. 19, pp.1952-1961.
3. John F. Cyranski: *Quantum Measurement as a communication with Nature,* 1978, Found. Phys. 8, pp. 805-823.
4. H. Ioannidou: *Quantum-Mechanical Measurement as a Communication Process,* 1979, in M.H. Hamza, S.G. Tzafestas, Eds. "Advances in Measurement and Control", Vol.1, pp.104-106, Acta Press.

5. T.L. Fine: *Theories of Probability,* 1973, Academic Press.
6. J.von Neumann: *Mathematical Foundations of Quantum Mechanics,* 1955, Princeton Univ. Press.
7. J. Rayski: *Controversial Problems of Measurements within Quantum Mechanics,* 1979, Found. Phys. pp.217-236.
8. W. Heisenberg: *Physics and Philosophy,* 1958, Harper and Bros., New York.
9. Alfred Landē: *New Foundations of Quantum Mechanics,* 1965, Cambridge Univ. Press.
10. P. Mittelstaedt: *Philosophische Probleme der modernen Physik,* 1968, Bibl. Inst. Mannheim.
11. J.von Neumann: ref. (6), pp. 419-421.
12. E. Wigner: *Remarks in the Mind-Body Question* , in E.Wigner " Symmetries and Reflections" 1967, The M.I.T Press.
13. A. Einstein, B. Podolski and N. Rosen: *Can Quantum-Mechanical Description of Physical Reality Be Considered Complete*? 1935, Phys. Rev. 47, pp.777-780.
14. L. Ballentine: *The Statistical Interpretation of Quantum Mechanics,* 1970, Revs. of Mod. Phys., 42, pp. 358-381.

ON SOME FUNCTIONAL EQUATIONS FROM ADDITIVE AND NONADDITIVE MEASURES - II

Pl. Kannappan

Department of Pure Mathematics, University of Waterloo, Waterloo, Ontario, Canada, N2L 3G1

A functional equation connected with additive measures such as Shannon's entropy etc. and nonadditive measures such as entropy of degree β etc. is solved. Generally functional equations of this type are solved for all pairs (m,n) or for some special pairs (2,3) etc. when the function (s) involved is continuous. Here, the functional equation is solved for some fixed pair (m,n), under the condition of measurability of the functions, showing thereby that the solutions indeed depend upon the pair (m,n).

1. Introduction

Various Characterizations are known to characterize the Shannon entropy $H_n(P) = -\sum p_i \log p_i$, which is an additive measure and the entropy of degree β, $H_n^\beta(P) = a(\sum p_i^\beta - 1)$ which is a non-additive measure. Several functional equations arising from the sum representation of these measures is the following one:

$$\sum_{i=1}^{n} \sum_{j=1}^{n} f(x_i y_j) = \sum_{i=1}^{m} f(x_i) + \sum_{j=1}^{n} f(y_j) + c\sum_i f(x_i) \sum_j f(y_j) \tag{1.1}$$

$$(x_i, y_j \in [0,1], \quad \sum x_i = 1 = \sum y_j)$$

($c = 0$ leads to H_n, $c \neq 0$ leads to H_n^β). Starting from Chaundy and McLeod [5], (1.1) with $c = 0$ or $c \neq 0$ has been studied by many authors in [1,2,3,4,6,7,8] under various hypothesis, such as (1.1) holding for all m,n, and f is continuous, (1.1) (with $c = 0$) holding for $m = 2$, $n = 3$ and f is measurable etc. Only recently, (1.1) was studied in [9] for fixed m and n, under the condition of measurability of f and it is natural to expect

D. G. Lainiotis and N. S. Tzannes (eds.), Advances in Communications, 45–50.

the solutions of (1.1) to depend upon m and n, which is indeed the case.

In this paper, along the same lines, we study the functional equation

$$\sum_{i=1}^{m} \sum_{j=1}^{n} f(x_i y_j) = \sum_{i=1}^{m} g(x_i) + \sum_{j=1}^{n} h(y_j) + \sum_{i=1}^{m} k(x_i) \sum_{j=1}^{n} \ell(y_j) \tag{1.2}$$

$$(x_i, y_j \in [0,1], \quad \sum x_i = 1 = \sum y_j)$$

which is a generalization of (1.1); which gives rise to more measures and which also helps to solve functional equations arising from other measures such as directed divergence etc. As a matter of fact, we solve the functional equation (1.2) holding for some (arbitrary but) fixed pair (m,n) when the functions involved are Lebesque measurable using different techniques and show that the solutions indeed depend upon the pair (m,n) and obtain type of solutions of (1.2) which are different in nature from that of (1.1).

2. Auxiliary results

In this section are listed two known results from [9,10] which are made use of in the next section to obtain the solutions of (1.2). Let $\Gamma_n = \{P = (p_1, \ldots, p_n)\colon \ p_i \geq 0, \ \sum p_i = 1\}$; $I = [0,1]$, $I_1 =]0,1]$. We follow the convention $0 \log 0 = 0$, $0^\beta = 0$, $1^\beta = 1$.

Result 1. [9] Let $F\colon I \times I \to R$ (reals), be measurable in each variable and satisfy the functional equation

$$\sum_{i=1}^{m} \sum_{j=1}^{n} F(x_i, y_j) = 0 \tag{2.1}$$

$(X = (x_i) \in \Gamma_m, \ \ Y = (y_j) \in \Gamma_n)$ holding for some fixed pair of integers $m,n \geq 3$. Then F is given by

$$F(x,y) = F(0,y)(1-mx)+F(x,0)(1-ny) \tag{2.2}$$
$$-F(0,0)(1-mx)(1-ny).$$

Result 2. [10] Let $F,G,H,K,L\colon S \to C$, satisfy

$$F(xy) = G(x)+H(y)+K(x)L(y) \tag{2.3}$$

where S is an arbitrary Abelian Semigroup which has a fixed element 'a' such that $a \cdot x = b$ is solvable in x for every $b \in S$, C, the set of complex numbers. Then the general solutions of

(2.3) are the following:

$$(s_1)\begin{cases} F(x) = \phi(x)+\alpha_1, \quad G(x) = \phi(x)-\alpha_3K(x)+\alpha_2+\frac{\alpha_1}{2}, \\ H(x) = \phi(x)+\frac{\alpha_1}{2}-\alpha_2, \quad K, \text{ arbitrary}, \quad L(x) = \alpha_3; \end{cases}$$

$$(s_2)\begin{cases} F(x) = \alpha_1\psi(x)+\phi(x)+\alpha_2, G(x) = \alpha_3\psi(x)+\phi(x)+\alpha_4 \\ H(x) = \alpha_5\psi(x)+\phi(x)+\alpha_6, \quad K(x) = \alpha_7\psi(x)+\alpha_8, \\ L(x) = \alpha_9\psi(x)+\alpha_{10}, \end{cases}$$

with $\alpha_1 = \alpha_7\alpha_9, \ \alpha_3+\alpha_7\alpha_{10} = 0 = \alpha_5+\alpha_8\alpha_9,$

$\alpha_2 = \alpha_4+\alpha_6+\alpha_8\alpha_{10},$

$$(s_3)\begin{cases} F(x) = \alpha_1\phi(x)^2+\alpha_2\phi(x)+\phi_1(x)+\alpha_3, G(x) = a_1\phi(x)^2+\phi_1(x)+\alpha_4 \\ H(x) = \alpha_1\phi(x)^2+\alpha_5\phi(x)+\phi_1(x)+\alpha_6, \quad K(x) = 2\alpha_1\phi(x)+\alpha_7, \\ L(x) = \phi(x)+\alpha_8 \end{cases}$$

with $\alpha_2 = 2\alpha_1\alpha_8 = \alpha_5+\alpha_7, \ \alpha_3 = \alpha_4+\alpha_6+\alpha_7\alpha_8,$

where ϕ, ϕ_1 respectively ψ satisfy

$$\phi(xy) = \phi(x)+\phi(y), \tag{2.4}$$

$$\psi(xy) = \psi(x)\psi(y), \tag{2.5}$$

and α_i's constants.

3. Solutions of the functional equation (1.2)

Let $f,g,h,k,\ell: I \to R$, be measurable and satisfy the functional equation (1.2), for a fixed pair m,n (≥ 3).

By defining

$$F(x,y) = f(xy)-yg(x)-xh(y)-k(x)\ell(y), \tag{3.1}$$

for $x,y \in I$, it follows that F is measurable in each variable and by (1.2) F satisfies (2.1) so that by Result 1, (2.2) holds, where by (3.1),

$$F(x,0) = \beta_1-\beta_3x-\beta_5k(x),$$

$$F(0,y) = \beta_1-\beta_2y-\beta_4\ell(y),$$

$$F(0,0) = \beta_1 - \beta_4\beta_5, \quad \text{with}$$

$$\beta_1 = f(0),\ \beta_2 = g(0),\ \beta_3 = h(0),\ \beta_4 = k(0)\ ,$$

$$\beta_5 = k(0). \tag{3.2}$$

Thus,

$$f(xy)-yg(x)-xh(y)-k(x)\ell(y) = (\beta_1-\beta_2 y-\beta_4\ell(y))(1-mx)$$

$$+(\beta_1-\beta_3 x-\beta_5 k(x))(1-ny)-(\beta_1-\beta_4\beta_5)(1-mx)(1-ny),$$

which can be rewritten as

$$\frac{f(xy)-\beta_1}{xy} + nm\beta_1 - nm\beta_4\beta_5 = \frac{g(x)-\beta_2+n\beta_5(k(x)-\beta_4)}{x} + m\beta_2$$

$$+ \frac{h(y)-\beta_3+m\beta_4(\ell(y)-\beta_5)}{x} + n\beta_3 + \frac{k(x)-\beta_4}{x}\,\frac{\ell(y)-\beta_5}{y}$$

for $x,y \in I_1$, which in turn by defining

$$\begin{cases} F(x) = \dfrac{f(x)-\beta_1}{x} + mn\beta_1 - mn\beta_4\beta_5 \\[2ex] G(x) = \dfrac{g(x)-\beta_2+n\beta_5(k(x)-\beta_4)}{x} + m\beta_2 \\[2ex] H(x) = \dfrac{h(x)-\beta_3+m\beta_4(\ell(x)-\beta_5)}{x} + n\beta_3 \\[2ex] K(x) = \dfrac{k(x)-\beta_4}{x}\ , \quad L(x) = \dfrac{\ell(x)-\beta_5}{x}\ , \quad \text{for } x \in I_1, \end{cases} \tag{3.3}$$

can be put in the form (2.3). Thus by Result 2, the solutions of (2.3) are given either by (s_1) or by (s_2) or by (s_3). The measurability of f,g,h,k,ℓ implies the measurability of F,G,H,K,L which in turn implies the measurability of ϕ,ϕ_1 satisfying (2.4) and ψ satisfying (2.5). Hence, ϕ,ϕ_1,ψ appearing in $(s_1),(s_2)$, (s_3) are of the form

$$\phi(x) = a \log x,\ \phi_1(x) = b \log x,\ \psi(x) = x^{\beta-1}, \tag{3.4}$$

where a,b,β are real constants.

Thus, the solution of (1.2), corresponding to the solution (s_1) of (2.3), from (3.3) and (3.4), has the form

$$(a_1)\begin{cases} f(x) = ax\log x + (\alpha_1 - mn\beta_1 + mn\beta_4\beta_5)x + \beta_1 \\ g(x) = ax\log x + (\alpha_2 + \frac{\alpha_1}{2} - m\beta_2 - \alpha_3 K(x) - n\beta_5 K(x))x + \beta_2 \\ h(x) = ax\log x + (\frac{\alpha_1}{2} - \alpha_2 - n\beta_3 - m\alpha_3\beta_4)x + \beta_3 \\ k(x) = xK(x) + \beta_4,\ \ell(x) = \alpha_3 x + \beta_5 \end{cases}$$

where $K(x)$ is arbitrary. It is easy to see from (3.2), that (a_1) also holds for $x = 0$. Thus (a_1) constitutes a solution of (1.2), where K is an arbitrary function and a, α_i's, β_i's are arbitrary constants.

Similarly, from the solutions (s_2) and (s_3) of (2.3), (3.3), (3.4) and (3.2), we obtain the following solutions of (1.2):

$$(a_2)\begin{cases} f(x) = \alpha_1 x^\beta + ax\log x + (\alpha_2 - mn\beta_1 + mn\beta_4\beta_5)x + \beta_1 \\ g(x) = (\alpha_3 - n\alpha_7\beta_5)x^\beta + ax\log x + (\alpha_4 - m\beta_2 - n\alpha_8\beta_5)x + \beta_2 \\ h(x) = (\alpha_5 - m\alpha_9\beta_4)x^\beta + ax\log x + (\alpha_6 - m\alpha_{10}\beta_4 - n\beta_3)x + \beta_3 \\ k(x) = \alpha_7 x^\beta + \alpha_8 x + \beta_4,\ \ell(x) = \alpha_9 x^\beta + \alpha_{10}x + \beta_5 \end{cases}$$

with $\alpha_1 = \alpha_7\alpha_9, \alpha_3 + \alpha_7\alpha_9 = 0 = \alpha_5 + \alpha_8\alpha_9,\ \alpha_2 = \alpha_4 + \alpha_6 + \alpha_8\alpha_{10}$;

and

$$(a_3)\begin{cases} f(x) = \alpha_1 a^2 x\log^2 x + (\alpha_2 a + b)x\log x + (\alpha_3 - mn\beta_1 + mn\beta_4\beta_5)x + \beta_1 \\ g(x) = \alpha_1 a^2 x\log^2 x + (b - 2na\beta_5\alpha_1)x\log x + (\alpha_4 - m\beta_2 - n\beta_5\alpha_7)x + \beta_2 \\ h(x) = \alpha_1 a^2 x\log^2 x + (a\alpha_4 + b - ma\beta_4)x\log x + (\alpha_6 - n\beta_3 - m\alpha_8\beta_4)x + \beta_3 \\ k(x) = 2\alpha_1 ax\log x + \alpha_7 x + \beta_4 \\ \ell(x) = ax\log x + \alpha_8 x + \beta_5 \end{cases}$$

with $\alpha_2 = 2\alpha_1\alpha_8 = \alpha_5,\ \alpha_3 = \alpha_4 + \alpha_6 + \alpha_7\alpha_8$, where a, b, α_i's, β_i's are arbitrary constants.

Thus, we have proved the following theorem:

Theorem. Let $f, g, h, k, \ell\colon I \to R$ be measurable. Then these functions satisfy the functional equation (1.2), for a fixed pair m, n (≥ 3) if and only if they are given either by (a_1) or by (a_2) or by (a_3).

It is easy to verify that (a_1), (a_2), (a_3) indeed are solutions of (1.2).

Remark. These type of equations were studied either for all m,n or for some special pairs m,n like (2.3), (3.3) etc. This is the first time that these type of equations are solved for fixed pair m,n showing thereby that the solution indeed depend upon m,n also and also has more solutions of different nature hitherto known. The methods employed even though involves different techniques, they are simple and direct. These type of equations (solutions) may yield more measures and are helpful in characterizing Shannon's entropy, entropy of degree β etc.

References

(1) Aczél, J. and Daróczy, Z.: (1975) On measures of information and their characterizations, Academic Press, New York.

(2) Aczél, J. and Daróczy, Z.: (1963) Charakterisierung der Entropien positiver orderung und der Shannonschen Entropie, Acta Math. Acad. Sci. Hung. 14, pp.95-121.

(3) Behara, M.: (1978) Additive and nonadditive measures of entropy with applications, McMaster University report.

(4) Behara, M. and Nath, P.: (1973) Additive and nonadditive entropies of finite measurable partitions, Probability and Information Theory, Vol. 2, Springer Verlag, V296, pp.102-138.

(5) Chaundy, T.W. and McLeod, J.B.: (1960) On a functional equation, Edin. Math. Notes 43, pp.7-8.

(6) Daróczy, Z.: (1971) On the measurable solutions of a functional equation, Acta Math. Acad. Sci. Hungar. 22, pp.11-14.

(7) Kannappan, Pl.: (1972) On Shannon's entropy, directed divergence and inaccuracy, Z. Wahr. Verw. Geb. 25, pp.95-100.

(8) Kannappan, Pl.: On a generalization of sum form functional equation - I (to appear).

(9) Kannappan, On some functional equations from additive and nonadditive measures - I (to appear in Proc. Edin. Math. Soc.)

(10) Vincze, E.: (1962) Eine allgemeinere Methode in der Theorie der Funktional gleichungen, I, Publ. Math. Debrecen, V9, pp.149-163.

INFORMATION AND COMPLEX SYSTEMS

J. Dufour , G. Gilles

Université de Lyon I , France
Laboratoire d'Automatique

ABSTRACT

The systems that Man has to analyse are becoming more and more complex. This leads to elaborate an "intensity" of the various relationships between the numerous characteristic variables in order to get a hierarchy between the relationships. This measurement presents an informational feature.
After a recall on the basic concepts concerning the informational entropy, their extension to complex systems is presented. Those tools are used for a proposal of partitions procedure in both static and dynamic cases. Finally, it is shown how to evaluate the partitions quality according to the Information decomposition formulas issued from recent researches.

1. COMPLEX SYSTEMS ANALYSIS

Man is concerned with more and more complex systems that he needs to deeply analyse and eventually to control : natural complex systems such as living organisms, economic systems, industrial processes for which higher and higher performances are required.
The analysis and the control of complex systems cannot be correctly achieved by using classical methods. That is why, for more than 10 years, the evolution of Automatic Control is strongly devoted to the elaboration of new methods suited to complex systems.
The possibility of more precise analysis of complex systems is given by the improvement of computer science and the huge technological progress in the field of digital computers.
Modellisation and control of complex systems require the systematic use of all the Information gathered on the system (phenomeno-

D. G. Lainiotis and N. S. Tzannes (eds.), Advances in Communications, 51–58.

logical knowledge, data information, ...). Then it becomes necessary to develop methods which allow structurising the informations, initially too numerous, and synthesizing them into different forms (graphs,...) in order to use them more easily.
In this way, it is important to point out the main mechanisms that occur in a complex system with respect to the weaker ones by measuring what we can call an "intensity"of linkages between variables or groups of variables. Then, in order to reach better commands of a complex system, it becomes necessary to decompose it into subsystems as weakly coupled as possible.

2. THE FUNDAMENTAL CONCEPTS OF INFORMATIONAL ENTROPY AND OF TRANSINFORMATION

Measuring the intensity of a relationship between 2 variables x_a and x_b or between 2 groups of variables in a system has a definite informational aspect since this involves detecting, in the information concerning a variable x_b, the information coming from a variable x_a ; information related to other variables appears as noise.
We will briefly recall some elements of the Information theory introduced by Shannon (1) and developped by Mc. Gill (2), Ashby (3) and Conant (4).
Most of the time, the observations of the same variable x_a are made by a sampling with a constant period : $(x_a^1, x_a^2, ..., x_a^k, ...$ $x_a^N)$, x_a^k being the kth observation of the ath variable. However,in a more general manner, any variable x_a can be either continuous or discrete ; in all cases, its value range has to be reduced to a certain number of classes, which implies a value range quantification in the continuous case and a possible value regrouping in the discrete case (5).
Let therefore $I_a = \{1, 2, ..., i, ..., c_a\}$ be the set of x_a variable classes and let $n_a(i)$ be the number of the event happenings "x_a belongs to the ith class". The frequencies $n_a(i)/N$ thus defined can tend towards the $p_a(i)$ probabilities when N tends to infinity, but in all what follows the existence of such probabilities is not necessary.
The informational entropy of a variable x_a can be calculated from N observations of x_a with the help of the expression :

$$H(x_a) = \log N - \frac{1}{N} \sum_{i=1}^{c_a} n_a(i) \log n_a(i)$$

Whereas in the probabilistic domain this function can be considered as an a priori indetermination measurement of the variable x_a, leaving aside this domain in the present application, it can be interpreted as a measurement of the variability of x_a. As a matter of fact, this function is inferiorly limited by zero, value that is reached when x_a has all its values which belong to the same class, superiorly limited by 1, value corresponding to an uniformly distributed variable.

The conjoint entropy of the 2 variables x_a and x_b is defined in the same way by :

$$H(x_a,x_b) = \log N - \frac{1}{N}\sum_{i=1}^{c_a}\sum_{j=1}^{c_b} n(i,j)\log n(i,j)$$

$n(i,j)$ being the number of event happenings "x_a belongs to the ith class of I_a "and" x_b belongs to the jth class of J_b".
The conditional entropy : $H_{x_a}(x_b)$ can be defined by fixing the variable x_a.
This definition leads to the property :

$$H(x_a,x_b) = H(x_a) + H_{x_a}(x_b) = H(x_b) + H_{x_b}(x_a)$$

The preceeding concepts allow the introduction of the global trans-information of x_a to x_b :

$$T(x_a : x_b) = H(x_a) + H(x_b) - H(x_a, x_b)$$

This is a non negative measurement of the relationship intensity between x_a and x_b. It takes a zero value if and only if x_a and x_b are statistically independant and it has its maximum value when one variable depends only on the other.

3. EXTENSION TO COMPLEX SYSTEMS

The extension of these concepts to complex systems does not present any difficulties. As a matter of fact, let $I_1, I_2, \ldots, I_a, \ldots, I_p$ be the class sets of the variables $x_1, x_2, \ldots, x_a, \ldots, x_p$ which characterise the system S. Let us call $n(i,j,\ldots,r)$ the number of event happenings "x_1 belongs to the ith class of I_1 and x_2 belongs to the jth class of I_2 and ... and x_p belongs to the rth class of I_p". Then the entropy of the system S is expressed by :

$$H(x_1,x_2,\ldots,x_a,\ldots,x_p) = -\sum_{i\in I_1}\sum_{j\in I_2}\ldots\sum_{r\in I_p}\frac{n(i,j,\ldots,r)}{N} \cdot \log\left\{\frac{n(i,j,\ldots,r)}{N}\right\}$$

The entropy of a group of variables $(x_1, x_2,\ldots,x_p)$ can also be defined in the equivalent following way by using the conditional entropies :

$$H(x_1, x_2,\ldots,x_p) = H(x_1) + H_{x_1}(x_2) + \ldots + H_{x_1\ldots x_{p-1}}(x_p)$$

The global transinformation of the system S generalises the trans-information of 2 variables :

$$T(x_1 : x_2 \ldots : x_a : \ldots x_p) = H(x_1)+H(x_2)+\ldots+H(x_a)+\ldots+H(x_p) - H(x_1, x_2,\ldots,x_a,\ldots,x_p)$$

It measures the intensity of all the linkages between all the variables in the system S.
Other particular types of transinformations are often used for

complex systems :

- $T[(x_1,x_2,\ldots,x_{p-1}) : x_p]$ measures the linkage intensity between x_p and the group of variables $(x_1, x_2,\ldots,x_{p-1})$; the linkages between the variables $x_1, x_2,\ldots,x_{p-1}$ are not taken into account.
- $T[(x_1,x_2,\ldots,x_r) : (x_{r+1}, \ldots, x_p)]$ measures the linkage intensity between the 2 groups of variables $(x_1,x_2,\ldots,x_r)$ and $(x_{r+1}, \ldots, x_p)$; the other linkages between the variables $x_1, x_2,\ldots, x_r$ and between $x_{r+1},\ldots,x_p$ are not taken into account.
- $T_{\underline{Z}-\{x_a,x_b\}}(x_a : x_b)$ measures the <u>direct</u> transinformation between the 2 variables x_a and x_b. The other variables of the system S characterised by the vector $\underline{Z} = (x_1 \ldots x_a \ldots x_p)^t$ being fixed, the apparent linkages between x_a and x_b via the other variables are not taken into account.

4. THE CONCEPT OF COMPLEXITY

The concept of informational entropy $H(\underline{Z})$ of a system allows specifying the commonly used, more or less vague, idea of complexity of a system. Ashby (3) proposed the use of the informational entropy $H(\underline{Z}) = H(x_1,\ldots,x_a,\ldots,x_p)$ as a measurement of the complexity of a system described by a vector $\underline{Z}$ of characteristic variables $(x_1,\ldots,x_a,\ldots,x_p)$. In fact, intuitively, we see that the more numerous are the variables of a system, the more important is the number of classes of each variable and the higher can be the entropy value. But the difficulty is due to the fact that complexity is not an internal property of a system but depends on its relation with an observer.

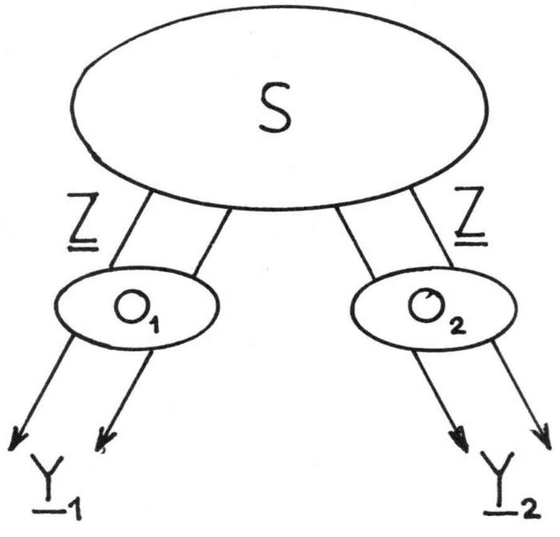

Fig. 1 Observations of a complex system

Consider a system S described by a characteristic vector $\underline{Z}$ and observed by 2 different operators O_1 and O_2 respectively restoring the vectors $\underline{Y_1} = (y_{11},\ldots,y_{1a},\ldots,y_{1n})$ and $\underline{Y_2} = (y_{21},\ldots, y_{2a}, \ldots, y_{2m})$ with $m, n \leq p$. It can be easily demonstated (6) that :

$$H(\underline{Z}) \geq H(\underline{Y}_1) \text{ and } H(\underline{Z}) \geq (\underline{Y}_2).$$

Moreover, if one operator O_1 restores a vector $\underline{Y_1}$ whose dimension is greater than that of $\underline{Y}_2$, i.e. $\underline{Y_1} = (\underline{Y}_2, y_o)$, then $H(\underline{Y_1}) \geq H(\underline{Y_2})$, which means that a more precise observation of a system leads to a loss of complexity.
Other measurements of complexity, also entropy-based, can be used in order to take into account the evolution of the system knowledge (4).

5. FORMULAS OF INFORMATION DECOMPOSITION

5.1. Partition into subsystems

A data analysis of the linkage between variables taken in pairs leads to the construction of a coupling matrix. The introduction of the classes defined previously has a direct influence on the results and it leads to use reduced transinformations relatively insensitive to the number of classes of each variable. We propose:

$$t(x_a : x_b) = \frac{T(x_a : x_b)}{H(x_b)}$$

which becomes 1 when there is an application of the value range of x_a in that of x_b. However, it is not totally insensitive to the number of classes ; this influence is calculated in the reference (5).
Having obtained the linkage matrix by the above mentioned procedure, the most difficult problem is to propose one or more partitions using the couplings matrix. For a relatively high number of variables, this cannot be done at first sight. It is therefore necessary to be helped by the decomposition techniques of valuated graphs. Moreover, as several partitions into subsystems can be proposed for the initial complex system, an important problem is to measure the "quality" of the partitions in order to select the best one with respect to a decoupling criterion. Formulas of information decomposition will allow this measurement in the static and dynamic cases.

5.2. Static partitions

Consider a system S characterised by a vector $\underline{Z}$ and a particular partition of S into q subsystems SS_m themselves characterised by vectors $\underline{Z}_m$.
Conant (6) has demonstrated the following formula of information decomposition :

$$T(\underline{Z}) = \sum_{m=1}^{q} T(\underline{Z}_m) + T(\underline{Z}_1 : \underline{Z}_2 : \dots : \underline{Z}_m : \underline{Z}_q)$$

which means that the "activity" inside the whole system S is equal to the sum of the activities inside each subsystem SS_m augmented by the interactions between all the subsystems.
So, one can get a partition into subsystems as well decoupled as $T(\underline{Z}_1 : \dots : \underline{Z}_m : \dots : \underline{Z}_q)$ is small with respect to $T(\underline{Z})$

5.3. Dynamic partitions

In order to measure the importance of a dynamic linkage between 2 variables x_a and x_b, the transinformation coefficient is expressed from the variable x_a shown by the sequence $(x_a^1, x_a^2, \dots, x_a^{N-1})$ and from the variable x_b shown by the sequence $(x_b^2, x_b^3, \dots, x_b^N)$. We will call x_b' such a variable having undergone a time delay of one sampling period. In this case,

$$T(x_a : x_b') = H(x_a) + H(x_b') - H(x_a, x_b')$$

measures the linkage intensity between x_a taken at a given time and x_b taken at the following sampling time ; it is a linkage measurement in the direction $x_a \rightarrow x_b$ and in general has not the symmetric characteristic.
From the linkage matrix made with the help of this coefficient, several partitions can be suggested and different procedures can be proposed in order to test their validity.
Conant (6) suggested the following heuristic procedure :
$\underline{Z}_m = (x_{m_1}, \dots, x_{m_\ell}, \dots, x_{m_{r_m}})^t$ being the characteristic vector associated with the subsystem SS_m, we compute :
$T_{wm} = T(x_{m_1} : \dots : x_{m_\ell} : x_{m_{r_m}} : x'_{m_1} : \dots : x'_{m_\ell} : \dots : x'_{m_{r_m}})$
which signifies the dynamic activity inside the subsystem SS_m.
We also compute :
$T_{bmn} = T\,[(\underline{Z}_m, \underline{Z}'_m) : (\underline{Z}_n, \underline{Z}'_n)]$ which measures the dynamic interactions between the subsystems SS_m and SS_n
and $T_b = T[(\underline{Z}_1, \underline{Z}'_1) : \dots : (\underline{Z}_m, \underline{Z}'_m) : \dots : (\underline{Z}_q, \underline{Z}'_q)]$
which takes into account all the interactions between the subsystems.
A partition is then said to be valid if the following inequalities are satisfied : $T_{wm} + T_{wn} \gg T_{bmn} \quad \forall\, m,n\ ,\ m \neq n$

$$\sum_{m=1}^{q} T_{wm} \gg T_b$$

Richetin (8) thought that it is difficult to precise the concept of independance between the subsystems if the structure of the system is not defined. So, he assumed that the evolution of the system is described by state equations and he got formulas of information decomposition in the dynamic case. We will not develop them because they can appear as a particular case $\ell = 0$ of our work.

Then, Dufour (6) gave a more general representation of the evolution of the system by means of equations involving the vector $\underline{Z}$ of the "characteristic" variables :

$$x_a(k+1) = f_a[\underline{Z}(k), \underline{Z}(k-1),\ldots,\underline{Z}(k-\ell),\ldots,\underline{Z}(k-n)] \quad a = 1 \text{ to } p$$

By calculating the generalised dynamic transinformation :

$$T_D = \sum_{a=1}^{p} T[(\underline{Z}(k),\ldots,\underline{Z}(k-\ell),\ldots,\underline{Z}(k-n)) : x_a(k+1)]$$

he got the following formula of dynamic information decomposition:

$$T_D = \sum_{m=1}^{q} \Big(\sum_{x_a \in \underline{Z}_m} T[\underline{Z}_m(k) : x_a(k+1)]\Big) +$$

$$+ \sum_{\ell=1}^{n} \Big(\sum_{m=1}^{q} \sum_{x_a \in \underline{Z}_m} T_{\underline{Z}(k)\ldots\underline{Z}(k-\ell+1)}[\underline{Z}_m(k-\ell) : x_a(k+1)]\Big) +$$

$$+ \sum_{m=1}^{q} \Big(\sum_{x_a \in \underline{Z}_m} T_{\underline{Z}(k)}[\underline{C}_m(k) : x_a(k+1)]\Big) +$$

$$+ \sum_{\ell=1}^{n} \Big(\sum_{m=1}^{q} \Big(\sum_{x_a \in \underline{Z}_m} T_{\underline{Z}(k)\ldots\underline{Z}(k-\ell+1)\underline{Z}(k-\ell)}[\underline{C}_m(k-\ell):x_a(k+1)]\Big)\Big)$$

where $\underline{C}_m(k)$ = "$\underline{Z}(k) - \underline{Z}_m(k)$" represents the vector of the variables of $\underline{Z}$ outside of $\underline{Z}_m$.
The practical meaning of this formula appears more clearly in the following reduced form :

$$T_D = T_{DI} + \sum_{\ell=1}^{n} T_{D\ell I} + T_{DB} + \sum_{\ell=1}^{n} T_{D\ell B}$$

whichs signifies that the dynamic activity inside the whole system S is equal to the sum of the dynamic activities inside (subscript I) each subsystem SS_m augmented by the dynamic interactions between (subscript B) all the subsystems.

6. APPLICATIONS AND CONCLUSION

The computation of transinformations can be considered as a basic tool very useful in the methods elaborated for analysing complex systems. It has been fruitfully used in differents kinds of applications, mainly involving industrial processes and macroeconomic systems (6)(10).
Nevertheless, we must not forget that the transinformation does not explain the causality of the phenomena, that the use of this coefficient for all the possible combinations for analysing large-scale complex dynamic systems leads to a huge amount of computations and that several technical problems occur in concrete applications(9).

The structural analysis methods based on Information theory allow establishing a hierarchy between the relationships in a complex system (internal mechanisms and input-output relationships) and thus they point out the most significant control variables. They allow dynamic analysis of relationships and thus precise the memory effects in the system. One of the most important advantages of the transinformation is not requiring linearity assumptions. So, the associated methods can lead to the analysis of non linear linkages, and also linkages involving more than 2 variables, by using $t[(x_a, x_b):x_c]$ for example. Of course, one important practical application of those methods is their use for a proposal of partitions of a complex system into subsystems and mainly the possibility of measuring the quality of those partitions, which can lead to choice criteria very useful for a future hierarchical control. Finally, those methods can appear as a help to modelling. By comparison of the results obtained from the real data on the system and from a model, they point out the sharp features and the relative weaknesses of the modellisation. So they appear as a way to improve a model. Undoubtless, a precise informational analysis of complex systems allows a better overall knowledge of them and can lead to better commands.

REFERENCES

(1) Shannon, C.E. : A mathematical theory of communication, 1948, The Bell syst. tech. J., 17, pp. 623-656

(2) McGill, W.J. : Multivariate information transmission, 1954, Psychometrika, 19, pp. 97-116

(3) Ashby, W.R. : Measuring the internal information exchange in a system, 1965, Cybernetica, 1, pp. 5-22

(4) Conant, R.C. : Information transfer in complex systems with applications to regulation, 1968, Ph. D. thes., Univ. of Illinois

(5) Dufour, J., Gilles, G., Foulard, C. : Analyse structurale et partition des systèmes dynamiques complexes à l'aide de la théorie de l'Information ; étude de l'influence du nombre de classes, 1976, C.R.A.S., t. 282, pp. 543-546

(6) Dufour, J. : Méthodes et méthodologie d'analyse de systèmes complexes. Application aux procédés industriels et aux systèmes macroéconomiques, 1979, thes. Dr. Etat, Univ. Lyon, France

(7) Conant, R.C. : Detecting subsystems of a complex system, 1972, I.E.E.E. trans. syst. man. cyb., pp. 550-553

(8) Richetin, M. : Analyse structurale des systèmes complexes en vue d'une commande hiérarchisée, 1975, thes. Dr. Etat, Univ. Toulouse, France

(9) Dufour, J. : Analyse structurale et partition des systèmes complexes. Application aux systèmes macroéconomiques, 1977, R.A. I.R.O. Aut., 11, pp. 335-354

(10) Dufour, J., Gilles, G. : Application of some concepts of the information theory to structural analysis and partition of macroeconomic large scale systems, 1977, IFAC workshop 'Information and systems', Compiègne, France, Pergamon press

THE EVOLUTION OF ENTROPY AND OF OTHER STATISTICAL PARAMETERS OF WRITTEN GREEK

F.Koursoubas and G.Kokkinakis

Laboratory of Wire Communications, School of Engineering,
University of Patras, Patras, Greece.

ABSTRACT

The entropy of written Greek in various historical periods has been calculated as an approximation of N-gram entropies according to the Shannon formula. For this purpose, the frequencies of letters, digrams and the most common words, were measured in each case from selected texts with the aid of a computer.

Five texts covering a period of 3000 years were used: Texts from Homer (9th century B.C.), from Plato (5th century B.C.), from Klemes of Alexandria (2nd century A.D.), from Kornaros (17th century A.D.), and from several books and newspapers of modern Greek.

Comparisons of the statistical parameters of written Greek with those of other languages are also made, and conclusions are drawn concerning their evolution. In addition, several tests on predictability of modern Greek based on the found statistics have been performed, and the results are evaluated.

I. INTRODUCTION

This paper deals with the statistics of the written Greek language, especially with the entropy and the redundancy, from ancient times to our days.

It is known, that the entropy of a language is a measure of the information contained in each letter (or symbol) of a text in the language. It is given by the average number of binary digits

D. G. Lainiotis and N. S. Tzannes (eds.), Advances in Communications, 59–68.

(bits) per letter, when the language is encoded in the most efficient way. In this case only the meaningfull combinations of letters, words and sentences of the language are considered, while all possible but meaningless combinations are neglected.

The entropy which is found when n adjacent symbols of a text are considered is called the N-gram entropy F_N. According to C.E.Shannon the N-gram entropy is given by the formula (1,2):

$$F_N = -\sum_{i,j} p(b_i,j)\log_2 p_{b_i}(j) =$$

$$= -\sum_{i,j} p(b_i,j)\log_2 p(b_i,j) + \sum_i p(b_i)\log_2 p(b_i)$$

where b_i is a group of N-1 letters [(N-1)-gram]
j is an arbitrary letter following b_i
$p(b_i,j)$ is the probability of the N-gram b_i,j
$p_{b_i}(j)$ is the conditional probability of letter j after the group b_i, and is given by $p(b_i,j)/p(b_i)$

From this formula the maximum possible entropy H_m of a language is obtained, assuming that all the letter combinations are used and all letters occur with equal probability in the text:

$$H_m = F_o = \sum_{i=1}^{v} p(i)\log_2 p(i) = \sum_{i=1}^{v} \log_2 p(i)$$

with v the amount of letters (or symbols) of the language.

F_o is 4.70 bits per letter in English, German and other languages with 26 letters, 4.56 bits per letter in Greek which has 24 letters. If the space is included as an additional letter, F_o increases to 4.76 bits per letter in English and 4.64 bits per letter in Greek.

Taking in to account the probability with which each letter appears in the text, the entropy is reduced, for instance in English, to

$$F_1 = -\sum_{i=1}^{26} p(i)\log_2 p(i) = 4{,}14 \text{ bits per letter}$$

Considering, further, the probability of appearance in the text of digrams, that is combinations of two letters, of trigrams,etc., the entropy is further reduced, due to the constraints of structure imposed in the text. With N increasing, the N-gram entropy F_N takes more and more from the statistics of the language. So it is possible to estimate the entropy H of a language as an approximation of the N-gram entropies. That is

$$H = \lim_{N\to\infty} F_N$$

This procedure has been applied for the estimation of the Greek language's entropy.

The redundancy on the other hand is a measure of the efficiency with which a language is used. It is given by the difference between the maximum possible entropy and the real entropy in per cent of the maximum entropy:

$$R = \frac{H_m - H}{H_m} 100\%$$

A big redundancy means that a big amount of possible combinations of letters, words, sentences etc. remain unused.

II. THE STATISTICS OF THE GREEK LANGUAGE

To apply the above described method for the entropy's estimation the statistics of a language must be known. In several languages some statistics were available, for example the frequencies of letters and digrams, the most frequent syllables, words etc. Using such statistics and several approximations for the parameters not available the entropy of written English was estimated by Shannon and that of German by Küpfmüller (2,3).

In Greek no statistics of this kind existed, and so we had to undertake the necessary measurements from the beginning. This was accomplished with the aid of a computer. Programs for measuring the frequencies of letters, digrams, trigrams and of the most common words were established and several texts were selected and given in to the computer.

In order to get a first approach of the evolution of entropy and of other statistical parameters, texts of five different periods were analysed. That is texts from Homer's Iliad (9th century B.C.), from Plato's Symposium (5th century B.C.), from Klemes' of Alexandria Exhortative Speach to Greeks (2nd century A.D.), from Kornaros Erotokritos (17th century A.D.) and texts from several books and newspapers of modern Greek. The texts had a length from 12000 to 14000 symbols (letters and spaces) with exception of the modern Greek for which in some cases texts of nearly 1 million symbols were also used.

From the measured frequencies of letters etc., and the given formulas the entropies F_1, F_2 and F_3 were calculated, also with the aid of a computer. In order to assess the error contained in that values, due to the limited number of text symbols taken into account, the entropies F_1 and F_2 of the modern Greek were measured also in a text of 980000 symbols. The error between the values in both texts, i.e. in the text with 12500 and in the text with 980000 symbols, was .030% and .044% correspondingly.

To obtain the entropy F_{word} we plot on a log-log paper the frequency of occurance of the words measured, against their frequency rank, as it is shown in Fig.1 for the text of Homer. In each case the values have been approximated with a straight line, so that the probability of the n_{th} most frequent word may be taken roughly from the formula

$$p_n = \frac{k}{n^{tana}}$$

where tana is the slope of the line and k is the probability for n=1. We used then the formula,

$$F^* = -\sum_{1}^{n} p_n \log_2 p_n \text{ bits per word}$$

n was taken from the equation

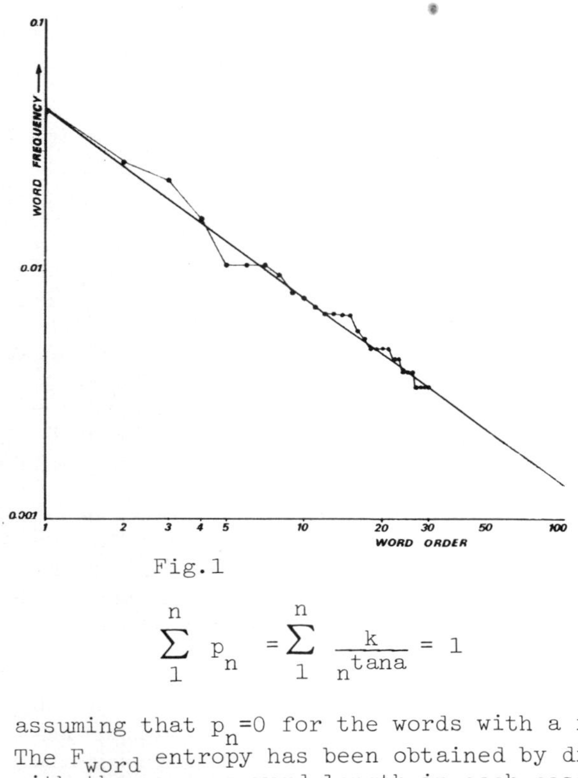

Fig.1

$$\sum_{1}^{n} p_n = \sum_{1}^{n} \frac{k}{n^{tana}} = 1$$

assuming that p_n=0 for the words with a rank higher than n (2). The F_{word} entropy has been obtained by dividing the F^*-entropy with the average word length in each case.

Table 1 gives the calculated F_N-entropies from the 5 texts. The value which F_N approximate when N goes to infinity is the entropy of the language. A calculation of that value must take into account combinations of words and phrases in long texts. Because that can not be practically done by measuring word and sentence

Table 1

	HOMER	PLATO	KLEMES	KORNAROS	MODERN GREEK
F_o (24 letter)	4.558	4.558	4.558	4.558	4.558
F_o (25 letter)	4.64	4.64	4.64	4.64	4.64
F_1 (24 letter)	4.089885	4.04737	4.076616	4.080366	4.083083
F_1 (25 letter)	4.045938	4.011512	4.056942	4.013937	4.075071
F_2	3.330454	3.097159	3.157458	3.135322	3.228751
F_3	2.608071	2.402081	2.5284	2.477356	2.532526
F_{word}	2.0124	2.0643	1.7163	1.9743	1.8247

combinations, we used the following method to estimate the entropy of modern Greek.

A person was given a text of modern Greek, long enough to get an idea of the described topic, for example a text with 300 words, and was asked to predict the next trigram of the text. If the predicted trigram was not correct, the person was told the correct trigram and tried to find out the next trigram. The test was repeated many times with several persons and different texts. From the average number of correctly predicted trigrams relative to the entire trigram numbers of the texts it became evident that approximately 65% of the trigrams of a common text in modern Greek can be predicted correctly by a person, due to the information contained in the previous text. That means that the entropy of trigrams can be reduced by a factor of .35, when the long range statistics of the language are considered. So we get an approximate value of .9 bits per letter for the modern Greek Entropy.

As it was impossible to perform the same test with the texts of the ancient periods, we estimated their entropy by extrapolating the F_N values under consideration of the relations existing between the entropies of the same order in the five texts.

From the estimated entropies H the redundancy R of the language was calculated in each case and is given in table 2.

Table 2

	HOMER	PLATO	KLEMES	KORNAROS	MODERN GREEK
H	1.2	0.80	0.85	1.0	0.90
R	0.7368	0.8245	0.8135	0.7807	0.8026

III. EVALUATION OF THE RESULTS

a. Entropy and redudancy

In Fig.2 the entropies of the five texts are presented together with their mean values. Table 3 gives the mean and the variance of the entropies.

From table 3 and Fig.2 we see that:

a. The variance σ is relatively small for each N-gram entropy, which means that the dispersion of values about the mean is small. In other words, no major variations of the information content of the language can be noted.

b. The dispersion of the values of each N-gram entropy diminishes with time. It seems that changes of the information content are geting smaller with time.

c. The variance increases with the rank of the N-gram entropy. In the entropy of letters the differences between the 5 texts are

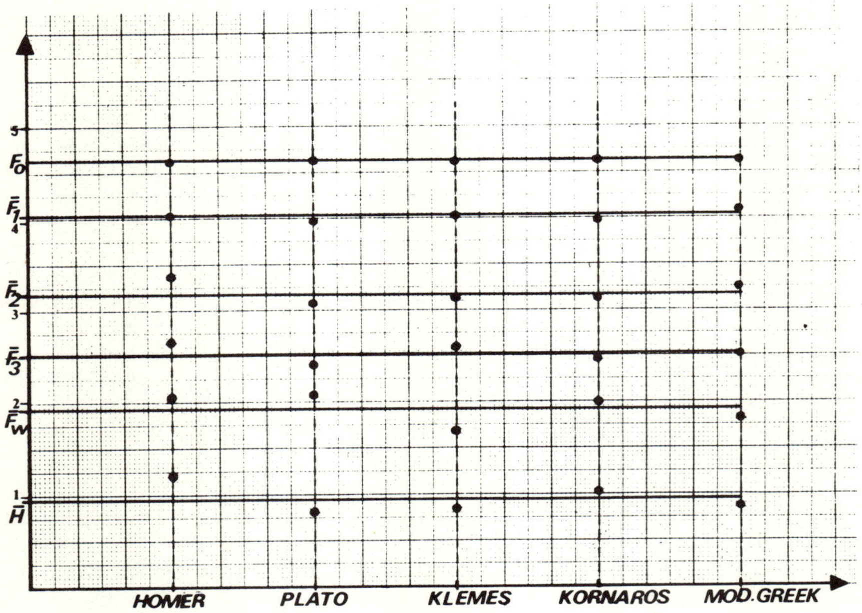

Fig. 2

Table 3

$\bar{F}_1$	$\bar{F}_2$	$\bar{F}_3$	$\bar{F}_{WORD}$	$\bar{H}$
4.044	3.186	2.519	1.918	0.95
σ_1^2	σ_2^2	σ_3^2	σ_{WORD}^2	σ^2
0.0080	0.0085	0.010	0.0135	0.025

smaller than in the entropy of digrams, trigrams, words etc. That was expected because the differentiation of the texts increases as the number of symbols in the N-grams is increased.

The fluctuation of the entropy reflects the changes of the language with time arising from changing circumstances. In general changes of the cultural system of people, as a consequence of economic, social, religious, etc. changes, change part of the language and therefore its entropy. Perhaps it would be possible to relate the entropy changes of language with progress or decadence of human societies. The results from the 5 periods examined contain some encouraging indications, but they are yet too limited to lead to conclusions.

b. The length of the words

Table 4 shows the average word length in letters in the five texts. We note a continuous increase of the word length from 4.761 letters in Homer to 5.815 in modern Greek. An exception is the length in Kornaros Erotokritos, which is 4.367 letters. But Erotokritos is a poem in which necessarily many articles, prepositi-

Table 4

HOMER	PLATO	KLEMES	KORNAROS	MODERN GREEK
4.7614	4.8517	5.2080	4.3670	5.8159

ons, conjunctions, etc. are used, that is small words which reduce the average word length of the text.

c. The frequency of vowels, consonants, initial and final letters of words in the text.

Table 5 shows the frequencies of vowels and their order in the 5 texts. Since Plato's time "A" is the most frequent letter in written Greek, with a frequency above 10%,"Ω" is the less frequent letter in all texts, with a frequency under 3.7%.

Table 5

HOMER	PLATO	KLEMES	KORNAROS	MODERN GREEK
E 0.122507	A 0.115138	A 0.111744	A 0.130072	A 0.120927
A 0.107179	I 0.102437	O 0.096260	O 0.097283	I 0.100598
I 0.98911	E 0.102334	E 0.093364	I 0.092572	O 0.094044
O 0.091853	O 0.095518	I 0.092683	E 0.087500	E 0.088059
H 0.039020	Y 0.048430	Y 0.049152	H 0.049728	H 0.050632
Y 0.037306	H 0.037794	H 0.036886	Y 0.037138	Y 0.042557
Ω 0.029441	Ω 0.036865	Ω 0.035182	Ω 0.017120	Ω 0.022894
Σ 0.526217	Σ 0.538516	Σ 0.515271	Σ 0.511413	Σ 0.519711

The last row of the table 5 gives the sum of frequencies and indicates that in all cases more than 51% of the letters are vowels. From the same row a small decrease of the total vowel frequency with time can be seen, with exception of Plato.

Table 6 shows the frequencies of the 5 most common consonants and their order in each text. The consonants N,T,Σ are from Homer's time at the top, and form more than 20% of the text. The consonants Z,Ψ are the least frequent. The 10 most frequent consonants form more than 41% of the text in each case.

Table 6

HOMER	PLATO	KLEMES	KORNAROS	MODERN GREEK
N 0.07572	N 0.08498	N 0.09242	T 0.07436	T 0.08114
Σ 0.07158	T 0.08023	T 0.07547	N 0.07092	Σ 0.08049
T 0.05495	Σ 0.06071	Σ 0.07343	Σ 0.06811	N 0.07304
P 0.03992	M 0.03201	Π 0.03288	K 0.04827	P 0.04171
Λ 0.03599	Π 0.03149	P 0.03288	P 0.04202	K 0.03989

d. Comparison of modern Greek with other languages.

Table 7 gives the frequencies for some letters of modern Greek and that of six other languages.
Table 8 gives the vowel percentages in the same languages, while table 9 gives the percentages of the consonants L,N, R,S,T (4).

Table 7

	ENGLISH	GERMAN	FRENCH	SPANISH	PORTUGUESE	ITALIAN	GREEK	
A	7.81	5	9.42	12.69	13.50	11.74	10.84	A
B	1.28	2.5	1.02	1.41	0.50	0.92	0.55	B
C	2.93	1.5	2.64	3.93	3.50	4.50	1.56	Γ
D	4.11	5	3.38	5.58	5	3.73	1.87	Δ
E	13.05	18.5	15.87	13.15	13	11.79	8.33	E

Table 8

ENGLISH	GERMAN	FRENCH	SPANISH	PORTUGUESE	ITALIAN	GREEK
33%	34%	34%	31%	29%	30%	26%

Table 9

ENGLISH	GERMAN	FRENCH	SPANISH	PORTUGUESE	ITALIAN	GREEK
40%	40%	45%	47%	48%	48%	51%

From these tables we see that:

a. "A" is the most frequent letter in Greek, while "E" is first in all other languages.

b. The percentage of vowels in Greek is considerably higher than that in other languages.

A comparison of the values in all tables shows that, concerning the single letter frequencies, the percentage of vowels and the percentage of L,N,R,S,T. Greek is closer to Italian followed by the Portuguese, and Spanish.

The average word length of the considered languages is given in table 10. With an average of 5.8 letters per word the Greek words are bigger than the words of the other languages (2,3,4).

Table 10

ENGLISH	GERMAN	FRENCH	SPANISH	ITALIAN	GREEK
4.5	5.53	4.3	4.4	4.5	5.8

Table 11 gives the N-gram entropies, the estimated entropy and the redundancy of English, German and Greek. In comparison with English and German the N-gram entropies of Greek are lower, and so is also the approximated value of the entropy. Correspondingly, the redundancy of Greek is with 80.26% higher than that of English and German (2,3).

Table 11

	F_o	F_1	F_2	F_3	F_{word}	H	R
ENGLISH	4.76	4.03	3.32	3.1	2.14	1.0	75%
GERMAN	4.76			2.8	2.0	1.6	66%
GREEK	4.64	4.07	3.22	2.53	1.82	0.90	80.26%

e. Some tests on predictability of Greek.

In addition to the test mentioned before by which the next trigram of a text was predicted by a person, we have performed two other tests in order to assess the information contained on meaning, syntax, orthography, etc., of a text.

In the first test we have eliminated approximately half of the letters of a text with 250 symbols, by puting a dash in the place of the middle letter of each trigram. The text was given to several persons, who were asked to complete the failing letters. In one case the persons were also given a text, which preceded the reduced text, so they knew somehow the topic discussed in that text. In a second case the topic of the reduced text was completely unknown to the persons.

In the first case nearly 99% of the unknown letters were found correctly, in the second nearly 95%.

After that the same texts were completed using the trigram frequencies of modern Greek. For each unknown trigram a trigram was selected from the trigram-table, which had the same first and last letter, as the unknown trigram, and the highest frequency among similar trigrams. From the letters inserted in the texts in this way nearly 50% were correct.

The difference between the results of both tests is 45% to 49%. This means that a machine which would have to perform the prediction of a text in the described way and with the same success as a person, should get nearly 50% of the information required from the meaning, the syntax, the orthografy, etc., of the text.

VI. CONCLUSION

The Greek language possesing written texts extending in a period of 3000 years offers a unique opportunity to study the evolution of statistics and especially of the entropy of a language.

This has been done in a first stage by analysing only five texts covering the whole period in order to get a first approach of this evolution.

The results have shown fluctuations of the entropy and of other statistical parameters, which could be perhaps related to the progress or decadence of culture in the various historical periods. But it is evident that more texts of each period and other periods should be examined in order to get more accurate results and more valid conclusions.

As for the statistics of modern Greek many peculiarities have been obtained, which differenciate it from other modern languages.

This comes from the origin of the Greek language which is an independent branch of the Indo-European family of languages, and of the long period of its evolution.

REFERENCES

1. Shannon C.E., "A Mathematical Theory of Communication", Bell System Technical Journal, v.27, pp.379,623, 1948.
2. Shannon C.E., "Prediction and Entropy of Printed English", Bell System Technical Journal, v.30, p.50, 1951.
3. Küpfmüller K., "Die Entropie der deutschen Sprache", Fernmeldetechnische Zeitschrift, v.7, p.265, 1954.
4. Fouche'Gaines H., "Gryptanalysis", Dover Publications, Inc. New York, 1956.

STATISTICAL AND NON STATISTICAL MEASURES OF INFORMATION

K. Ferentinos and T. Papaioannou

Statistics Unit, University of Ioannina

ABSTRACT

Measures of information are characterized as parametric, non parametric and entropy-type. Statistical measures of information are nonnegative, strongly or weakly additive, have the maximal information property and are invariant under sufficient transformations of the data. The various measures of information are examined in the light of the above properties and characterizations. They are also examined for convexity, loss of information and appearance in Cramér-Rao type inequalities. Methods for obtaining parametric measures of information from the non parametric ones and from information matrices are also presented.

1. INTRODUCTION

The concept of measure or amount of information is of fundamental importance in statistics, probability theory and communication theory. Perhaps more fundamental is the concept of information itself. In general the word "information" means the amount by which the uncertainty of an event is reduced as a result of an experiment. A historical account of the development of the concepts of entropy, communication theory and their relation to statistical information is given by Kendall in (7), Csiszar in (3) and specially for the concept of statistical information by Papaioannou and Kempthorne in (11).

Several functions have been proposed in the literature as measures of information and each definition enjoys certain axiomatic and/or heuristic properties. A convenient way to differentiate among the various measures of information is to classify them in

D. G. Lainiotis and N. S. Tzannes (eds.), Advances in Communications, 69–76.

the following three categories:parametric, non parametric and entropy-type measures of information. Parametric measures of information measure the amount of information supplied by the data about the unkown parameters $\underset{\sim}{\theta}$ and are functions of $\underset{\sim}{\theta}$.In this case the best known measure is Fisher's measure of information. Non parametric measures express the amount of information supplied by the data for discriminating in favor of a distribution f_1 against another f_2, or measure the distance or affinity between f_1 and f_2. The best known measure of this type is the Kullback-Leibler measure.Measures of entropy give the amount of information contained in a distribution, that is the amount of uncertainty concering the outcome of an experiment. The classical measures of this type are Shannon's and Renyi's.

The purpose of the paper is (i) to present in an unified way the properties of measures of information and in particular to characterize the statistical measures of information (Section 2), (ii) to give the main measures of information and their properties (Section 3) and (iii) to present new parametric measures of information (Section 4).

2. PROPERTIES OF MEASURES

The measures of information proposed in the literature have various properties which establish their applicability in various sciences. The improtance of an amount which has been proposed or can be proposed as a measure of information emanates from these properties. Many properties have appeared in the literature in various contexts. Not all measures satisfy these properties. If one excludes certain basic ones it appears that there is no agreement as to which is the desirable minimal set of properties to be satisfied by a measure of information.

Let I be a measure of information of whatsoever category. Let X be a r.v. and T(X) any statistic. The most important of the properties of I are summurized as follows [cf.(2),(8),(11),(15),(16)].

(1) Non-negativity: $I_X \geq 0$

(2) Additivity:

$$I_{X,Y}=I_X+I_Y, \qquad I_{X,Y}=I_X+I_{Y|X}, \qquad I_{X,Y} \leq I_X+I_Y .$$

(weak if X ind Y) (strong) (subadditivity)

(3) Conditional inequality: $I_Y \geq I_{Y|X}$

(4) Maximal information : $I_X \geq I_{T(X)}$

(5) Invariance under sufficient transformations: $I_X=I_{T(X)}$, iff T(X) is a sufficient statistic.

(6) Convexity: Let $\alpha_1+\alpha_2=1$, $\alpha_1,\alpha_2 \geq 0$ and f_1,f_2 pdf's, then

$$I_X(\alpha_1 f_1+\alpha_2 f_2) \leq \alpha_1 I_X(f_1)+\alpha_2 I_X(f_2) .$$

(7) Loss of information: Let G be the set of all partitions of R^k and I_g the measure of information of a random variable g, then $\sup_g I_g = I_X$, where $g \in G$.

(8) If the experiment E_X is sufficient for the experiment E_Y (denoted by $E_X \succeq E_Y$) with the same parameter space Θ according to the Blackwell's definition, then $I_X \geq I_Y$.

Other properties or characterizations of measures of information are as follows: the invariance under parametric transformations, appearance in inequalities of Cramér-Rao type, the maximum and minimum value of them, etc. [cf. (1), (4), (5), (9), (10), (12), (13), (14)].

The properties nonnegativity, additivity (weak), maximal information and invariance under sufficient transformations express the concept of statistical information in a more characteristic way. The above characterization in general is not accepted in all studies of statistical inference. There are theoretical examples for which the above properties and specially the weak additivity property are not valid. These examples are more pathological than substantial. For more details see (11) and (7). For convenience and statistical deontology the measures of information which satisfy the above four properties will be characterized as statistical measures of information.

3. THE MAIN MEASURES OF INFORMATION AND THEIR PROPERTIES

Several functions have been proposed in the literature as measures of information. In this section we present the main measures of information and their properties in a systematic way. Thus Tables 1,2,3 contain the main measures of information and Table 4 their properties. In Table 4 the symbol + means that the corresponding property is satisfied, while the symbol - means the opposite. Symbol ⊕ means that the property is partly satisfied. Some of the properties in Table 4 have been proved in the literature. Others are the result of our work. Details will appear elsewhere.

4. NEW PARAMETRIC MEASURES OF INFORMATION

The construction of measures of information is not an easy problem. Ideally one would like to find measures which satisfy all of the properties of Section 2 for any family of probability distributions. Such measures do not exist. So one has to be less demanding and require the fulfillment of as many properties as possible. The non-negativity, maximal information and invariance under sufficient transformations properties constitute the minimum property requirement of a measure of information. Thus our target is

Table 1 :Parametric measures

Name of measure	Symbol Definition
Fisher (1925)	$I_X^F(\theta)=\begin{cases} E_\theta[\frac{\partial}{\partial\theta}\ln f(x,\theta)]^2 \\ \|E_{\underset{\sim}{\theta}}[\frac{\partial}{\partial\theta_i}\ln f(x,\underset{\sim}{\theta})\frac{\partial}{\partial\theta_j}\ln f(x,\underset{\sim}{\theta})]\|_{kxk} \end{cases}$
Mathai (1967)	$I_X^{Mat}(\theta)=[E_\theta\|\frac{\partial}{\partial\theta}\ln f(x,\theta)\|^\alpha]^{1/\alpha}$, $\alpha\geq 1$
Vajda (1971)	$I_X^V(\theta) = E_\theta\|\frac{\partial}{\partial\theta}\ln f(x,\theta)\|^\alpha$, $\alpha\geq 1$
Boekee (1975)	$I_X^{Bo}(\theta) =[E_\theta\|\frac{\partial}{\partial\theta}\ln f(x,\theta)\|^{s/(s-1)}]^{s-1}$, $1<s<\infty$

Table 2: Non parametric measures

Name of measure	Symbol Definition
Bhattacharyya (1943)	$I_X^B(f_1,f_2) =-\ln E_{f_2}[f_1/f_2]^{1/p}$, $1/p+1/q=1$
Kullback-Leibler(1951)	$I_X^{KL}(f_1,f_2)= E_{f_1}[\ln(f_1/f_2)]$
Renyi (1961)	$I_X^R (f_1,f_2)= \frac{1}{\alpha-1}\ln E_{f_2}[f_1/f_2]^\alpha$, $\alpha>0$, $\alpha\neq 1$
Kagan (1963)	$I_X^{Ka}(f_1,f_2)= E_{f_1}[1-(f_2/f_1)]^2$
Csiszar (1963)	$I_X^C (f_1,f_2)= E_{f_2}[f(f_1/f_2)]$, f convex
Matusita (1967)	$I_X^M (f_1,f_2)=[2[1-E_{f_2}(f_1/f_2)]^{1/2}]^{1/2}$
Vajda (1971)	$I_X^V (f_1,f_2)= E_{f_1}\|1-(f_2/f_1)\|^\alpha$, $\alpha\geq 1$

Table 3 :Some entropy-type measures

Name of measure	Symbol Definition
Shannon (1948)	$H_X^S = E[-\ln p_i]$
Renyi (1961)	$H_X^R = \frac{1}{1-\alpha}\ln E(p_i^{\alpha-1})$, $\alpha>0$, $\alpha\neq 1$

Table 4 :Properties of measures

	Properties→ / Measures ↓	Non negativity	Additivity (weak)	Maximal information	Invariance	Convexity	Loss of information	Information in experiment	Cramér-Rao type inequal.
	$I_X^F(\theta)$	+	+	+	+	+	⊕	+	+
Parametric	$I_X^{Mat}(\theta)$	+	⊕	+	+	+	+	+	+
	$I_X^V(\theta)$	+	⊕	+	+	+	+	+	+
	$I_X^{Bo}(\theta)$	+	⊕	+	+	+	+	+	+
	$I_X^B(f_1,f_2)$	+	+	+	+	+	+	+	–
	$I_X^{KL}(f_1,f_2)$	+	+	+	+	+	+	+	–
	$I_X^R(f_1,f_2)$	+	+	+	+	⊕	+	⊕	–
Non Parametric	$I_X^{Ka}(f_1,f_2)$	+	+	+	+	+	+	+	–
	$I_X^C(f_1,f_2)$	⊕	⊕	+	+	+	+	+	–
	$I_X^M(f_1,f_2)$	+	–	+	+	–	+	+	–
	$I_X^V(f_1,f_2)$	+	⊕	+	+	+	+	+	–
Entropy-type	H_X^S	⊕	+	–	–	–	–	+	–
	H_X^R	⊕	+	–	–	⊕	–	+	–

the determination of amounts which satisfy these three properties. Then we eximine the fulfillment of other properties as well. In this section we are interested in parametric measures of information and in methods of constructing such measures.

The pure parametric measures of information are the measures of Fisher, Mathai, Vajda and Boekee. If θ is k-variate Fisher's measure is a kxk matrix. In general information matrices are positive semidefinite real symmetric matrices, which satisfy the maximal information and invariance under sufficient transformations properties for every $\theta \in \Theta$. These matrices provide a way to construct new one-dimensional measures of information by considering special functions of their eigenvalues. In particular we have proved that every increasing function of the eigenvalues of an information matrix can be taken as a measure of information. For example if we start with the Fisherian matrix, the quantities

$$\det[I_X^F(\theta)],\ tr[I_X^F(\theta)],\ \lambda_i[I_X^F(\theta)],\ \sum_i w_i\lambda_i[I_X^F(\theta)] \quad w_i>0,$$

constitute reasonable parametric measures of information.Other information matrices can be derived form the parametric measures of Mathai, Boekee and Vajda. In a similar manner increasing functions of their eigenvalues can yield one-dimensional measures of information. The details of this methodology are presented elsewhere.

Let $I_X(f_1,f_2)$ be any non parametric measure of information, e.g. the measures of Renyi, Kullback-Leibler, Kagan etc. Non parametric measures of information provide two ways of constructing parametric measures of information as follows:

(a). Let f_1,f_2, be of the same functional form but differ according to the values of a parameter θ. Then we define as one-dimensional parametric measure of information the quantity:

$$I_X(\theta)=\begin{cases} \liminf\limits_{t\to 0} \dfrac{1}{t^2} I_X[f(x,\theta),f(x,\theta+t)] & \text{if } \theta \text{ univariate} \\ \liminf\limits_{\|t\|\to 0} \dfrac{1}{\|t\|^2} I_X[f(x,\theta),f(x,\theta+t)] & \text{if } \theta \text{ k-variate} \end{cases}$$

For example, if $I_X(f_1,f_2)$ is the Kagan measure of information and θ univariate, then

$$I_X^{Ka}(\theta) = \liminf_{t\to 0} \frac{1}{t^2} E_\theta\left[1- \frac{f(x,\theta+t)}{f(x,\theta)}\right]^2 .$$

In a similirar manner one can construct parametric measures of information derived from the measures of information of Renyi, Kullback-Leibler, Bhattacharyya, Matusita, Csiszar etc. We have proved that these measure satisfy several of the basic properties of Section 2. Details are given elsewhere. If θ is k-variate the non parametric measures of information can produce interesting parametric matrices in an analogous way. We have shown:these matrices are not necessarily information matrices. The above measure was firstly introduced by Kagan in (6).

(b). Let $k(\theta)$ be a one-to-one transformation of the parametric space Θ onto itself with $k(\theta)\neq\theta$. The quantity

$$I_X[\theta,k(\theta)] = I_X[f(x,\theta),f(x,k(\theta))]$$

can be considered as a parametric measure of information. If $I_X(f_1,f_2)$ is the Kullback-Leibler measure then

$$I_X^{KL}[\theta,k(\theta)] = E_\theta[\ln \frac{f(x,\theta)}{f(x,k(\theta))}].$$

This definition was firstly used by Papaioannou and Kempthorne in (11).

The above parametric measures of information, which are derived from the non parametric, are defined without any requlatity condition. If we impose the regularity conditions of the Fisherian theory of information [cf. (11)] these measures become linear functions of Fisher's measure. This provides evidence to conjecture that, if we require the information about θ contained in data from regular families of distributions to enjoy certain properties-axioms (non-negativity, maximal information and invariance), the quantity

$$I_X^F(\theta) = E_\theta[\frac{\partial}{\partial\theta} \ln f(x,\theta)]^2$$

is unique up to a positive constant. This is analogous with Shannon's entropy.

REFERENCES

(1) Aczél, J., and Daróczy, Z.:1975, *On Measures of Information and their Characterizations.*Academic Press. New York.

(2) Boekee, D.: 1975, An extension of the Fisher information measure. *Colloquia Mahtematica Societatis Janos Bolyai,* 16 Budabest 1977, pp. 113-123.

(3) Csiszar, I.: 1974, Information measures: A critical survey. *Trans. of the Seventh Prague Conference,* Prague 1974, Vol. A, pp. 73-85.

(4) Ferentinos, K., and Papaioannou, T.: 1979, Loss of information due to groupings. *Trans.of the Eighth Prague Conference,* Prague 1978,Vol. C, pp. 87-94.

(5) Goel, P., and DeGroot, M.: 1979, Comparison of experiments and information measures.*The Annals of Statistics* 7, pp.1066-1077.

(6) Kagan, M.: 1963, On the theory of Fisher's amount of information. (in Russian). *Dokl. Acad. Nauk. SSSR* 151, pp. 277-278.

(7) Kendall, M. G.: 1973, Entropy, probability and information. *Int. Stat. Rev.* 41, pp. 59-68.

(8) Kullback, S.: 1959, *Information Theory and Statistics.* J.Wiley, New York.

(9) Mathai, M., and Rathie, P.: 1975, *Basic Concepts in Information Theory and Statistics.* Wiley Eastern Limited - New Delhi.

(10) Nemetz, T.: 1977, *Information Type Measures and their Applications to Finite Decision Problems.* Lecture Notes. Reprint No. 7 and 8 1977. Reprint of the Math. Inst. of the Hung. Acad. of Sciences.

(11) Papaioannou, P., and Kemthorne, O.: 1971, On statistical information theory and related measures of information. *Aerospace Res. Lab.* ARL 71-0059. Ohio.

(12) Pinsker, M.: 1964, *Information and Information Stability of Random Variables and Processes.* Holden-Day, Inc. (translated by A. Feinstein from the Russian, Published in Moscow 1960).

(13) Renyi, A.: 1961, On measures of entropy and information. *4th Berkeley Symp.* Vol. 1, pp. 547-561.

(14) Shannon, E., and Weaver, W.: 1949, *The Mathematical Theory of Communication.* Urbana, Illinois, Univ. of Illinois Press.

(15) Stam, A.: 1959, *Some Mathematical Properties of Quantities of Information.* Ph. D. Thesis. Delft Univ. of Technology, Delft, the Netherlands, 1959.

(16) Vajda, I.: 1973, χ^{α}-divergence and generalized Fisher's information. *Trans. of the Sixth Prague Conference, Prague 1971*, Academia Prague 1973, pp. 873-886.

STORAGE VERSUS PERFORMANCE EXCHANGE IN PROBABILISTIC QUESTION-ANSWERING SYSTEMS

Judea Pearl and Alain Crolotte

School of Engineering and Applied Science
University of California, Los Angeles

ABSTRACT

In this paper we study the trade-offs between the required storage space and the quality of answers produced by Probabilistic Question-Answering (PQA) systems.

If the correct answer to a query is 'YES' and the system, instead, issues the estimate that the query has a probability π of being true, then the economic loss to the user can be measured by a distortion function which decreases monotonically with π. A system will be called <u>elastic</u> if a drastic memory saving may be achieved by tolerating a small level of average distortion.

The main result reported in this paper is that for a large class of distortion measures (i.e., measures exceeding $(1 - \pi)^\alpha$ $\alpha > 0$), if a binary QA system (true-false answers) is inelastic then the corresponding probabilistic QA system must also be inelastic. This result implies, for example, that a PQA system which is designed to answer all binary questions on an arbitrary dataset is inelastic. Similarly a PQA system which admits singly conjunctive questions (e.g., "Are both x and y in the dataset?") is also inelastic.

SUMMARY

A binary Question-Answering (QA) system, after consulting a summary of the input data stored in its memory, produces true-false answers to a given set of binary queries. Probabilistic Question-Answering (PQA) systems permit the answers to be cast in

D. G. Lainiotis and N. S. Tzannes (eds.), Advances in Communications, 77–80.

probabilistic phrasing, (e.g., "Yes, but I am only 80 percent sure"). The use of a PQA system is justified when the size of the data exceeds the system's storage capacity, in which case, storage space is economized at the expense of answers' quality.

Whereas the user of a binary QA system with insufficient storage must be resigned to the fact that a certain fraction of the answers will be utterly erroneous, the user of a PQA system also obtains estimates of each answer's reliability. An answer such as, "Yes, Mr. Smith is still employed by the X Corporation" when in fact he is no longer with X, is far more damaging than, "Yes, Mr. Smith is probably still with X, but I am only 85 percent sure". The basic difference is that in the first case, the user, acting blindly on the basis of wrong information, may make critical decisions to his detriment. In the second case, the user is alerted to proceed with caution (e.g., consult additional information sources, monitor progress carefully, etc.) and is able to judge, given the probability estimates, what level of precautionary effort is warranted in any specific situation.

If the correct answer to a given query is 'YES' ($i = 1$) and the system, instead, issues the estimate that the query has a probability π of being true ($0 \leq \pi \leq 1$), then the economic loss to the user is assumed to be expressed by a distortion function $\delta(1, \pi) = \delta(0, 1 - \pi)$. Similarly, we assume that the overall mean degradation in answer validity is measured by:

$$D = \frac{1}{Q} \sum_{q=1}^{Q} \delta(i_q, \pi_q)$$

where Q is the total number of admissible queries, $i_q \in \{0, 1\}$ represents the correct answer to query q implied by a given dataset, and π_q represents the corresponding probabilistic estimates issued by the system.

In this paper we study the trade-offs between the required storage space and the quality of answers produced by QA systems. Our main objective has been to devise simple tests for determining whether tolerating a low level of answers' indefiniteness may result in drastic savings of memory space.

We treat a PQA system as a communication channel whereby the content of the system's memory is viewed as a channel-code connecting the input messages (datasets) and the output messages (answers to queries) reproduced at the receiving end. This permits us to lower bound the memory requirement by Shannon's rate-distortion function R(D) if an average distortion less than D has to be insured.

The ability of a PQA system to convert an amount D of validity degradation into a large memory saving is measured by the ratio R(D)/R(0). If this ratio goes to zero as M, the size of the input ensemble, tends to infinity, the system will be called elastic, as a drastic saving in memory might be feasible by tolerating a small level of indefiniteness. A contrario, if R(D)/R(0) does not tend to zero (inelastic system), no coding scheme exists which drastically reduces memory requirements with only small degradation in answers validity.

Criteria for testing the elasticity of binary QA systems were established in (1) and (2) using the proportion of erroneous answers as a distortion measure. A previous analysis of storage-fidelity exchange in probabilistic QA systems demonstrated a very weak (inelastic) exchange for systems admitting only logically independent queries under both logarithmic and quadratic distortion functions (3). This paper extends the analysis to PQA systems involving arbitrary (e.g., logically dependent) query sets by studying the asymptotic behavior of a lower bound $R_L(D)$ to R(D).

The main result reported in this paper is the following: For distortion measures of the type $\delta(1, \pi) \geq (1 - \pi)^{\alpha}$, $\alpha > 0$, if the inelasticity of a binary QA system (true-false answers) can be established on the basis of such a lower bound, then the corresponding probabilistic QA system must also be inelastic. The class of distortion measure to which this theorem applies encompasses all reproducing scoring rules (5) used in practice, such as the (truncated) logarithmic, quadratic, and spherical scoring rules.

The proof of this theorem is rather involved and depends heavily on the asymptotic bounds developed in (2). The details of the proof will be reported elsewhere (4).

The theorem stated above implies, for example, that a PQA system which admits all binary questions on an arbitrary dataset is inelastic. Similarly a PQA system which admits singly conjunctive questions (e.g., "Are both x_i and x_j in the dataset?") is also inelastic. The results in (3) are special cases of this theorem since the inelasticity of a binary QA system with logically independent queries and probability of error distortion criterion, follows directly from the fact that the query set is non-redundant (1).

Another implication of our result applies to fuzzy systems. Regardless of how one chooses to quantify uncertainties, be it linguistically or numerically, the theorem above states that the relative savings in storage requirements due tolerating indefinite answers is basically identical to the savings realized by tolerating a given percentage of errors in true-false answers. For this statement to hold, though, we must insist that the quality of fuzzy answers be subject to numerical quantification using the function $\delta(i_q, A_q)$ where A_q now stands for the linguistic answer assigned to query q. $\delta(i_q, A_q)$ should also satisfy the reasonable condition $\delta(1, A_q) > 0$ for all answers A_q which express doubts regarding the truth of proposition q.

REFERENCES

(1) Crolotte, A. and Pearl, J., "Bounds on Memory Versus Error Trade-Offs in Question-Answering Systems," IEEE Transactions on Information Theory, Vol. IT-25, No. 2, pp. 193-202, March 1979.

(2) Crolotte, A. and Pearl, J., "Elasticity Conditions for Storage Versus Error Exchange in Question-Answering Systems," to be published in IEEE Transactions on Information Theory, November 1979.

(3) Pearl, J., "On Summarizing Data Using Probabilistic Assertions," IEEE Transactions on Information Theory, Vol. IT-23, No. 4, pp. 459-465, July 1977.

(4) Pearl, J. and Crolotte, A., "Storage-Space Versus Answers-Validity in Probabilistic Question-Answering Systems," UCLA-ENG-CSL-7933, University of California, Los Angeles, May 1979. Submitted for publication to IEEE Transactions on Information Theory.

(5) Shufford, E., Albert, A., and Massengill, H., "Admissible Probability Measurements Procedures," Psychometrica, Vol. 31, pp. 125-145, 1966.

SPECTRAL PROPERTIES OF BIPOLAR LINE CODES WITH ZERO EXTRACTION

Gianfranco L. Pierobon and Silvano G. Pupolin

University of Padova, Padova, Italy
Department of Electrical Engineering

ABSTRACT

The paper deals with bipolar line codes with zero extraction, like as B6ZS and HDB3, in which long sequences of zeros are converted into non zero ternary sequences. In order to evaluate code performances, a spectral analysis of the coded sequence is adequate; in the paper this is performed, in general, on the basis of a natural encoder model as a finite-state sequential machine driven by a variable-length word sequence.

1. INTRODUCTION

In PCM digital transmission systems wide use is made of encoding schemes with the following targets: 1) reduction of dc wandering, 2) reduction of intersymbol interference effects, 3) self-clocking capability, and 4) efficient error monitoring. To evaluate the encoding performances with respect to items 1), 2), and 3) a spectral analysis of the encoded message is adequate. As a matter of fact, 1) is achieved by a zero mean encoded message, 2) by a small low-frequency content of the spectral density, 3) by a nonvanishing spectral density, even in the presence of a zero input sequence.

In this paper we deal with *bipolar line codes with zero extraction*, which improve the classical bipolar code, in that they allow to maintain self-clocking capability even in the presence of long zero sequences.

In the literature the spectral analysis of these codes has been considered only for particular cases *(1,2)*, sometimes by assuming an input stream with equally likely zeros and ones, in any case with a cumbersome amount of computations. The reason lies essentially

D. G. Lainiotis and N. S. Tzannes (eds.), Advances in Communications, 81–88.

on the model usually considered for the line encoder, namely a sequential machine model operating on a symbol level. However, since bipolar codes with zero extraction have different encoding rules for single symbols and for strings of n zeros, a sequential machine model operating on a level of words with *variable length* (VL) appears to be a more natural one.

2. BIPOLAR CODES WITH ZERO EXTRACTION

2.1 *Bipolar with n Zero Substitution codes*

In these codes *(3)*, known as BnZS, a binary input stream is converted into a ternary output stream according to the following rules: a) binary 1's are converted into +1 and −1 alternately, b) binary 0's are converted into ternary 0's, apart from strings of n consecutive 0's, which are encoded into non zero ternary *substituting words* $\underline{w}^+$ or $\underline{w}^-$; c) the substituting word is $\underline{w}^+$ ($\underline{w}^-$) if the last binary 1 was encoded into +1(−1).

About the choice of the substitution words, we note that: 1) in order to assure a correct decoding, they must imply some violation of the alternation law; 2) 1+ and −1 must be balanced in $\underline{w}^+$ and $\underline{w}^-$, so that the mean of the encoded sequence is vanishing also in the presence of long zero sequences. Moreover, it appears natural to choose $\underline{w}^- = -\underline{w}^+$.

A suitable VL word encoding model can be used for the BnZS codes, if we regard the incoming binary stream as framed into VL words satisfying a *prefix condition*, namely into the words

$$\underline{\alpha}_1 = |1| \quad \underline{\alpha}_2 = |01| \ldots \underline{\alpha}_n = |00\ldots01| \quad \underline{\alpha}_{n+1} = |00\ldots00| \,. \tag{1}$$

2.2 *High Density Bipolar codes of order n*

These codes *(4)*, known as HDBn, have the following rules: a) binary 1's are encoded into +(−) if the last encoded ternary non zero digit is −(+); b) binary 0's remain unchanged apart from strings of n+1 consecutive 0's, which are converted into substitution words 00...0V or B0...0V, where B(V) stands for a ternary non zero digit that satisfies (violates) the alternation law a); c) the sequence of violations V have to satisfy the alternation law a).

It is clear that a prefix framing into words (1) will be adequate also for HDBn codes.

3. ENCODER MODEL

We consider the general VL encoding model depicted in fig. 1, where the encoding operation is split into *framing*, *word encoding*, and *deframing*. The source emits a sequence of symbols a_k, at a fixed

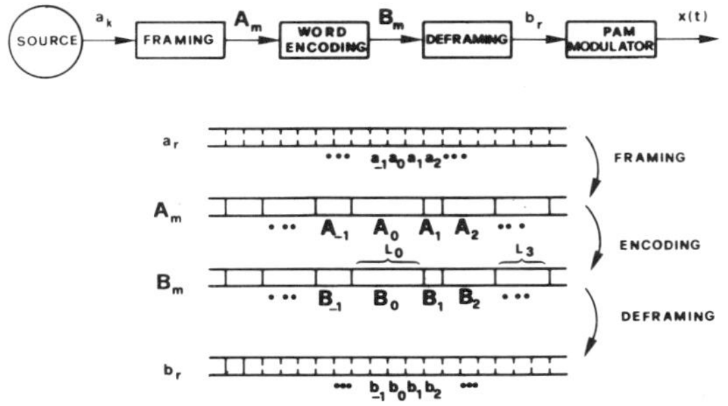

Fig. 1 - Model of a variable-length encoder

rate; these symbols are then framed into blocks to form a sequence $\underline{A}_m$ of VL *sourcewords*, which are encoded into another sequence $\underline{B}_m$ of VL *codewords*. Thereafter, the codewords are deframed into the sequence b_r of their symbols, which are sequentially transmitted by a PAM signal x(t).

The word encoding is assumed to be a finite state sequential machine $\mathcal{M}=\{\mathcal{A},\mathcal{B},\mathcal{S}, g,\underline{h}\}$, where: $\mathcal{A}=\{\underline{\alpha}_1,\ldots,\underline{\alpha}_K\}$ is the *sourceword operation*, i.e. the set of the VL words obtained by framing operation; $\mathcal{B}$ is the *codeword dictionary*; $\mathcal{S}=\{1,\ldots,I\}$ is the set of *machine states*; g is the *state transition function*, which assigns the next state in terms of the present one and of the input word: $S_{m+1} = g(S_m,\underline{A}_m)$; $\underline{h}$ is the *output function*, which assigns the output word in terms of the present state and of the input word: $\underline{B}_m=\underline{h}(S_m,\underline{A}_m)$.

We note that in the output function it is marginally implied the *length function*: $L_m=\ell(S_m,\underline{A}_m)$, which gives the length of the output word. However, we shall assume that codeword lengths are state independent, i.e. $L_m=\ell(\underline{A}_m)$, in accordance with the fact that synchronization imposes a fixed ratio between lengths of sourcewords and codewords.

In the following we find it convenient to use a matrix encoder specification. Namely, for each sourceword $\underline{\alpha}_u$ we define the binary square matrix $\underline{E}_u \triangleq |e_u(i,j)|$, $1\leq i,j\leq I$, with $e_u(i,j)=1$ if $g(i,\underline{\alpha}_u)=j$ and $e_u(i,j)=0$ otherwise; in words: $e_u(i,j)=1$ only if the sourceword $\underline{\alpha}_u$ forces the encoder to pass from state i to state j. Moreover, we index the codewords on the basis of the state-sourceword pairs by setting $\underline{\beta}_{iu} \triangleq \underline{h}(i,\underline{\alpha}_u)$, i.e. $\underline{\beta}_{iu}$ is the codeword produced by the sourceword $\underline{\alpha}_u$, when the encoder is in the state i. Then, the code-words produced by the sourcewords $\underline{\alpha}_u$, having a common length $\ell_u \triangleq \ell(\underline{\alpha}_n)$, may be stored into the $I\times\ell_u$ matrix:

$$\underline{\beta}_u \triangleq \begin{vmatrix} \underline{\beta}_{1u} \\ \cdots \\ \underline{\beta}_{Iu} \end{vmatrix} , \qquad 1\leq u\leq K . \tag{2}$$

Finally, we denote by $\mathcal{L}$ the set of the codeword lengths, and, for each $\lambda \in \mathcal{L}$, we define the set $U_\lambda \stackrel{\Delta}{=} \{u : \ell(\underline{\alpha}_u) = \lambda\}$ of the indexes of the sourcewords that give rise to codewords with a common length λ.

BnZS codes. As input dictionary $\mathcal{A}$ we assume the set of the VL words (1). Then, from the analysis of the encoding rules in 2.1, one recognizes two encoder states, corresponding to the sign of the last non zero ternary digit encoding a binary 1: $\mathcal{S}=\{+,-\}$. Then, by inspection, we get the following matrix representation:

$$\underline{\beta}_u = \begin{vmatrix} -\underline{\alpha}_u \\ +\underline{\alpha}_u \end{vmatrix}, \quad 1 \le u \le n; \qquad \underline{\beta}_{n+1} = \begin{vmatrix} \underline{w}^+ \\ \underline{w}^- \end{vmatrix}$$
$$\underline{E}_u = \begin{vmatrix} 0 & 1 \\ 1 & 0 \end{vmatrix}, \quad 1 \le u \le n; \qquad \underline{E}_{n+1} = \begin{vmatrix} 1 & 0 \\ 0 & 1 \end{vmatrix} \tag{3}$$

The transition diagram corresponding to matrices $\underline{E}_u$ is depicted in fig. 2a). It may be noted that the state number and the transition diagram structure do not depend on the order n of the code.

HDBn codes. For these codes, we can choose four different states (fig. 2b)), i.e. $\mathcal{S}=\{++,+-,-+,--\}$ where the first (second) sign is the sign of the last non zero encoded symbol (of the last violation). Then, the matrix representation is

$$\underline{\beta}_u = \begin{vmatrix} -\underline{\alpha}_u \\ -\underline{\alpha}_u \\ +\underline{\alpha}_u \\ +\underline{\alpha}_u \end{vmatrix}, \quad 1 \le u \le n+1; \qquad \underline{\beta}_{n+2} = \begin{vmatrix} - & 0 & \dots & 0 & - \\ 0 & 0 & \dots & 0 & + \\ 0 & 0 & \dots & 0 & - \\ + & 0 & \dots & 0 & + \end{vmatrix}$$
$$\underline{E}_u = \begin{vmatrix} 0 & 0 & 1 & 0 \\ 0 & 0 & 0 & 1 \\ 1 & 0 & 0 & 0 \\ 0 & 1 & 0 & 0 \end{vmatrix}, \quad 1 \le u \le n+1; \qquad \underline{E}_{n+2} = \begin{vmatrix} 0 & 0 & 0 & 1 \\ 1 & 0 & 0 & 0 \\ 0 & 0 & 0 & 1 \\ 1 & 0 & 0 & 0 \end{vmatrix} \tag{4}$$

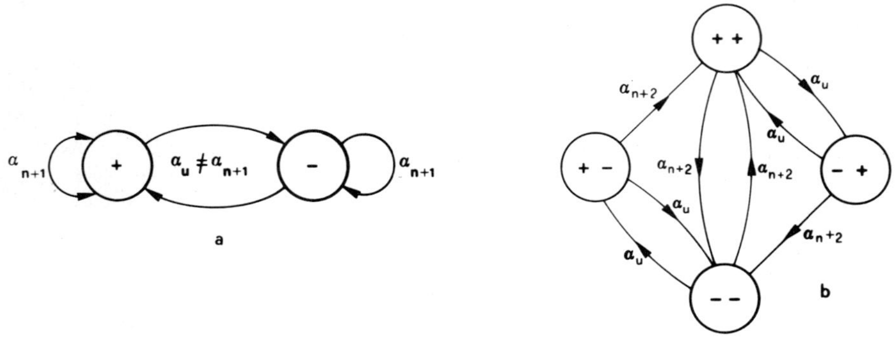

Fig. 2 - State transition diagram of BnZS (a) and HDBn (b) codes

4. SPECTRAL ANALYSIS OF THE ENCODED MESSAGE

4.1 Spectral analysis of variable length vector processes

In order to relate the second order analysis of the codeword sequence $\underline{B}_m$ to the one of the deframed symbol process b_r, we have to involve the distance in symbols between codewords $\underline{B}_m$ and $\underline{B}_n$, namely:

$$\tau_{mn} \stackrel{\Delta}{=} \sum_{i=m}^{n-1} L_i \ , \quad n > m \ ,$$

whereas $\tau_{mn} \stackrel{\Delta}{=} 0$, if m=n, and $\tau_{mn} \stackrel{\Delta}{=} -\tau_{nm}$, if n<m. Then, by using the special purpose expectation $E\{\underline{X};\mathscr{E}\} \stackrel{\Delta}{=} E\{\underline{X}|\mathscr{E}\}Pr\{\mathscr{E}\}$, we define the *cross correlation matrix functions between words of fixed lengths at a fixed symbols' distance:*

$$\underline{R}_{\lambda\mu}(k) \stackrel{\Delta}{=} E\{\underline{B}'_m\underline{B}_n;\ L_m=\lambda,\ L_n=\mu,\ \tau_{mn}=k\} \ , \tag{5}$$

where, for a given m, the index n of the codeword, if any, at a fixed distance k, is itself random. Correspondingly, we can define the $\lambda x\mu$ *cross spectral density matrix functions*:

$$\underline{W}_{\lambda\mu}(f) \stackrel{\Delta}{=} \Sigma_k\, T\,\underline{R}_{\lambda\mu}(k)z^k, \quad z \stackrel{\Delta}{=} \exp(-j2\pi fT) \ , \tag{6}$$

($T=1/f_c$ is the symbol period).

Now as it is shown in *(6)*, the second order analysis of the deframed symbol sequence b_r can be performed in terms of (6). Namely, it turns out that the spectral density of the deframed symbol sequence b_r is given by:

$$W_b(f) = \frac{1}{L}\, \Sigma_{\lambda\mu}\ \underline{V}^*(z)\underline{W}_{\lambda\mu}(f)\underline{V}_\mu(z) \tag{7}$$

where asterisk denotes transpose conjugate, $\underline{V}_\lambda(z) \stackrel{\Delta}{=} |z, z^2, \ldots, z^\lambda|$, and $L \stackrel{\Delta}{=} E\{L_m\}$ is the average codeword length.

In conclusion, the spectral analysis of the coded sequence is performed via the evaluation of the cross correlation matrices $\underline{R}_{\lambda\mu}(k)$.

4.2 Second order analysis of the codeword sequence

We suppose that 1) the *symbol source* generating a_k is *stationary and memoryless* and 2) the *framing operation is memoryless*. Then, also the sourceword sequence $\underline{A}_m$ is stationary and memoryless, so that it is statistically specified by the probabilities $q_u \stackrel{\Delta}{=} Pr\{\underline{A}_m=\underline{\alpha}_u\}$, which can be evaluated from the statistics of a_k and from the framing rules.

Now, the length sequence L_m turns out to be stationary and memoryless, specified by probabilities:

$$p_\lambda \triangleq \Pr\{L_m=\lambda\} = \Sigma_{U(\lambda)} q_u \; ; \tag{8}$$

as follows from the identity between events $\{L_m=\lambda\}=\bigcup_{U(\lambda)}\{\underline{A}_m=\underline{\alpha}_u\}$. By probabilities (8), we can evaluate $L=\Sigma\lambda p_\lambda$.

Next, by virtue of the sequential machine nature of the encoder, it follows that the state sequence S_m is a stationary Markov chain with transition probability matrix, $\underline{\Pi}=\Sigma_u q_u \underline{E}_u$. For a proof, see *(6)*. We shall denote by $\underline{\pi}$ the vector of the absolute state probabilities $\pi_i \triangleq \Pr\{S_m=i\}$, which is the normalized solution of the linear equation system $\underline{\pi}=\underline{\pi}\,\underline{\Pi}$. We denote by $\underline{D}$ the diagonal matrix $\underline{D}=\mathrm{diag}\{\underline{\pi}\}$ and by $\underline{\Pi}_\infty$ the matrix having all its rows equal to $\underline{\pi}$.

As it is shown in *(6)*, we get:

$$\begin{aligned} \underline{R}_{\lambda\mu}(\underline{0}) &= \delta_{\lambda\mu}\, \Sigma_{U(\lambda)}\, q_u \underline{\beta}'_u\, \underline{D}\underline{\beta}_u \\ \underline{R}_{\lambda\mu}(k) &= \underline{K}_\lambda \underline{Y}(k-\lambda)\underline{J}_\mu \qquad k>0 \\ \underline{R}_{\lambda\mu}(k) &= \underline{J}'_\lambda \underline{Y}(-k-\mu)\underline{K}'_\mu \qquad k<0 \end{aligned} \tag{9}$$

where $\underline{K}_\lambda=\Sigma_{U(\lambda)}\, q_u\underline{\beta}'_u\, \underline{D}\, \underline{E}_u$ and $\underline{J}_\lambda \triangleq \Sigma_{U(\lambda)}\, q_u\underline{\beta}_u$, whereas $\underline{Y}(k)\triangleq|\Pr\{S_n=j, \tau_{mn}=k, S_m=i\}|$. The matrix function $\underline{Y}(k)$ is the solution of the difference matrix equation *(6)*:

$$\underline{Y}(k) = \Sigma_\lambda\, \underline{H}_\lambda \underline{Y}(k-\lambda)\,, \quad k>0\,, \tag{10}$$

with initial conditions $\underline{Y}(0)=\underline{I}$ (identity matrix) and $\underline{Y}(k)=\underline{0}$, for $k<0$. The matrices $\underline{H}_\lambda$ are defined as

$$\underline{H}_\lambda = \Sigma_{U(\lambda)}\, q_u \underline{E}_u\,, \tag{11}$$

and can be evaluated in terms of source and encoder specifications. In particular, $\underline{Y}(k)$ converges to $\underline{Y}(\infty)\triangleq\frac{1}{L}\,\underline{\Pi}_\infty$ and its z-transform is given by

$$\underline{Y}(z) \triangleq \Sigma_k\, \underline{Y}(k)z^k = (\underline{I}-\Sigma_\lambda z^\lambda \underline{H}_\lambda)^{-1}\,. \tag{12}$$

4.3 Spectral analysis of the deframed symbol process

By using eqs. (7), (6), (9), and (12), after a cumbersome algebra one gets the spectral density of the deframed symbol sequence b_r:

$$W_b(f) = \frac{T}{L}\,\{\Sigma_\lambda \underline{V}^*_\lambda(z)\underline{R}_{\lambda\lambda}(0)\underline{V}_\lambda(z) + 2\mathrm{Re}\left[\underline{K}^*(z)(I-\Sigma_\lambda z^\lambda K_\lambda)^{-1}\underline{J}(z)\right]\} \tag{13}$$

with $\underline{K}(z)\triangleq\Sigma_\lambda\, z^{-\lambda}\underline{K}'_\lambda\underline{V}_\lambda(z)$ and $\underline{J}(z)=\Sigma_\lambda\, \underline{J}_\lambda\underline{V}_\lambda(z)$.

Owing to the polynomial structure of matrices in (13), the spectral density reduces, in its closed form, to a rational function in $z=\exp(-j2\pi fT)$. Since the codes dealt with in the paper have no spectral lines, we do not consider here the line extraction problem (for a detailed discussion of the general case see *(6)*).

5. SPECTRAL DENSITY OF BIPOLAR CODES WITH ZERO EXTRACTION

By using the above theory, we have found simple general expressions of the spectral densities of BnZS and HDBn codes, for any value of n and of the probability of 0 in the input symbol sequence: $q \triangleq \Pr\{a_r = 0\}$.

In fig. 3a) it is shown the spectral density of a B6ZS code for three values of q. In fig. 3b) two B6ZS codes are compared with the classical bipolar code, for equally likely binary digits; it can be noted that the choice of the substitution word has a

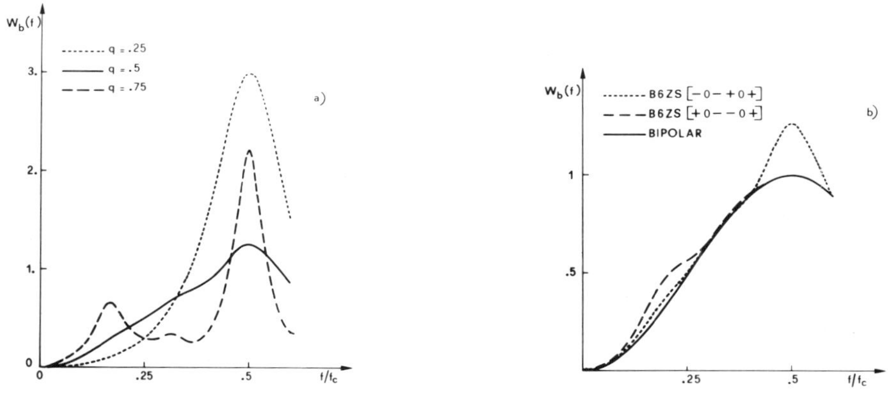

Fig. 3-a) Spectrum of a B6ZS code - b) Comparison between B6ZS and bipolar codes for q=1/2.

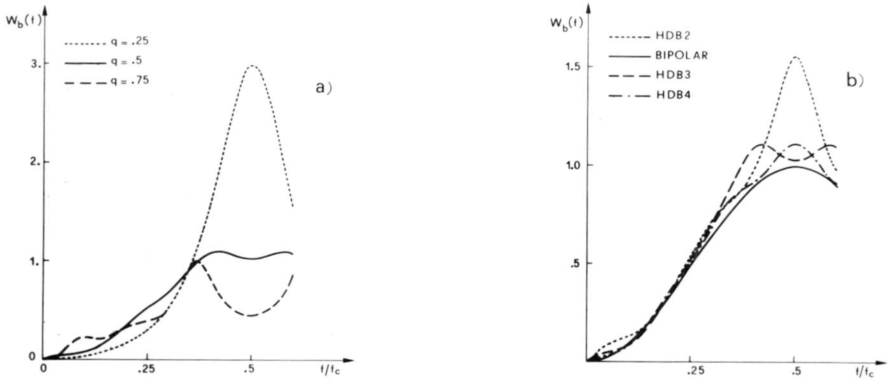

Fig. 4-a) Spectrum of the HDB3 code - b) Comparison between HDBn and bipolar codes for q=1/2.

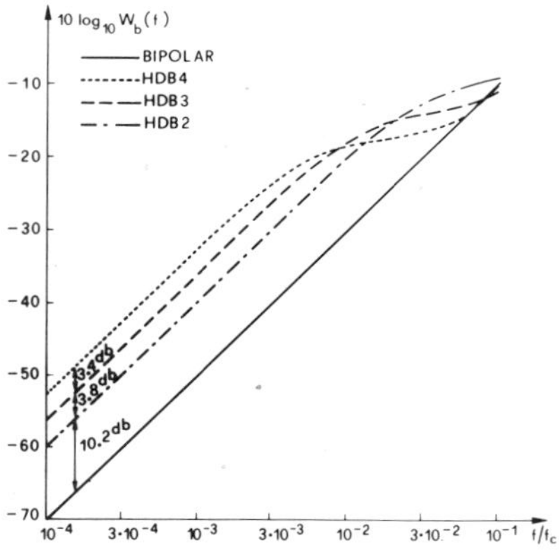

Fig. 5 - Low-frequency power content of bipolar codes.

remarkable relevance in shaping the spectral density. The B6ZS codes exhibit a larger power than the bipolar code. This is a penalty to be paid in order to assure self-clocking capability.

In fig. 4a) the spectral density of the HDB3 code is depicted for three values of q. In fig. 7b) a comparison is made among the HDB2, HDB3, HDB4, and bipolar codes; the difference between HDBn and bipolar code density reduces as n increases. Finally, in fig. 5 several codes are compared on the basis of their low frequency behaviour for q=1/2.

We note that HDBn codes have a larger amount of power density at small frequencies than the bipolar code ; this implies a further penalty with respect to intersymbol interference effects produced by low frequency cutoff. On the contrary there is no appreciable difference between the bipolar code and the BnZS codes.

ACKNOWLEDGEMENTS

The work was supported by Italian CNR under grant CT 78.00693.07 Thanks are due to Adriano Rossi for his programming support.

REFERENCES

(1) Breant, P.; 1974, *Cables et Transmission*, pp. 3-36
(2) Desblache, A.; 1971, *IBM Memo.*, CS209.
(3) Johannes, V.I., Kaim, A.G., and Walzman T.; 1969, *IEEE Trans. on Comm. Techn.*, pp. 303-310.
(4) Croisier, A.; 1970, *IBM J. Res. Develop.*, pp. 354-367.
(5) Cariolaro, G.L., and Pierobon, G.L.; 1977, *IEEE Trans. on Inf. Th.*, pp. 243-253.
(6) Pierobon, G.L., and Pupolin, S.; 1979, *Int. Rep. Univ. di Padova*, UPee 79/11.

ANALYSIS AND DESIGN OF AN ERROR CORRECTION CODE FOR A LINEAR DELTA MODULATION SYSTEM

G.Kourelis, N.C.David and V.Makios

Laboratory of Electromagnetics, School of Engineering, University of Patras, Patras, Greece.

ABSTRACT

The purpose of this paper is to examine the behavior of the demodulator (receiver) of a linear delta modulation system, in the presence of transmission errors and indicate a method for improving the performance of the demodulation system whenever a transmission error has been detected. According to the error correction code adopted here, the transmitter sends beyond the binary representation of the analog input signal, another two bits indicative of the parity and order of "zeros" in a sixteen bit word. This additional information makes the receiver capable of detecting an error and provides the demodulator with the capability of correcting it. A detailed block diagram of the circuitry implementation of the code is then presented and experimental results are given.

I. INTRODUCTION

Techniques for correcting transmission errors in digitally encoded signals have been reported in the literature in the past by D.J.Connor (1), N.Rydbeck and C.E.W.Sundberg (2), N.Scheinberg and D.L.Schilling (3), and others (4),(5). Specifically Scheinberg and D.L.Schilling have presented algorithms for correcting transmission errors for digital delta modulation systems. In this paper we examine the behavior of a linear DM system in the presence of transmission errors and introduce a simple algorithm for correcting them.

D. G. Lainiotis and N. S. Tzannes (eds.), Advances in Communications, 89–95.

II. THE EFFECT OF TRANSMISSION ERRORS ON A LINEAR DELTA MODULATION SYSTEM.

The block diagram of a linear digital DM system is shown in Fig.1.

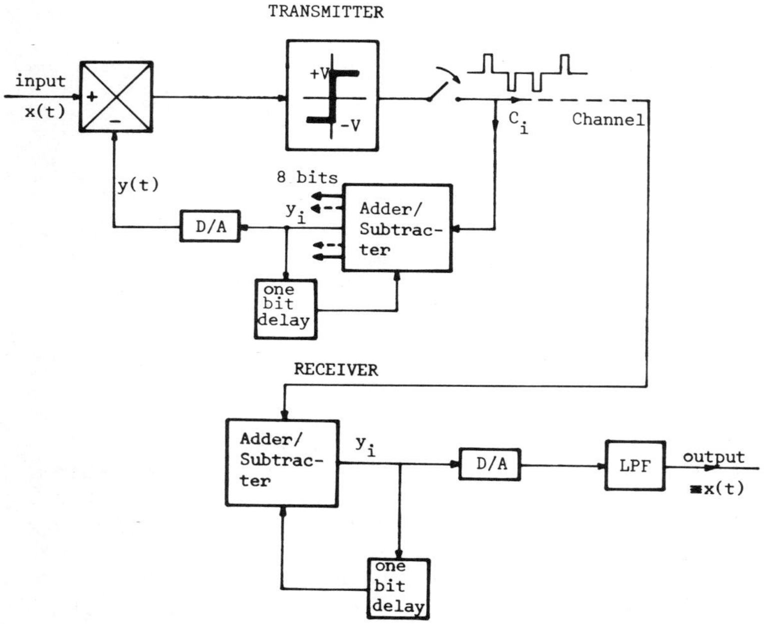

Fig.1. Block diagram of a Linear DM system.

When c_i is at a high state (logical "one") y increases by one step Δ, and when c_i is at a low state (logical "zero") y decreases by one step Δ, in a manner such that y always approximates the input signal. Fig.2 shows the response of the DM system in an errorless transmission of c_i (solid line) and in the presence of one transmission error (dotted line) for an arbitrary input waveform x(t). From Fig.2 we observe that in the presence of a transmission error there is a permanent dc shift equal to 2Δ. Thus when an error is detected the demodulator should proceed as follows. If the number of zeros sent, is less (larger) than the number received the demodulator should increase (decrease) the dc level of the demudulated signal by 2Δ.

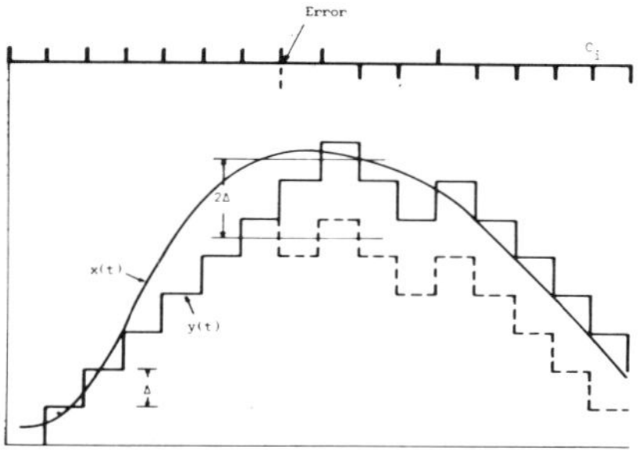

Fig.2. Response of the Linear DM system to an arbitrary input signal

III. ERROR CORRECTION CODE

Every n bits, we send another two bits which indicate the following:

1. The first bit (FB) indicates whether the sent sequence of the n bits contains odd or even number of zeros in the following manner:

   Even number of zeros : FB = 0

   Odd number of zeros : FB = 1

2. The second bit (SB) is "zero" whenever the number of zeros is 4K or 4K+1, K=0,1,2,... and "one" whenever the number of zeros is 4K+2 or 4K+3, K=0,1,2,...

We assume that only one error may be introduced every n-bits and that the correction bits are always correct. Upon receiving the n-bits the receiver compares the number of "zeros" with the first bit and if:

a. The number of zeros is even and the first bit 0, or
b. The number of zeros is odd and the first bit 1,

it instructs the demodulator to proceed to the decoding procedure.
If:

a. The number of received zeros (n_r) is even and the first bit is "one" (FB=1), or
b. The number of received zeros (n_r) is odd and the first bit is "zero" (FB=0) an error has been detected.

Thus if FB=1, the number of "zeros" transmitted (n_t), is odd, and if FB=0, the number of "zeros" transmitted (n_t) is even. So if the receiver has received $n_r=2m+2$ "zeros" then:

$$n_t=2m+1 \quad \text{or} \quad n_t=2m+3 \;, \quad m=0,1,2....$$

In addition to the above if SB=0, the n_t will be such that:

$$n_t=2m+1=4K+1 \quad \text{or} \quad n_t=2m+3=4K+1$$

and if SB=1, the n_t will be such that:

$$n_t=2m+1=4K+3 \quad \text{or} \quad n_t=2m+3=4K+3$$

A similar situation holds when, $n_r=2m+1$

III.1. Design of the correction algorithm

We choose n=16. At the receiver there is a counter which counts the "zeros" contained in every arriving 16-bit word. Obviously if the LSB of the counter after the arrival of the 16-bit word is at a high state ($Q_0'=1$) an odd number of zeros has been counted (i.e. n_r is odd). Similarly if $Q_0'=0$, n_r is even,

Thus if $FB=Q_0'$ no error has been detected,

and if $FB\neq Q_0'$ an error has been detected.

In the first case ($FB=Q_0'$) the demodulator is instructed to proceed to the decoding. We now investigate for different cases when an error has been detected i.e. $FB\neq Q_0'$.

1. FB=1, $Q_0'=0$.
   In this case the probable states of the counter at the receiver are shown in Table 1. From the above table we observe the following which can be summarized in the following Table 2.

Table 1

n_r	Q_3'	Q_2'	Q_1'	Q_0'	n_t/SB=0	n_t/SB=1
0	0	0	0	0	1	impos.
2	0	0	1	0	1	3
4	0	1	0	0	5	3
6	0	1	1	0	5	7
8	1	0	0	0	9	7
10	1	0	1	0	9	11
12	1	1	0	0	13	11
14	1	1	1	0	13	15

Table 2

SB	Q_1'	n_t
0	0	n_r+1
0	1	n_r-1
1	0	n_r-1
1	1	n_r+1

Thus setting $C=SB\oplus\bar{Q}_1'$ we may say that the step size should increase by 2Δ if C=1 and becrease by 2Δ when C=0.

2. FB=0, $Q_0'=1$. In this case the probable states of the counter are shown in Table 3. From table 3 we can get in summary the following Table 4.

Table 3

n_r	Q_3'	Q_2'	Q_1'	Q_0'	n_t/SB=0	n_t/SB=1
1	0	0	0	1	0	2
3	0	0	1	1	4	2
5	0	1	0	1	4	6
7	0	1	1	1	8	6
9	1	0	0	1	8	10
11	1	0	1	1	12	10
13	1	1	0	1	12	14
15	1	1	1	1	impos.	14

Table 4

SB	Q_1'	n_t
0	0	n_r-1
0	1	n_r+1
1	0	n_r+1
1	1	n_r-1

Thus for $C=\overline{SB\oplus Q_1'}$ we may say that the step size should increase by 2Δ if C=1 and decrease by 2Δ when C=0. Concluding we have:

if FB=1 then $C=SB\oplus SB'$

and if FB=0 then $C=\overline{SB\oplus SB'}$

The correction (Addition or Subtraction) at the receiver is made with the arrival of the 8th bit, so that the d.c. shift introduced by an error be as short in time as possible.
Thus the situation of the 8-th bit is shown in Table 5.

Table 5

C_8	C	Step $FB=Q_0'$	Step $FB\neq Q_0'$
0	0	$-\Delta$	$-\Delta-2\Delta$
0	1	$-\Delta$	$-\Delta+2\Delta$
1	0	$+\Delta$	$+\Delta-2\Delta$
1	1	$+\Delta$	$+\Delta+2\Delta$

IV. BLOCK DIAGRAM AND EXPERIMENTAL RESULTS

In Fig.3 is shown the block diagram of the linear digital DM System with the error correction code implemented. The modulator

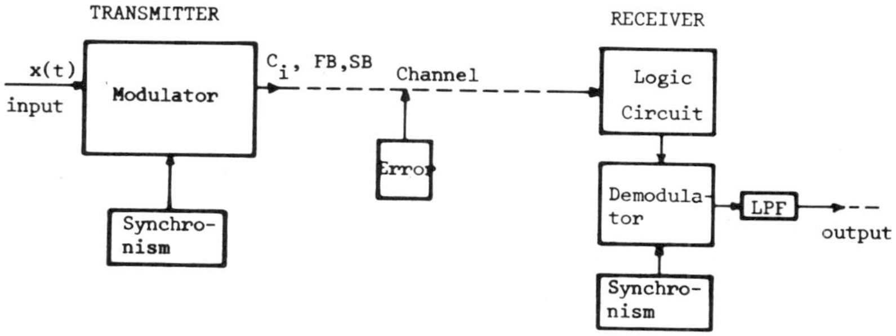

Fig.3. Modified block diagram of the Linear DM system with the error correction circuit added.

produces the digital sequence c_i and the two correction bits which are transmitted through the information channel. In this channel also an error is introduced every 16 bits. The logic circuit in the receiver produces the needed control signal (according to the tables given in section III) and drives accordingly the inputs of the adder in the demodulator to reach the desired correction.

Now we give the experimental results. At the input of the transmitter a pulse train (x(t)) of 7×10^2Hz was applied. In the upper trace of the oscillogram of Fig.4(a) the step re-

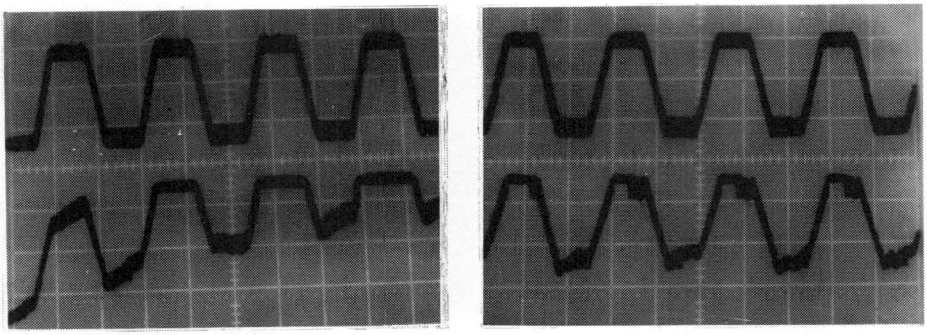

(a) (b)

Fig.4. Effects of channel errors on a delta-modulated encoded pulse train.
(a) without using the correction code.
(b) with the use of the correction code.
($f_{pulse}=7\times10^2$Hz, $f_{sampling}=30\times10^2$Hz).

sponse of the modulator is shown, and the lower trace shows the output of the receiver after introducing an error per 16 bits. In fig.4(b) the upper trace also shows the output of the modulator, and the lower trace shows the output of the demodulator after an error was introduced and the error was corrected, using the described method. We observe that the desired correction is fulfilled and the designed method gives satisfactory results.

V. CONCLUSIONS

In this paper we have given a new method for improving the performance of a linear digital DM system whenever a transmission error has been detected. The experimental results prove that the receiver is capable, in a satisfactory manner, of detecting an error and provides the demodulator with the capability of correcting it.

ACKNOWLEDGMENT

The authors would like to thank the staff of the Electronics Laboratory of the University of Patras and Mr.P.Prokopiou for their valuable technical help.

REFERENCES

1. Connor, D.J., IEEE Trans. Commun., vol COM-21, pp.695-706, June 1973.
2. Rydbeck, N., and Sundberg, C.E.W, IEEE Trans. Commun., vol. COM-24, pp.59-65, Jan. 1976.
3. Scheinberg, N., and Schilling, D.L., IEEE Trans. Commun., vol. COM-24, pp.1064-1069, Sept. 1976.
4. Yan, H., and Donaldson, R.W., IEEE Trans. Commun., vol. COM-20, pp.281-290, June 1972.
5. Dostis, I., Bell Syst. Tech. J., vol 44, pp.2227-2243, Dec. 1965.

THE ENTROPY OF WRITTEN ITALIAN

Renato M. Capocelli and Luigi M. Ricciardi

Istituto di Scienze dell'Informazione, Università di Salerno, Salerno, Italy

ABSTRACT

The entropies to the fourth order and the word entropy, with and without space symbol (blank), of an Italian text are calculated. The obtained value of the fourth order entropy shows that a redundancy of about 53% is exhibited by the text.

Since Shannon's pioneering work on the structure and the information content of written English (1) several papers have appeared that provide an evaluation of entropies of other languages (2) ÷ (8). The practical motivation for these works ultimately rests on the possibility of achieving compression for storage and time saving purposes. In this work the results of a study on the statistical association of chacacters in an Italian classical text are presented. This is an extention of (6) since the analysis is extended to tetragrams and the space symbol is also looked upon as a character. Furthermore, as claimed by the author himself, the results of (6) for digrams and particularly for trigrams are not reliable because of the insufficient size of the considered texts (only 10,000 characters). To be able to obtain a meaningful statistics also for tetragrams, the size of the text had to be chosen far larger than that of (6). The statistics were thus derived from the analysis of the first six chapters of I Promessi Sposi (The Betrothed) by A.Manzoni. As remarked in (6), the choice of the text is not expected to affect much the numerical results. This is indeed stressed by the observed excel-

D. G. Lainiotis and N. S. Tzannes (eds.), Advances in Communications, 96–100.

lent agreement of our numerical results with those of (6) in the cases when the comparison can be performed. The total sample consisted of 177, 419 characters forming 31, 336 words. The mean length L of a word was 5.6618 characters. The number of primitive characters was 22 (Italian alphabet plus blank). Hence the zero order entropy is $H_0=\log_2 22 \simeq 4.4594$ bits/character.

As is well known, the entropy per character H(S) of a text, regarded as a sequence of letters emitted by a stationary source, is given by

$$H(S) = \lim_{n\to\infty} n^{-1} H(X_1 X_2 \ldots X_n) = \lim_{n\to\infty} H(X_n | X_1 X_2 \ldots X_{n-1}),$$

where $n^{-1} H(X_1 X_2 \ldots X_n)$ is the entropy per character for blocks of n characters and $H(X_n | X_1 X_2 \ldots X_{n-1})$ is the conditional entropy of the next character when the n-1 preceding ones are known. From

$$H(X_n | X_1 X_2 \ldots X_{n-1}) = H(X_1 X_2 \ldots X_n) - H(X_{n-1} | X_1 X_2 \ldots X_{n-1})$$

$$- \ldots - H(X_2 | X_1) - H(X_1)$$

we have:

$$K_n \equiv n^{-1} H(X_1 X_2 \ldots X_n) \geq H_n \equiv H(X_n | X_1 X_2 \ldots X_{n-1}).$$

Successive approximations to H(S) may thus be obtained by considering only the statistical structure of sequences extending over 1,2,...,N characters, H_N coinciding with H(S) whenever no statistical influences extend beyond the N-th successive character.

It is perhaps useful to recall here that entropy is a measure of the information carried on the average by each character of a text. If this is coded by means of binary digits, the entropy gives the least number of binary digits required per character of the original text.

Our statistical study shows that the single character probabilities vary from $5963 \cdot 10^{-6}$ to $1766 \cdot 10^{-4}$. The average amount of information per character for single characters H_1, i.e. the first order entropy, is 3.9608 bits/character. The digram probabilities vary from $5636 \cdot 10^{-9}$ to $4798 \cdot 10^{-5}$; the number of distinct digrams found in the text was 285. The second order entropy H_2 and the entropy per character for digrams K_2 were 3.1005 bits/character and 3.5306 bits/character, respectively. The trigram probabilities of occurrence were found to range between $5636 \cdot 10^{-9}$ and $9576 \cdot 10^{-6}$. The distinct trigrams were 2,336 and we found $H_3 = 2.5863$ bits/character and $K_3 = 3.2159$ bits/character. Finally, 10,376 tetragrams were found. Their probabilities vary between $5636 \cdot 10^{-9}$ and $6803 \cdot 10^{-6}$, whereas $H_4 = 2.0984$ bits/character and $K_4 = 2.9365$ bits/character. Since the uncertainty on a character decreases as the range of possible statistical influences is made to increase, H_4 can be taken as an upper limit to the text entropy H(S). A better upper limit, very close to H(S), may be obtained by evaluating the word-character entropy $H_W(S)$. Indeed, due to the strong internal structure of natural languages $H_W(S)$ corresponds to the entropy of n-grams whose length is relevantly larger than the mean words length (1). In our case we have found $H_W(S) = 1.7116$ bits/character roughly corresponding to H_8 or to K_{11}.

A similar set of calculations has been carried out excluding the blank, giving a 21 character alphabet for which the zero order entropy is $H_o = \log_2 21 \simeq 4.3923$. The obtained results (in bits/character) are summarized below:

$$\overline{H}_1 = 3.9935 \quad \overline{H}_2 = 3.3845 \quad \overline{H}_3 = 2.9920 \quad \overline{H}_4 = 2.4802$$

$$\overline{K}_2 = 3.6890 \quad \overline{K}_3 = 3.4566 \quad \overline{K}_4 = 3.2127 \quad \overline{H}_W(S) = 2.0788$$

Note that each of the above values is larger than the corresponding value for the blank included case. The ratios $\overline{H}_n/H_n$ and $\overline{K}_n/K_n$ approach the limit L/(L-1) as n increases since the blank becomes completely redundant (1). In our case this limit is practically attained for the word-character entropies.

In conclusion, the above results show that a redundancy of roughly 53% is exhibited by the text so that a compression of about 47% can be achieved by a compression scheme based upon the tetragrams dictionary. As is well known, the redundancy measures the amount of constraints imposed on the text due to its statistical structure. Of course, compression schemes based upon longer n-grams (or upon words) yield a larger compression at the prize of an increased size dictionary on the text due to its statistical structure. Of course, compression schemes based upon longer n-grams (or upon words) yield a larger compression at the prize of an increased size dictionary.

We would also like to mention that an alternative approach

to the compression problem relies upon an optimal parsing of the text into a reasonably limited number of highly cohesive sequences of charcters with strong intervals statistical influences (9). Such possibility arises when some structural regularities are observed in the strings of the text and when certain types of constraints exist. This certainly occurs for natural languages. By substituting the original words by new shorter strings of characters carrying the same amount of information a relevant reduction in the storage and transmission costs may then be obtained. The problem of information compression becomes thus a problem of feature extraction. These features should be suitable groups of characters typical of the text under consideration. Once they have been identified, a minimum-redundancy coding problem is faced.

In (9) an algorithm to find a set of natural features (i.e. such that there are special symbols denoting the end of the features) has been proposed. An application of this algorithm to our text has yielded 473 features consisting of 6 characters, 81 digrams, 299 trigrams and 95 tetragrams, with a mean length of 1.76 characters. A high degree of compression can thus be achieved if the text is first parsed in terms of these features and then coded. The ensuing amount of compression is indeed roughly the one we would get by parsing the text in terms of trigrams with the advantage of using a reasonably small dictionary. The entropy per character of a feature was approximately found to be 3.27 bits/character. Hence, each character of a feature carries the same amount of information as a symbol of a trigram even though the expected features' length is about a half of that of trigrams'.

ACKNOWLEDGEMENTS

Work supported in part by the Group for Mathematical Information Sciences (GNIM) of the National Council for Research.

REFERENCES

(1) C.E.Shannon, "Prediction and entropy of printed English", Bell.Syst.Tech.J., vol.30,pp.50-64,1951.

(2) F.M.Badran, "Entropy and internal information of printed Arabic", in Trans.Third Prague Conf.on Inform.Theory, pp.73-81, Prague,1963.

(3) G.A.Barnard,III, "Statistical calculation of word entropies for four western languages",IEEE Trans.Inform.Theory, vol.1, pp.49-53,1955.

(4) H.Hansson, "The entropy of the Sweedish language", in Trans. Second Prague Conf.on Inform.Theory, 215-217,Prague,1960.

(5) K.Kuepfmueller, "Die Entropie der deutschen Sprache",F.T.Z. VII, p.265 and foll.,1954.

(6) R.L.Manfrino, "L'entropia della lingua italiana ed il suo calcolo", Alta Frequenza,vol.XXIX, 4-29,1960.

(7) R.L.Manfrino, "Printed Portuguese (Brazilian) entropy statistical calculation", IEEE Trans.Inform.Theory (Abstract), vol.16,p.122, 1960.

(8) M.A.Wanas, A.I.Zayed, M.M.Shaker and E.H.Taha, "First-second and third-order entropies of Arabic text", IEEE Trans.Inform. Theory,vol.22,p.123,1976.

(9) R.M.Capocelli and L.M.Ricciardi, "A heuristic approach to feature extraction and compression for written languages", in Proc. 4-th Int'l Joint Conference on Pattern Recognition, Kyoto,pp.364-368,1978.

CUTOFF RATES WHEN DECODING WITH APPROXIMATE CHANNEL STATISTICS

Jim K. Omura

University of California, Los Angeles, California

ABSTRACT

We derive expressions for the "cutoff rate" of memoryless channels and certain channels with finite memory when the decoder uses approximate channel statistics. For channels with memory we examine both the use of ideal interleavers and deinterleavers with conventional decoders and the use of optimum maximum likelihood decoders that take advantage of the channel memory.

1. INTRODUCTION

Traditionally coding analysis begins with some assumed discrete memoryless channel (DMC) which is available for coding and do not consider how the DMC was created. Wozencraft and Jacobs [1], Wozencraft and Kennedy [2] and more recently Masey [3] and Lee [4] have taken the broader view that, when using coding, the modulation/demodulation system design which creates the DMC should be based on a "cut off rate" R_0 criterion instead of the usual "symbol error probability" criterion. The R_0 criterion recognizes that the real goal of the modulation system is to create the best DMC as seen by the coding system. This R_0 parameter depends only on the DMC and is independent of specific codes. It gives a region of rates where it is possible to operate with arbitrarily small probability of error. Moreover, this rate also is the "R_{comp}" of sequential decoding [1], i.e., the rate above which the average number of decoding steps per decoded digit becomes infinite. In practice, sequential decoders can comfortably operate at rates R near R_0.

D. G. Lainiotis and N. S. Tzannes (eds.), Advances in Communications, 101–110.

In this paper derive expressions for the cut off rate of memoryless channels and certain channels with finite memory when the decoder uses approximate channel statistics. For channels with memory we examine both the use of ideal interleavers and deinterleavers with conventional decoders and the use of optimum maximum likelihood decoders that take advantage of the channel memory. Although we limit this presentation to discrete alphabets, there is no basic changes in the form of the cut off rate expressions with more general channel output alphabets. This work is an extension of Viterbi and Omura [5] to the case where the decoder uses approximate channel statistics.

2. DISCRETE MEMORYLESS CHANNEL (DMC)

Consider a DMC with input alphabet $\mathscr{X}$, output alphabet $\mathscr{Y}$, and conditional probabilities $\{p(y|x): x\varepsilon\mathscr{X}, y\varepsilon\mathscr{Y}\}$. The space of channel input sequences of length N is denoted $\mathscr{X}_N$ while the space of channel output sequences is denoted $\mathscr{Y}_N$. The memoryless condition is

$$p_N(\underline{y}|\underline{x}) = \prod_{n=1}^{N} p(y_n|x_n)$$

for every $\underline{x} = (x_1,\ldots,x_N)\varepsilon\ \mathscr{X}_N$ and $\underline{y}(y_1,\ldots,y_N)\varepsilon\ \mathscr{Y}_N$.

Consider now a block code of rate R and block length N consisting of codewords $\underline{x}_1,\underline{x}_2,\ldots,\underline{x}_M\varepsilon\mathscr{X}_N$ where $M = e^{RN}$. In the case where there is complete knowledge of the channel statistics the minimum probability of error decision rule (equal probable codewords) is the maximum likelihood decision rule where the channel output space $\mathscr{Y}_N$ consists of decision regions*

$$\Lambda_m = \{\underline{y}:\ p_N(\underline{y}|\underline{x}_m) \geq p_N(\underline{y}|\underline{x}_{m'})\ \text{all } m' \neq m\}$$

$$m = 1,2,\ldots,M$$

where if $\underline{y}\varepsilon\Lambda_m$ then the decoder chooses codeword $\hat{\underline{x}} = \underline{x}_m$.

Assume codeword $\underline{x}_1$ is sent. The probability of error is then

$$P_{E_1} = \sum_{\underline{y}\varepsilon\Lambda_1^c} p_N(\underline{y}|\underline{x}_1), \qquad \Lambda_1^c = \text{complement of } \Lambda_1$$

$$\leq \sum_{m=2}^{M} \sum_{\underline{y}\varepsilon\Lambda_m} p_N(\underline{y}|\underline{x}_1) \sqrt{\frac{p_N(\underline{y}|\underline{x}_m)}{p_N(\underline{y}|\underline{x}_1)}}$$

$$\leq \sum_{m=2}^{M} \sum_{\underline{y}} \sqrt{p_N(\underline{y}|\underline{x}_1)p_N(\underline{y}|\underline{x}_m)}$$

where the first inequality is due to the condition

$$\text{if } \underline{y}\varepsilon\Lambda_m \text{ then } 1 \leq \sqrt{\frac{p_N(\underline{y}|\underline{x}_m)}{p_N(\underline{y}|\underline{x}_m)}}$$

by definition of Λ_m and the second inequality is due to the extension of the summation from $\underline{y}\varepsilon\Lambda_m$ to all $\underline{y}\varepsilon\,\mathscr{Y}_N$.

We now randomly select codes by independently choosing codewords according to probability

$$q_N(\underline{x}) = \prod_{n=1}^{N} q(x_n) \tag{1}$$

where $q(\cdot)$ is some arbitrary probability function on $\mathscr{X}$. Averaging $P_{E_1|}$ over this code ensemble where $\bar{P}_{E_1}$ denotes this average,

we have the bound

$$\bar{P}_{E_1} \leq \sum_{m=2}^{M} \prod_{n=1}^{N} \left[\sum_{x} \sum_{x'} q(x)q(x')\left(\sum_{y} \sqrt{p(y|x)p(y|x')}\right)\right]$$

$$= (M-1)\left[\sum_{x} \sum_{x'} q(x)q(x')\left(\sum_{y} \sqrt{p(y|x)p(y|x')}\right)\right]^N$$

$$\leq e^{-N[R_0(q)-R]}$$

where we used $M - 1 \leq M = e^{RN}$ and defined the function

$$R_0(q) = -\ln\left[\sum_{x} \sum_{x'} q(x)q(x')\left(\sum_{y} \sqrt{p(y|x)p(y|x')}\right)\right]$$

$$= -\ln \sum_{y}\left(\sum_{x} q(x)\sqrt{p(y|x)}\right)^2$$

Since the choice of $q(\cdot)$ is arbitrary we can select it to minimize the bound. We define the channel cut off rate as

$$R_0 = \max_{q} R_0(q)$$

$$= - \min_q \ln \sum_y \left(\sum_x q(x) \sqrt{p(y|x)} \right)^2 \tag{2}$$

By symmetry all the code averaged conditional error probabilities are the same and so the error probability

$$P_E = \frac{1}{M} \sum_{m=1}^{M} P_{E_m}$$

when averaged over the code ensemble gives

$$\bar{P}_E \leq e^{-N[R_0 - R]}$$

Clearly there exists at least one code of rate R and block-length where

$$P_E(\mathscr{C}) \leq \bar{P}_E \leq e^{-N[R_0 - R]}$$

This is commonly known as the union Bhattacharyya bound. It is a simple bound which shows that R_0 is an achievable rate with arbitrarily small error probabilities. Although with more complex bounds we can show that higher rates are achievable (up to channel capacity) R_0 easy to evaluate and closer to practically achievable rates with coding.

3. DMC WITH APPROXIMATE CHANNEL STATISTICS

Suppose the decision regions given above are based on approximate channel conditional probabilities denoted $\{p^*(y|x): x \varepsilon \mathscr{X}, y \varepsilon \mathscr{Y}\}$. Then the decision regions become

$$\Lambda_m = \{\underline{y}: \; p_N^*(\underline{y}|\underline{x}_m) \geq p_N^*(\underline{y}|\underline{x}_{m'}) \text{ all } m' \neq m\}$$

$$m = 1,2,\ldots M$$

where

$$p_N^*(\underline{y}|\underline{x}) = \prod_{n=1}^{N} p^*(y_n|x_n)$$

In this case we use [6]

$$\text{if } \underline{y} \varepsilon \Lambda_m \text{ then } 1 \leq \left[\frac{p_N^*(\underline{y}|\underline{x}_m)}{p_N^*(\underline{y}|\underline{x}_1)} \right]^\lambda, \; \lambda \geq 0$$

to obtain a bound to the conditional error probability

$$P_{E_1} = \sum_{\underline{y} \in \Lambda_1^c} p_N(\underline{y}|\underline{x}_1)$$

$$\leq \sum_{m=2}^{M} \sum_{\underline{y} \in \Lambda_m} p_N(\underline{y}|\underline{x}_1) \left[\frac{p_N^*(\underline{y}|\underline{x}_m)}{p_N^*(\underline{y}|\underline{x}_1)} \right]^{\lambda}$$

$$\leq \sum_{m=2}^{M} \sum_{y} p_N(\underline{y}\ \underline{x}_1) \left[\frac{p_N^*(y|x_m)}{p_N^*(\underline{y}|\underline{x}_1)} \right]^{\lambda}$$

Averaging this over the same code ensemble (2), we obtain

$$\bar{P}_{E_1} \leq e^{-N[R_0(\lambda,q)-R]}$$

where

$$R_0(\lambda,q) = -\ln\left\{\sum_{x}\sum_{x'} q(x)q(x') \sum_{y} p(y|x) \left[\frac{p^*(y|x')}{p^*(y|x)}\right]^{\lambda}\right\} \tag{3}$$

The natural expression for the cut off rate for this approximate channel statistics case is thus

$$R_0 = \sup_{\lambda>0} \max_{q} R_0(\lambda,q) \tag{4}$$

where the last equality follows because $R_0(\lambda,q)$ is a convex function of λ and $R_0(\lambda,q) \to 0$ for $\lambda \to 0$ and $R_0(\lambda,q) \to -\infty$ for $\lambda \to \infty$. For the special case where $p(*y|x)$ for all $x \varepsilon \mathscr{X}$ and $y \varepsilon \mathscr{Y}$, $\lambda = 1/2$ achieves the maximum and we get back to our original definition for the cut off rate R_0 as given in (2).

4. FINITE MEMORY CHANNEL (FMC)

We now examine finite memory channels of the form where the coding channel consists of a known deterministic part described by

$$z_n = F(x_n, x_{n-1}, \ldots, x_{n-L+1})$$

$$= F(x_n, \underline{S}_n) \qquad n = 1,2,\ldots \tag{5}$$

where

$$\underline{S}_n = (x_{n-1},\ldots,x_{n-L+1})$$

is the channel state at time n and L is the constraint length of the channel memory. The deterministic memory is followed by a DMC defined by the conditional transition probabilities $\{p^*(y|z)\}$. We assume the decoder knows the deterministic finite memory part of the channel described by the function $F(\cdot,\cdot)$ but only knows approximately the DMC conditional transition probabilities, $\{p^*(y|x)\}$. This type of FMC is typical of channels with intersymbol interference and we shall later discuss bandlimited nonlinear satellite channels of this form.

5. FMC WITH IDEAL INTERLEAVING

In practice, channels with memory are often made to become approximately memoryless with the use of interleavers and deinterleavers. This avoids the need to develop special codes for each type of FMC's. The price of this approach is some loss in achievable coded performance however. We assume the ideal interleaver and deinterleaver create an equivalent DMC where the channel conditional probability is

$$\hat{p}(y|x) = \frac{1}{|S|}\sum_{\underline{S}} p(y|F(x,\underline{S}))$$

where $|S|$ is the total number of channel states. Thus the ideal interleaving is assumed to create a channel where the state $\underline{S}$ consisting of L-1 past channel input symbols are all equal probable during any particular use of the channel.

If we now assume the decoder bases its decision rule on $p^*(y|x)$, we have the cut off rate given by

$$R_0 = \max_{\lambda\geq 0}\ \max_{q}\ R_0(\lambda,q) \tag{6}$$

where

$$R_0(\lambda,q) = -\ln\left\{\sum_x\sum_{x'} q(x)q(x')\sum_y \hat{p}(y|x)\left[\frac{p^*(y|x')}{p^*(y|x)}\right]^{\lambda}\right\} \tag{7}$$

$$= -\ln\left\{\frac{1}{|S|}\sum_{\underline{S}}\left(\sum_x\sum_{x'} q(x)q(x')\sum_y p(y|F(x,\underline{S}))\right.\right.$$

$$\left[\frac{p^*(y|x')}{p^*(y|x)}\right]^{\lambda}\Bigg)\Bigg\}$$

lets assume that λ^* and q^* are such that

$$R_0(\lambda^*,q^*) = \max_{\lambda\geq 0} \max_{q} R_0(\lambda,q)$$

then it is easy to show that

$$e^{-R_0} = \frac{1}{|S|} \sum_{\underline{S}\in S} e^{R_0(\lambda^*,q^*,\underline{S})} \tag{8}$$

where

$$R_0(\lambda^*,q^*,\underline{S}) = -\ln\left\{\sum_{x}\sum_{x'} q^*(x)q^*(x') \sum_{y} p(y|F(x,\underline{S})) \left[\frac{p^*(y|x')}{p^*(y|x)}\right]^{\lambda^*}\right\} \tag{9}$$

This last relationship is of a universal character and the cut off rate R_0 can be interpreted to be that of a "sum channel" [7] where for each channel input one of $|S|$ DMC's is chosen with equal probability.

6. FMC WITH APPROXIMATE MAXIMUM LIKELIHOOD DECODING

Suppose now we have the FMC channel and the code $\mathscr{C}$ of rate R and blocklength N consisting of codewords $\underline{x}_1,\underline{x}_2,\ldots,\underline{x}_M\varepsilon\mathscr{X}_N$ where $M = e^{RN}$. When $\underline{x}\varepsilon\mathscr{X}_N$ enters the FMC, the conditional probability of the output sequence $\underline{y}\varepsilon\mathscr{Y}_N$ is given by

$$p_N(\underline{y}|\underline{x}) = \prod_{n=1}^{N} p(y_n|F(x_n,\underline{S}_n))$$

where $z_n = F(x_n,\underline{S}_n)$ and $\underline{S}_0$ is some given initial state.

The decoder has knowledge of the function $F(\cdot,\cdot)$, but only of $p^*(y|z)$. Hence the decoder uses the conditional probability

$$p_N^*(\underline{y}|\underline{x}) = \prod_{n=1}^{N} p^*(y_n|F(x_n,\underline{S}_n))$$

to define the decision regions

$$\Lambda_m = \{\underline{y}: p_N^*(\underline{y}|x_m) \geq p_N^*(\underline{y}|\underline{x}_{m'}) \text{ for all } m' \neq m\} \quad m = 1,2,\ldots,M.$$

Again, assuming codeword $\underline{x}_1$ is sent we have the union Bhattacharyya bound [6]

$$P_{E_1} \leq \sum_{m=2}^{M} \sum_{\underline{y}} p_N(\underline{y}|\underline{x}_1) \left[\frac{p_N^*(\underline{y}|\underline{x}_m)}{p_N^*(\underline{y}|\underline{x}_1)}\right]^{\lambda}$$

$$= \sum_{m=2}^{M} \prod_{n=1}^{N} \left(\sum_{y} p(y|Fx_{1n},\underline{S}_{1n})) \left[\frac{p^*(y|F(x_{mn},\underline{S}_{mn}))}{p^*(y|F(x_{1n},\underline{S}_{1n}))}\right]^{\lambda} \right) \quad (10)$$

Now as before we randomly select codes by independently choosing codewords according to probability

$$q_N(\underline{x}) = \prod_{n=1}^{N} q(x_n)$$

where $q(\cdot)$ is some arbitrary probability function on $\mathscr{X}$. Averaging P_{E_1} over this code ensemble yields the bound

$$\bar{P}_{E_1} \leq M \sum_{\underline{x}} \sum_{\underline{x}'} \prod_{n=1}^{N} q(x_n)q(x_n') \left(\sum_{y} p(y|F(x_n,\underline{S}_n)) \left[\frac{p^*(y_n|F(x_n',\underline{S}_n'))}{p^*(y_n|F(x_n,\underline{S}_n))}\right]^{\lambda} \right)$$

We now define pair states $\hat{\underline{S}}_n = (\underline{S}_n,\underline{S}_n')$ and pair inputs $\hat{\underline{x}}_n = (\underline{x}_n,\underline{x}_n')$. Let $\underline{\Delta}_1,\underline{\Delta}_2,\ldots,\underline{\Delta}_\Gamma$ be the $\Gamma = |\mathscr{X}|^{2(L-1)}$ possible distinct pair states. Here $|\mathscr{X}|$ is the size of the channel input alphabet and is the constraint length of the channel memory. Define

$$Q(\underline{x},\hat{\underline{S}};q,\lambda) = q(x)q(x') \left(\sum_{y} p(y|F(x,\underline{S})) \left[\frac{p^*(y|F(x',\underline{S}))}{p^*(y|F(x,\underline{S}))}\right]^{\lambda} \right) \quad (11)$$

and the $\Gamma\times\Gamma$ matrix $\underline{B}(q,\lambda) = \{b_{ij}\}$ where

$$b_{ij} = \begin{cases} Q(\underline{x},\underline{\Delta}_j);q,\lambda), & \text{if state } \underline{\Delta}_i \text{ can be reached from state } \underline{\Delta}_j \text{ for some } \underline{x} \text{ input otherwise} \\ 0, & \end{cases} \quad (12)$$

$$i,j = 1,2,\ldots L = |\mathscr{X}|^{2(L-1)}.$$

Then

$$\bar{P}_{E_1} \leq M \sum_{\hat{\underline{x}}_1} \sum_{\hat{\underline{x}}_2} \ldots \sum_{\hat{\underline{x}}_N} \prod_{n=1}^{N} Q(\hat{\underline{x}}_n,\hat{\underline{S}}_n;q,\lambda)$$

$$= M[1 \;\; 1 \; \ldots \; 1]B(q,\lambda)^N \begin{bmatrix} 1 \\ 0 \\ \vdots \\ 0 \end{bmatrix}$$

where we assume initial state $\hat{\underline{S}} = \underline{\Delta}_1$.

Although the matrix $B(q,\lambda)$ can be quite large, it is generally very sparse with only $|\mathscr{X}|^2$ nonzero elements per row and column. Hence computing $\underline{B}(q,\lambda)^N$ can be done for large values of $\Gamma = |\mathscr{X}|^{2(L-1)}$. The matrix $\underline{B}(q,\lambda)$ is also nonnegative and irreducible. The Perron-Frobenius Theorem [8] states that such a matrix has a real maximum eigenvalue $\rho(q,\lambda)$ and an associated positive left eigenvector. Defining $\alpha > 0$ to be the largest component of the left eigenvector divided by its smallest component we have inequalities

$$\frac{1}{\alpha}\rho(q,\lambda)^N \leq [1 \; 1 \; \ldots \; 1]\underline{B}(q,\lambda)^N \begin{bmatrix} 1 \\ 0 \\ \vdots \\ 0 \end{bmatrix} \leq \alpha\rho(q,\lambda)^N$$

and thus

$$\bar{P}_{E_1} \leq \alpha M \rho(q,\lambda)^N = e^{-N[R_0(q,\lambda)-R]}$$

where we define

$$R_0(q,\lambda) = -\ln \rho(q,\lambda).$$

The natural definition for the cut off rate here is then

$$\begin{aligned} R_0 &= \sup_{\lambda \geq 0} \max_q R_0(q,\lambda) \\ &= \sup_{\lambda \geq 0} \max_q \{-\ln \rho(q,\lambda)\} \end{aligned} \tag{13}$$

Approximations to $\rho(q,\lambda)$ can be obtained by computing powers of the sparse matrix $\underline{B}(q,\lambda)$ [9].

REFERENCES

(1) Wozencraft, J. M., and Jacobs, I. M., Principles of Communication Engineering, John Wiley & Sons, New York, 1965.

(2) Wozencraft, J. M., and Kennedy, R. S., "Modulation and Demodulation for Probabilistic Coding," IEEE Trans. Inform. Th., Vol. IT-12, No. 3, July 1966, pp. 291-297.

(3) Massey, J. L., "Coding and Modulation in Digital Communications," Proceedings International Zurich Seminar on Digital Communications, Switzerland, March 12-15, 1974.

(4) Lee, L., "On Optimal Soft-Decision Demodulation," IEEE Trans. Inform. Thy., Vol. IT-22, No. 4, July 1976, pp. 437-444.

(5) Viterbi, A. J., and Omura, J. K., Digital Communication and Coding, McGraw-Hill, New York, 1979.

(6) Stiglitz, I. G., "Coding for a Class of Unknown Channels," IEEE Trans. on Inform. Th., Vol. IT-12, April 1966, pp. 189-195.

(7) Gallager, R. G., Information Theory and Reliable Communication, John Wiley & Sons, New York, 1968, pp. 525.

(8) Bellman, R., Introduction to Matrix Analysis, McGraw-Hill, New York, 1970, pp. 288.

(9) Divsalar, D., "Mismatched Viterbi Receivers for ISI Channels," Ph.D. dissertation, University of California, Los Angeles, December 1978.

CHARACTERIZATION AND PERFORMANCE EVALUATION OF A (56,28) CODE

George Young and V.K. Bhargava
Teleglobe Canada
680 Sherbrooke Street West
Montreal, Canada
H3A 2S4

V.K. Bhargava
Depart. of Electrical Eng.
Concordia University
Montreal, Canada
H3G 1M8

ABSTRACT

We characterize a (54,27) code in terms of the incidence matrix of the difference set defined on $GF(3^3)$. It is known that all the weights in this code will be of the form 4t or 4t-1. The minimum distance of the code is found to be 11 and its weight distribution is computed. The (54,27) d = 11 is extended to a (56,28) d = 12 code. This code is known to be self-dual with all weights of the form 4t. The performance of the (56,28) d = 12 code in conjunction with QPSK modulation over a satellite channel is evaluated.

1. INTRODUCTION

Coding of digital information before transmission offers the possibility of removing errors introduced by a noisy transmission medium. Block coding breaks up a digital transmission stream into equal segments called blocks of length n bits. Each block is formed from k information bits according to some rule and the resulting set of 2^k codewords is called an (n,k) block code.

It is convenient to think of codewords as vectors from the vector space V_n of all n-tuples. A linear block code is one in which the codewords form a k-dimensional subspace of V_n. A kxn matrix which has, as its rows, basis vectors of the subspace is called the generator matrix of the code. The codeword corresponding to a particular k information bits is found by pre-multiplying the generator matrix by the 1xk information vector.

D. G. Lainiotis and N. S. Tzannes (eds.), Advances in Communications, 111–116.

This paper describes a (54,27) error-correcting linear block code, and the procedure used to obtain the weight distribution of the code. As a method of verifying the accuracy of the results, the code was extended to a (56,28) code, and the weight distribution of the extended code was compared with previously tabulated results (1). The performance of the (56,28) code in conjunction with QPSK modulation over a satellite channel is evaluated.

2. THE (54,27) CODE

The code is specified by a generator matrix of the form

$$G = [I_{27}, A].$$

Here I_{27} is a 27x27 identity matrix and A is the "incidence matrix" of a (27,14,7) group difference set described below.

Let $p = 3^3 = 27$. In $GF(3^3)$ with $x^3 = x+2$

$$D = (0, x, x+2, 2x^2+x+2, x^2+2x+2, x+1, x^2+x+2, 2, 2x^2, 2x^2+x, 2x^2+2x+2, x^2+1, 2x^2+2x, 2x^2+1)$$

is a (27,14,7) group difference set. The incidence matrix is given by

$$A = \begin{bmatrix} P & Q & R \\ R & P & Q \\ Q & R & P \end{bmatrix}$$

and each of the P,Q, and R is comprised of nine 3x3 circulants which are given by:

$$P = \begin{bmatrix} P_1 & P_2 & P_3 \\ P_3 & P_1 & P_2 \\ P_2 & P_3 & P_1 \end{bmatrix} \text{ with } P_1 = \begin{bmatrix} 1 & 0 & 1 \\ 1 & 1 & 0 \\ 0 & 1 & 1 \end{bmatrix} \quad P_2 = \begin{bmatrix} 1 & 1 & 1 \\ 1 & 1 & 1 \\ 1 & 1 & 1 \end{bmatrix} \text{ and } P_3 = \begin{bmatrix} 0 & 0 & 0 \\ 0 & 0 & 0 \\ 0 & 0 & 0 \end{bmatrix},$$

$$Q = \begin{bmatrix} q_1 & q_2 & q_3 \\ q_3 & q_1 & q_2 \\ q_2 & q_3 & q_1 \end{bmatrix} \text{ with } q_1 = \begin{bmatrix} 0 & 1 & 0 \\ 0 & 0 & 1 \\ 1 & 0 & 0 \end{bmatrix} \quad q_2 = \begin{bmatrix} 0 & 0 & 1 \\ 1 & 0 & 0 \\ 0 & 1 & 0 \end{bmatrix} \text{ and } q_3 = \begin{bmatrix} 0 & 0 & 1 \\ 1 & 0 & 0 \\ 0 & 1 & 0 \end{bmatrix}$$

and

$$R = \begin{bmatrix} r_1 & r_2 & r_3 \\ r_3 & r_1 & r_2 \\ r_2 & r_3 & r_1 \end{bmatrix} \text{ with } r_1 \begin{bmatrix} 1 & 1 & 0 \\ 0 & 1 & 1 \\ 1 & 0 & 1 \end{bmatrix} \quad r_2 = \begin{bmatrix} 1 & 0 & 1 \\ 1 & 1 & 0 \\ 0 & 1 & 1 \end{bmatrix} \text{ and } r_3 = \begin{bmatrix} 1 & 0 & 1 \\ 1 & 1 & 0 \\ 0 & 1 & 1 \end{bmatrix}$$

Note the cyclic and repetitive nature of the matrix A which will be exploited to compute the weight distribution of the code.

2.1. WEIGHT DISTRIBUTION DETERMINATION

The weight distribution of a code is very important from many points of view. For communications engineers, its importance lies in the exact determination of the post-decoding bit error rate (2). In addition it can be used to ascertain the error-correcting capability of the code.

The determination of the weight distribution of a code is a straightforward if lengthy procedure. All output codewords are generated, which is to say all possible modulo-2 combinations of the rows of the generator matrix are formed, the number of ones in each codeword is counted, and the quantity of output codewords of each weight is totalled.

The above procedure would seem to lend itself immediately to a solution by computer program, and the Control Data computer at Concordia University was used. Assembler language programming was chosen for its ease of handling binary numbers and logic operations, and its relatively high execution speed compared to higher level languages such as Fortran.

A direct method, which simply generated all code vectors, and counted the ones in each codeword, would require an estimated three and one-half hours of central processor time to complete the determination of the entire weight distribution. The cyclic nature of the generator matrix was examined, and methods were found which reduced the required computer time to under half an hour. Details of the methods will not be presented here but, can be obtained from the authors. The (54,27) code generated by G has the following weight distribution (with A(i) denoting the number of code words of weight i):

A(0)	=	1
A(11)	=	1,404
A(12)	=	5,031
A(15)	=	129,312
A(16)	=	315,198
A(19)	=	2,734,992
A(20)	=	4,786,236
A(23)	=	16,185,312
A(24)	=	20,906,028
A(27)	=	29,006,824
A(28)	=	27,970,866
A(31)	=	16,185,312
A(32)	=	11,633,193

A(35)	=	2,734,992
A(36)	=	1,443,468
A(39)	=	129,312
A(40)	=	48,492
A(43)	=	1,404
A(44)	=	351

Note that all the weights in this code are of the form 4t or 4t-1. Further observe that A(2i-1) = A(54-2i+1). This phenomenon has been observed in many other classes of codes (3,4).

3. EXTENSION TO (56,28) CODE

Adding an overall parity bit to the (54,27) code results in a (55,27) code with the following weight distribution:

A(0)	=	1
A(12)	=	6,435
A(16)	=	444,510
A(20)	=	7,521,228
A(24)	=	37,091,340
A(28)	=	56,977,690
A(32)	=	27,818,505
A(36)	=	4,178,460
A(40)	=	177,804
A(44)	=	1,755

We now add a row of 55 ones to the generator matrix of the (55,27) code. This, results in a (55,28) code with the following weight distribution:

A(0) & A(55)		1
A(11) & A(44)	=	1,755
A(12) & A(43)	=	6,435
A(15) & A(40)	=	177,804
A(16) & A(39)	=	444,510
A(19) & A(36)	=	4,178,460
A(20) & A(35)	=	7,521,228
A(23) & A(32)	=	27,818,505
A(24) & A(31)	=	37,091,340
A(27) & A(28)	=	56,977,690

Finally we again add an overall parity bit to the (55,28) code. This results in a (56,28) code with the following weight distribution:

A(0)	& A(56)	=	1
A(12)	& A(44)	=	8,190
A(16)	& A(40)	=	622,314
A(20)	& A(36)	=	11,699,688
A(24)	& A(32)	=	64,909,845
A(28)		=	113,955,380

The above mentioned (56,28) code is the external self-dual code whose weight distribution (but not the characterization) was reported in (1). To summarize, the generator matrix of the (56, 28) external self-dual code is:

$$G' = \left[I_{28}, \begin{matrix} 0 & j \\ j^T & A \end{matrix} \right]$$

where j=(1,1,...,1) is a row of 27 ones and A is the incidence matrix of a (27,14,7) group difference set defined on $GF(3^3)$.

4. PERFORMANCE EVALUATION OF THE (56,28) CODE

The performance of the (56,28) code in conjunction with QPSK modulation over a satellite channel was evaluated using the method described in (5). The following is a summary of the Net Coding Gain (NCG) at various Bit Error Rates (BER):

BER	10^{-2}	10^{-3}	10^{-4}	10^{-5}
NCR (in dB)	4.09	4.55	4.91	5.07

For details, the reader is referred to (6).

ACKNOWLEDGEMENTS

This research was supported by Natural Sciences and Engineering Research Council of Canada and by FCAC-Quebec.

REFERENCES

(1) Mallows,C.L., and Sloane,N.J.A.: 1973, "An upper bound for self-dual codes", Information and Control,22, pp.188-200.

(2) Lin,S.: 1970,"An introduction to error-correcting codes", Prentice Hall Inc., Englewood Cliffs, New Jersey.

(3) Karlin,M., Bhargava,V.K., and Tavares,S.E.: 1978, "A note on extended quaternary quadratic residue codes and their binary images", Information and Control, 38, pp. 148-153.

(4) Bhargava,V.K.: 1977,"Odd weight symmetry in some binary codes", IEEE Trans. on Info. Theory, 23, pp. 518-520.

(5) Huntoon,Z., and Michelson,A.M.: 1977,"On the computation of the probability of post-decoding error events for block codes", IEEE Trans. on Info. Theory, 23, pp. 399-403.

(6) Avni,M., and Bhargava,V.K.,: 1979, "On improving the bit error rate of various modulation systems by block coding" submitted for publication.

WEIGHT DISTRIBUTION OF SOME CYCLIC CODES OF MACWILLIAMS

V.K. Bhargava and C. Nguyen

Dept. Elec. Eng.
Concordia University
1455 de Maisonneuve W.
Montreal, Canada H3G 1M8

Canadian Marconi Company
2442 Trenton Avenue
Montreal, Quebec
Canada

ABSTRACT

Mrs. MacWilliams recently gave methods for the decomposition of cyclic codes of block lengths 3p, 5p, and 7p into systematic quasi-cyclic form. In this paper, using her results, we construct and enumerate the complete weight distribution of many cyclic codes. Among the many codes, exhibiting remarkable gaps in their weight distributions are the (87,28,24), (51,16,16) and the (91, 12,36) codes with weights divisible by four, and a (85,16,32) code with weights divisible by eight. We also report on a (52,13, 16) quasi-cyclic subcode (of the (65,13,25) cyclic code) with all weights divisible by four. It is not known if this code is cyclic as well.

1. INTRODUCTION

An (mp,p) cyclic code is generated by a polynomial g(x) of degree p(m-1), which divides $x^{mp}-1$. It is also known that the cyclic code can be specified as the row space of

$$G = \begin{bmatrix} g(x) \\ xg(x) \\ \cdot \\ \cdot \\ \cdot \\ x^{p-1}g(x) \end{bmatrix}$$

D. G. Lainiotis and N. S. Tzannes (eds.), Advances in Communications, 117–122.

Mrs. F.J. MacWilliams (1) recently examined conditions under which the matrix G can be re-arranged in the form

$$G' = [Ip, C_1, C_2, \ldots, C_{m-1}]$$

where each C_i is a pxp circulant matrix over GF(2), of the form

$$C_i \quad \begin{bmatrix} a_o & a_1 & \cdots & a_{p-1} \\ a_{p-1} & a_o & \cdots & a_{p-2} \\ \cdots & \cdots & \cdots & \cdots \\ a_1 & a_2 & \cdots & a_o \end{bmatrix}$$

The algebra of pxp circulant matrices over GF(2) is isomorphic to that of polynomials modulo x^p-1 over GF(2) with the following one-to-one correspondence:

$$C = \begin{bmatrix} a_o & a_1 & \cdots & a_{p-1} \\ a_{p-1} & a_o & \cdots & a_{p-2} \\ \cdots & \cdots & \cdots & \cdots \\ a_1 & a_2 & \cdots & a_o \end{bmatrix} \leftrightarrow c(x)= a_o+a_1x+\ldots+a_{p-1}x^{p-1}$$

Thus we shall often speak of the circulant C as being specified by the "associated" polynomial c(x). We shall also represent the polynomials in their octal form. For instance, the octal number five has the binary form 101 corresponding to the polynomial $1+x^2$ and represents the circulant

$$\begin{bmatrix} 1 & 0 & 1 \\ 1 & 1 & 0 \\ 0 & 1 & 1 \end{bmatrix}$$

The code generated by G' is referred to as a rate 1/m systematic quasi-cyclic code, In her paper, Mrs. MacWilliams has demonstrated that the possibility of decomposing a (mp,p) cyclic code into a (mp,p) systematic quasi-cyclic code depends on the length of the cyclotonnic cosets mod p (say k) and on the length of cyclotonnic cosets mod mp (say k'). The advantages of having the code in systematic quasi-cyclic form are (i) both the code and the dual code are specified by the same circulants (or the same polynomials associated with the first row of the circulants),(ii) the decoding method proposed by Karlin (2), may be applied to these codes, and (iii) ease in computing the weight distribution and in putting the weight distribution in factored form.

In this paper we compute the weight distribution of a number of cyclic codes for which MacWilliams has given a decomposition in systematic quasi-cyclic form and for which p is at most 29 (the present limit of our computer programme).

2. CYCLIC CODES OF BLOCK LENGTH 3P

Here the decomposition is possible if k is divisible by two (1). For $p=8m\pm3$ for which 2 is a primitive root the codes are exactly those described by Bhargava, Séguin and Stein (3). The weight distributions for $p=5,11,13,17$ and 19 were reported in (3). For $p=29$, the weight distribution of the (87,29) code generated by $c_1(x)=6364221362$ and $c_2(x)=5413556414$ is as follows (here $A(i)$ denotes the number of code words of weight i):

A(0) & A(87) =	1
A(24) & A(63) =	71,253
A(28) & A(59) =	613,872
A(31) & A(56) =	1,238,793
A(32) & A(55) =	5,618,025
A(35) & A(52) =	17,276,112
A(36) & A(51) =	24,701,040
A(39) & A(48) =	62,560,134
A(40) & A(47) =	65,315,250
A(43) & A(44) =	91,040,976

If we "expurgate" the above to a (87,28) cyclic code, we see that all weights are divisible by four.

The decomposition is possible for $p=8m\pm3$ when 2 is not a primitive root, but the forms of the circulants C_1 and C_2 is not simple. The construction does not work for $p=8m-1$. Further for $p=8m+1$, the decomposition works only if k is even, but again the forms of the circulants C_1 and C_2 are not simple (1). However for $p=17$, the weight distribution of the (51,17) code generated by $c_1(x) = 551454$ and $c_2(x) = 626324$ is as follows:

A(0) & A(51) =	1
A(16) & A(35) =	1,530
A(19) & A(32) =	5,661
A(20) & A(31) =	8,161
A(23) & A(28) =	24,480
A(24) & A(27) =	25,704

Clearly the "expurgated" (51,16,16) code will have all weights dividible by four.

The duals of these codes, the (3p,2p) rate 2/3 quasi-cyclic codes are not good. In fact it can be shown that they have minimum distance of at most six (3).

3. CYCLIC CODES OF BLOCK LENGTH 5p

For this case the decomposition works only if k is a multiple of four. Consequently, p must have the form 8m-3 or 8m+1, p greater than five (1). The only prime numbers satisfying these conditions and which are within the computing range of our programme are 13 and 17.

For p=13, the weight distribution of the (65,13) code generated by $c_1(x)=3575$, $c_2(x)=37464$, $c_3(x)=26374$ and $c_4(x)=13734$ is as follows:

A(0)	&	A(65)=	1
A(25)	&	A(40)=	195
A(26)	&	A(39)=	390
A(27)	&	A(38)=	520
A(28)	&	A(37)=	455
A(29)	&	A(36)=	585
A(30)	&	A(35)=	780
A(31)	&	A(34)=	390
A(32)	&	A(33)=	780

For p=17, the weight distribution of the (85,17) code generated by $c_1(x)=42412$, $c_2(x)=41144$, $c_3(x)=540014$ and $c_4(x)=602202$ is as follows:

A(0) =	A(85) =	1
A(21)=	A(64) =	85
A(32)=	A(53) =	3,570
A(37)=	A(48) =	23,800
A(40)=	A(45) =	38,080

The above code can be "expurgated" to a (85,16,32)code with all weights divisible by eight.

3.1.1 The (52,13,16) Quasi-Cyclic Subcode

While examining the quasi-cyclic subcodes of the (65,13,25) cyclic code, the authors found the following (52,13,16) quasi-cyclic code generated by $c_1(x)=3575$, $c_2(x)=37464$ and

$c_3(x)=26374$:

A(0)	& A(52) =	1
A(16)	& A(36) =	13
A(20)	& A(32) =	1,092
A(24)	& A(28) =	2,990

This is the first known example of a rate 1/4 systematic quasi-cyclic codes with weights divisible by four.

4. CYCLIC CODES OF BLOCK LENGTH 7p

In this case a necessary condition for the decomposition to work is that $p=6t+1$, and p is greater than seven. The circulant matrices however have simple forms only for the case $p=8m\pm3$ for which two is a primitive root.

For $p=13$, the weight distribution of the (91,13) code generated by $c_1(x)=4264$, $c_2(x)=61104$, $c_3(x)=75134$, $c_4(x)=5403$, $c_5(x)=63744$ and $c_6(x)=5667$ is as follows:

A(0)	& A(91) =	1
A(36)	& A(55) =	364
A(39)	& A(52) =	728
A(40)	& A(51) =	546
A(43)	& A(48) =	1,365
A(44)	& A(47) =	1,092

If we "expurgate" the code to a (91,12) code we see that all weights are divisible by four.

5. CONCLUSIONS

In this paper we have presented the weight distribution of a number of low rate 1/m cyclic codes in quasi-cyclic form. Many of these codes can be suitably expurgated to form constant weight codes. This coding technique has recently been used over the Raleigh fading channel (4). Also, advantage can be taken of the MacWilliams identities to compute the minimum distance (and possibly the complete weight distribution) of the rate (m-1)/m dual code.

ACKNOWLEDGEMENTS

This research was supported by the Natural Sciences and Engineering Research Council of Canada and by FCAC-Quebec.

REFERENCES

(1) MacWilliams,F.J.: 1979,"Decomposition of cyclic codes of block lengths 3p,5p,7p", IEEE Trans. on Info. Theory, 25,

(2) Karlin,M.: 1970,"Decoding of circulant codes", IEEE Trans. on Info. Theory, 16,pp.797-802.

(3) Bhargava,V.K., Séguin,G.E., and Stein,J.M.: 1978,"Some (mk,k) cyclic codes in quasi-cyclic form", IEEE Trans. on Info. Theory, 24, pp. 630-632.

(4) Pieper,J.F., Proakis,J.G., Reed,R.R., and Wolf,J.K.: 1978, "Design of Efficient Coding and Modulation for a Rayleigh Fading Channel", IEEE Trans. on Info. Theory, 24,pp. 457 - 469.

ON WEAKLY PREFIX SUBSEMIGROUPS OF A FREE SEMIGROUP

Renato M. Capocelli

Laboratorio di Cibernetica del CNR - 80072 Arco Felice, Naples, Italy.

ABSTRACT. *A remarkable family of free subsemigroups of a free semigroup, the family of weakly prefix subsemigroups, is considered. An algorithm for obtaining the minimal weakly prefix subsemigroup containing a given finite subset is proposed.*

INTRODUCTION.

In this paper weakly prefix subsemigroups of a free semigroup and some of their properties are considered. A weakly prefix subsemigroup Z of a free semigroup X^+ is a subsemigroup of X^+ satisfying the condition: for all $a \in Z, x, y \in X^+$; $ax, xy, yx \in Z$ imply $x, y \in Z$. Weakly prefix subsemigroups and their bases, weakly prefix codes, have been the object of recent investigations (4), (5). They appear to be a remarkable family of free subsemigroups exibiting properties of relevant interest in information transmission. In particular, in the finite case their bases coincide with codes having finite decipherability delay.

The optimal factorization of a subsemigroup into a weakly prefix subsemigroup is also considered. It is shown that the intersection of the family of weakly prefix subsemigroups containing a given subset A of X^+ is itself a weakly prefix subsemigroup. It provides the minimal weakly prefix subsemigroup containing A. In the finite case, a procedure for constructing the basis F of such a subsemigroup is proposed. Obviously if A is a weakly prefix code F coincides with A, and viceversa. Therefore the procedure for obtaining F provides a criterion to test whether a finite code is decipherable with finite delay. When A^+ is not a weakly prefix subsemigroup of X^+, F provides

D. G. Lainiotis and N. S. Tzannes (eds.), Advances in Communications, 123–129.

somewhat optimal code, with the finite decipherability property, to deal with A-messages. We remark that the proposed procedure is very simple and based explicitly on certain fundamental properties of codewords and their sequences.

WEAKLY PREFIX SUBSEMIGROUPS.

Let X be a finite nonempty set and let X^+ and X^* be the *free semigroup* and the *free monoid* generated by X, respectively. We call *letters* the elements of X, *words* the elements of X^+ and denote by $l(w)$ the *length* of the word $w \in X^+$ and by X^n the n-fold concatenation of X with itself. Given $p, r, w \in X^+$; if $pr = w$ then p is a *prefix* of w and r is a *suffix* of w. Let A be a subset of X^+. A is a *uniquely decipherable code*, or simply a *code*, iff A^+ is a free subsemigroup of X^+ having A as basis. A code A has *finite decipherability delay* iff there exists an integer p such that if $x \in X^+$, $l(x) \geq p$, and $xy \in A^+$, then x has a decomposition $x = x_1 x_2$ such that whenever $xz \in A^+$, $x_1 \in A^+$ and $x_2 z \in A^*$.

Given a subset A of X^+, a sequence of words $w_1, w_2, \ldots, w_n$ ($w_i \in A$) is called an *L(linked) sequence* of prefix s_o and suffix s_n (3) if

$$
\begin{array}{lll}
w_1 = s_o s_1 & & \\
w_2 = s_1 s_2 & \text{or} & s_1 = w_2 s_2 \\
w_3 = s_2 s_3 & \text{or} & s_2 = w_3 s_3 \qquad [1] \\
\ldots\ldots\ldots & & \\
w_n = s_{n-1} s_n & \text{or} & s_{n-1} = w_n s_n, \quad (s_i \in X^+).
\end{array}
$$

Given an L-sequence, let us denote by a_k the kth prefix of [1] (i.e., if $w_{k+1} = s_k s_{k+1}$, then $a_k = s_k$; if $s_k = w_{k+1} s_{k+1}$, then $a_k = w_{k+1}$) and let us consider the sequence

$$s_o \; a_1 \; a_2 \; \ldots \; a_k \; \ldots \; a_{n-1} \; s_n \qquad [2]$$

obtained by concatenating to s_o the prefixes a_k of [1] and s_n. It can be immediately seen that, in general, either

$$s_o a_1 \ldots a_{n-1} \in A^+ \quad \text{and} \quad a_1 a_2 \ldots a_{n-1} s_n \in A^+, \text{ or}$$

$$s_o a_1 \ldots a_{n-1} s_n \in A^+ \text{ and } a_1 \ldots a_{n-1} \in A^+ .$$

In particular, if s_o, $s_n \in A^+$ the above provide two different factorizations of [2].

Further let us say that:

A word w_1 is *divided* by the set A if $w_1 = s_o s_1$ with $s_o \in A$, and if there exists an L-sequence *(dividing sequence)* $w_1 w_2 \ldots w_n \in A^+$ of prefix s_o and suffix $s_n \in A$.

A word w_1 *generates a loop* in A if $w_1 = s_o s_1$, $s_o \in A$, and if there exists an L-sequence *(loop sequence)* $w_1 w_2 \ldots w_n \in A^+$ of prefix s_o and suffix $s_n = s_i$ $(i \neq 0)$.

A word w_1 *belongs to a loop* in A if $w_1 = s_o s_1$, $s_o \notin A^+$, and if there exists an L-sequence *(cyclic sequence)* $w_1 w_2 \ldots w_n \in A^+$ of prefix s_o and suffix $s_n = s_o$.

Theorem 1. *A finite subset A of X^+ is a code iff no words of A are divided by A.*

Theorem 2. *A finite code A has finite decipherability delay iff no codewords of A generate a loop in A.*

Theorem 3. *A finite code A is synchronizable iff no codewords of A belong to a loop in A.*

We say that a code A is *weakly prefix* iff for any $a \in A^+$ and $x,y \in X^+$ the following relations $ax \in A^+$, $yx \in A^+$, $xy \in A^+$ imply $x,y \in A^+$.

It is easy to prove the following Theorem.

Theorem 4. *A finite code A has finite decipherability delay iff it is weakly prefix.*

Proof. If A does not have a finite decipherability delay there exists a loop sequence and therefore one has

$$s_o a_1 \ldots a_{i-1} \in A^+$$

$$s_o a_1 \ldots a_{i-1} s_i \in A^+$$

$$s_i (a_{i+k+1} \ldots a_{n-1} s_n a_{i+k+1} \ldots a_{n-1}) \in A^+$$

$$(a_{i+k+1} \ldots a_{n-1} s_n a_{i+k+1} \ldots a_{n-1}) s_n \in A^+$$

with $s_o \in A$ and $s_i = s_n = a_i \ldots a_{i+k} \notin A^+$.

Conversely, if A is not weakly prefix (that is there exist the sequences $\alpha = ax \in A^+$, $\beta = xy \in A^+$, $\gamma = yx \in A^+$ with $a \in A^+$ and $x,y \notin A^+$), from

$\alpha = a/x$
$\beta = x/y$
$\gamma = y/x$ with $a, \alpha, \beta, \gamma \in A^+$, $x \notin A^+$,

we can immediately deduce the existence of a loop sequence so that A does not have a finite delay of decipherability.

Since a maximal finite code either is prefix or does not have finite decipherability delay (8) (a code is maximal if it is no proper subset of any other code in the same alphabet), from Theorem 4 we obtain the following

Corollary 1. *A finite weakly prefix complete code is a prefix code.*

A subsemigroup Z of X^+ is said to be *weakly prefix* iff for all $a \in Z$ and $x,y \in X^+$ the relations $ax \in Z$, $xy \in Z$, $yx \in Z$ imply $x,y \in Z$.

The following statements may be easily proved.

i) A subsemigroup Z of X^+ is weakly prefix iff its minimal set of generators $Z \setminus Z^2$ is a weakly prefix code.

ii) If Z is a weakly prefix subsemigroup of X^+ then it is free in X^+.

iii) If B is a weakly prefix subsemigroup of X^+ and Z a weakly prefix subsemigroup of B, Z is a weakly prefix subsemigroup of X^+.

It is also easy to prove the following Theorems:

Theorem 5. *The family of weakly prefix subsemigroups is closed under intersection.*

Proof. Let Z_n, $n \in N$, be a nonempty family of weakly prefix subsemigroups of X^+ and set $Z = \cap Z_n$. Z is a subsemigroup of X^+. Further for any $x,y \in X^+$; $a, ax, xy, yx \in Z$ imply $x,y \in Z$. That is, Z is a weakly prefix subsemigroup of X^+.

Theorem 6. *The family of weakly prefix subsemigroups (weakly prefix codes) has the Zorn property.*

Proof. Let Z_n, $n \in N$, be a chain of weakly prefix subsemigroups and set $Z = \underset{n}{\cup} Z_n$. Z is a subsemigroup of X^+. Further suppose that $ax, xy, yx \in Z$ with $a \in Z$ and $x,y \in X^+$, there exists then an element of the chain, Z_m, such that $ax, xy, yx \in Z_m$ with $a \in Z_m$ and $x,y \in X^+$. Since Z_m is weakly prefix it follows that $x,y \in Z_m \subset Z$. Therefore Z is a weakly prefix subsemigroup.

Theorem 7 (6). *Let A be the basis of $A^+ \subseteq X^+$. The following statements are equivalent.*

(1) A^+ is a weakly prefix subsemigroup.

(2) For all $h \in X^+$, $a \in A^+$
$ah \in A^+$ and $hA^+ \cap A^+h \neq \emptyset$ imply $h \in A^+$.

(3) For all $a \in A^+$; $u,v \in X^+$
$av \in A^+$, $uv \in A$ imply $vAu \cap A^+ = \emptyset$.

(4) For all $u \in X^*$, $v \in X^+$, $a \in A^+$
$av \in A^+$, $x, uv \in A$ with $l(x) < l(v)$ imply
$vA^*u \cap xA^* = \emptyset$.

MINIMAL WEAKLY PREFIX SUBSEMIGROUPS.

As it has been shown in the previous Section the family of weakly prefix subsemigroups of X^+ is closed under intersection. Therefore, for any $A \subset X^+$, the intersection of all weakly prefix subsemigroups containing A provides the minimal weakly prefix subsemigroup containing A. Let be F the basis of such a minimal subsemigroup. Similarly to what has been proved elsewhere for free (very pure) subsemigroups of X^+ (1), (7) it is easy to see the following Theorem and its consequent Corollary.

Theorem 7. *Any element of the basis F of the minimal weakly prefix subsemigroup containing A is a left factor (right factor) of some word of A.*

Proof. Suppose that for some $f \in F$ $fF^* \cap A = \emptyset$. Consider the set $Z = (F \setminus \{f\})f^*$; Z is a weakly prefix code in the alphabet F and, since F is a weakly prefix code in the alphabet X, is a weakly prefix code in the alphabet X. Therefore $A \subset Z^+ \subset F^+$ which contradicts the assumption that F^+ is the minimal weakly prefix subsemigroup containing A.

Corollary 2. *If A is finite then Card (F) $\leq$ Card (A). In particular if A is not a weakly prefix code Card (F) $\leq$ Card (A) - 1.*

Proof. Since $A \subset F^+$ and F freely generates F^+ we can introduce the map $g : A \to F$ defined for all $x \in A$ as $g(x) = f$ iff $x \in fF^*$. From the previous Theorem it follows that g is surjective so that in the finite case Card (F) $\leq$ Card (A). If A is not a weakly prefix code there exist $x,y \in A$, $x \neq y$, such that $g(x) = g(y)$; therefore g is not injective.

If A is a finite subset of X^+ it is possible to provide an algorithm for obtaining the basis F of the minimal weakly prefix subsemigroup containing A.

Indeed let A be an irreducible finite subset of X^+ and let us operate on it according to the following procedure.

1. Put A-elements in a stack called W and set $C \leftarrow \emptyset$, $F \leftarrow \emptyset$.

2. If W is empty continue; otherwise go to 4.

3. If C = F stop; F is the basis of F^+ . Otherwise go to 5.

4. Pot the first element of W, call it w and set $F \leftarrow F \cup \{w\}$.

5. Set $C \leftarrow F$ and test if each element h of C either is divided by C or generates a loop in C; then, if $h = s_0 s_1$ ($s_0 \in C$) is divided by (or generates a loop in) C, set $F \leftarrow C \cup \{s_1\} \setminus \{h\}$.

6. Go to 2.

It can be easily proved the following

Theorem 8. *The set F obtained by operating on A with the above procedure is the basis of the minimal weakly prefix subsemigroup containing A.*

Proof. Obviously, F is a weakly prefix code and $F^+ \supseteq A^+ \supset A$. Indeed the elements of F are either elements of A or obtained by factorizing elements of A (the words that, so to say, break the finite decipherability of A). Furthermore, F^+ is the minimal weakly prefix subsemigroup containing A. Indeed any other factorization of A in a weakly prefix code can produce only elements of F or factors of them. Therefore in order to generate a weakly prefix subsemigroup $Z^+ \supset A$, any other basis Z must contain, or provide by concatenation, all elements of F. That is, $Z^+ \supseteq F^+$ and $Z^+ = F^+$ iff $Z = F$.

If A is a weakly prefix code no codewords are divided or generate a loop in A and the minimal weakly prefix subsemigroup F^+ containing A is the subsemigroup generated by A itself; namely F = A and vice-versa. Hence

A necessary and sufficient condition for a finite code A to be weakly prefix is that F = A.

If A^+ is not a weakly prefix subsemigroup, F provides the "most convenient" weakly prefix code to handle the A-messages.

REFERENCES

(1) Berstel, J., Perrin, D., Perrot, J.F. and A. Restivo: 1978, Sur le Theoreme du Defaut, J. of Algebra (to appear)

(2) Capocelli, R.M.: 1975, A necessary and sufficient condition for unique decipherability, Notices of A.M.S. 22, A-714.

(3) Capocelli, R.M.: 1979, A note on uniquely decipherable Codes, IEEE Trans. Information Theory, 25, pp. 90-94.

(4) Capocelli, R.M.: 1979, On optimal factorization of free semigroups into free subsemigroups, Semigroup Forum (to appear)

(5) Capocelli, R.M.: 1979, A combinatorial Property of codes having finite deciphering delay, Preprint.

(6) Capocelli, R.M.: 1979, On some properties of weakly prefix codes, Preprint.

(7) De Luca, A. and A. Restivo : 1979, Synchronization and Maximality for very pure subsemigroups of a free semigroup, Lectures Notes in Computer Sciences, 74, pp. 363-371.

(8) Schützenberger, M.P.: 1966, On a question concerning certain free submonoids, Journal Combinatorial Theory, 1, pp. 437-442.

A CELLULAR APPROACH FOR THE IMPLEMENTATION OF DIGITAL SIGNAL PROCESSING ALGORITHMS

G. D. Papadopoulos*, K. Pekmestzi**

* Professor of Applied Electronics, University of Patras
** Research Scientist, Nuclear Research Center Demokritos

Abstract: In this paper a unified cellular circuit approach is presented for the implementation of digital filters and of the Fast Fourier Transform (FFT). Specifically, the second order section and the butterfly computation of the FFT are realized by using two's complement serial arithmetic in conjuction with the pipeline technique. The proposed method yields a cascade cellular structure with small circuit complexity and well suited for LSI realization. Finally, this cellular approach is functionally complete, that is, in addition to the arithmetic operations, it also provides the filter delays and the rounding.

I. INTRODUCTION

Digital signal processing algorithms are implemented mainly by software, due to the large number of computations involved. However, in many applications, as in communication systems, there is need for real-time processing. With the recent advances in large-scale integration, however, an inexpensive special-purpose hardware realization is now possible. Some interesting work has been done in this direction. Jackson et al (1) proposed a filter structure well suited for LSI by introducing the serial pipeline multiplier. In Peled's and Liu's realization (2) the partial results of a filter are precomputed and stored in a ROM. Thus, no multiplier is required. The same technique has been also applied for the FFT realization by the same authors (3).

In this work a cellular realization of a second order section is presented as well as of the butterfly computation. These two form the basic algorithms for digital filtering and FFT computa-

D. G. Lainiotis and N. S. Tzannes (eds.), Advances in Communications, 130–137.

tions, respectively. The operations are performed with the pipeline technique in order to obtain high speed processing. The proposed designs are functionally complete, that is, the hardware perfoms the rounding and the control, in addition to the arithmetic operations.

II. PIPELINE IMPLEMENTATION OF THE OPERATION AX+BY+Z

A very useful funtional for digital signal processing is shown in Eq. 1:

$$W = AX+BY+Z \tag{1}$$

In this section the serial-pipeline implementation of this operation with cellular circuits is presented.

The quantities X, Y, Z represent the variables and A, B the constant coefficients. The variables magnitude is scaled to be less than one. The coefficients' magnitude is smaller than one for the fast Fourier transform and for digital filter applications it is assumed to be less than two. In accordance with the above restrictions the variables and the constants are represented in two's complement form as shown in Equs. 2,

$$X = x^0 + \sum_{K=1}^{N-1} x^k 2^{-k} \tag{2}$$

$$A = \{a^0 + \sum_{j=1}^{M-1} a^j 2^{-j}\}2^v \qquad v=0,1$$

where v distinguishes between the FFT (v=0) and digital filtering applications (v=1).

A similar representation is valid for Y, Z, W, B. Substitution of A and B in Equ. 1 results in Equ. 3,

$$W = -(Xa^0 + Yb^0)2^v + Z + \sum_{j=1}^{M-1} (Xa^j + Yb^j)2^{-j+v} \tag{3}$$

The term Xa^j+Yb^j is computed serially in the jth order cell of the circuit shown in Fig. 1. This circuit accepts the variables, X, Y at the first cell in serial form and produces the relult W at the last cell also in serial form.

The circuit of the cell used is shown in Fig. 2. As can be seen it consists of two full adders that add the partial products W_j^k shown in Equ. 4 with the PS_i and PC_i terms to produce the outputs PS_o and PC_o.

$$W_j^k = (X^k a^j + Y^k b^j)2^{-k-j+v} \tag{4}$$

To synchronized the output PS_o of the jth cell with the quantities added in the j-1 cell, a delay unit is inserted after the PS_i input in each cell. The PC_o output of the J-1 cell, which is a carry, has the same weight as the product of the previous cell and is connected directly to PC_i of the j-th cell. The carry C_2 is saved for the next step in the same cell. The carry C_1 and the sum S_2 are the outputs PC_o and PS_o respectively.

Different length two's complement numbers addition requires sign-extension. For the circuit discussed the S_1 sum produces a number of length smaller by one bit in comparison with the numbers added in the second full-adder. The required sign-extension of S_1 is performed by the switch SW1 one bit after the MSB of the variables.

It is clear that a space interval is required for the above operation and this is achieved by inserting a zero bit after the MSB of each variable. At the same time SW1 truncates PS_o of the next operation, so that mixing of the results from two distinct operations is avoided. The truncated quantities are smaller than 2^{-N+1} except for the truncation at the last cell where a term of weigth 2^{-N+v} is truncated. Consequently, for digital filter applications, the last truncation must be inhibited. In order to retain the cellular form, a second zero bit is inserted before the LSB of the next variable, so that the last truncation is performed on a step earlier and the truncated term is of weight 2^{-N}.

As can be seen from Equ. (3) the last cell must perform subtraction. This is realized by inserting inverters at the inputs PS_i and PC_i of the last two cells. This modification permits the addition of different sign digits as is shown in Fig. 5. The inverter that corresponds to the PC_o output of the last cell turns out to be a bonus because it provides automatic rounding for the operation of the next set of variables. When the inputs to the last cell are the most significant bits of X and Y . This zero is converted to a one of weight 2^{-N} by means of the inverter and is added as a rounding quantity. Thus, truncation is converted into rounding. The truncation, the sign-extension and the zero bit insertion is performed by a control signal equally delayed as X, Y.

III. APPLICATIONS FOR DIGITAL SIGNAL PROCESSING (DSP)

The basic algorithms in DSP are the second order section algorithm and the butterfly computation algorithm. The implementation of these algorithms with the proposed cellular pipeline technique is presented next.

i) Second Order Section

The canonic form realization of a second order section is described by Equs. 5:

$$z_n = x_n + b_1 z_{n-1} + b_2 z_{n-2}$$
$$y_n = z_n + a_1 z_{n-1} + a_2 z_{n-2} \qquad (5)$$

Two circuits such as that of Fig. 1 can be used for the realization of Equs. (5) simultaneously with a single control signal and the same common variables. Such an implementation is shown in Fig. 2; one is in the top compartment and the other in the bottom in a mirror image arrangement, with common delays for the variables and the control. The overall delay from the time of arrival of the LSB of the variables until the formation of the LSB of the result in the circuit of Fig. 3 is equal to 2M-2 delay units where M is the number of digits of the coefficients. This delay is considered as the delay of the filter and, thus, we can connect the z_n output with the z_{n-1} input. For synchronization of this feedback connection the following relationship must hold,

$$2M-2 = N+2 \qquad (6)$$

This relationship does not limit the application because in most cases the variables have more bits than the coefficients. The output z_n is also connected to the additive input of the top compartment which yields the output y_n. The input x_n is connected to the additive input of the bottom compartment and the z_{n-2} variable is obtained by the z_{n-1} line before the last cell. It must be pointed out that the filter of Fig. 3 is functionally complete, that is, it offers arithmetic operations, rounding, data delay and control.

ii) Butterfly Computation

The basic operation for a radix-2 decimation-in-time FFT algorithm is the so-called butterfly in which two complex numbers X, Y are combined to give two numbers X′, Y′ via the operation, shown in Equs. 7.

$$X' = X + Ye^{-2\pi k/N}$$
$$Y' = X - Ye^{-2\pi k/N} \qquad (7)$$

By considering the real and the imaginary parts of each number the following equivalent equations are obtained,

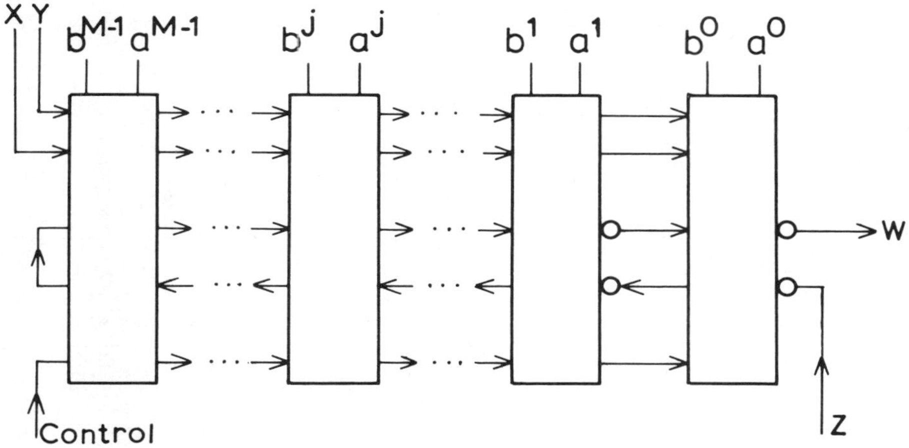

Fig. 1 The operation W = AX+BY+Z

$$
\begin{aligned}
X'_r &= X_r + (\cos \frac{2\pi k}{N} Y_r + \sin \frac{2\pi k}{N} Y_i) \\
X'_i &= X_i + (-\sin \frac{2\pi k}{N} Y_r + \cos \frac{2\pi k}{N} Y_i) \\
Y'_r &= X_r - (\cos \frac{2\pi k}{N} Y_r + \sin \frac{2\pi k}{N} Y_i) \\
Y'_r &= X_i - (-\sin \frac{2\pi k}{N} Y_r + \cos \frac{2\pi k}{N} Y_i
\end{aligned}
\qquad (8)
$$

The first two equations can be realized with a similar circuit as in Fig. 3 but without the feedback connections. The last two equations require the subtraction of the terms in parenthesis instead of addition and are realized with a different last cell, which does not include inverters, as is shown in Fig. 4. The sign of each term that is added at the last two cells is shown in Fig. 5. It can be seen that the terms added at the second of the last two cells are of positive sign and consequently the full adders do not require any modification.

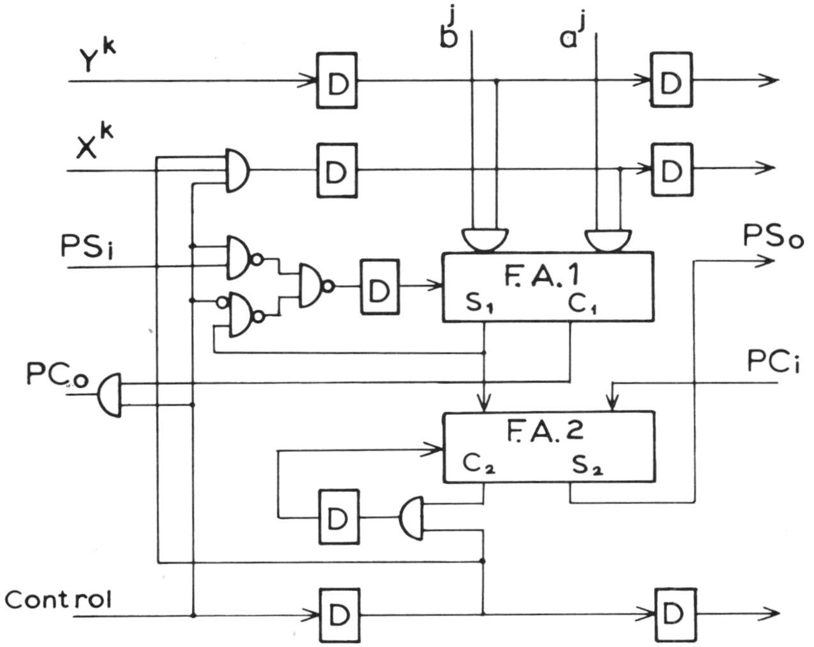

Fig. 2 The basic cell

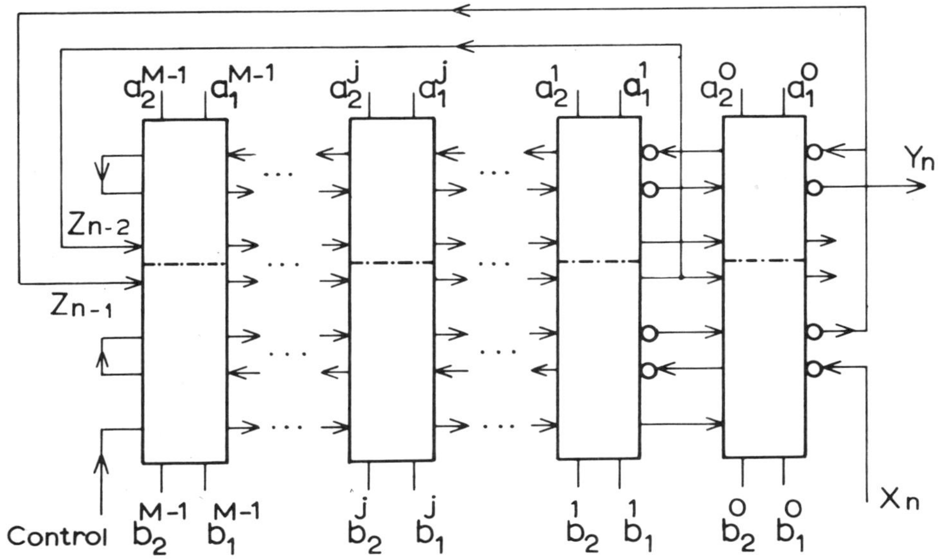

Fig. 3. The proposed second order section

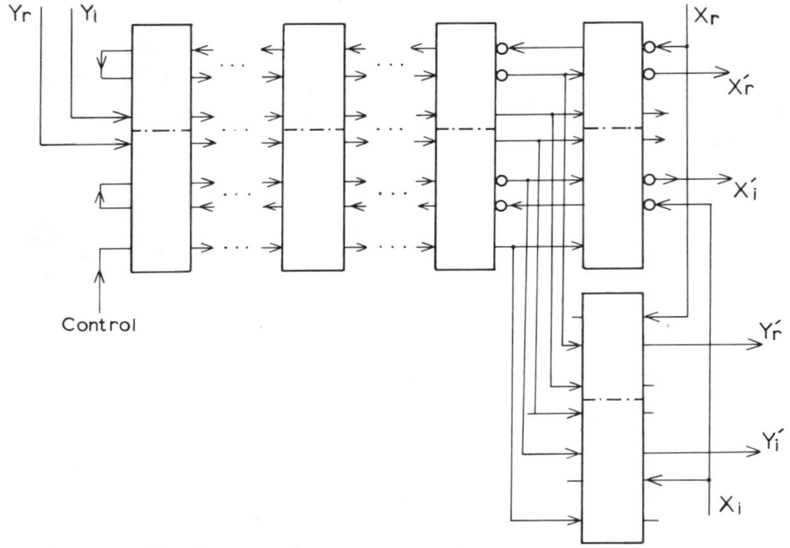

Fig. 4 The butterfly computation

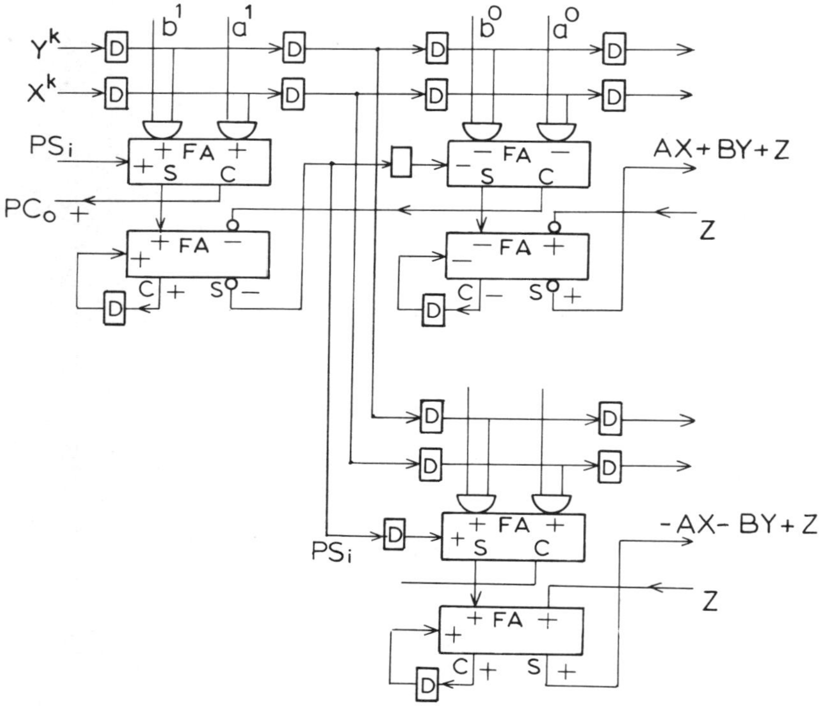

Fig. 5 Signed additions at the last two cells

IV. DISCUSSION

The proposed cellular pipeline realization of digital signal processing algorithms has the following features:

i) Small circuit complexity. Each cell contains 120 gates while the equivalent realization with four serial pipeline multipliers requires 200 gates.
ii) High speed processing due to the pipelining. The maximum bit rate is $f_{max} = 1/10\tau_g$ where τ_g is the one gate delay.
III) Each cell is functionally complete and, consequently, this structure is expandable to any length to achieve the required accuracy and permits the design by using only one component as the building block.

Due to these advandages the integrated circuit realization of a module can be realized with Emitter function Logic (EFL) (4) that has been proven very efficient for the realization of serial pipeline multipliers (5), (6). Such a chip will need 28 pins and will permit data processing rates of 20 Mbits/sec with a 300 mWatt power dissipation.

REFERENCES

(1) L. B. Jackson, J. F. Kaiser, and H. S. McDonald, "An approach to the implementation of digital filters," IEEE Trans. Audio Electroacoust., vol. AU-16, pp. 413-421, Sept. 1968.
(2) A. Peled, B. Liu, "A new hardware realization of digital filters," IEEE Trans. on Acoustics, Speech and Signal Processing, ASSP-22 No. 6, pp. 456-462, December 1974.
(3) B. Liu, A. Peled, "A new hardware realization of high-speed Fast Fourier Transformers," IEEE Trans. on Acoustics, Speech and Signal Processing, ASSP-23, No. 6, pp. 543-547, Dec.1975.
(4) Z. E. Skokan, "Emitter Function Logic-Logic Family for LSI," IEEE Journal of Solid-State Circuit, vol. SC-8 No. 5, October 1973.
(5) J. Kane, "A low-poer, bipolar, two's complement serial pipeline multiplier chip, " IEEE Journal of Solid-State Circuits, vol. SC-11 No. 5, October 1976.
(6) G. L. Baldwin, B. L. Morris, D. B. Morris, D. B. Fraser and A. R. Tretola, "A modular high-speed serial pipeline multiplier for signal processing," IEEE Journal of Solid-State Circuits, vol. SC-13 No. 3, June 1978.

A METHOD FOR OBTAINING SECOND-ORDER DIGITAL FILTER STRUCTURES WITH PRESCRIBED POLE-ZERO QUANTIZATION ERRORS

G. J. Karayannopoulos and C. C. Halkias

Research Assistant
Department of Electrical Engineering
National Technical University of Athens

Professor of Electronics
Department of Electrical Engineering
National Technical University of Athens

ABSTRACT

This paper proposes a method for designing digital filter structures having parameter quantization errors with prescribed effect on the poles and zeros. This generalized method, which is valid only for second-order digital filters, is being investigated in special cases and several conclusions are carried out. This method is able to provide digital filter realizations characterized by one-to-one correspodence between the coefficients of the transfer function and the parameters of the multipliers. Several examples indicate the efficiency of the proposed method.

This paper is part of the first author's senior thesis.

INTRODUCTION

The implementation of a digital filter using either special-purpose digital hardware or a computer involves facing an important problem related to the number representation method and the corresponding word length. This is a drastical factor for the computational complexity and it plays an important role to the final determination of the cost and the speed. Therefore the word length cannot be as large as desirable and thus several quantization methods must be used for representing the parameters with a limited number of bits. The coefficient of the digital filter are some of these parameters and their quantization affects the standard characteristics of the filter (tranfer function, etc). The effects of the quantization procedure on these characteristics depend upon the structure of the filter. A basic idea we can use, in order to reduce the errors due to quantization, is to use alternative

D. G. Lainiotis and N. S. Tzannes (eds.), Advances in Communications, 138–145.

realization methods. This paper deals with a theory that can be used to provide such methods having prescribed effects of the quantization on the poles of a second-order digital filter.

THEORY

Consider the transfer function denominator polynomial of a second order digital filter:

$$D_1(z) = 1 - a_1 z^{-1} + a_0 z^{-2} \tag{1}$$

Any alternative method for realizing this filter characterized by one-to-one correspodence between coefficients and multipliers (two coefficients and thus two multipliers), e.g. K and L, will have a transfer function denominator polynomial of the form:

$$D_2(z) = 1 - f(K,L) z^{-1} + g(K,L) z^{-2} \qquad (2),$$

that is, a transformation of the form:

$$\begin{aligned} a_1 &= f(K,L) \\ a_0 &= g(K,L) \end{aligned} \tag{3}$$

holds. The idea is to substitute the coefficients a_1 and a_0 by expressions of some other variables K and L. We want the above functions f(K,L) and g(K,L) to have the form of a polynomial:

$$\sum_{i=1}^{N} \sum_{j=1}^{M} c_{ij} K^i L^j \tag{4}$$

where c_{ij} are positive or negative numbers having a short word length so that they can be realized by adders and binary point shifters (avoiding the use of additional multipliers and the necessity of quantizing these numbers). For instance, if

$$f(K,L) = 2.25\, K^2 - 3\, L$$

this expression involves, except the multipliers K and L, 5 adders and 3 binary point shifters (because

$$2.25 = 2^1 + 2^{-2} \quad \text{and} \quad 3 = 2^1 + 2^0 \;)$$

and only K and L must be quantized.
Suppose now that the poles of this filter are $re^{\pm j\theta}$ (r = magnitude and θ = phase). Let us define Δa_1 and Δa_0 the quantization errors in a_1 and a_0 respectively and Δr and $\Delta\theta$ the change of the magnitude and the phase of the poles respectively due to the quantization. Using Newton's formula we have the following approxi-

mate relations of Δr, $\Delta\theta$, and $\Delta a_1, \Delta a_0$:

$$\Delta r = \frac{\Delta a_0}{2r} \tag{5}$$

and

$$\Delta\theta = \frac{\Delta a_0}{2r^2 \tan\theta} - \frac{\Delta a_1}{2r\sin\theta} \tag{6}$$

When realizing a_0 and a_1 with expressions of K and L (see equation (3)) the latter variables have to be quantized. The equivalent change of a_0 and a_1 due to the quantization errors of K and L, respectively ΔK and ΔL, can be written:

$$\Delta a_0 = \frac{\partial g}{\partial K}\Delta K + \frac{\partial g}{\partial L}\Delta L \tag{7}$$

and

$$\Delta a_1 = \frac{\partial f}{\partial K}\Delta K + \frac{\partial f}{\partial L}\Delta L \tag{8}$$

Substituting (7) and (8) into equations (5) and (6) we have:

$$\Delta r = \frac{1}{2r}\left(\frac{\partial g}{\partial K}\Delta K + \frac{\partial g}{\partial L}\Delta L\right) \tag{9}$$

and

$$\Delta\theta = \left(\frac{\partial g/\partial K}{2r^2 \tan\theta} - \frac{\partial f/\partial K}{2r\sin\theta}\right)\Delta K + $$

$$+ \left(\frac{\partial g/\partial L}{2r^2 \tan\theta} - \frac{\partial f/\partial L}{2r\sin\theta}\right)\Delta L \tag{10}$$

We assume now that we want:

$$\Delta r = h'_{11}(r,\theta)\Delta K + h'_{12}(r,\theta)\Delta L \tag{11}$$

$$\Delta\theta = h'_{21}(r,\theta)\Delta K + h'_{22}(r,\theta)\Delta L \tag{12}$$

where $h'_{11}(r,\theta)$, $h'_{12}(r,\theta)$, $h'_{21}(r,\theta)$ and $h'_{22}(r,\theta)$ are desired functions of r and θ representing the effects of ΔK and ΔL on Δr and $\Delta\theta$. In other words, equations (11) and (12) are our specifications.

Comparing equations (9) and (10) to equations (11) and (12) we conclude that (11) and (12) are valid iff:

$$\frac{\partial g}{\partial K} = 2r\, h'_{11}(r,\theta) \tag{13}$$

$$\frac{\partial g}{\partial L} = 2r\, h'_{12}(r,\theta) \tag{14}$$

$$\frac{\partial f}{\partial K} = 2\cos\theta\; h'_{11}(r,\theta) - 2r\sin\theta\; h'_{21}(r,\theta) \tag{15}$$

$$\frac{\partial f}{\partial L} = 2\cos\theta\; h'_{12}(r,\theta) - 2r\sin\theta\; h'_{22}(r,\theta) \tag{16}.$$

The forms of equations (13) to (16) are suitable to work , because they are relations involving r, θ and <u>not</u> f,g,K and L. From (1) we have:

$$r = \sqrt{a_0} \tag{17}$$

$$\theta = \arccos\left(\frac{a_1}{2\sqrt{a_0}}\right) \tag{18}$$

Using these relations equations (13) to (16) can be written:

$$\frac{\partial g}{\partial K} = 2\sqrt{a_0}\; h_{11}(a_0,a_1) \tag{19}$$

$$\frac{\partial g}{\partial L} = 2\sqrt{a_0}\; h_{12}(a_0,a_1) \tag{20}$$

$$\frac{\partial f}{\partial K} = \frac{a_1}{\sqrt{a_0}}\, h_{11} \pm \sqrt{(4a_0 - a_1^2)}\; h_{21} \tag{21}$$

$$\frac{f}{L} = \frac{a_1}{\sqrt{a_0}}\, h_{12} \pm \sqrt{(4a_0 - a_1^2)}\; h_{22} \tag{22}$$

where

$$h_{11}(a_0,a_1) = h'_{11}(r,\theta),\; h_{12} = h'_{12},\; h_{21} = h'_{21} \text{ and } h_{22}=h'_{22} \tag{23}$$

The double sign in (21) and (22) originates from the symmetry of the two comples conjugate poles. Any sign can be selected. Finally, according to the initial choise of (3), equations (19) to (22) can be written (eliminating for simplicity the arguments):

$$\frac{\partial g}{\partial K} = 2\sqrt{g}\; h_{11}(g,f) \tag{24}$$

$$\frac{\partial g}{\partial L} = 2\sqrt{g}\; h_{12}(g,f) \tag{25}$$

$$\frac{\partial f}{\partial K} = \frac{f}{\sqrt{g}} h_{11}(g,f) \pm \sqrt{(4g-f^2)}\; h_{21}(g,f) \tag{26}$$

$$\frac{\partial f}{\partial L} = \frac{f}{\sqrt{g}} h_{12}(g,f) \pm \sqrt{(4g-f^2)}\; h_{22}(g,f) \tag{27}$$

We have a system of partial differential equations in two unknown functions of two variables. All we have to do is to solve it for the two functions. However, this is not easy at all. Therefore we shall work on some applications and examine some special cases. Functions h_{ij} i,j=1,2 can be functions of g,f and/or K,L. Notice that the following must be valid:

$$\frac{\partial h_{11}}{\partial L} = \frac{\partial h_{12}}{\partial K} \qquad \frac{\partial h_{21}}{\partial L} = \frac{\partial h_{22}}{\partial K}$$

Let us examine an example. suppose that we want:

$$h_{11}=1,\ h_{12}=0,\ h_{21}=\left(\frac{-1}{2r\sin\theta} + \frac{1}{2r^2\tan\theta}\right)$$

and

$$h_{22}= \frac{1}{2r\sin\theta}$$

For this selection we see that for $\theta \rightarrow \frac{\pi}{4}$ and $r \simeq \cos\theta$, $\Delta\theta$ can become very small. This set of selections gives:

$$\frac{\partial g}{\partial K} = 1, \frac{\partial g}{\partial L} = 0, \frac{\partial f}{\partial K} = 1, \frac{\partial f}{\partial L} = 1$$

which is easy to be solved and the solution is

$$a_0 = g(K,L) = K$$
$$a_1 = f(K,L) = K + L$$

and thus the transfer function denominator polynomial is

$$D_3(z) = 1 - (K + L)z^{-1} + Kz^{-2}$$

This is a realization scheme proposed by Wells ([1]). This example shows that we can find a system that can be easily solved selecting h_{11}, h_{12}, h_{21}, h_{22} properly, so that Δr and $\Delta\theta$ be small for our r and θ . However, intuition may be necessary to get a solvable system of partial differential equations. Let us see another example. Suppose that we want the possible pole locations on the z-plane to be distributed in such a way that their density is constant. If

$$re^{j\theta} = p + jq$$

taking the differentials of both parts we have:

$$\Delta r\, e^{j\theta} + jr\, e^{j\theta}\, \Delta\theta = \Delta p + j\,\Delta q \qquad \text{or}$$

$$\Delta p = \Delta r\cos\theta - r\sin\theta \cdot \Delta\theta$$
$$\Delta q = \Delta r\sin\theta + r\cos\theta \,.\, \Delta\theta$$

Substituting equations (11) and (12) into the above equations we have:

$$\Delta p = (h_{11}\cos\theta - r\sin\theta\, h_{21})\Delta K + (h_{12}\cos\theta - r\sin\theta\, h_{22})\Delta L$$

$$\Delta q = (h_{11}\sin\theta + r\cos\theta\, h_{21})\Delta K + (h_{12}\sin\theta + r\cos\theta\, h_{22})\Delta L$$

The desired situation (constant density) can be met by the condition:

$\Delta p = \Delta K$ and $\Delta q = \Delta L$

because $K = r\cos\theta$ and $L = r\sin\theta$. These conditions reduce to a system of simple equations:

$$h_{11}\cos\theta - r\sin\theta\, h_{21} = 1$$

$$h_{12}\cos\theta - r\sin\theta\, h_{22} = 0$$

$$h_{11}\sin\theta + r\cos\theta\, h_{21} = 0$$

$$h_{12}\sin\theta + r\cos\theta\, h_{22} = 1$$

Solving it for h_{ij} we have:

$$h_{11} = \cos\theta \ , \ h_{12} = \sin\theta \ , \ h_{21} = -\frac{\sin\theta}{r} \ , \ h_{22} = \frac{\cos\theta}{r}$$

Using equations (17) and (18) and substituting into (24), (27) we obtain:

$$\frac{\partial g}{\partial K} = -f \ , \ \frac{\partial g}{\partial L} = \sqrt{(4g-f^2)} \ , \ \frac{\partial f}{\partial K} = -2 \ , \ \frac{\partial f}{\partial L} = 0$$

from which simply:

$$f = -2K \text{ and } g = K^2 + L^2$$

which is the realization proposed by Rader and Gold ([2]).
Let us see now some very interesting cases. We first notice that equations (24) and (25) are of a simpler form than equations (26) and (27) and thus we can work on them in order to reduce Δr, because r is a function of g only. Let us define:

$$u = \frac{\partial g}{\partial K} \tag{29}$$

$$v = \frac{\partial g}{\partial L} \tag{30}$$

If $u = cv$ (31),
then from (24) and (25):

$$h_{11} = c h_{12} \tag{32}$$

where c is a constant. Equation (32) must be valid so that (31) be valid. In this case taking the total differential of g we have

$$dg = u dK + v dL \tag{33}$$

or

$$dg = 2\sqrt{g}\, h_{12}\ (c dK + dL)$$

or

$$\frac{dg}{2\sqrt{g}\ h_{12}} = c dK + dL \tag{34}$$

Selecting now h_{12} so that h_{12} and $h_{11}=ch_{12}$ be small enough, from (34) we can calculate g as a function of K and L. Attention must be paid to the selection of h_{12} in terms of g, because the latter has to be in the form of (4) and

$$\frac{\partial h_{12}}{\partial K} = c\,\frac{\partial h_{12}}{\partial L} \tag{35}$$

Finally, let us examine the case of the existence of a relation of the form:

$$a_1 = x(a_0) \quad \text{or } f = x(g) \tag{36}$$

Defining $y = \frac{\partial x}{\partial g}$ and considering the above definitions we have:

$$\frac{\partial f}{\partial K} = uy \tag{37}$$

$$\frac{\partial f}{\partial L} = vy \tag{38}$$

Equations (25) to (28) can form:

$$F = (u-2\sqrt{g}\,h_{11})^2 + (v - 2\sqrt{g}\,h_{12})^2 = 0 \tag{39}$$

and

$$G = [uy - \frac{x(g)}{\sqrt{g}}\,h_{11} \pm (4g-x^2(g))^{1/2} h_{21}]^2 +$$

$$+ [vy - \frac{x(g)}{\sqrt{g}}\,h_{12} \pm (4g-x^2(g))^{1/2}\,h_{22}]^2 = 0 \tag{40}$$

We have now a system of first order partial differential equations. This system is compatible iff (see [3]):

$$J \triangleq \frac{\partial(F,G)}{\partial(u,v)} = \begin{vmatrix} \frac{\partial F}{\partial u} & \frac{\partial F}{\partial v} \\ \frac{\partial G}{\partial u} & \frac{\partial G}{\partial v} \end{vmatrix} \neq 0 \tag{41}$$

Expressing the above partial derivatives in terms of u,v,g,x,y and h_{ij}'s we have that (41) is equivalent to:

$$(u - 2\sqrt{g}\,h_{11})[vy - \frac{x}{\sqrt{g}}h_{12} \pm (4g - x^2)^{1/2} h_{22}] \neq$$
$$\neq (v - 2\sqrt{g}\,h_{12})[uy - \frac{x}{\sqrt{g}}h_{11} \pm (4g - x^2)^{1/2} h_{21}] \qquad (42)$$

assuming that $y \neq 0$ at $g = a_0$.

CONCLUSIONS

We have proposed a method for obtaining second-order digital filter structures with prespecified quantization effects on the poles of the transfer function.

REFERENCES

[1] : H. Wells : "Rader M.T.I. : A Comparison of 2nd-order Digital-Filter Realizations when the Stopband is very Narrow", Electronics Letters, January 8, 1976, vol. 12, No 1, pp 3-4.

[2] : C. Rader and R. Gold: "Effects of Parameter Quantization on the Poles of a Digital Filter", Proceedings of the IEEE, May 1967, pp 688-689.

[3] : Ian N. Sneddon : "Elements of Partial Differential Equations", Mc Graw-hill, 1957, p 67.

NORMALIZED GRADIENT ALGORITHMS FOR ADAPTIVE RECURSIVE DIGITAL FILTERS

Dimitris Manolakis

University of Athens, Department of Physics,
Division of Electronics
104, Solonos Str., Athens
GREECE

ABSTRACT

Adaptive non-recursive digital filters have been extensively studied in the literature. However in some applications there is a strong need for adaptive recursive digital filters since they can produce both poles and zeros in their system function. Although the error surface for these filters in non-quadratic, the steepest-descent algorithm can be used for the adaptation. A problem arising then is that its convergence properties depend on the input signal level. To avoid this problem the use of some normalized gradient algorithms is proposed. The performance of these algorithms is investigated via simulation.

1. INTRODUCTION

Adaptive non-recursive digital filters have been extensively studied in the literature (1,2). The error surface, for these filters, is quadratic, their stability properties have been well understood and their performance in many applications, is very good. These are some of the reasons behind their wide-spread use.

However, as these filters have a finite impulse response, i.e. they can produce only zeros with poles at z=0 in the filter's system function, their capability is limited in some applications. Such a case is met e.g. when the desired response has a rational spectrum. In such cases the use of an adaptive recursive digital filter (ARDF) is indicated.

D. G. Lainiotis and N. S. Tzannes (eds.), Advances in Communications, 146–155.

Although the error surface for these filters is, in general, non-quadratic the method of steepest - descent can be used for the adaptation. A problem which arises then is that the convergence properties depend on the input signal level. This dependence is undesirable in practice, especially when the various signals have unknown or time varying variances.

To avoid this problem some modified algorithms ("normalized" gradient algorithms) are suggested.

This paper is organized as follows: Sections 2, 3 and 4 include a statement of the problem and a description of the optimum and adaptive filter solution. The proposed algorithms are described in section 5.The last two sections are reffered to the experimental results. The investigations are limited to ARDFs having a direct form-1 structure.

2. STATEMENT OF THE PROBLEM

The time domain input-output relation for a Nth order recursive digital filter is given by

$$y(n) = \sum_{k=0}^{M} a_k x(n-k) + \sum_{k=1}^{N} b_k y(n-k) \quad (M \leq N) \tag{1}$$

If d(n) is the desired response (see fig.1) then the error e(n) is

$$e(n) \triangleq d(n) - y(n) \tag{2}$$

Let $P[e(n)]$ be an appropriately defined performance criterion. The problem is to determine the set of filter coeffients for which the given criterion takes its minimum value.

The solution to this problem depends on the amount of the a-priori information at hand. If there is "enough" a-priori information, the design of an optimum filter is feasible, at least in principle. If there is not "enough" a-priori information available or in a non-stationary environment, the use of an adaptive filter is the only feasible solution.

3. OPTIMUM FILTER SOLUTION (WIENER SOLUTION)

In this case, the a-priori information at hand is that the signals x(n) and d(n) are (at least) jointly wide-sense stationary stochastic processes with zero means and known second order statistics. The criterion of performance used is the $P[e^2(n)] \triangleq E[e^2(n)]$, where E denotes the expectation value.

By using matrix notation equ.(1) can be written as

$$y(n) = \underline{A}^T \underline{X}(n) + \underline{B}^T \underline{Y}(n) \quad (3)$$

where

$$\underline{A} \triangleq [a_0, a_1, \ldots, a_M]^T \text{ etc.}$$

(T denotes the traspofe of a matrix).

By using eqs. (2) and (3) we have after some manipulations:

$$E[e^2(n)] = E[d^2(n)] + \underline{A}^T R_{XX} \underline{A} + \underline{B}^T R_{YY} \underline{B} - 2\underline{A}^T \underline{R}_{dX} - 2\underline{B}^T \underline{R}_{dY} + 2\underline{A}^T R_{XY} \underline{B} \quad (4)$$

where $R_{XX} \triangleq E[\underline{X}(n)\underline{X}^T(n)]$ is the input signal correlation matrix, $\underline{R}_{dX} \triangleq E[d(n)\underline{X}(n)]$ is the desired response-input crosscorrelation vector and R_{YY}, $\underline{R}_{dY}$, R_{XY} are similarly defined.

As it has been shown in (3), $E[e^2(n)]$ is not a convex function of the filter coefficients. Thus in general does not posses a unique minimum. The optimum set of coefficients is the one which corresponds to the global minimum of (4).

However in such a case there is no guarantee that this global minimum can always be found either analytically or by numerical means. This is due to the fact that the conditions

$$\underline{\nabla}_{\underline{A}} E[e^2(n)] = \underline{0} \quad \text{and} \quad \underline{\nabla}_{\underline{B}} E[e^2(n)] = \underline{0} \quad (5)$$

which characterize an extremum of (4) are only necessary and not sufficient ($\underline{\nabla}_{\underline{A}}$ denotes the gradient operator with respect to vector $\underline{A}$). These equations are highly non-linear and their solution, even if it is very difficult, is in principle possible. However in practice second-order statistics are never known and then arises the need for an adaptive filter.

4. ADAPTIVE FILTER SOLUTION (LMS ADAPTIVE FILTERS)

An adaptive filter is the only feasible solution to the problem under discussion either when second-order statistics are non explicitly known or in the case of time-varying statistics.

An adaptive recursive digital filter is described by the relation:

$$y(n) = \underline{A}^T(n)\underline{X}(n) + \underline{B}^T(n)\underline{Y}(n) \quad (6)$$

where the filter coefficient vectors are now time-depended.

The problem is to find a coefficient-adjustment mechanism (adaptive algorithm) resulting in an adaptive filter which tries to minimize a given criterion of performance during its operation. In the previous case the criterion $E[e^2(n)]$ was taken in order to remove the stochastic aspect. However in a hardware realization an adaptive filter has only one ensemble element to operate on. This make necessary the use of estimates instead of expectation values.

Estimates based on the LMS hypothesis, (1), have led to adaptive transversal filters with good performance. According to this hypothesis, local values are good estimates of expected values e.g. $E[e^2(n)] \simeq e^2(n)$.

The performance criterion which is chosen to be minimized here is:

$$P[e(n)] = e^2(n) \tag{7}$$

This is obviously a stochastic criterion function. By choosing the steepest-descent algorithm for seeking a minimum (at least local) on the error surface, the following adaptation algorithm results:

$$\begin{aligned} \underline{A}(n+1) &= \underline{A}(n) + k_1 \cdot \nabla_{\underline{A}(n)} e^2(n) \\ \underline{B}(n+1) &= \underline{B}(n) + k_2 \cdot \nabla_{\underline{B}(n)} e^2(n) \end{aligned} \tag{8}$$

where k_1 and k_2 are negative constants which determine the convergence properties of the algorithm.

By using for the gradients appearing in equ. (8) the approximation given in ref.(4) the following (LMS) algorithm for ARDFs results:

$$\begin{aligned} \underline{A}(n+1) &= \underline{A}(n) - 2 \cdot k_1 \cdot e(n) \cdot \underline{X}(n) \\ \underline{B}(n+1) &= \underline{B}(n) - 2 \cdot k_2 \cdot e(n) \cdot \underline{Y}(n) \end{aligned} \tag{9}$$

Algorithms based on other approximations can be found in (5, 6).

5. NORMALIZED GRADIENT ALGORITHMS

As it is well known the steepest descent algorithm is searching for a minimum by trying at each iteration, a descent in the direction of the negative gradient. The length of such a descent depents both on the magnitude of the gradient vector and a constant, which determines the stability and the convergence properties of the algorithm.

As it seems from eqs.(9), in the case of an ARDF the length of the descent depends not only on the constants of convergence but on the input signal level and the error e(n) as well. This kind of dependence makes the control of the convergence properties of adaptive filter, via a proper selection of k_1 and k_2, difficult in case the signals have unknown or time varying variances (1, 7). This is of great importance in many cases of practical interest as e.g. in the case of adaptive prediction of non-stationary or quasistationary signals.

In order to be able to control the algorithm's convergence properties in a more efficient way, via the constants of convergence, some "normalized" gradient algorithms are suggested.

Algorithms using a square-error criterion

The two algorithms, which follow, use $e^2(n)$ as criterion of performance.

Algorithm 1

This algorithm is described by the equations:

$$\underline{A}(n+1) = \underline{A}(n) - 2 \cdot k_1 \cdot e(n) \frac{\underline{X}(n)}{\|\underline{X}(n)\|}$$

$$B(n+1) = B(n) - 2 \cdot k_2 \cdot e(n) \frac{\underline{Y}(n)}{\|\underline{Y}(n)\|} \tag{10}$$

where $\|\underline{X}\|$ is the Euclidean norm of the vector $\underline{X}$.

As it is obvious from the above equations the length of the descent depends only on the constants of convergence and the error e(n). However the error e(n) depends on the input signal level which means that still there exists a dependence of the convergence properties on the input signal level. Of course this dependence is not so strong as in the algorithm (9).

Algorithm 2

The idea behind this algorithm is in the learning identification method suggested by Nagumo and Noda(11). The algorithm is described by the equation:

$$\underline{C}(n+1) = \underline{C}(n) - 2\cdot k\cdot e(n)\cdot\frac{\underline{Z}(n)}{\|\underline{Z}(n)\|^2} \tag{11}$$

where

$$\underline{C}(n) \triangleq [\underline{A}^T(n),\underline{B}^T(n)]^T \text{ and } \underline{Z}(n) \triangleq [\underline{X}^T(n),Y^T(n)]^T$$

In this case the convergence properties are independent of the input signal level. This can be easily seen for the case of identification examples in the next section.

As it is obvious the use of normalization in the previous algorithms increases their complexity in comparison to the algorithm (9).

Algorithms using an absolute error criterion.

The criterion used in this case is

$$P[e(n)] \triangleq |e(n)| = e(n)\cdot \mathrm{sgn}e(n) \tag{12}$$

where sgne(n) means the sign of e(n).

Algorithm 3

By following the procedure of section 4 it can be easily found that the adaptation algorithm (based again on the method of steepest descent) is

$$\begin{aligned} \underline{A}(n+1) &= \underline{A}(n) - 2\cdot K_1\cdot \mathrm{sgn}e(n)\cdot\underline{X}(n) \\ \underline{B}(n+1) &= \underline{B}(n) - 2\cdot k_2\cdot \mathrm{sgn}e(n)\cdot\underline{Y}(n) \end{aligned} \tag{13}$$

(The factor 2 is included only for similarity with (12)).

To avoid the dependence of the convergence properties on the input signal level, the algorithm (13) is modified as:

$$\begin{aligned} \underline{A}(n+1) &= \underline{A}(n) - 2\cdot k_1\cdot \mathrm{sgn}e(n)\cdot\frac{\underline{X}(n)}{\|\underline{X}(n)\|} \\ \underline{B}(n+1) &= \underline{B}(n) - 2\cdot k_2\cdot \mathrm{sgn}e(n)\frac{\underline{Y}(n)}{\|\underline{Y}(n)\|} \end{aligned} \tag{14}$$

This algorithm is similar to that of equs. (10), but in algorithm (14) is only the choise of constants of convergence which determines the convergence properties of the algorithm.

This can be easily shown from eqs. (14), since the length of the descent, at each iteration, is always two times the constant of convergence.

6. EXPERIMENTAL RESULTS

To investigate the performance of the proposed algorithms, simulation of a system identification problem was made. An ARDF was used as a learning model for the identification of an "unknown" fixed recursive digital filter (plant) having the same number of feedback and feedforward coefficients (see fig. 2). The input x(n) is a zero mean, uniformly distributed,white noise stochastic process.

The following secon-order processes,suggested in the literature (8,9), were used in the simulations:

$$H_1(z) = \frac{0.05 - 0.4\, z^{-1}}{1 - 1.1314\, z^{-1} + 0.25\, z^{-2}} ,$$

$$H_2(z) = \frac{z^{-1} + 0.5\, z^{-2}}{1 - 1.5\, z^{-1} + 0.7\, z^{-2}} .$$

The first plant has a non-oscilating unit-sample response whereas the second has an oscilating one. The criterion used for the description of performance of this identification method is a quantity Q,defined as:

$$Q(n) = 10 \log_{10} \frac{\|\underline{C} - \underline{C}(n)\|^2}{\|\underline{C}\|^2}$$

where $\underline{C}$ is the plant's parameter vector. Although the measure Q can not be used in practice it is very usefull in analytical studies and studies by simulation.

Simulations have shown that with algorithms 1 and 2 the ARDF coefficients always converge to the fixed filter coefficients. The same happens with algorithm 3 provided that the constants of convergence are much smaller than in algorithms 1 and 2. This is because for algorithms 1 and 2 the length of the descent tends to zero near to a minimum whereas in algorithm 3 is fixed. This results to a fluctuation of ARDF coefficients around the values of the fixed filter coeffiecents when k_1 and k_2 are not small enough. The indepen-

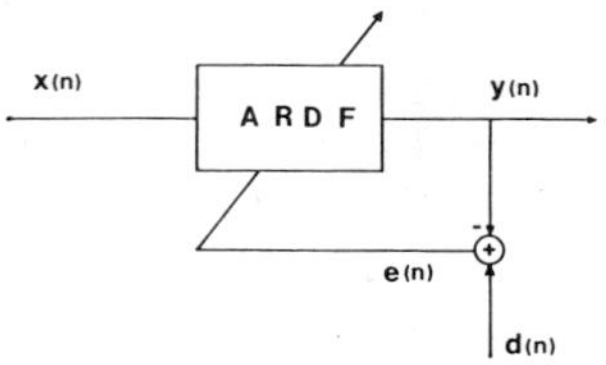

Fig. 1 Adaptive recursive digital filter.

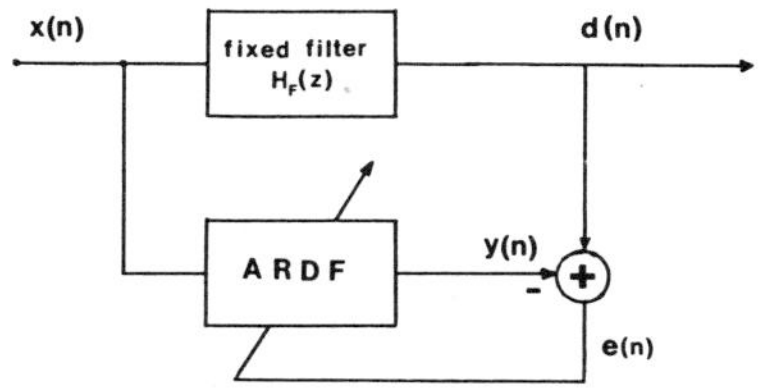

Fig. 2 System identification experiment.

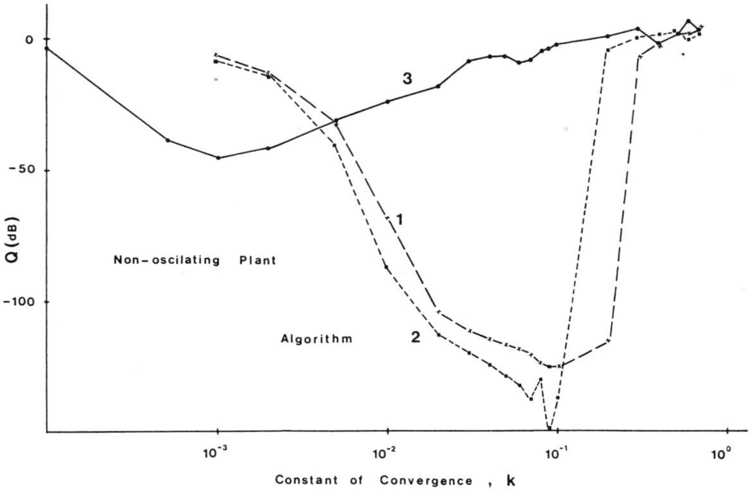

Fig. 3 Q versus k curves for the $H_1(z)$ plant.

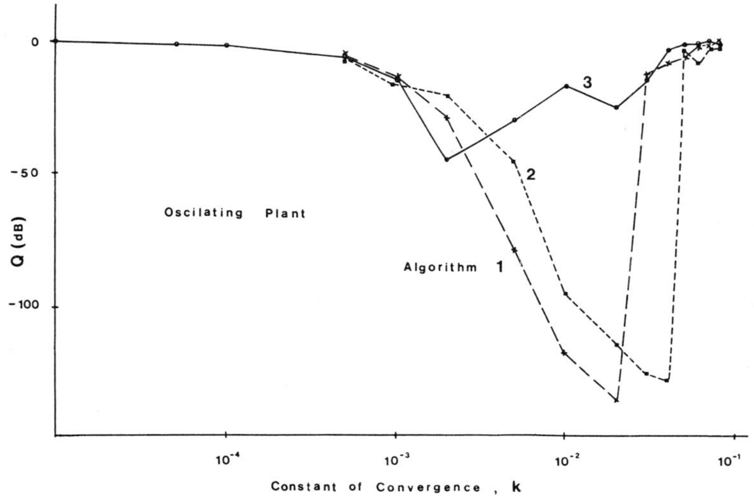

Fig. 4 Q versus k curves for the $H_2(z)$ plant.

dence of the convergence properties of algorithms 1 and 3 on the input signal level was apparent in the simulations.

As it has been seen, using the simpler to calculate absolute value norm instead of the Euclidean in algorithms 2 and 3 does not make much difference. This norm is used in algorithm 3.

Another thing important to mension is the trade-off between the rate of convergence and the residual mean square error due to the fluctuation of the filter coefficients arrount their optimum values. This error dominates for large values of k's. Choosing smaller k's reduces this error. However making k's very small to reduce even more the residual mean square error does not help because the adaptation stops when the correction term in the various algorithms becomes smaller in magnitude than the least significant digit of a given coefficient (of course this depends on the computer wordlength) (10). This is indicated in figs. 3 and 4. The same figures show that $0 < k < 2$ ($k \triangleq k_1 = k_2$ were used), a result which has been proved for adaptive transversal filters (1).

7. CONCLUSIONS

Three new algorithms for adaptive recursive digital filters were presented in this paper. Although these algorithms require relatively more computational effort than the algorithm (9) they have the advantage that their convergence properties are independent of the input signal level. Experimental results demostrating the performance, of the proposed algorithms, in a system identification experiment are included. Further investigations on these algorithms, including their application in other signal processing problems are being pursued.

ACKNOWLEDGEMENT

The author would like to thank Prof. C. Caroubalos for his encouragement and support.

This work was sponsored by the National Hellenic Research Foundation. Special thanks are also due to the National Observatory of Athens for making available its computer facilities to the author.

References

1. Widrow,B. and J.McCool "Principles and Applications of Adaptive filters: A Tutorial review", Proc. of International Specialist Seminar on "The Impact of New Technologies on Signal Processing" 20-24 Sept. 1976, Aviemore, Scotland.

2. Widrow,B. et all. "Adaptive Noise Cancelling:Principles and Applications", Proc. IEEE, Dec.1975,pp. 1692-1716.

3. Mantey, P.E., "Convergent Automatic-Synthesis procedures for Sampled-Data Networks with Feedback", Stanford Electronics Laboratories, Stanford University, Stanford CA; TR no. 6773-1, October,1964.

4. Feintuch, P.L. "An Adaptive Recursive LMS Filter", Proc. IEEE, Nov. 1976, pp. 1622-1624.

5. White, S.A., "An Adaptive Recursive Filter", in Proc. 9th Asilomar Conf. Circuits, Systems and Computers, pp. 21-25, Nov. 1975.

6. Stearns, S.D. et all, "On Adaptive Recursive Filtering", In Proc. Tenth Asilomar Conf. Circuits and Systems, 1976, pp. 5-10.

7. Rabiner,L. et all, "FIR system Modeling and Identification in the presence of Noise and with Band-Limited Inputs" IEEE trans. ASSP, vol.26, Aug. 1978, pp. 319-333.

8. Eykhoff,P. System Identification, Wiley, 1974.

9. Johnson, C.R. and M.G. Larimore, Comments on and Additions to "An Adaptive Recursive LMS Filter", Proc. IEEE, Sept. 1977, pp. 1399-1402.

10. Gitlin, R.D. and S.B.Weinstein, "On the Required Tap-Weight Precision for Digitally Implemented, Adaptive, Mean-Squared Equalizers" BSTJ, Vol.58,No. 2. Febr. 1979, pp. 301-321.

11. Nagumo,J. and A.Noda, "A learning method for system identification", IEEE Trans. AC, vol.12,1967, pp. 282-287.

A CLASS OF SECOND-ORDER DIGITAL FILTERS CHARACTERIZED BY VERY SMALL POLE DISPLACEMENTS DUE TO PARAMETER QUANTIZATION

G.J. Karayannopoulos and C.C. Halkias

Research Assistant
National Technical University
of Athens

Professor of Electronics
National technical University
of Athens

ABSTRACT

Second-order digital filters having poles, whose phase or magnitude are almost not affected by parameter quantization, are investigated. The results of this investigation are locii of poles corresponding to digital filters, which, when they are realized by the or the Rader-Gold methods have pole magnitude or phase errors very close to zero. Several examples of such filters are also presented.

This paper is part of the first author's senior thesis.

INTRODUCTION

As it is already known, the realization of a digital filter involves some important procedures strongly related to the final characteristics of the filter. One of them is the coefficient quantization, a procedure whose necessity originates from the limited word length we are forced to use. This procedure causes errors in the standard characteristics of the digital filter, such as the transfer function or in other words its poles and zeros, and even the stability of the filter. It is therefore obvious that coefficient quantization errors may be of paramount importance, not always of course, but at least in some specific cases, and thus solutions to the appropriate arising problems must be found.

Since the effects of these errors on the characteristics of the digital filter are related to the structure we use to realize the filter, one way to solve the above problems is to find alternative realization methods and to work on their quantization errors. The present paper deals with classes of digital filters of second order characterized by very small quantization errors in the magni-

D. G. Lainiotis and N. S. Tzannes (eds.), Advances in Communications, 156–164.

tude or the phase of the poles of the transfer function. We shall see that any realization method corresponds to an appropriate class of this kind, characterized by very small (approximately equal to zero) quantization errors, even if the errors in the coefficients are not zero, that is, even if the quantization strongly affects the coefficients. First of all we shall formulate the general case and then we shall proceed to special applications of the theory.

GENERAL THEORY

We shall examine the idea of these classes of digital filters only for the case of second-order filters with zero errors in the magnitude or phase of the poles. Therefore from now on we shall restrict our interest in the transfer function denominator polynomial only.

For the case of the standard realization method this polynomial is:

$$D_1(z) = 1 - a_1 z^{-1} + a_0 z^{-2} \quad (1)$$

where a_1 and a_0 are the multipliers used for the realization. The above mentioned alternative realization methods are based on the following relationships:

$$a_1 = f(K,L) \quad (2)$$

$$a_0 = g(K,L) \quad (3)$$

If $re^{\pm j\theta}$ are the two complex conjugate poles of the filter, then

$$r = \sqrt{a_0} \quad (4)$$

$$\theta = \text{arc cos} \left(\frac{a_1}{2\sqrt{a_0}}\right) \quad (5)$$

Substituting equations (2) and (3) into (4) and (5) we have r and θ as functions of K and L in the form:

$$r = h_1(K,L) \quad (6)$$

$$\theta = h_2(K,L) \quad (7)$$

Variables K and L are the multipliers of the digital filter structure based on (2) and (3). Suppose now that they are quantized. If Q(K) and Q(L) are their new values respectively, we define the following corresponding quantization errors in K and L:

$$\Delta K = Q(K) - K \quad (8)$$

$$\Delta L = Q(L) - L \quad (9)$$

Using equations (6) and (7) and applying Newton's formula we can approximately estimate the errors in r and θ , Δr and $\Delta\theta$ respectively, in the form:

$$\Delta r = \frac{\partial h_1}{\partial K}\Delta K + \frac{\partial h_2}{\partial L}\Delta L \qquad (10)$$

$$\Delta\theta = \frac{\partial h_1}{\partial K}\Delta K + \frac{\partial h_2}{\partial L}\Delta L \qquad (11)$$

It is obvious that, if $\Delta K = \Delta L = 0$, that is if the quantization does not affect K and L, Δr and $\Delta\theta$ will be equal to zero too. However there exist some specific cases, where Δr or $\Delta\theta$ are equal to zero even if ΔK and/or ΔL are differnt than zero. This happens whenever the partial errors due to the quantization of each of K and L are cancelling eachother. These specific cases can be found either from (6) and (7) or from (10) and (11).
After the quantization procedure the magnitude and the phase of the poles in their new locations will be:

$$r' = h_1\,[\,Q(K), Q(L)\,] \qquad (12)$$

and

$$\theta' = h_2\,[\,Q(K), Q(L)\,] \qquad (13)$$

Thus the exact values of Δr and $\Delta\theta$ are:

$$\Delta r = r' - r = h_1[Q(K), Q(L)] - h_1(K,L) \qquad (14a)$$

$$\Delta\theta = \theta' - \theta = h_2[Q(K), Q(L)] - h_2(K,L) \qquad (14b)$$

It is now obvious that $\Delta r = 0$ iff:

$$h_1[Q(K), Q(L)] = h_1(K,L) \qquad (15a)$$

and $\Delta\theta = 0$, iff

$$h_2[Q(K), Q(L)] = h_2(K,L) \qquad (15b)$$

Equations (15a) and (15b) are relations between Q(K), Q(L), K and L, while using (8) and (9) we can have relations between ΔK, ΔL, K and L.
However, it is sometimes useful to obtain simpler expressions, because (15a) and (15b) might be too complicated. This can be achieved from (10) and (11).
Thus, we have: $\Delta r = 0$ iff:

$$\frac{\partial h_1}{\partial K}\Big/\frac{\partial h_1}{\partial L} = -\frac{\Delta L}{\Delta K} \qquad (16)$$

and $\Delta\theta = 0$, iff:

$$\frac{\partial h_2}{\partial K} \Big/ \frac{\partial h_2}{\partial L} = -\frac{\Delta L}{\Delta K} \tag{17}$$

Expressing now the above partial derivatives in terms of K and L we obtain conditions in terms of K, L, ΔK and ΔL.
Since equations (10) and (11) are approximations of Δr and $\Delta\theta$, the above conditions (equations (16) and (17)) can guaranty that Δr or $\Delta\theta$ will be different than but very close to zero.

APPLICATIONS TO SEVERAL REALIZATION METHODS

We first apply the above theory to the standard structure characterized by the transfer function denominator polynomial of equ. (1). In this case no transformation between (a_0, a_1) and (K,L) holds, that is, (2) and (3) are:

$$a_0 = K$$

$$a_1 = L$$

Equations (4), (5), (14a) and (14b) provide the following exact conditions:

$$\Delta r = 0, \text{ iff } Q(a_0) = a_0 \tag{18}$$

and

$\Delta\theta = 0$, iff

$$\frac{a_1}{2\sqrt{a_0}} = \frac{Q(a_1)}{2\sqrt{Q(a_0)}} \tag{19}$$

If:

$$\Delta a_0 = Q(a_0) - a_0$$

$$\Delta a_1 = Q(a_1) - a_1$$

equation (19) can be written:

$$\frac{a_1}{2\sqrt{a_0}} = \frac{a_1 + a_1}{2\sqrt{\Delta a_0 + a_0}} \tag{20}$$

Equation (18) has no interest because it is obvious.
Estimating Δr and $\Delta\theta$ in the form of equations (10) and (11) where K and L are substituted by a_0 and a_1 respectively we obtain:

$$\Delta r = \frac{\Delta a_0}{2\sqrt{a_0}} \tag{21}$$

$$\Delta\theta \simeq \frac{\Delta a_0}{2r^2\tan(\theta)} - \frac{\Delta a_1}{2r\sin(\theta)} \tag{22}$$

(see also [1]).
Condition (18) can be provided from (21) as well. On the other hand $\Delta\theta$ can be written:

$$\Delta\theta = \frac{1}{2r\ \sin(\theta)}\ (\Delta a_0 \frac{\cos\theta}{r} - \Delta a_1)$$

or taking equations (4) and (5) into account:

$$\Delta\theta = \frac{1}{2r\ \sin(\theta)}\ (\Delta a_0 \frac{a_1}{2a_0} - \Delta a_1)$$

Thus the approximate condition is:

$$\Delta\theta = 0 \quad \text{if} \quad \frac{\Delta a_0}{2a_0} = \frac{\Delta a_1}{a_1} \tag{23}$$

We can draw the locii of the points (a_0, a_1) characterized by (23) on a plane having a_0 and a_1 as coordinates, taking into account the specific quantization method (either truncation or rounding) we are using. For the case of fixed point arithmetic and truncation with 2 bits (just for the sake of simplicity of the figures) the appropriate locii are indicated in figure 1 (see next page).

It is very interesting to see that condition (23) can be derived from condition (20) through some approximation. In fact, from (20) we have:

$$\frac{a_1^2}{2a_0} = \frac{(\Delta a_1)^2 + a_1^2 + 2a_1\Delta a_1}{2a_0 + 2\Delta a_0}$$

and considering that $(\Delta a_1)^2 << a_1^2 + 2a_1\Delta a_1$ we have:

$$\frac{a_1^2}{2a_0} \simeq \frac{a_1^2 + 2a_1\Delta a_1}{2a_0 + 2\Delta a_0}$$

which can be reduced to:

$$\frac{\Delta a_0}{2a_0} = \frac{\Delta a_1}{a_1}$$

which is equation (23).

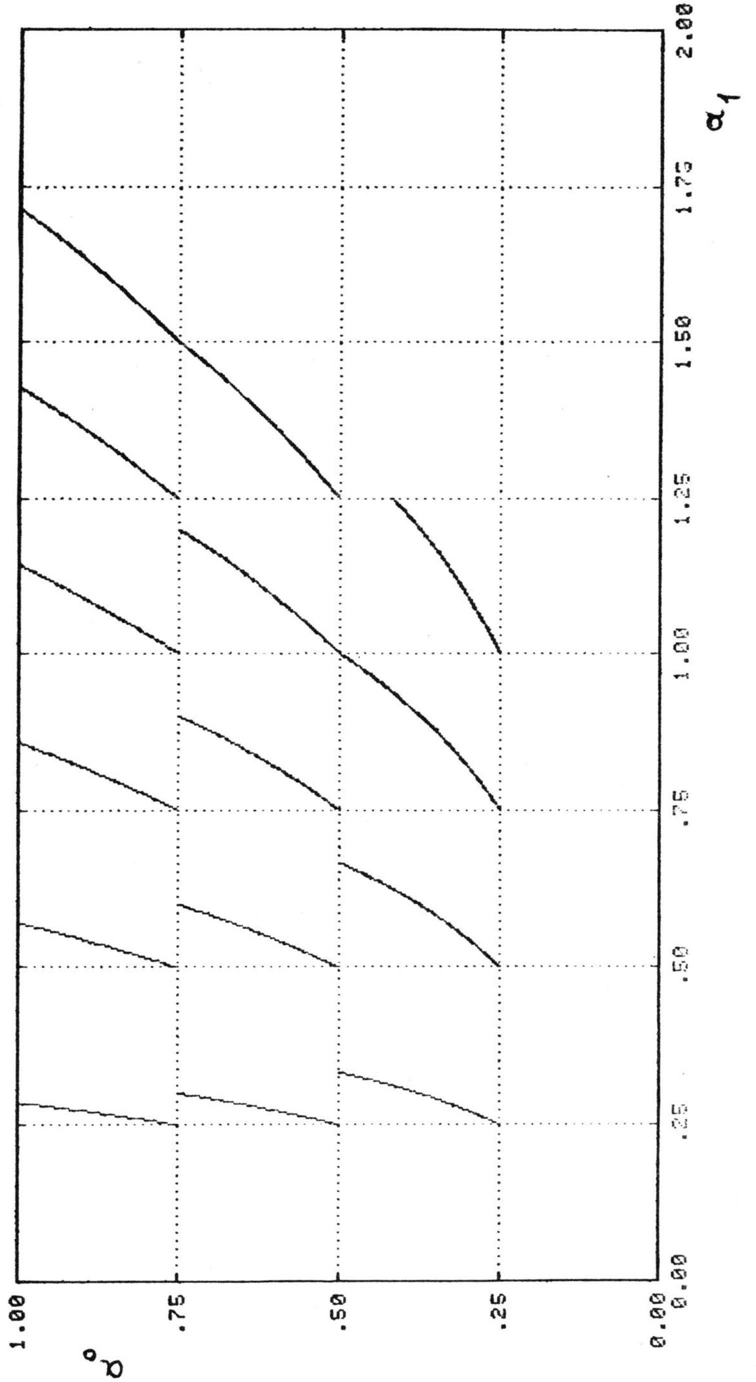

Figure 1

Another realization method has been proposed by Rader and Gold ([1]). This realization is based on the coefficient transformations:

$$a_0 = K^2 + L^2 \tag{24}$$

$$a_1 = - 2K \tag{25}$$

so that the transfer function denominator polynomial becomes:

$$D_2(z) = 1 - 2Kz^{-1} + (K^2 + L^2)z^{-2} \tag{26}$$

where K and L are the multipliers now. Obviously:

$$r = \sqrt{K^2 + L^2} \tag{27}$$

and

$$\theta = \arccos(\frac{2K}{\sqrt{K^2 + L^2}}) \tag{28}$$

From the above equations it is simple to derive that the actual errors are:

$$\Delta r = r' - r = \sqrt{Q(K)^2 + Q(L)^2} - \sqrt{K^2 + L^2}$$

and

$$\Delta\theta = \theta' - \theta = \arccos(\frac{2Q(K)}{\sqrt{Q(K)^2+Q(L)^2}}) - \arccos(\frac{2K}{\sqrt{K^2 + L^2}})$$

From these equations we get the conditions so that $\Delta r = 0$ or $\Delta\theta = 0$ (exactly, and not approximately):

$$\Delta r = 0 \text{ if } : \quad (K+\Delta K)^2 + (L+\Delta L)^2 = K^2 + L^2 \tag{29}$$

$$\Delta\theta = 0 \text{ if } : \frac{2K+2\quad K}{\sqrt{(K+\quad K)^2+(L+\quad L)^2}} = \frac{2K}{\sqrt{K^2 + L^2}} \tag{30}$$

Similarly we can find the conditions so that $\Delta r \simeq 0$ or $\Delta\theta \simeq 0$. Applying equations (17a) and (17b) and taking into account that equ. (6) and (7) are now equations (27) and (28), we have:

$$\Delta r = 0, \text{ if } \quad \frac{\Delta K}{\Delta L} = -\frac{\sin\theta}{\cos\theta}$$

and

$$\Delta\theta = 0, \text{ if } \quad \frac{\Delta K}{\Delta L} = \frac{\cos\theta}{\sin\theta}$$

or, since $K=r\cos\theta$ and $L=r\sin\theta$ (31)

$$\Delta r = 0, \text{ if } \quad \frac{\Delta K}{\Delta L} = -\frac{L}{K} \tag{32}$$

and

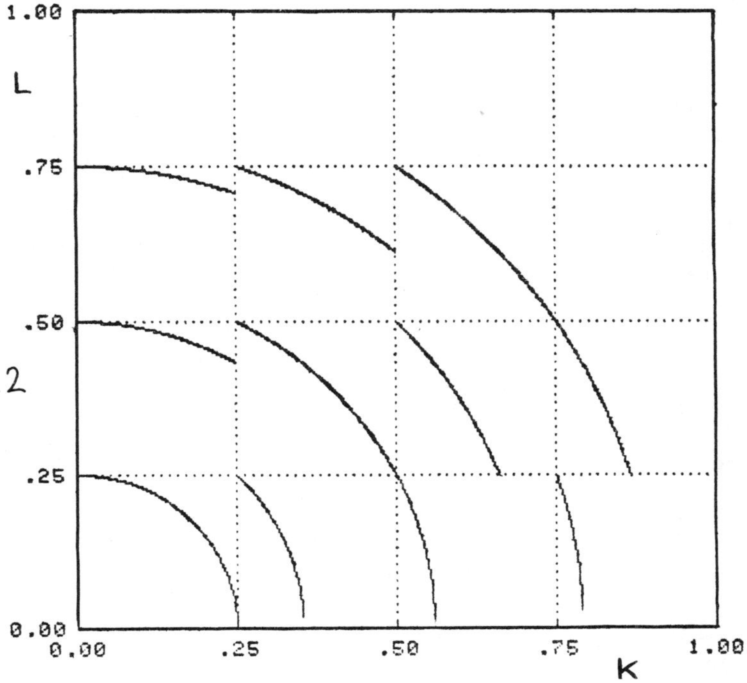
1.00
L
.75
.50
.25
0.00
0.00
.25
.50
.75
1.00
K

Fig. 2

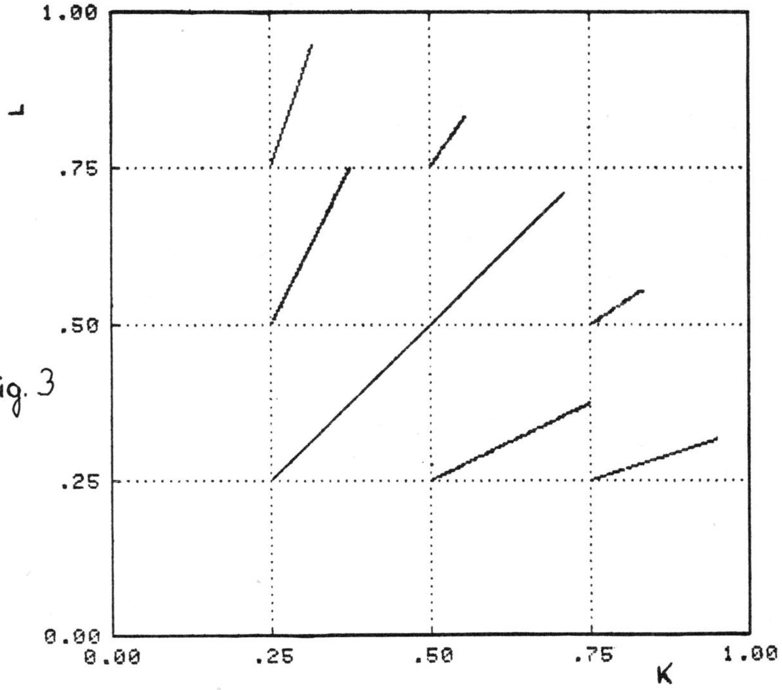
1.00
L
.75
.50
.25
0.00
0.00
.25
.50
.75
1.00
K

Fig. 3

$\Delta\theta = 0$, if

$$\frac{\Delta K}{\Delta L} = \frac{K}{L} \tag{33}$$

The appropriate locii on the (K,L) plane can be seen in figures 2 (for $\Delta r=0$) and 3 (for $\Delta\theta=0$) for the same kind of arithmetic and quantization as previously (fixed point, truncation).
From figure 2 we see that because of the negative sign of equ. (32) Q(K) or Q(L) is not equal to the lowest quantization level but the highest one, that is, quantizing K (or L) we truncate it to m number of bits and add 2^{-m} (m is the number of fractional bits).
Equations (32) and (33) can be similarly derived from equations (29) and (30). In fact, equation (29) can be written:

$$K^2+(\Delta K)^2+2K\Delta K+L^2+(\Delta L)^2+2L\Delta L = K^2 + L^2$$

or, since $(\Delta K)^2$ and $(\Delta L)^2$ are very small in value:

$$2K\Delta K + 2L\Delta L = 0$$

from which we obtain equation (32).
Similarly, equation (30) can be written:

$$\frac{K^2+(\Delta K)^2+2K\Delta K}{K^2+2K\Delta K+L^2+2L\Delta L+(\Delta K)^2+(\Delta L)^2} = \frac{K^2}{K^2 + L^2} \qquad \text{or}$$

$$\frac{K^2+2K\Delta K}{K^2+2K\Delta K+L^2+2L\Delta L} = \frac{K^2}{K^2 + L^2}$$

which can be reduced to (33), as it is easy to prove.

A general theory of finding classes of second-order digital filters having pole magnitude or phase error very close to zero, is proposed and applied to certain digital filter structures. This method examines where the coefficients of the multipliers have to move, being quantized, so that the poles move on a circle ($\Delta r=0$) or on a radius ($\Delta\theta=0$).

REFERENCE

[1] : C. Rader and R. Gold: "Effects of Parameter Quantization on the Poles of a Digital Filter", Proceedings of the IEEE, May, 1967, pp. 688-689.

IMPROVEMENT OF THE RECONSTRUCTION AND CONTROL OF THE COMPRESSOR OUTPUT

C.A. Caroubalos, Ch. Metaxaki-Kossionides.
S.A. Athenaios, D.C. Kotsis.
University of Athens, Department of Physics,
Division of Electronics.
104, Solonos str., Athens-Greece.

ABSTRACT

The problem of applying the well known Zero Order Predictor is the loss of useful information when the "aperture" γ is several quantization steps; this problem can be reduced by a more sophisticated refilling of the supressed samples. The adaptive method proposed here is based on an "adaptive-sinc" refilling completed by an "important points" insertion. This method leads to improvement of the accuracy of the reconstructed signal before the transmission, especially in the low frequencies of the spectrum. At the same time it gives the possibility to control either the accuracy of the reconstructed receiving signal (in absence of channel er rors) or the Compression Ratio in order to better exploit the available capacity of the transmission channel.

1. Introduction

Among the different methods of Data Compression the "prediction method" is very well known (1,2). This method is succesfully applied in many cases, especially when the data source is causal (2). One of the simplest "prediction" algorithms is the "Zero Order Predictor" (ZOP) based on the suppression of every sample that differs from the last transmitted sample less than a given tolerance (aperture) γ.

The performance of different variations of the ZOP algorithms (fixed, floating aperture, etc.) has been

D. G. Lainiotis and N. S. Tzannes (eds.), Advances in Communications, 165–172.

studied, in order to obtain the optimum values of compression parameters (1,3). However some aspects of this prediction technique, seem not to have been investigated enough. Two of them are:

1. The improvement of the reconstruction algorithm in order to obtain better accuracy as well as the development of proper methods, leading to the control of the accuracy before transmission of the samples.

2. The control of the compression ratio (CR= number of original samples/ number of transmitted samples)through the compressor parameters, for optimal use of the available capacity of the transmission channel.

In order to achieve improvement of the accuracy as well as the control of CR, adaptive methods must be used.

For the D/A conversion of a sampled signal, a sinc interpolator is commonly used. But in the case of a compressed signal, we must proceed to the refilling of the space of the suppressed samples before the D/A conversion. This refilling can simply be effected by repeating the last not suppressed sample. But in this way an important part of useful imformation is lost, especially when the tolerance γ is several times the quantization step, and generally the so reconstructed signal takes a stepped shape.

A more efficient way for the refilling would be the use of a sinc interpolator too, but this one must be an adaptive sinc as the number of missing samples varies with time (4).

Considering points 1 and 2 above, a more sophisticated adaptive method for the "compression-reconstruction" is proposed, by which the compression of the signal and the control of the compressor output is possible.

This method consists of (a) a ZOP (b) an adaptive sinc-trial reconstruction, (c) a test of the obtained accuracy and (d) in case of failure an enrichment of the compressor output by inserting "important samples" taken from the original sampled source.

The method has been applied on ECG signals and on solar radio bursts, both sampled on Nyquist rate. Although the sources have different properties, the application of the method results in compression ratios (CR) of the same order 4-10, with rmse 0.9-2.2. provided that the corresponding singals are normalised approximately to the same valueof initial signal to noise ratio.

2. THE ADAPTIVE METHOD.

The proposed adaptive "compresion-reconstruction" method, shown in fig. 1 consists of:

a. A first step of fixed ZOP compressor, leading to an initial CR value.

b. A second step of trial refilling using an adaptive sinc interpolator. The adaptive element of this interpolator is the time scale Tk of the sinc, which varies according to the number k of the sequentially suppressed samples.

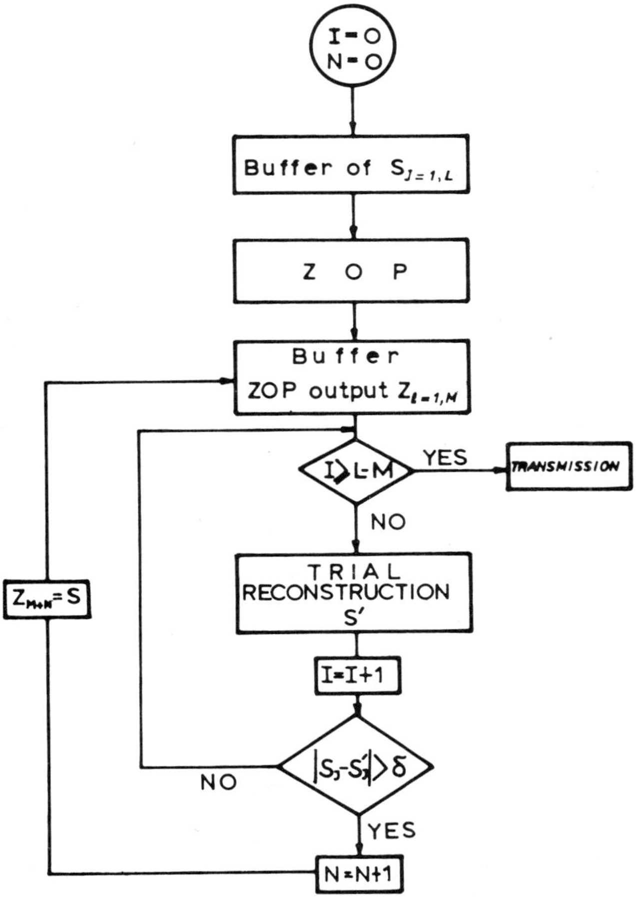

Fig. 1 The Diagram of "Compression-reconstruction method".

In the time domain the adaptive-sinc attempts the refilling based on variable "subsampling" the frequency of which varies according to k. In the frequency domain this corresponds to the use of only the lower frequencies of the spectrum. But as the signal is not at the same time apropriately filtered the adaptive-sinc-refilling often fails, due to aliasing, resulting in a point outside the ZOP aperture γ. The refilling will be succesful only when the aliasing spectrum does not affect the truespectrum. This occurs only when the number k is small and the neighbouring part of the waveform, has not intensive high frequency components. To avoid this problem and given that an adaptive filtering would make the system much slower and complicated the following step (c), which inserts "importat points"(5) is added:

(c) Every sample computed by the adaptive-sinc is compared to its corresponding original sample of the source. If it differs from it more than a given tolerance δ≤γ it is not accepted and is replaced by the original sample of the source. This last sample is considered as an "important point" which is added in the ZOP output buffer. When this operation is completed, for all missing points, the content of the so enriched buffer, is transmitted.

In this way a reconstruction whith controlled accuracy is guaranteed. Since better accuracy (lower δ) results in smaller CR a compromise must be made between accuracy and compression ratio.

If the channel capacity is limited the CR value is more important. In this case we can start whith a high value of ZOP tolerance γ (high value of CR) and then decreasing δ progressively we can gain in accuracy until the limited value of CR is reached.

3. RESULTS AND DISCUSSION

The results of applying the proposed method on solar radio burst signals are shown in Fig.2. The original signal (2.a) was compressed with a ZOP aperture γ=4 leading to CR=6.8. The following remarks can be made:

(a) Comparing 2.c and 2.d we see improvement of the reconstruction by the use of adaptive-sinc. The artificial stepped shape has been removed and the peaks are well reproduced without any introduction of shifts. The e(t) profile which represents the difference between original and reconstructed signal is also reduced from 2.c

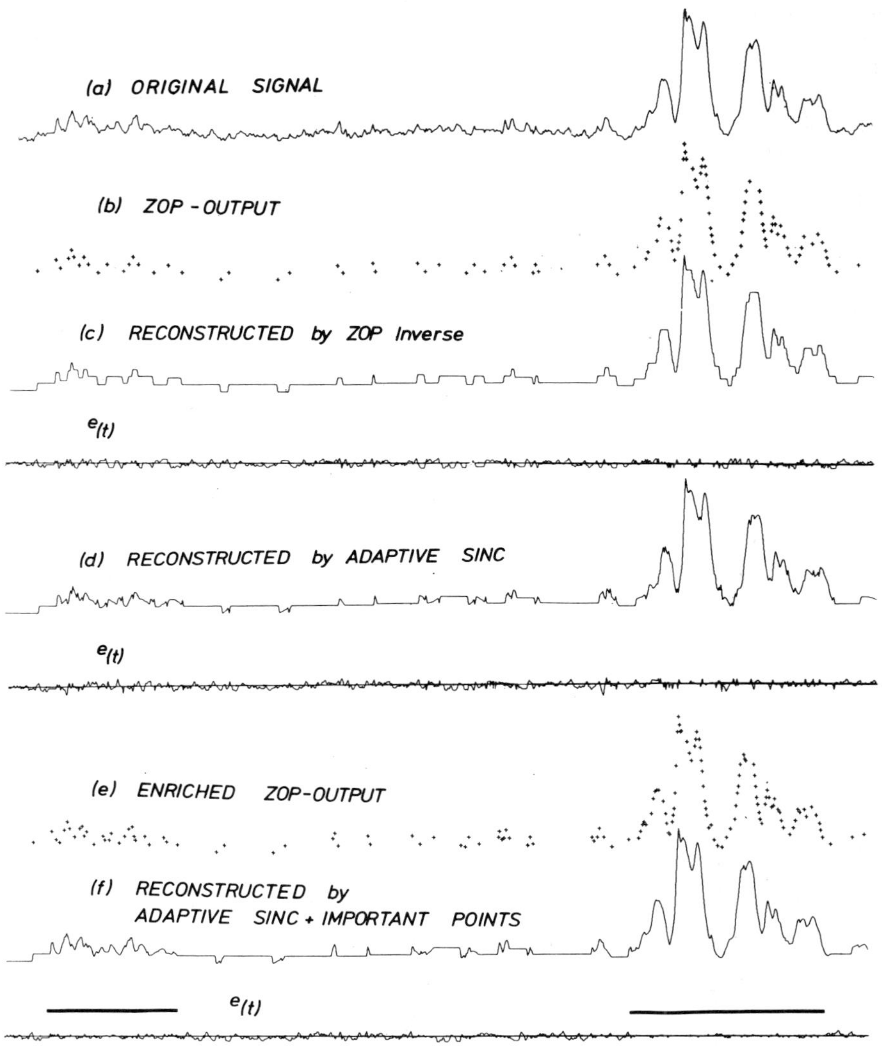

Fig.2 Application of the compression-reconstruction method on solar radio bursts. The successive error profiles e(t) show the strong selective improvement during the active periods.

to 2.d, especially near the peaks; the correspoding rmse values are 1,9 and 1,68. This improvement is exclusively due to the use of adaptive-sinc whithout any loss in the CR value . However, since it does not permit any further improvement or control of the compressor output it can not be considered as satisfactory.
(b) Comparing figures 2.d and 2.f we see a clear improvement in the reconstruction obtained by the insertion of important points, with simultaneous reduction of CR value.

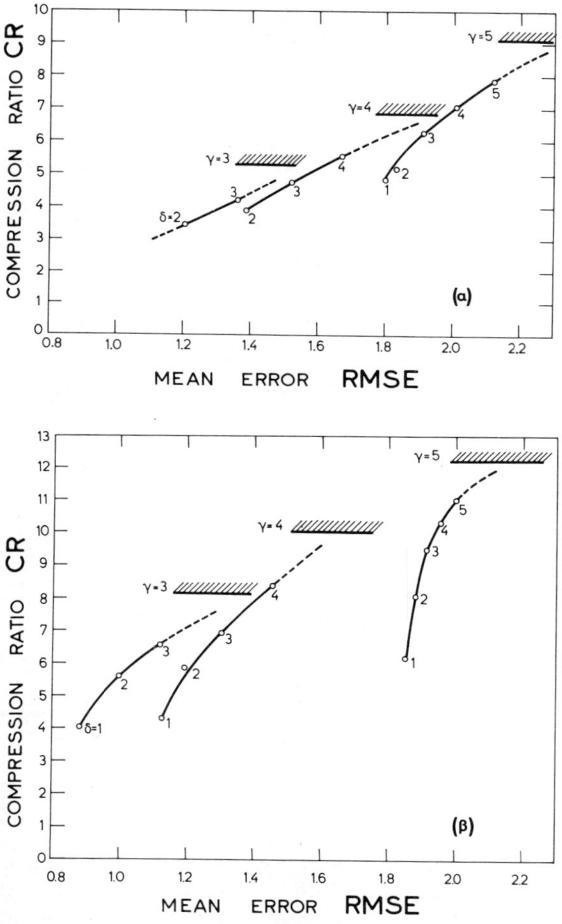

Fig. 3. The control action of parameter δ on CR and rmse, for different values of γ.(a) radio bursts (b) ECG.

It must be emphasized that this improvement is very strong, especially during the active periods (underlined segments) which are the more important. Furthermore, this improvement can be controlled by the choice of parameter δ (for 2.f δ=3). Concerning the position of important points insertion, we can see in 2.e that the important points are mainly accumulated in the region of smaller fluctuations while the peaks are well reproduced mainly by the combined action of ZOP+adaptive sinc.

Figures 3.a, 3.b present the control action of the parameter δ for different values of ZOP aperture γ, for radio bursts and ECG signals respectively . It seems that the reduction of rmse is not fast enough when δ decreases, but the rmse value of the total signal does not represent the strong improvement during the active periods, as it is already noted.

CONCLUSIONS

The proposed method of compression-reconstruction has the following advantages: first it gives the possibility for the control of compressor output and second it leads to a strong improvement of the reconstructed signal during the intervals of activity. It is to note that other methods in the bibliography (5,6,7) lead to enhanced error near the peak activity whith comparative or even smaller CR values.

The limit of the method for real time applications has not been tested and it will not be probably fast enough for broad spectrum signals.

This work was partially supported by the Greek National Foundation of Research.

REGERENCES

1. Andrews,C.A. et al "Adaptive Data Compression",Proc. IEEE,Vol.55, No.3, Mar. 1967, pp. 267-277.

2. Davisson,L.D. "The theoretical Analysis of Data Compression Systems",Proc. IEEE, Vol.56, No.2,Febr.1968.

3. Capellini,V. et al,"A Study on Data Compression Techniques", Rep. ESTEC/Cont.902/1970AA, Inst. di onde Electromagnetiche, Univ. Firenze.

4. Caroubalos C.et al,"La compression des données Radioastronomiques transmisés par satellite et la reconstitution du signal, "Rap. du Depart. Spatiale No 60/stereo 1968, Observatoire de Meudon.

5. Sandman, A.M. et al, "A Solution to the bandwith Compression- Important points", IEEE Trans. on Commun., Sept. 1976, pp. 1022-1023.

6. Ahmed, N. et al, "Electrocardiographic Data Compression via Orthogonal Transforms", IEEE Trans. on Biom. Engineering, Vol.22, No.6, Nov.75, pp. 484-487.

7. Lohn, D.C. et al "Efficient Digitization methods for Electrocardiograms", Proc. of Joint Automatic Control Conference, 1977, pp. 1386-1391.

TWO-DIMENSIONAL DIGITAL FILTERING.
BASIC CONSIDERATIONS AND SUBROUTINES FOR IMPLEMENTATION*

W.Waung and A.N.Venetsanopoulos J.M. Costa

University of Toronto
Toronto, Ontario, Canada

Bell Northern Research
Ottawa,Ontario,Canada

ABSTRACT

This paper considers filtering of two-dimensional data. The mathematics governing these filters are outlined. Different mathematical transforms useful in simplifying calculations and techniques to simplify computation are discussed. A subrountine (written in FORTRAN) to perform two-dimensional fast Fourier transforms of a real-valued array is briefly explained. An example of two-dimensional data processing is also given.

I. INTRODUCTION

During the past few decades, the rapid development of digital computers has made such devices available to the general public. Advances in this field have also greatly increased the capabilities of such machines [1], especially in the areas of storage capacity and speed of computation. These improvements have resulted in improvements in the retrieval of data in digital form. However with data stored digitally, it becomes more convenient to perform data processing in a digital fashion. This creates new problems, but at the same time offers degrees of freedom not available before [2-4].

In many cases, the data represent multidimensional information. Examples of such include contour maps, two-dimensional images, position versus time data in seismology etc. Of these data sets, two-dimensional information is usually most easily

* This work was supported by the Natural Sciences and Engineering Research Council of Canada, under grant No. A-7397.

D. G. Lainiotis and N. S. Tzannes (eds.), Advances in Communications, 173–182.

understood and thus generally most useful. Data sets of three or more dimensions are harder to visualize.

Some of the applications of two dimensional data processing include restoration of Images, enchancement of images and pattern recognition.

There are many other uses and examples [2] but those mentioned give a good indication of the importance and scope of such processing.

This paper is divided into two major section. The first section deals with an introduction to digital filtering and the mathematics involved. Brief mention is made of number theoretic transforms and discrete Fourier transforms. In the second, a subroutine to implment two-dimensional fast Fourier transforms is presented and briefly explained. Finally, an example of usage of the mentioned subroutine is given.

II. DIGITAL FILTERING

As stated in the introduction, digital data processing is particularly useful due to the availability of inexpensive digital computers. However the idea of processing data gives rise to notions of complicated algorithms. Viewed in a different manner this processing becomes easy to understand. One can usually model the process in terms of a digital filter. Then the output of such a filter is simply the processed result of the input to the filter [7]. The problem is then converted from that of the design of algorithms, to that of the design of a filter which will achieve the desired processing. Such a problem requires a different solution for each particular situation and is not discussed in this paper. There are many available references describing how to design digital filters for restoration, enhancement etc. [5-6].

The resulting filters can have either infinite (IIR) or finite impulse response (FIR). However, IIR filters possess certain properties that complicate their implementation. One problem is that of stability. It is generally, difficult to ensure bounded-input; bounded-output (BIBO) stability in multidimensional IIR filters. Yet FIR filters are always stable in the BIBO sense. (See Appendix A). This paper discusses FIR filters only.

It is well known that the output of a filter is the convolution of the filter impulse response and the input sequence. In the case of a finite input sequence, filtered by a FIR filter, we have

$$y(n_1,n_2)=\sum_{m_1=0}^{M_1-1}\sum_{m_2=0}^{M_2-1} x(m_1,m_2)h(n_1-m_1,n_2-m_2)$$

where $x(n_1,n_2)$ is the input sequence of length M_1M_2, i.e.

$$x(n_1,n_2) = 0 \qquad \begin{array}{l} n_1 < 0 \text{ or } n_1 > M_1 \\ n_2 < 0 \text{ or } n_2 > M_2 \end{array}$$

$h(n_1,n_2)$ is the impulse response of length $M_1'M_2'$
and $y(n_1,n_2)$ is the output sequence of length N_1N_2
where $N_1 = M_1'-1+M_1$
$N_2 = M_2'-1+M_2$

However, doing this convolution directly requires large number of operations. For each output sequence of length N_1,N_2, approximately $N_1^2N_2^2$ multiplications and additions are needed. There are faster and more efficient methods of achieving the same result. These are usually accomplished by the use of some mathematical transforms, so that a convolution can be replaced by simple multiplication.

III. MATHEMATICAL TRANSFORMS

The alternative to direct convolution is the use of mathematical transforms. The solution may require fewer operations in the transform domain, thus rendering it more practical.

Such an approach became practical when Cooly and Tudey first introduced a fast algorithm for performing the discrete Fourier transform and its inverse. A general DFT like transform in two-dimensions is

$$X(k_1,k_2) = \sum_{n_1=0}^{N_1-1}\sum_{n_2=0}^{N_2-1} x(n_1,n_2)\alpha_1^{n_1k_1}\alpha_2^{n_2k_2} \tag{1}$$

where α_1 is a primitive N_1th root of unity
α_2 is a primitive N_2th root of unity.

The inverse transform is

$$x(n_1,n_2) = (N_1N_2)^{-1}\sum_{k_1=0}^{N_1-1}\sum_{k_2=0}^{N_2-1} X(k_1,k_2)\alpha_1^{-n_1k_1}\alpha^{-n_2k_2} \tag{2}$$

The expressions (1) and (2) can be shown to form a transform pair.

Note now that the transform of the output sequence is the point by point multiplication of the transforms of the two input sequences.

$$Y(k_1,k_2) = \sum_{n_1} \sum_{n_2} y(n_1,n_2)\alpha^{n_1k_1} \alpha_2^{n_2k_2}$$

$$= \sum_{n_1} \sum_{n_2} \sum_{m_1} \sum_{m_2} x(m_1,m_2)h(n_1-m_1,n_2,m_2)\alpha^{n_1k_1} \alpha^{n_2k_2}$$

$$= X(k_1,k_2)H(k_1,k_2) \tag{3}$$

Therefore the output sequence can be obtained from the inverse transform of the point by point multiplication of the transforms of the input sequence. And

$$y(n_1,n_2)=(N_1,N_2)^{-1} \sum_{k_1=0}^{N_1-1} \sum_{k_2=0}^{N_2-1} X(k_1,k_2)H(k_1,k_2)\alpha_1^{-n_1k_1} \alpha_2^{-n_2k_2} \tag{4}$$

By using fast transform algorithms, 1-dimensional transforms can be obtained using approximately $N\log_2 N$ multiplications and additions. Therefore the number of operations required to implement the convolution in the transform domain is approximately $3(N\log_2 N)+N$. For N large, this is much smaller than the N^2 operations required for direct convolution. Convolution by operations in the transform domain is usually preferred.

IV. NUMBER THEORETIC TRANSFORM (NTT)

In number theoretic transforms, all operations are done in modular arithmetic. That is, additions and multiplications are done modulo M (mod M) where M is an integer called the modulus. The allowable numbers are the set of integers 0,1,2,...M-2,M-1. This set is called a field or a ring. In modular arithmetic all numbers must be members of the above finite set. This is accomplished by a residue reduction operation such that

$$((x)) = X - rM \tag{5}$$

where r is an integer and $0 \leq ((x)) \leq M-1$.

For a number in modular arithmetic to have multiplicative inverse, the modulus M must be a prime number. Therefore, the general NTT of a sequence of N_1N_2 integers

$$x(n_1,n_2);\ n_1 = 0,1,2,\ldots,\ N_1 - 1,\ n_2 = 0,1,\ldots,N_2 - 1$$

where modulus M is another sequence of N_1N_2 integers

$$X(k_1,k_2) = \sum_{n_1=0}^{N_1-1} \sum_{n_2=0}^{N_2-1} x(n_1,n_2)\alpha_1^{n_1k_1}\alpha_2^{n_2k_2} \quad (6)$$

where α_1 is relatively prime to M_1 and of order N_1 (i.e. $\alpha_1^{N_1} \equiv 1$)
α_2 is mutually prime to M_2 and of order N_2
and N^{-1} and N_2^{-1} (multiplicative inverses N_1 and N_2 respectively) exist,
$i-\alpha_1^i$, is mutually prime to M_1 for all $i = 1,2,\ldots, N-1$
$i-\alpha_2^i$, is mutually prime to M_2 for all $i = 1,2,\ldots, N-1$.

However, only NTT's with highly composite N_1 and N_2 can support an efficient FFT-like algorithm [8-9]. It is then necessary to find suitable α_1, α_2, M_1 and M_2 for such transforms. Unfortunately only a few classes of NTT's satisfy the above requirements. Two of these are the Fermat number transforms (FNT) and the Mersenne number transforms (MNT). FNT's use a Fermat number F_t as a modulus M, that is

$$F = F_t = 2^{2^t} + 1 \quad (7)$$

MNT's, on the other hand, use a modulus of a Mersenne number

$$M = M_P = 2^P - 1 \quad (8)$$

where P is a prime number.

Number theoretic transforms offer the advantages of integer arithmetic. All operations and intermediate steps result in integers which can be represented exactly in digital machines. Thus all roundoff and truncation errors are limited to those resulting from initial quantization of the input sequences. However, this advantage is offset by the restrictions on size of the modulus N and α. These difficulties make NTT's a subject suitable to individuals with advanced background in modular arithmetic and transforms. For those interested, the topic is discussed in [8-9].

V. DISCRETE FOURIER TRANSFORMS

This discrete Fourier transform (DFT) is an extension of the Fourier Transform. Following the general form of mathematical transforms, we have

$$DFT(x(n_1,n_2)) = X(k_1,k_2) = \sum_{n_2=0}^{N_1-1} \sum_{n_2=0}^{N_2-1} x(n_1,n_2)\exp(-j\frac{2\pi}{N_1} n_1k_1) \exp(-j\frac{2\pi}{N_2} n_2k_2) \quad (9)$$

$$\mathrm{IDFT}(X(k_1,k_2))=x(n_1,n_2)=(N_1N_2)^{-1}\sum_{k_1=0}^{N_1-1}\sum_{k_2=0}^{N_2-1}X(k_1,k_2)\exp(j\frac{2\pi}{N_1}n_1k_1)\exp(j\frac{2\pi}{N_2}n_2k_2) \quad (10)$$

This transform assumes a periodicity such that

$$(n_1+rN_1,n_2+\ell N_2) = x(n_1,n_2) \quad r,\ell = \ldots -1,0,1,2\ldots \quad (11)$$

The convolution property applies for DFT also such that

$$y(n_1,n_2) = \mathrm{IDFT}\{\mathrm{DFT}(x(n_1,n_2))\times \mathrm{DFT}(h(n_1,n_2))\} \quad (12)$$

However to avoid erroneous results from the assumed periodicity, the DFT's should be performed with

$$N_1 = M_1 + M_1' - 1$$
$$N_2 = M_2 + M_2' - 1 \quad (13)$$

where $x(n_1,n_2)$ is a sequence of length $M_1 \times M_2$
$h(n_1,n_2)$ is a FIR of length $M_1' \times M_2'$
$y(n_1,n_2)$ is a sequence of length $N_1 \times N_2$

To be able to use the fast Fourier transform algorithm developed by Cooley and Tukey [10-11], N_1 and N_2 must be powers of 2 (radix 2). This restriction is much easier to satisfy than those on NTT's. Further work on these algorithms has also made it possible to have a radix 4 implementation. This method has been also extended to radix 3 FFT algorithms [12] and other extensions [13], which allow more flexibility in the choice of transform size.

VI. SOFTWARE AVAILABLE

We shall now present the various subprograms that are currently available at the University of Toronto's Computer Centre (UTCC). These are subroutines written mainly by J.Costa a Ph.D. student in Electrical Engineering.

One often used two-dimensional FFT subroutine is discussed in some detail. This is the program FFT2R and it operates on two dimensional data that is purely real. This type of data is encountered most often in the real world therefore making this subroutine of special interest.

1. FFT2R

In general, a two-dimensional sequence of data is presented by a set of complex numbers. Using the Fortran programming language, this would mean $2N^2$ storage locations are required to record an NxN matrix of points. That is, each complex number would have been stored as a number pair of real and imaginary parts of magnitude and argument. However, much of the data collected in actual situations involve only real numbers. This means that only N^2 memory locations are required. In general, the transform of such a set of real data, is a set of complex numbers, which requires $2N^2$ storage points. For an algorithm that produces the transform in place, it would seem that storage requirement is not reduced even if the input vector is purely real. Further examination reveals that the transform of a two-dimensional real matrix is characterized by certain symmetry properties. The two-dimensional transform is defined as

$$X(k_1,k_2) = \sum_{n_1=0}^{N_1-1} \sum_{n_2=0}^{N_2-1} x(n_1,n_2) e^{-j\frac{2\pi}{N_1} k_1 n_1} e^{-j\frac{2\pi}{N_2} k_2 n_2}$$

where n_1 and n_2 are indices of the matrix with dimension N_1 and N_2 respectively

$x(n_1,n_2)$ is the two-dimensional input sequence and $X(k_1,k_2)$ is the two-dimensional transformed sequence. Now examine

$$X(N_1-k_1,N_2-k_2) = \sum_{n_1=0}^{N_1-1} \sum_{n_2=0}^{N_2-1} x(n_1 n_2) e^{-j\frac{2\pi}{N_1}(N_1-k_1)n_1} e^{-j\frac{2\pi}{N_2}(N_2-k_2)n_2}$$

$$= \left[\sum_{n_1=0}^{N_1-1} \sum_{n_2=0}^{N_2-1} x^*(n_1,n_2) e^{-j\frac{2\pi}{N_1}k_1 n_1} e^{j\frac{2\pi}{N_2}k_2 n_2} \right]$$

$$= X^*(k_1,k_2)$$

where * indicates the complex conjugate of a number. The above symmetry property, together with the assumed periodicity property means that the transform of a NxM real sequence will result in $(\frac{NM}{2} + 2)$ distinct numbers, four of which are real numbers. These numbers can therefore be stored in the same memory space used to store NM real numbers. This observation is used in the subroutine FFT2R to conserve storage requirements. However, slightly more than NM storage locations were used to reduce the processing complexity.

The actual memory allocation techniques are best explained by way of an illustration. Suppose we wish to obtain the DFT of a real two-dimensional sequence of data, a(x,y). That is we want

$$A(x,y) = DFT[a,(x,y)]. \tag{14}$$

The pionts a(x,y) are stored in a FORTRAN array as follows.

$$X(I,J)\ ;\quad 1 \leq I \leq M, \qquad 1 \leq J \leq N \tag{15}$$

Therefore X(1,1) is the value of a (0,0) - the value at the origin. To illustrate this M and N is chosen to equal 8, i.e. M=N=8. Generalization to the other powers of 2 is simple.

The real function a(x,y), $-\frac{M}{2} < x \leq \frac{M}{2}$, $-\frac{N}{2} < y \leq \frac{N}{2}$ is stored as shown in Figure 1.

a(0,0) → X(1,1)	a(0,1) → X(1,2)	a(0,2) → X(1,3)	...a(0,-1)→X(1,8)
a(1,0) → X(2,1)	a(1,1) → X(2,2)	a(1,2) → X(2,3)	...a(1,-1)→X(2,8)
a(2,0) → X(3,1)	a(2,1) → X(3,2)	a(2,2) → X(3,3)	...a(2,-1)→X(3,8)
a(3,0) → X(4,1)	a(3,1) → X(4,2)	a(3,2) → X(4,3)	...a(3,-1)→X(4,8)
a(4,0) → X(5,1)	a(4 1) → X(5,2)	a(4,2) → X(5,3)	...a(4, 1)→X(5,8)
a(-3,0)→ X(6,1)	a(-3,1) → X(6,2)	a(-3,2) → X(6,3)	...a(-3,-1)→X(6,8)
a(-2,0)→ X(7,1)	a(-2,1) → X(7,2)	a(-2,2) → X(7,3)	...a(-2,-1)→X(7,8)
a(-1,0)→ X(8,1)	a(-1,1)→ X(8,2)	a(0,2) → X(8,3)	...a(-1,-1)→X(8,8)

Figure 1. Storege allocation of (8,8) data sequence

VII. EXAMPLE

In Figure 2 to 4 an example achieving two dimensional filtering of a noisy input is presented. Below are a set of print-outs showing pictorially the input sequence, filter characteristic and the filtered output.

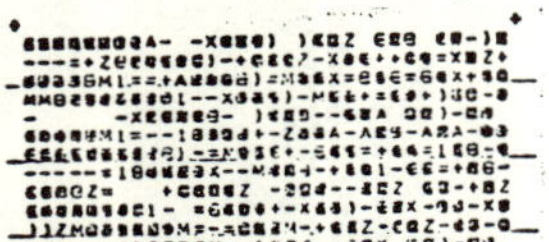

Figure 2
Input sequence

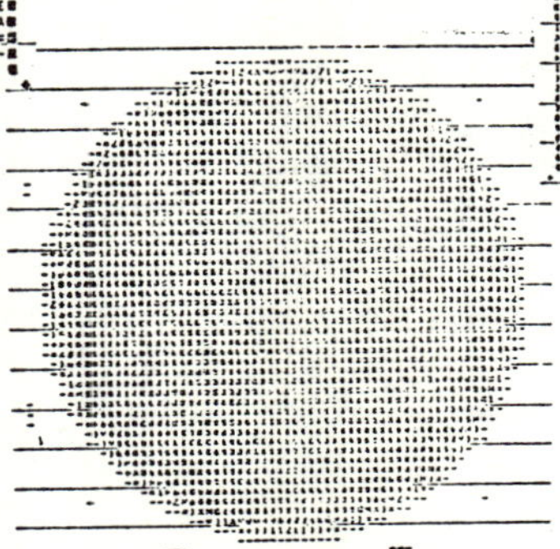

Figure 3 Filter characteristic

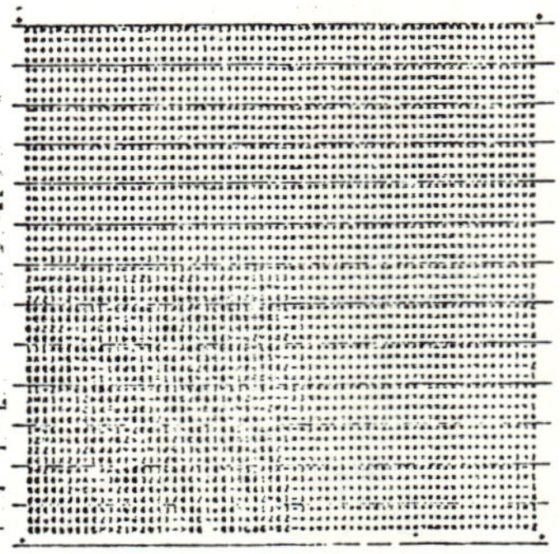

Figure 4
Filtered output

VIII. CONCLUSIONS

This paper has dealt with digital filtering which is a very powerful tool in data processing. The emphasis has been placed on two-dimensional data, as it is both useful and conceptually easy to visualize. Details on how such filtering can be achieved by software was presented, with special discussion on a two-dimensional FFT program available at UTCC. It is hoped that the reader will find out more about these subroutines and make use of them. The usefulness and power of digital filtering can only be fully appreciated by actual experience with such methods.

APPENDIX A - BIBO stability of two-dimensional FIR filters

A two-dimensional finite impulse response (FIR) filter is defined by an impulse response sequence h(m,n) such that

$$h(m,n) = 0 \text{ for } m < 0 \text{ or } m > N_1 \text{ or } n < 0 \text{ or } n > N_2 \qquad (a.1)$$

gives an input sequence x(m,n), the output of the filter is

$$y(k,\ell) = \sum_{m=0}^{N_1} \sum_{n=0}^{N_2} h(m,n)\, x(k-m, \ell-n) \qquad (a.2)$$

The magnitude of the output sequence is therefore

$$|y(k,\ell)| = \left| \sum_{m=0}^{N_1} \sum_{n=0}^{N_2} h(m,n)\, x(k-m,\ell-n) \right|$$

$$\leq \sum_{m=0}^{N_1} \sum_{n=0}^{N_2} |h(m,n)|\,|x(k-m,\ell-n)| \qquad (a.3)$$

If the input sequence is bounded such that

$$|x(m,n)| \leq B \text{ for all } m,n \qquad (a.4)$$

then it follows that

$$|y(k,\ell)| \leq B \sum_{m=0}^{N_1} \sum_{n=0}^{N_2} |h(m,n)| < \infty \qquad (a.5)$$

Therefore, for a bounded input, the output of a two-dimensional FIR filter is always bounded. Thus bounded-input-bounded-output (BIBO) stability is guaranteed for all FIR filters.

REFERENCES

1. Stanley, W.D.: 1975, Digital Signal Processing, Reston Publishing Co.
2. Ackroyd, M.H.: 1973, Digital Filters, Butterworths, London.
3. Bogner, R.E. and Constantinides, A.G. (editor): 1975, Introduction to Digital Filtering, Wiley.

4. Rabiner, L.R. and Gold, B.: 1975, Theory and Application of Digital Signal Processing, Prentice Hall.
5. Hunt, B.R. and Andrews, H.C.: 1977, Digital Image Restoration, Prentice Hall.
6. Rosenfeld, A. and Kak, A.C.: 1969, Digital Picture Processing, Academic Press, N.Y.
7. Oppenheim, A.V. (editor): 1969, Papers on Digital Signal Processing, MIT Press.
8. Agarwal, R.C. and Burrus, C.S.: 1974, "Fast Convolution Using Fermat Number Transforms with Application to Digital Filtering", IEEE Transactions on Acoustic Speech and Signal Processing, ASSP-22, pp. 87-97.
9. Dubois, E. and Venetsanopoulos, A.N.: 1978, "The Discrete Fourier Transform Over Finite Rings with Application to Fast Convolution", IEEE Trans. on Computers, C-27, 7, pp. 586-593.
10. Brigham, E.O.: 1974, The Fast Fourier Transform, Prentice-Hall.
11. Cooley, J.W. and Tukey, J.W.: 1965, "An Algorithm for Machine Calculation of Complex Fourier Series", Math. Computation,19, pp. 297-301.
12. Venetsanopoulos, A.N. and Dubois, E.: 1968, "A Radix 3 FFT with No Multiplications in the 3 Point DFT's", Trans. on Acoustics Speech and Signal Processing, ASSP-26, 3.
13. Rabiner, L.R. and Rader, C.M.: 1973, Digital Signal Processing, IEEE Press.
14. Hamming, R.W.: 1977, Digital Filters, Prentice-Hall, N.J.
15. Waung, W., Costa, J.M. and Venetsanopoulos, A.N.: 1979, "Two Dimensional Digital Signal Processing - Introductory Theory and Subroutines for Implementation", Communications Tech. Report No. 79-1, University of Toronto, Toronto, Canada

FREQUENCY RESPONSE OF MULTI-DIMENSIONAL LINEAR PHASE FIR DIGITAL FILTERS

N.C. Hu

Shanghai University of Technology, CHINA

In this paper, the frequency responses of multi-dimensional linear phase FIR digital filters are discussed. The general expressions of symmetric, anti-symmetric and hybrid-symmtric frequency response modes in the multi-dimensional cases are obtained. The results show, the total number of the frequency response modes in the K-dimensional spaces to be 2^{2K}, in which the number of basic modes is $(K+1)^2$, while the other modes may be abtained from the basic modes by the substitutions of the frequency coordinates.

1. INTRODUCTION

Linear phase FIR digital filters are widely applied in the signal processing techniques, for the one-dimensional case, reference (2) has discussed the frequency response of filters and found four symmetric and anti-symmetric basic frequency response modes. With the development of digital signal processing techniques, the analysis and design of the high-dimensional digital filters have gradually attracted attention. The purpose of this paper is in the multi-dimensional cases to discuss the frequency response modes of linear phase FIR digital filters, to find the mode number in the different dimensions and the general expressions of the frequency response.

2. BASIC NOTIONS

Let $h(n_1,\cdots,n_K)$ be the K-dimensional orthogonal sequences within the region $0 \leqslant n_i \leqslant N_i-1$ $(i=1,2,\cdots,K)$, its Z-transform may

D. G. Lainiotis and N. S. Tzannes (eds.), Advances in Communications, 183–191.

be written as

$$H(z_1,\cdots,z_K)=\sum_{n_1=0}^{N_1-1}\cdots\sum_{n_K=0}^{N_K-1} h(n_1,\cdots,n_K)\prod_{i=1}^{K} z_i^{-n_i} \tag{1}$$

the relative Fourier transform is

$$H(e^{jw_1},\cdots,e^{jw_K})=\sum_{n_1=0}^{N-1}\cdots\sum_{n_K=0}^{N-1} h(n_1,\cdots,n_K)\exp(-j\sum_{i=1}^{K} n_i w_i) \tag{2}$$

this is a function of which the period is 2 in the every dimension of the K-dimensional spaces, i.e. we have

$$H(e^{jw_1},\cdots,e^{jw_K})=H(e^{j(w_1+2\pi m_1)},\cdots,e^{j(w_K+2\pi m_K)}) \tag{3}$$

here $m_1,\cdots,m_K$ are integers. Formula (2) may also be rewritten as

$$H(e^{jw_1},\cdots,e^{jw_K})=|H(e^{jw_1},\cdots,e^{jw_K})|\, e^{j\theta(w_1,\cdots,w_K)} \tag{4}$$

where $|H(e^{jw_1},\cdots,e^{jw_K})|$ and $e^{j\theta(w_1,\cdots,w_K)}$ are the module and phase of H respectively.

In the linear phase case we have

$$\theta(w_1,\cdots,w_K)=-(d_1w_1+\cdots+d_Kw_K)=-\sum_{i=1}^{K} d_i w_i \tag{5}$$

where d is the group delay factor in the i-th dimension. Now the formula (2) can be rewritten in the following form

$$H(e^{jw_1},\cdots,e^{jw_K})=|H(e^{jw_1},\cdots,e^{jw_K})|\exp(-j\sum_{i=1}^{K} d_i w_i) \tag{6}$$

from formula (2) and (6) we find

$$\tan(\sum_{i=1}^{K} d_i w_i)=\frac{\sum_{n_1=0}^{N_1-1}\cdots\sum_{n_K=0}^{N_K-1} h(n_1,\cdots,n_K)\sin(\sum_{i=1}^{K} n_i w_i)}{\sum_{n_1=0}^{N_1-1}\cdots\sum_{n_K=0}^{N_K-1} h(n_1,\cdots,n_K)\cos(\sum_{i=1}^{K} n_i w_i)}$$

$$=\frac{\sum_{n=0}^{N=1}\cdots\sum_{n=0}^{N-1} h(n_1,\cdots,n_K)\sin(\sum_{i=1}^{K} n_i w_i)}{h(0,\cdots,0)+\sum_{n_1=0}^{N_1-1}\cdots\sum_{n_K=1}^{N_K-1} h(n_1,\cdots,n_K)\cos(\sum_{i=1}^{K} n_i w_i)} \tag{7}$$

There exist two possible solutions to Eq.(7), the first possibility is that

$$\tan\left(\sum_{i=1}^{K} d_i w_i\right)=0 \tag{8}$$

and from Eq.(7) it has been found

$$h(n_1,\cdots,n_K)\begin{cases}\neq 0, & n_1=\cdots=n_K=0\\ =0, & \text{elsewhere}\end{cases} \tag{9}$$

in this case, there is only one impulse, i.e. h(o, ,o), therefore, it has not general significance; the other possible case is when

$$\tan\left(\sum_{i=1}^{K} d_i w_i\right)\neq 0 \tag{10}$$

from Eq.(7) we can find

$$\sum_{n_1=0}^{N_1-1}\cdots\sum_{n_K=0}^{N_K-1} h(n_1,\cdots,n_K)\sin\left\{\sum_{i=1}^{K}(d_i-n_i)w_i\right\}=0 \tag{11}$$

this result has been obtained from the linear phase condition of Eq.(5). In the general case, the linear phase condition may be rewritten as

$$\theta(w_1,\cdots,w_K)=\sum_{i=1}^{K}(\beta_i-d_i w_i) \tag{5)a$$

where β_i is the phase delay factor in the i-th dimension. hence Eq. (11) becomes

$$\sum_{n_1=0}^{N_1-1}\cdots\sum_{n_K=0}^{N_K-1} h(n_1,\cdots,n_K)\sin\left\{\sum_{i=1}^{K}\left((d_i-n_i)w_i-\beta_i\right)\right\}=0 \tag{12}$$

Similarly, Eq.(2) becomes

$$H(e^{jw_1},\cdots,e^{jw_K})=\sum_{n_1=0}^{N_1-1}\cdots\sum_{n_K=0}^{N_K-1} h(n_1,\cdots,n_K)\exp\left[-j\sum_{i=1}^{K}(n_i w_i-\beta_i)\right] \tag{6)a$$

Eq.(12) may be solved when β_i are the integral times of $\frac{\pi}{2}$, so that we can let

$$\beta_i=R_i(\pi/2) \qquad (i=1,2,\cdots,K) \tag{13}$$

here R_i are integers. when R_i is even, the $h(n_1,\cdots,n_K)$ that satisfy Eq.(12) are symmetric around $(N_i-1)/2$, and when R_i is odd, they are anti-symmetric. To simplify the description, we will add one small point over R_i or N_i to indicate they are odd, and two points over them when they are even,soch as $\dot{R}_2$ indicates R_2 is odd, $\ddot{N}_4$ indicates N_4 is even etc.. In the following we will be further discusing from the two-dimensional case, then find the general expressions of the frequency response.

3. TWO-DIMENSIONAL FREQUENCY RESPONSE

For the two-dimensional case, Eq.(12) becomes

$$\sum_{n_1=0}^{N_1-1} \sum_{n_2=0}^{N_2-1} h(n_1,n_2)\sin\left[(d_1-n_1)w_1+(d_2-n_2)w_2-(\beta_1+\beta_2)\right]=0 \qquad (12)_2$$

from Eq.(13) we have $\beta_1=R_1(\pi/2)$, $\beta_2=R_2(\pi/2)$, here R_1,R_2 may be odd or even respectively, thus, there are four possible different combinations:

(1) $\ddot{R}_1,\ddot{R}_2$; (2) $\dot{R}_1,\dot{R}_2$; (3) $\dot{R}_1,\ddot{R}_2$; (4) $\ddot{R}_1,\dot{R}_2$.

In the cases (1) and (2) $\beta_1+\beta_2$ is the integral times of π, therefore Eq.(12) may be rewritten as

$$\sum_{n_1=0}^{N_1-1} \sum_{n_2=0}^{N_2-1} h(n_1,n_2)\sin\left[(d_1-n_1)w_1+(d_2-n_2)w_2\right]=0 \qquad (12)_{2a}$$

The differences between the cases(3) and (4) are only in the interchange of w_1,w_2 and n_1,n_2 respectively. therefore, they may be considered in the same case. For instance, we need only discu cuss the case (3). Now Eq.(12) becomes

$$\sum_{n_1=0}^{N_1-1} \sum_{n_1=0}^{N_2-1} h(n_1,n_2)\cos\left[(d_1-n_1)w_1+(d_2-n_2)w_2\right]=0 \qquad (12)_{2b}$$

Corresponding to the three basic cases (1),(2) and (3) as shonw above, respectively, we can get three group basic conditions that satisfy Eq.(12), they are

(1) $d_1=(N_1-1)/2$
$d_2=(N_2-1)/2$
$h(n_1,n_2)=h(n_1,N_2-1-n_2)$
$=h(N_1-1-n_1,n_2)$ (14)a

(2) $d_1=(N_1-1)/2$
$d_2=(N_2-1)/2$
$h(n_1,n_2)=-h(n_1,N_2-1-n_2)$
$=-h(N_1-1-n_1,n_2)$ (14)b

(3) $d_1=(N_1-1)/2$
$d_2=(N_2-1)/2$
$h(n_1,n_2)=h(n_1,N_2-1-n_2)$
$=-h(N_1-1-n_1,n_2)$ (14)c

Similarly, when N and N are odd or even respectively, there are four different combinations,too,in which three combinations are basic, therefore, in the two-dimensional case, there are nine different basic solutions to Eq,(12) .

For example, in the following we shall discussing the case of $\ddot{R}_1$, $\ddot{R}_2$, $\dot{N}_1$, $\dot{N}_2$. Now Eq.(6)a becomes

$$H(e^{jw_1},e^{jw_2})=\sum_{n_1=0}^{N_1-1}\sum_{n_2=0}^{N_2-1} h(n_1,n_2)e^{-j[(n_1w_1+n_2w_2)-(\beta_1+\beta_2)]}$$

$$=(A_{11}+A_{21}+A_{10}+A_{01}+A_{00}+A_{02}+A_{20}+A_{12}+A_{22})e^{j(\beta_1+\beta_2)} \tag{15}$$

here

$$A_{11}=\sum_{n_1=0}^{(N_1-3)/2}\sum_{n_2=0}^{(N_2-3)/2} h(n_1,n_2)e^{-j(n_1w_1+n_2w_2)} \tag{16}$$

$$A_{21}=\sum_{n_1=(N_1+1)/2}^{N_1-1}\sum_{n_2=0}^{(N_2-3)/2} h(n_1,n_2)e^{-j(n_1w_1+n_2w_2)} \tag{16}$$

$$A_{10}=\sum_{n_1=0}^{(N_1-3)/2} h(n_1,\frac{N_2-1}{2})e^{-j[n_1w_1+(\frac{N_2-1}{2})w_2]} \tag{16}$$

$$A_{01}=\sum_{n_2=0}^{(N_2-3)/2} h(\frac{N_1-1}{2},n_2)e^{-j[(\frac{N_1-1}{2})w_1+n_2w_2]} \tag{16}$$

$$A_{00}=h(\frac{N_1-1}{2},\frac{N_2-1}{2})e^{-j[(\frac{N_1-1}{2})w_1+(\frac{N_2-1}{2})w_2]} \tag{16}$$

$$A_{02}=\sum_{n_2=(N_2+1)/2}^{N_2-1} h(\frac{N_1-1}{2},n_2)e^{-j[(\frac{N_1-1}{2})w_1+n_2w_2]} \tag{16}$$

$$A_{20}=\sum_{n_1=(N_1+1)/2}^{N_1-1} h(n_1,\frac{N_2-1}{2})e^{-j[n_1w_1+(\frac{N_2-1}{2})w_2]} \tag{16}$$

$$A_{12}=\sum_{n_1=0}^{(N_1-1)/2}\sum_{n_2=(N_2+1)/2}^{N_2-1} h(n_1,n_2)e^{-j(n_1w_1+n_2w_2)} \tag{16}$$

$$A_{22}=\sum_{n_1=(N_1+1)/2}^{N_1-1}\sum_{n_2=(N_2+1)/2}^{N_2-1} h(n_1,n_2)\ {}^{-j(n_1w_1+n_2w_2)} \tag{16}$$

Take n_1 and n_2 as the horizontal and vertical coordinates respectively, the distribution of the numerical regions of Eq. (15) may be represented by figure. There is a cross region through the center

$$h(d_1,d_2)=h(\frac{N_1-1}{2},\frac{N_2-1}{2}) \tag{17}$$

in which on the horizontal and vertical directions are only one

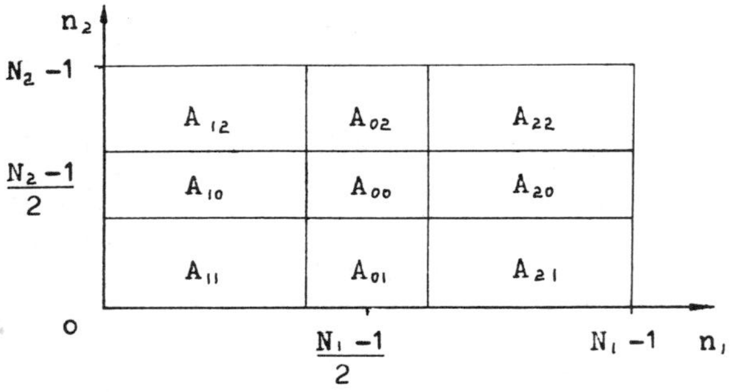

Fig. 2-D numerical region distribution for $\dot{N}_1, \dot{N}_2$.

row and one column (the center row and column) data, they are divided in five regions, namely $A_{10}, A_{01}, A_{00}, A_{02}, A_{20}$. In the four corners are four rectangular regions $A_{11}, A_{21}, A_{12}, A_{22}$.

Making suitable substitutions, after calculation, we find the actual expressions of Eq.(15) to be

$$H(e^{jw_1}, e^{jw_2}) = e^{-j\theta} \left\{ \sum_{n_1=0}^{\frac{N_1-1}{2}} \sum_{n_2=0}^{\frac{N_2-1}{2}} b(n_1, n_2) \cos(n_1 w_1) \cos(n_2 w_2) \right\} \quad (15)a$$

where

$$\theta = \left(\frac{N_1-1}{2}\right) w_1 + \left(\frac{N_2-1}{2}\right) w_2 - (\beta_1 + \beta_2) \quad (18)$$

is the phase angle. The term within the bracket of Eq.(15)a is the amplitude-frequency response for this case, which shows the cosine type distributions in the both w_1 and w_2 frequency directions. This is an all-symmetric mode.

Similarly, we can find all the nine basic frequency response modes of the two-dimensional digital filters. To simplify the description, let H represent $H(e^{jw_1}, e^{jw_2})$, thus, we have

(1) $\ddot{R}_1, \ddot{R}_2, \dot{N}_1, \dot{N}_2$:

$$H_1 = e^{-j\theta} \sum_{n_1=0}^{\frac{N_1-1}{2}} \sum_{n_2=0}^{\frac{N_2-1}{2}} b(n_1, n_2) \cos(n_1 w_1) \cos(n_2 w_2) \quad (19)$$

(2) $\ddot{R}_1, \ddot{R}_2, \ddot{N}_1, \ddot{N}_2$:

$$H_2 = e^{-j\theta} \sum_{n_1=1}^{\frac{N_1}{2}} \sum_{n_2=1}^{\frac{N_2}{2}} b(n_1, n_2) \cos\left[\left(n_1 - \frac{1}{2}\right) w_1\right] \cos\left[\left(n_2 - \frac{1}{2}\right) w_2\right] \quad (19)$$

(3) $\ddot{R}_1$, $\ddot{R}_2$, $\dot{N}_1$, $\ddot{N}_2$:

$$H_3 = e^{-j\theta} \sum_{n_1=0}^{\frac{N_1-1}{2}} \sum_{n_2=1}^{\frac{N_2}{2}} b(n_1,n_2)\cos(n_1 w_1)\cos\left[(n_2-\tfrac{1}{2})w_2\right] \tag{19}$$

(4) $\dot{R}_1$, $\dot{R}_2$, $\dot{N}_1$, $\dot{N}_2$:

$$H_4 = e^{-j\theta} \sum_{n_1=0}^{\frac{N_1-1}{2}} \sum_{n_2=0}^{\frac{N_2-1}{2}} b(n_1,n_2)\sin(n_1 w_1)\sin(n_2 w_2) \tag{19}$$

(5) $\dot{R}_1$, $\dot{R}_2$, $\ddot{N}_1$, $\ddot{N}_2$:

$$H_5 = e^{-j\theta} \sum_{n_1=1}^{\frac{N_1}{2}} \sum_{n_2=1}^{\frac{N_2}{2}} b(n_1,n_2)\sin\left[(n_1-\tfrac{1}{2})w_1\right]\sin\left[(n_2-\tfrac{1}{2})w_2\right] \tag{19}$$

(6) $\dot{R}_1$, $\dot{R}_2$, $\dot{N}_1$, $\ddot{N}_2$:

$$H_6 = e^{-j\theta} \sum_{n_1=0}^{\frac{N_1-1}{2}} \sum_{n_2=1}^{\frac{N_2}{2}} b(n_1,n_2)\sin(n_1 w_1)\sin\left[(n_2-\tfrac{1}{2})w_2\right] \tag{19}$$

(7) $\dot{R}_1$, $\ddot{R}_2$, $\dot{N}_1$, $\dot{N}_2$:

$$H_7 = e^{-j\theta} \sum_{n_1=0}^{\frac{N_1-1}{2}} \sum_{n_2=0}^{\frac{N_2-1}{2}} b(n_1,n_2)\sin(n_1 w_1)\cos(n_2 w_2) \tag{19}$$

(8) $\dot{R}_1$, $\ddot{R}_2$, $\ddot{N}_1$, $\ddot{N}_2$:

$$H_8 = e^{-j\theta} \sum_{n_1=1}^{\frac{N_1}{2}} \sum_{n_2=1}^{\frac{N_2}{2}} b(n_1,n_2)\sin\left[(n_1-\tfrac{1}{2})w_1\right]\cos\left[(n_2-\tfrac{1}{2})w_2\right] \tag{19}$$

(9) $\dot{R}_1$, $\ddot{R}_2$, $\dot{N}_1$, $\ddot{N}_2$:

$$H_9 = e^{-j\theta} \sum_{n_1=0}^{\frac{N_1-1}{2}} \sum_{n_2=0}^{\frac{N_2}{2}} b(n_1,n_2)\sin(n_1 w_1)\cos\left[(n_2-\tfrac{1}{2})w_2\right] \tag{19}$$

where Eq.(19) corresponding to previous Eq.(15)a; H_1, H_2, H_3, are all-symmetric modes, H_4, H_5, H_6, are anti-symmetric modes and H_7, H_8, H_9, are hybrid-symmetric modes. Clearly, according to four parameters R_1, R_2, N_1, N_2, are even or odd respectively, there are $2^4=16$ possible different odd-even combinations, which are equivalent to 16 frequency response modes, besides above 9 basic modes, the other 7 modes may be easy to obtained by suitable substitutions of the frequency coordinates.

4. FREQUENCY RESPONSES OF MULTI-DIMENSIONAL SPACES

Similar to the two-dimensional case, that for the multi-dimensional cases of K>2, according to 2K parameters $R_1,\cdots,R_K$ and $N_1,\cdots,N_K$ are odd or even respectively, there are $2^{2K}=4^K$ different odd-even combinations, which are equivalent to the same number frequency response modes. Using similar methods, we can obtain the expressions of their frequency responses. For example, when R_i are all even and N_i are all odd, the expresions are

$$H(e^{jw_1},\cdots,e^{jw_K})=e^{-j\theta_K}\sum_{n_1=o}^{\frac{N_1-1}{2}}\cdots\sum_{n_K=o}^{\frac{N_K-1}{2}} b(n_1,\cdots,n_K)\prod_{i=1}^{K}\cos(n_i w_i) \tag{20}$$

where

$$\theta_K=\sum_{i=1}^{K}\left(\frac{N_i-1}{2}\right)w_i-\sum_{i=1}^{K}\beta_i \tag{21}$$

This is an all-symmetric mode in the K-dimenaional space, the frequency response distributions are all cosine types for the every dimension.

To find the general expressions of the frequency response in the K-dimensional space, introdusing the odd-even functions $\dot{A}(x)$ and $\ddot{A}(x)$ (where x is the integer), their definition are

$$\dot{A}(x)=\begin{cases}1, & \text{for } \dot{x}\\ 0, & \text{for } \ddot{x}\end{cases} \tag{22)a}$$

$$\ddot{A}(x)=\begin{cases}1, & \text{for } \ddot{x}\\ 0, & \text{for } \dot{x}\end{cases} \tag{22)b}$$

Thus, the general expression of (6)a is

$$H=H(e^{jw_1},\cdots,e^{jw_K})$$

$$=e^{-j\theta_K}\sum_{n_1=\ddot{A}(N_1)}^{\frac{N_1-\dot{A}(N_1)}{2}}\cdots\sum_{n_K=\ddot{A}(N_K)}^{\frac{N_K-\dot{A}(N_K)}{2}} b(n_1,\cdots,n_K)\times$$

$$\times\prod_{i=1}^{K}\left\{\dot{A}(R_i)\sin\left[\left(n_i-\frac{\ddot{A}(N_i)}{2}\right)w_i\right]+\ddot{A}(R_i)\cos\left[\left(n_i-\frac{\ddot{A}(N_i)}{2}\right)w_i\right]\right\} \tag{23}$$

By means of the above general expression, for the different odd-even combinations we may obtain 4^K expressions of the frequency response, which in each dimension may take the forms of multi-term-sum of sine or cosine type respectively.

Clearly, from whether the $N_1,\cdots,N_K$ and $R_1,\cdots,R_K$ total 2K numbers are odd or even respectively, we can make $2^{2K}=4^K$ different odd-even combinations, therefore, the total mode number of the frequency responses of the K-dimensional linear phase FIR digital filter is also 4^K. On the other consideration, if the orders between the various frequency coordinates are neglected, for K numbers $N_1,\cdots,N_K$ or $R_1,\cdots,R_K$, there are $(K+1)^2$ different basic odd-even combinations, so that for the K-dimensional linear phase FIR digital filters, there are $(K+1)^2$ basic frequency response modes. When $K>1$, $4^K>(K+1)^2$, the other $4^K-(K+1)^2$ modes may be obtained from the $(K+1)^2$ basic modes by the substitution of the frequency coordinates.

5. CONCLUSION

From the above discussions we get the following results:

(1) The frequency response of multi-dimensional linear phase digital filters in each dimensions exist two possible modes: the symmetric mode (when the phase delay factor β is the even times of $\pi/2$) and the anti-symmetric mode (when the phase delay factor is the odd times of $\pi/2$).

(2) The frequency response of the K-dimensional linear phase FIR digital filters may be certainly expressed by the multiple term summation of the K-multiple productions of the sine or cosine functions.

(3) The total number of the frequency response modes of the K-dimensional linear phase FIR digital filters is $2^{2K}=4^K$, in which the number of basic modes is $(K+1)^2$, and the other modes may be obtained from the basic modes by the substitutions of the frequency coordinates.

REFERENCES

(1) L.R. Rabiner and R.W. Schafer: 1971, IEEE Trans. on Audio, AU-19, No.2, pp 200-207.
(2) L.R. Rabiner and B. Gold: Theory and Application of Digital Signal Processing. New Jersey, 1975.
(3) H.D. Helms: 1968, IEEE Trans. on Audio, AU-16, No.3, pp 336-342.

MULTI-DIMENSIONAL DIGITAL WINDOWS

N.C. Hu

Shanghai University of Technology, CHINA

The Digital Window Technique is one of the fundamental methods of digital filter design. This paper extends the principal properties of digital windows from one-dimensional case to multi-dimensional spaces; the basic properties of rectangular symmetric windows in the multi-dimensional spaces and the basic modes of multi-dimensional digital filters are discussed.

1. INTRODUCTION

In company with the development of digital integrated circuits, the one-dimensional digital filters have been widly employed in the computer signal processing techniques. In recent years, for two-dimensional case, some articles have been published[(1),(2)],but for higher dimensions references are lacking and the problem desires further study.

For low (one- or two-) dimensional digital filter designs, several methods are currently available, such as window design, frequency sampling design and optimal (minimax error) design, in the case of windows we also have different types. This paper deals with the extension of some properties of low-dimensional windows to high-dimensional spaces, the basic properties of rectangular symmetric windows in the multi-dimensional spaces and the basic modes of digital filters are discussed.

2. BASIC NOTIONS

For the linear time-invariant K-dimensional systems, if the

D. G. Lainiotis and N. S. Tzannes (eds.), Advances in Communications, 192–199.

input signal is sinusoidal

$$X(n_1,\cdots,n_K)=\prod_{i=1}^{K} e^{jw_i n_i} \qquad (-\infty\leqslant n_i\leqslant\infty) \tag{1}$$

then the output will be

$$\begin{aligned} Y(n_1,\cdots,n_K) &= \sum_{m_1=-\infty}^{+\infty}\cdots\sum_{m_K=-\infty}^{+\infty} h(m_1,\cdots,m_K)\cdot\prod_{i=1}^{K} e^{jw_i(n_i-m_i)} \\ &= \prod_{i=1}^{K} e^{jw_i n_i}\sum_{m_1=-\infty}^{+\infty}\cdots\sum_{m_K=-\infty}^{+\infty} h(m_1,\cdots,m_K)\prod_{i=1}^{K} e^{-jw_i n_i} \\ &= X(n_1,\cdots,n_K)\cdot H(e^{jw_1},\cdots,e^{jw_K}) \end{aligned} \tag{2}$$

in which the frequency response

$$H(e^{jw_1},\cdots,e^{jw_K})=\sum_{m_1=-\infty}^{+\infty}\cdots\sum_{m_K=-\infty}^{+\infty} h(m_1,\cdots,m_K)\prod_{i=1}^{K} e^{-jw_i m_i} \tag{3}$$

is K-multiple Fourier transform, their coefficients are

$$h(n_1,\cdots,n_K)=\frac{1}{(2\pi)^K}\underbrace{\int_{-\pi}^{\pi}\cdots\int_{-\pi}^{\pi}}_{K\text{-multiple}} H(e^{jw_1},\cdots,e^{jw_K})\prod_{i=1}^{K} e^{jw_i n_i}\,dw \tag{4}$$

Clearly $H(e^{jw_1},\cdots,e^{jw_K})$ is a K-multiple periodic function and

$$H(e^{jw_1},\cdots,e^{jw_K})=H(e^{j(w_1+q_1\pi)},\cdots,e^{j(w_K+q_K\pi)}) \qquad (q_1,q_2,\cdots,q_K=0,\pm2,\pm4,\cdots) \tag{5}$$

In formula (3), using the K-dimensional window (weighed function) $g(n_1,\cdots,n_K)$ may improve the convergence of the series and convenient to that approximated by the summation of finit terms, in which case (3) may be rewritten as

$$\bar{H}(e^{jw_1},\cdots,e^{jw_K})=\sum_{n_1=-\infty}^{+\infty}\cdots\sum_{n_K=-\infty}^{+\infty} h(n_1,\cdots,n_K)g(n_1,\cdots,n_K)\prod_{i=1}^{K} e^{-jw_i n_i} \tag{6}$$

the bar over H represents weighted by windows and the term coefficients of the series are

$$\bar{h}(n_1,\cdots,n_K)=h(n_1,\cdots,n_K)\cdot g(n_1,\cdots,n_K)$$

$$= \frac{1}{(2\pi)^K} \underbrace{\int_{-\pi}^{\pi} \cdots \int_{-\pi}^{\pi}}_{K\text{-multiple}} H(e^{jw_1}, \cdots, e^{jw_K}) \prod_{i=1}^{K} e^{jw_i n_i} dw \tag{7}$$

according to the convolution theorem, we therefore have

$$\bar{H}(e^{jw_1}, \cdots, e^{jw_K}) = H(e^{jw_1}, \cdots, e^{jw_K}) * G(e^{jw_1}, \cdots, e^{jw_K}) \tag{8}$$

in the equation

$$\bar{G}(e^{jw_1}, \cdots, e^{jw_K}) = \sum_{n_1=-\infty}^{+\infty} \cdots \sum_{n_K=-\infty}^{+\infty} g(n_1, \cdots, n_K) \prod_{i=1}^{K} e^{-jw_i n_i} \tag{9}$$

is the Fourier transform of $g(n_1, \cdots, n_K)$ i.e. the frequency response of the window.

Generally, the window $g(n_1, \cdots, n_K)$ must be approximated by the sum value of finite terms. Let $N_1, N_2, \cdots, N_K$ represent the term numbers of $g(n_1, \cdots, n_K)$ in the directions $n_1, \cdots, n_K$ respectively, then the formula (6) can be rewritten as

$$\hat{H}(e^{jw_1}, \cdots, e^{jw_K}) = \sum_{n_1=-\frac{1}{2}(N_1-1)}^{\frac{1}{2}(N_1-1)} \cdots \sum_{n_K=-\frac{1}{2}(N_K-1)}^{\frac{1}{2}(N_K-1)} h(n_1, \cdots, n_K) \prod_{i=1}^{K} e^{-jw_i n_i} \tag{10}$$

here "∧" over H represents weighted approximation. Hence we can rewrite formula (9) as

$$\hat{G}(e^{jw_1}, \cdots, e^{jw_K}) = \sum_{n_1=-\frac{1}{2}(N_1-1)}^{\frac{1}{2}(N_1-1)} \cdots \sum_{n_K=-\frac{1}{2}(N_K-1)}^{\frac{1}{2}(N_K-1)} g(n_1, \cdots, n_K) \prod_{i=1}^{K} e^{-jw_i n_i} \tag{11}$$

where $\hat{G}$ represents the frequency response that is weighted by finite term window. Since in practical applications we can only obtain the approximation of finite terms, hence we shall neglect the sign "∧" in our further discussions.

3. RECTANGULAR MULTI-DIMENSIONAL WINDOW

The general expressions of rectangular symmetric type windows in the K-dimensional spaces are

$$g_R(n_1, \cdots, n_K) = \begin{cases} 1, & -\frac{1}{2}(N_i - 1) \leqslant n_i \leqslant \frac{1}{2}(N_i - 1),\ i = 1, 2, \cdots, K \\ 0, & \text{elsewhere} \end{cases} \tag{12}$$

where $N_1, \cdots, N_K$ are integers, $g_R(n_1, \cdots, n_K)$ has the rectangular distribution in each dimension, the efficient region being a K-dimensional rectangular region of which the symmetric center is in the origine. From formula (11) we can find the correspanding frequency response to be

$$G_R(e^{jw_1},\cdots,e^{jw_K})=\sum_{n_1=-\frac{1}{2}(N_1-1)}^{\frac{1}{2}(N_1-1)}\cdots\sum_{n_K=-\frac{1}{2}(N_K-1)}^{\frac{1}{2}(N_K-1)}\left[\prod_{i=1}^{K}e^{-jw_i n_i}\right]$$

$$=\left[\sum_{n_1=-\frac{1}{2}(N_1-1)}^{\frac{1}{2}(N_1-1)}e^{-jw_1 n_1}\right]\times\cdots\times\left[\sum_{n_K=-\frac{1}{2}(N_K-1)}^{\frac{1}{2}(N_K-1)}e^{-jw_K n_K}\right]$$

$$=\prod_{i=1}^{K}\left[\sum_{n_i=-\frac{1}{2}(N_i-1)}^{\frac{1}{2}(N_i-1)}e^{-jw_i N_i}\right]$$

$$=\prod_{i=1}^{K}\left[\frac{e^{-jw_i\frac{1}{2}(N_i-1)}(1-e^{-jw_i N_i})}{1-e^{jw_i}}\right]$$

$$=\prod_{i=1}^{K}\left[\frac{\sin(\frac{1}{2}w_i N_i)}{\sin(\frac{1}{2}w_i)}\right] \qquad (13)$$

the above formula ahows the multi-dimensional frequency response in each dimension holds the properties similar to the one-dimensional case. Introdusing the normalized coefficient

$$p=(\prod_{i=1}^{K}N_i)^{-1} \qquad (14)$$

formula (13) becomes

$$\bar{G}_R(e^{jw_1},\cdots,e^{jw_K})=pG_R(e^{jw_1},\cdots,e^{jw_K})=\prod_{i=1}^{K}\left[\frac{\sin(\frac{1}{2}w_i N_i)}{N_i\sin(\frac{1}{2}w_i)}\right] \qquad (15)$$

When $w_1=w_2=\cdots=w_K=0$, i.e. at the origine of multi-dimensional space, we have

$$\bar{G}_{R0}=\bar{G}_R(e^0,\cdots,e^0)=1 \qquad (16)$$

here $\bar{G}_{R0}$ is called the main peak of the function $\bar{G}_R$.

If one of the K frequencies (e.g. w_i) is not equal to zero, and the other frequencies all equal to zero, we have

$$\bar{G}_{Ri}=\bar{G}_R(e^0,\cdots,e^0,e^{jw_i},e^0,\cdots,e^0)=\left[\frac{\sin(\frac{1}{2}w_i N_i)}{N_i\sin(\frac{1}{2}w_i)}\right] \qquad (17)$$

where $\bar{G}_{Ri}$ is called the main frequency response of i-th-dimensional frequency axis in the K-dimensional frequency space. Similar to the one-dimensional case, when we differentiate formula (17) respect to w_i and let the derivative equal to zero, we can find the necessary conditions of w_i where $\bar{G}_{Ri}$ will be extrema are

$$N_i \tan(\tfrac{1}{2} w_i) = \tan(\tfrac{1}{2} N_i w_i) \tag{18}$$

From the above equeation may be found the frequencies $w_{i\ell_i}$, in which the side lobe peaks are placed. The above equeation can generally be resolved by numerical methods. For the approximation, we may take

$$w_{i\ell_i} = (2\ell_i+1)\pi/N_i \tag{19}$$
$$(\ell_i = 1, 2, \cdots)$$

Setting (19) in to (17), we have

$$\bar{G}_{R_i} = (-1)^{\ell_i} / N_i \sin[(2\ell_i+1)\pi/2N_i] \tag{20}$$

when $\ell_i << N_i$, formula (19) may be simplify as

$$\bar{G}_{R_i} = (-1)^{\ell_i} / (\ell_i + \tfrac{1}{2}) \tag{21}$$

Similarly, when in the K frequencies where r frequencies are not equal to zero (not losing the generality, may let $w_1, \cdots, w_r$ be nonzero), and the all other frequencies are equal to zero, we have

$$\bar{G}_{R1\cdots r} = \prod_{i=1}^{r} \left[\frac{\sin(\frac{1}{2} N_i w_i)}{N_i \sin(\frac{1}{2} w_i)} \right] \tag{22}$$

where $\bar{G}_{R1\cdots r}$ is called the main frequency response of the r-dimensional sub frequency space that is formed by 1st, 2nd, ⋯, rth frequency axes in the K-dimensional frequency spaces. Because apart from the main peak, there is always

$$\frac{\sin(\frac{1}{2} N_i w_i)}{N_i \sin(\frac{1}{2} w_i)} < 1 \tag{23}$$

therefore side lobe amplitudes will be decreased when r is increased.

Differentiating expression (22) for $w_1, \cdots, w_r$ respectively, we can find

$$G_{R1\cdots r} = \prod_{i=1}^{r} \left[\frac{(-1)^{\ell_i}}{(\ell_i + \frac{1}{2})\pi} \right] \tag{24}$$
$$(\ell_i = 1, 2, \cdots)$$

Generally, when r=K, the peak amplitudes of all side lobes of the frequency response functions in the K-dimensional spaces can easily be obtained from formula (24). Let (24) have absolute value and take r=K, we have

$$|G_{R1\cdots r}| = \prod_{i=1}^{K} \left[\frac{1}{(\ell_i + \frac{1}{2})\pi} \right] = \frac{1}{\pi^K} \prod_{i=1}^{K} \left[\frac{1}{(\ell_i + \frac{1}{2})} \right] \tag{25}$$
$$(\ell_i = 1, 2, \cdots)$$

From (25) we can obtain the normalized peak amplitudes of all different order side lobes. For instance, the normalized peak amplitude of maximum side lobe (the one order first side lobe) is

$$|G_{R1}| = 2/3 = 0.212 \tag{26}$$

the attenuation relative to the main peak is

$$|G_{R1}|_{db} = 13.46db \tag{27}$$

4. PROPERTY IMPROVEMENT OF MULTI-DIMENSIONAL WINDOWS

To increase the attenuation of side lobes and improve the properties of filters, inthe high dimensional cases we can also find the new multi-dimensional windows by essential improvement of rectangular windows. For example, similar to the one-dimensional case, we can define the K-dimensional Hamming window as

$$g_H(n_1,\cdots,n_K) = \begin{cases} \prod_{i=1}^{K} \left[d_i + (1-d_i)\cos\left(\frac{2\pi n_i}{N}\right)\right], & -\frac{N_i-1}{2} \leq n_i \leq \frac{N_i-1}{2} \\ 0, \text{ elsewhere} \end{cases} \tag{28}$$

To find the frequency response function, it is first to prove the follwing theorem.

Theorem: If $x(t_1,\cdots,t_K)$ and $h(t_1,\cdots,t_K)$ are separable for $t_1,\cdots,t_K$, i.e. we can write

$$x(t_1,\cdots,t_K) = \prod_{i=1}^{K} x_i(t_i) \tag{29}$$

$$h(t_1,\cdots,t_K) = \prod_{i=1}^{K} h_i(t_i) \tag{30}$$

so that

$$x(t_1,\cdots,t_K) * h(t_1,\cdots,t_K) = \prod_{i=1}^{K} \left[x_i(t_i) * h_i(t_i)\right] \tag{31}$$

Proof:

$$\begin{aligned} x(t_1,\cdots,t_K) * h(t_1,\cdots,t_K) &= \int_{-\infty}^{+\infty}\cdots\int x(t_1,\cdots,t_K)h(t_1-\tau_1,\cdots,t_K-\tau_K)d\tau_1\cdots d\tau_K \\ &= \int_{-\infty}^{+\infty}\cdots\int \prod_{i=1}^{K} x_i(t_i)h_i(t_i-\tau_i)d\tau_i \\ &= \prod_{i=1}^{K} \int_{-\infty}^{+\infty} x_i(t_i)h_i(t_i-\tau_i)d\tau_i \\ &= \prod_{i=1}^{K} \left[x_i(t_i) * h_i(t_i)\right] \end{aligned}$$

the theorem is proved.

Use the similar method and the relationship of formula (26)

may be obtained the frequency response of the window (28) is

$$G_H(e^{jw_1},\cdots,e^{jw_K}) = \prod_{i=1}^{K}\left\{d_i\frac{\sin(\frac{1}{2}w_i n_i)}{\sin(\frac{1}{2}w_i)}+\frac{1}{2}(1-d_i)\frac{\sin[\frac{1}{2}(w_i-2\pi/N_i)N_i]}{\sin[\frac{1}{2}(w_i-2\pi/N_i)]}\right.$$

$$\left.+\frac{1}{2}(1-d_i)\frac{\sin[\frac{1}{2}(w_i-2\pi/N_i)N_i]}{\sin[\frac{1}{2}(w_i-2\pi/N_i)]}\right\} \quad (32)$$

Now the attenuation of the maximum side lobe (the one order but not the first side lobe) relative to the main peak is increased to -43db, therefore, improving the filter properties, as well as effectively preventing the Gibbs phenomenon in the frequency response.

5. BASIC MODES OF MULTI-DIMENSIONAL DIGITAL FILTERS

When the multi-dimensional window is found, we may be acorrding to the desirable filter properties (e.g. cut-off frequency,N , etc.) to design suitable multi-dimensional filters. In the one-dimensional case, there are four basic types of the filter, namely the low-pass, high-pass, band-pass and band-stop; in the multi-dimensional cases, generally, the filter can choose each one of the four basic types in each dimension, therefore there are 4 distinct combined filter modes. But the majority of these modes will be not have important practical significance. For example, in the two-dimensional case,the combined mode that in one dimension is band-stop and the other dimension is band-pass will have not significance. In general signal processing, the iso-type (i.e. where the filter type is the same in each dimension) combined filter modes have more practical important.

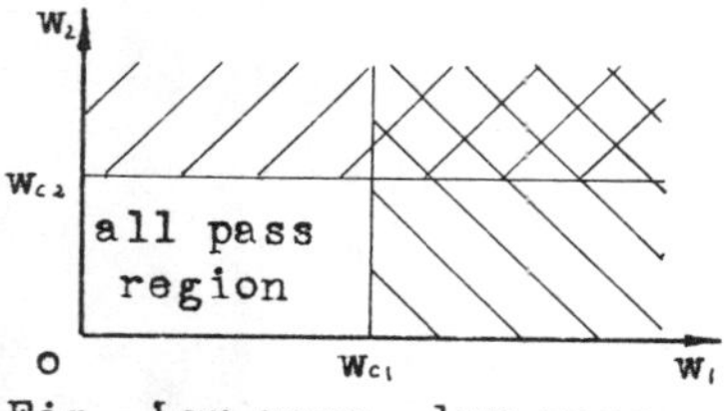

Fig. Low-pass ~ low-pass.

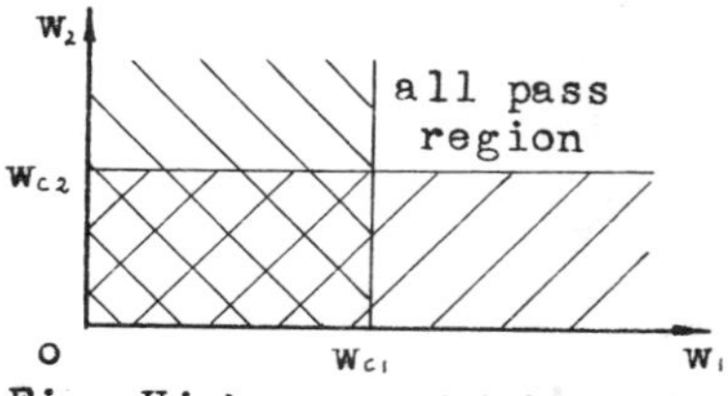

Fig. High-pass ~ high-pass.

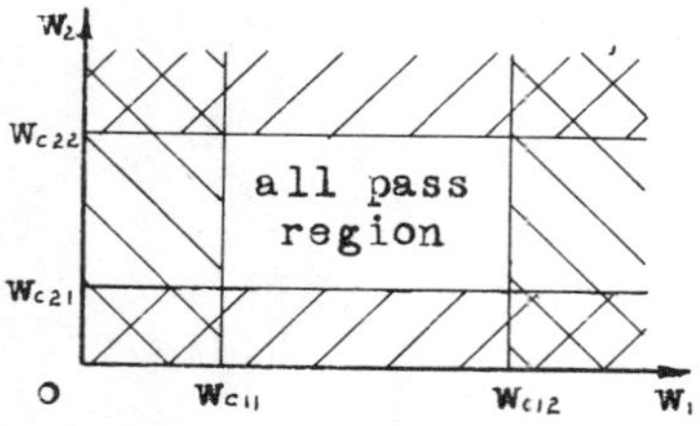

Fig. Band-pass ~ band-pass.

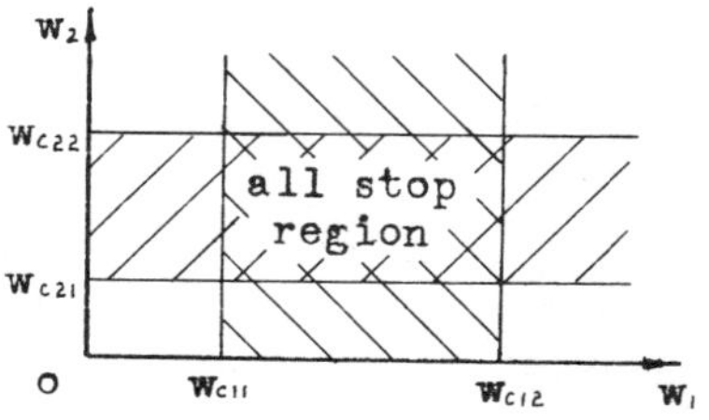

Fig. Band-stop ~ band-stop.

Four Figures show the graphical description of the pass region, partial pass region and stop region of four iso-type combined fel-ter modes in the two-dimensional case, In these figures only the positive frequency region are represented. Clearly, for the three-dimensional cases, the pass region, partial pass region and stop region of the combined filter modes must be described by the space region. For higher dimension cases, the filter regions will can not be described by these geometric methods.

Generally, for the K-dimensional combined filter modes, except the all pass region and all stop region, there are K-2 kinds of partial pass regions of where the stop orders are different. In these cases, the side lobe attenuations are proportionate to the number of the stop order.

6. CONCLUSION

We have discussed the fundamental properties of multi-dimensional windows and basic modes of the multi-dimensional digital filters. Similar to the one-dimensional case, the multi-dimensional digital filters also may be designed by frequency sampling techniques, optimal (minimax error) techniques and other methods, the properties of which shall be discussed in the other papers.

REFERENCES

(1) T.S. Huang: 1972, IEEE Trans. on Audio and Electroacoustics, AU-20, No.1, pp 88-89.
(2) T.V. Hu and L.R. Rabiner: 1972, IEEE Trans. on Audio and Electroacoustics, AU-20, No.4, pp 249-257.
(3) L.R. Rabiner and B. Gold: Theory and Application of Digital Signal Processing. New Jersey, 1975.

SECTION II

COMMUNICATIONS:

THEORY

AND

APPLICATIONS

CYCLO-ERGODICITY, PART I: APPLICATION TO SPECTRAL EXTRACTION

W.A. Gardner

University of California, Davis

We consider the problem of estimating the spectral density of an amplitude-modulated random process in the presence of an additive "noise" or "interference" process, and we propose several methods for obtaining consistent estimates. Consistent estimation of almost periodically time-varying probabilistic parameters such as mean and autocovariance is discussed also.

I. THE SPECTRAL EXTRACTION PROBLEM

Consider a zero-mean continuous-parameter process $X = \{X_t : t \in R\}$ that is composed of the sum of a sinusoid that is amplitude modulated by a stationary process W, and an "interference" or "noise" process Z that is uncorrelated with W and is not necessarily stationary:

$$X_t = Y_t + Z_t, \qquad Y_t = W_t \cos(\alpha_0 t + \phi).$$

We want to show that if α_0 is known and Z does not "contain the factor" $\cos(\alpha_0 t + \theta)$ (a precise statement of this condition is given in Section II), then the spectral densities of W and Y can be extracted from one sample path of X, regardless of Z.

The symmetrized autocorrelation function for X is

$$K_X(t,\tau) = E\{X_{t+\tau/2} X_{t-\tau/2}\} = \frac{1}{2} K_W(\tau)\cos(\alpha_0 \tau) + \frac{1}{2} K_W(\tau)\cos(2\alpha_0 t + 2\phi) + K_Z(t,\tau).$$

D. G. Lainiotis and N. S. Tzannes (eds.), Advances in Communications, 203–209.

The time-varying spectral density for X is (using $f_o \triangleq \alpha_o/2\pi$)

$$S_X(t,f) \triangleq \int_{-\infty}^{\infty} K_X(t,\tau)e^{-i2\pi f\tau}d\tau$$

$$= \frac{1}{4}S_W(f+f_o) + \frac{1}{4}S_W(f-f_o) + \frac{1}{2}S_W(f)\cos(2\alpha_o t+2\phi) + S_Z(t,f).$$

Use of conventional spectral estimation methods on X (e.g., bandpass-filter-squarer-lowpass-filter, or averaged or smoothed periodograms) yield estimates of the time-averaged spectral density [1], [2], defined by

$$\overline{S}_X(f) \triangleq \lim_{T\to\infty} \frac{1}{T}\int_0^T S_X(t,f)dt = \overline{S}_Y(f) + \overline{S}_Z(f)$$

$$= \frac{1}{4}S_W(f+f_o) + \frac{1}{4}S_W(f-f_o) + \overline{S}_Z(f).$$

If Z is a continuing process for which $\overline{S}_Z \neq 0$, and if Y and Z have spectral overlap (i.e., $\overline{S}_Y$ and $\overline{S}_Z$ do not have disjoint support), then although $\overline{S}_X$ can be estimated using conventional methods, estimates of $\overline{S}_Y$ and S_W cannot be obtained from the estimate of $\overline{S}_X$.

The purpose of Part I of our presentation is to show how novel time-varying spectral estimation methods can be used to obtain arbitrarily accurate estimates of S_W and $\overline{S}_Y$ (and $S_Y(t,f)$) from a sufficiently long record of X_t, regardless of spectral overlap between $\overline{S}_Y$ and $\overline{S}_Z$.

II. SOLUTION: GENERALIZED PERIODOGRAMS AND GENERALIZED SPECTRAL DENSITIES

We begin by observing that the time-averaged cross-spectral density of the frequency-shifted processes

$$X_t^+ \triangleq X_t e^{+i\alpha_o t} \qquad X_t^- \triangleq X_t e^{-i\alpha_o t} \tag{1}$$

which is defined by

$$\overline{S}_{X^+X^-}(f) \triangleq \lim_{T\to\infty} \frac{1}{T}\int_0^T S_{X^+X^-}(t,f)dt$$

$$S_{X^+X^-}(t,f) \triangleq \int_{-\infty}^{\infty} E\{X^+_{t+\tau/2}(X^-_{t-\tau/2})^*\}e^{-i2\pi f\tau}\, d\tau,$$

is given by

$$\bar{S}_{X^+X^-}(f) = \lim_{T\to\infty} \frac{1}{T}\int_0^T S_X(t,f)e^{-i2\alpha_o t}\,dt$$

$$= \frac{1}{4}S_W(f)e^{i2\phi} + \lim_{T\to\infty}\frac{1}{T}\int_0^T S_Z(t,f)e^{-i2\alpha_o t}\,dt. \qquad (2)$$

Assuming that $S_Z(t,f)$ does not contain a sinusoidally time-varying component with frequency $2\alpha_o$ (e.g., Z does not contain a component identical in form to Y), the second term in (2) is zero. Hence, conventional methods for cross-spectral estimation [3, chap. 9] will yield the desired estimate of S_W (as well as ϕ), provided that X possesses "appropriate" ergodic properties. For example, one can average a set of <u>co-periodograms</u> of X_t^+ and X_t^- defined by

$$P^{(T)}_{X^+X^-}(f) \triangleq \frac{1}{T}\tilde{X}_T^+(f)[\tilde{X}_T^-(f)]^*, \qquad \tilde{X}_T^{\pm}(f) \triangleq \int_{-T/2}^{T/2} X_t^{\pm}e^{-i2\pi ft}dt. \quad (3)$$

The bias of the estimate of $\bar{S}_{X^+X^-}$ obtained in this way $\to 0$ as $T\to\infty$, and the variance $\to 0$ as the number of co-periodograms averaged $\to\infty$, provided that X possesses an "appropriate" ergodic property.

Substitution of (1) into (3) yields an alternative interpretation of the co-periodogram of X_t^+, X_t^- as the α_o-<u>shift-periodogram</u> of X_t, which is defined by

$$P_X^{(T)}(f;\alpha_o) \triangleq \frac{1}{T}\tilde{X}_T(f-f_o)[\tilde{X}_T(f+f_o)]^* \equiv P^{(T)}_{X^+X^-}(f).$$

This reduces to the conventional periodogram for $\alpha_o = 0$. $P_X^{(T)}(f;\alpha_o)$ has been analyzed for the class of cyclostationary processes (e.g., Y herein) in [4, chap. VI].

In order to reveal what the "appropriate" ergodic property of X is, we consider the following alternative approach which yields an estimate of the α-<u>cyclic spectral density</u> of X, which is defined by

$$S_X^{\alpha}(t,f) \triangleq \lim_{T\to\infty}\frac{2}{T}\int_0^T S_X(u,f)\cos[\alpha(t-u)]du. \qquad (4)$$

For $\alpha = 0$, this reduces to the conventional time-averaged spectral density (except for a factor of 2). For the process X of interest here,

$$S_X^{2\alpha_o}(t,f) = \frac{1}{2}S_W(f)\cos(2\alpha_o t+2\phi). \qquad (5)$$

Therefore, we can indeed obtain an estimate of the desired S_W from an estimate of the $2\alpha_o$-cyclic spectral density of X. To obtain the latter estimate, we propose the α-cyclic spectrum analyzer diagrammed in Figure 1. For $\alpha=0$ this reduces to the conventional spectrum analyzer. For simplicity, the bandpass filters (BPFs) shown in Figure 1 are idealized; i.e., they have transfer functions defined by

$$H_{+}(\nu) = \begin{cases} 1/\sqrt{\Delta} & |\nu \pm (f+\alpha/4\pi)| \leq \Delta/2 \\ 0 & \text{otherwise,} \end{cases}$$

$$G(\nu) = \begin{cases} 1 & |\nu-\alpha/2\pi| \leq 1/T_{*} \leq \Delta/2 \\ 0 & \text{otherwise.} \end{cases}$$

If it is assumed that $S_X(t,f)$ has no sinusoidally time-varying components with frequencies β satisfying $|\alpha-\beta| \leq \Delta$ or $|2f-\beta| \leq \Delta$, except for $\beta=\alpha$, then it can be shown that (refer to Figure 1)

$$E\{S_t\} = \frac{1}{\Delta}\int_{-\Delta/2}^{\Delta/2} S_X^{\alpha}(t,f-\nu)d\nu. \tag{6}$$

(If the above hypothesis is not true, $E\{S_t\}$ will contain a β-cyclic spectral density term also.) Hence, we can use S_t as an estimate of $S_X^{\alpha}(t,f)$. For the process X of interest herein (with $\alpha=2\alpha_o$), (5) and (6) yield

$$E\{S_t\}= \frac{1}{\Delta}\int_{-\Delta/2}^{\Delta/2} S_W(f-\nu)d\nu[\cos(2\alpha_o t+2\phi)].$$

Hence, we can use S_t to obtain an estimate of S_W. The bias of this estimate $\to 0$ as $\Delta\to 0$. The variance $\to 0$ as $T_*\to\infty$ provided that X possesses an "appropriate" ergodic property, which is most easily characterized in terms of an ergodic property of the process V at the input to the output BPF shown in Figure 1. Specifically, the necessary and sufficient condition for variance $\{S_t\}\to 0$ as $T_*\to\infty$ is easily shown to be

$$E\{S_t\} = \frac{1}{\Delta}\int_{-\Delta/2}^{\Delta/2} S_W(f-\nu)d\nu[\cos(2\alpha_o t + 2\phi)]. \tag{7}$$

A process V satisfying (7) shall be said to be α-cyclo-ergodic in the mean. For $\alpha=0$, this reduces to the conventional property termed ergodicity in the mean. It can be shown that if X is a zero-mean Gaussian process V will be α-cyclo-ergodic in the mean for all α if $K_X(t,\tau) \to 0$ as $\tau\to\infty$.

III. CYCLE DETECTION AND FREQUENCY ESTIMATION

The key to the success of the spectral extraction methods proposed in Section II is knowledge of the value of the frequency $\alpha=2\alpha_o$. This may pose no problem in some applications, but may be crucial in others. Since the power in the output process S_t from the α-cyclic spectrum analyzer will $\to 0$ as $T_* \to \infty$ if and only if the α'-cyclic spectral density of X at frequency ν is zero for $|\alpha'-\alpha| \leq \Delta$ and $|2f-\alpha'| \leq \Delta$ for all ν in a neighborhood of f, then the power in S_t can be used to detect the presence of "cycles" in X, and to "tune" the frequency α to match an unknown or slowly varying frequency of a cycle in X (by maximizing the power in S_t).

When α is not perfectly tuned to the cycle frequency α' in X, it can be shown that although the variance $\{S_t\}$ decreases as T_* increases, the bias will increase. Hence, for a given frequency mismatch, there will exist an optimum integration time T_* that minimizes the mean squared error of the estimator for $S_X^{\alpha'}(t,f)$.

IV. ALMOST CYCLOSTATIONARY PROCESSES

When the only source of nonstationarity of a process X is cyclic phenomena, consisting of any number of cycles with arbitrary frequencies, the time-varying spectral density and other probabilistic parameters such as mean and autocovariance are almost periodic (AP) functions of the time-variable, t. Such processes are called almost cyclostationary (ACS), and are discussed at length in [5].

If an ACS process possesses "appropriate" α-cyclo-ergodic properties for every one of its frequencies , then in principle each cyclic component of an AP probabilistic parameter can be consistently estimated, and the estimates can be summed to provide a composite consistent estimate of the AP parameter. Therefore, although (as is well known) time-varying probabilistic parameters cannot, in general, be estimated from a single sample path of a process (i.e., without averaging over an ensemble of sample paths), if the time variation is AP, this well-known limitation does not apply. For example, the uncertainty principle for time-varying spectral densities [6] is violated by ACS processes possessing "appropriate" cyclo-ergodic properties.

Specifically, if X is ACS, then $S_X(t,f)$ has a Fourier series representation (assuming the series converges)

$$S_X(t,f) = \sum_{n=-\infty}^{\infty} S_X^{\alpha_n}(f) e^{i\alpha_n t} \tag{8}$$

$$S_X^{\alpha_n}(f) \triangleq \lim_{T\to\infty} \frac{1}{T} \int_0^T S_X(t,f) e^{-i\alpha_n t}\, dt.$$

Therefore,

$$S_X(t,f) = \overline{S}_X(f) + \sum_{n=1}^{\infty} S_X^{\alpha_n}(t,f) \tag{9}$$

$$S_X^{\alpha_n}(t,f) = 2|S_X^{\alpha_n}(f)| \cos(\alpha_n t + \arg\{S_X^{\alpha_n}(f)\}).$$

Hence, the α-shift periodogram can be used to estimate each of the terms in (8), or the α-cyclic spectrum analyzer can be used to estimate each of the terms in (9).

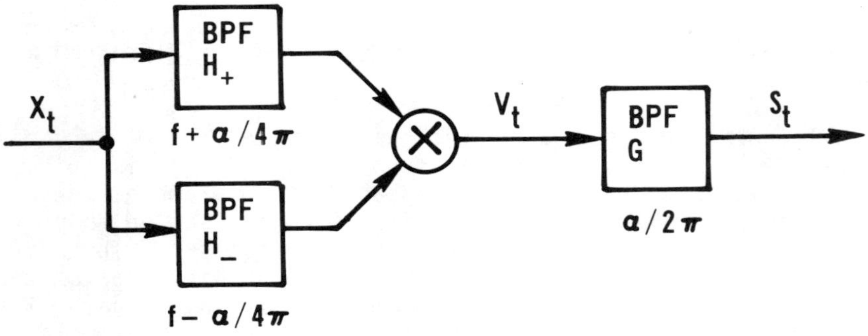

Figure 1. α − Cyclic Spectrum Analyzer

REFERENCES

1. Parzen, E.: 1962, "Spectral analysis of asymptotically stationary time series," Bull. Inst. Internat. Stat., 39, Liv. 2, pp. 87-103.

2. Kampé de Fériet, J., and Frenkiel, F.N.: 1962, "Correlations and spectra for non-stationary random functions," Mathematics of Computation, 16, no. 77, pp. 1-21.

3. Jenkins, G.M. and Watts, D.G.: 1968, Spectral Analysis and its Applications, San Francisco, Holden-Day.

4. Hurd, H.L.: 1969, "An investigation of periodically correlated stochastic processes," Ph.D. Dissertation, Duke Univ., Durham, North Carolina.

5. Gardner, W.A.: 1978, "Stationarizable random processes," IEEE Trans. on Information Theory, IT-24, pp. 8-22.

6. Priestley, M.B.: 1965, "Evolutionary spectra and nonstationary processes," J. Royal Statist. Soc., B, 27, pp. 204-237.

ASYMPTOTIC INFERENCE IN CONTINUOUS TIME MARKOV PROCESSES

M. G. Akritas, and G. G. Roussas

University of Wisconsin, Madison,
and University of Patras, Greece

ABSTRACT

In this paper, a certain class of continuous time parameter Markov processes is considered, whose probability law depends on a k-dimensional parameter. Then, under suitable regularity conditions, several asymptotic results are devived. These results are sufficient to allow us to draw statistical inferences about the stochastic processes in question.

Description of the model

Let Θ be an open subset of $\mathbb{R}^k$, $k \geq 1$, and for each $\theta \varepsilon \Theta$, let $\{X(t),\ t \varepsilon [0,\infty)\}$ be a Markov process of the completely discontinuous (or jump) type, where the r.v.'s $X(t)$, $t \varepsilon [0,\infty)$, are defined on the probability space (Ω, A, P_θ) and take values in the space S (a Borel subset of a finite dimensional Euclidean space).

Actually, the space Ω is taken to be the space $D([0,\infty))$ of all S-valued functions on $[0,\infty)$ which are right-continuous and have finite left-hand limits, and A is the Borel σ-field which is generated by the Skorokhod topology on $D([0,\infty))$. The process $\{X(t),\ t \varepsilon [0,\infty)\}$ is then the coordinate process, and suitable conditions secure that it is of the prescribed type.

Same conditions guarantee the right-continuity of the sample functions of the process $\{X(t),\ t \varepsilon [0,\infty)\}$. This fact leads to the following representation of the r.v.'s $X(t)$, $t \varepsilon [0,\infty)$, namely, there exist r.v.'s Z_n, n=1,2,... taking values in S, and r.v.'s ρ_n, n=1,2,... taking values in $\mathbb{R}^+ = (0,\infty)$ such that

$$X(t) = \begin{cases} Z_1\ , & \text{if } 0 \leq t < \rho_1 \\ Z_n\ , & \text{if } \rho_1 + \ldots + \rho_{n-1} \leq t < \rho_1 + \ldots + \rho_n,\ n \geq 2. \end{cases}$$

D. G. Lainiotis and N. S. Tzannes (eds.), Advances in Communications, 210–217.

The process $\{Z_n, n\geq 1\}$ is a Markov process with stationary transition probabilities. Also, $\rho_n<\infty$, $n\geq 1$, and $\rho_1+\ldots+\rho_n \xrightarrow[n\to\infty]{} \infty$ with probability one.

Formulation of a likelihood function

For any $T\geq 0$, let $\mathcal{A}_T=\sigma\{X(t), t\varepsilon[0,T]\}$ and let $P_{T,\theta}$ be the restriction of P_θ to $\mathcal{A}_T$. Then our purpose is to find a convenient expression for the likelihood function $\frac{dP_{T,\theta^*}}{dP_{T,\theta}}$ for each pair θ, $\theta^*\varepsilon\Theta$.

To this end, let

$$p_\theta(t;\xi,A) = P_\theta(X(s+t)\varepsilon A|X(s) = \xi)$$

be the (stationary) transition probabilities of the underlying process and let the function

$$q(.;\theta) : S \longrightarrow [0,\infty)$$

be defined as the (uniform in $\xi\varepsilon S$) limit

$$\frac{1-p_\theta(t;\xi,\{\xi\})}{t} \xrightarrow[t\to 0]{} q(\xi;\theta).$$

(In a similar way, one defines the function $q(.,.;\theta)$ on $S\times S_\xi$, where S_ξ is the σ-field of Borel subsets of $S-\{\xi\}$; that is,

$$q(\xi,A;\theta) = \lim_{t\to 0} \frac{P_\theta(t;\xi,A)}{t}.$$

The pair of functions $q(.;\theta)$ and $q(.,.;\theta)$ determines the process $\{X_t, t\varepsilon[0,\infty)\}$).

Let $\pi(.,.;\theta)$ be the (stationary) transition probabilities of the process $\{Z_n, n\geq 1\}$ and assume that, for each $\xi\varepsilon S$ and each pair $\theta,\theta^*\varepsilon\Theta$,

$$\pi(\xi,.;\theta) \approx \pi(\xi,.;\theta^*).$$

Also, let $\hat{\pi}(.;\theta)$ be the actual initial distribution of Z_1, and assume that, for each pair $\theta,\theta^*\varepsilon\Theta$,

$$\hat{\pi}(.;\theta) \approx \hat{\pi}(.;\theta^*).$$

Then set

$$\delta(\xi;\theta,\theta^*) = \frac{\hat{\pi}(d\xi;\theta^*)}{\hat{\pi}(d\xi;\theta)},$$

$$\delta(\xi,\eta;\theta,\theta^*) = \frac{\pi(\xi,d\eta;\theta^*)}{\pi(\xi,d\eta;\theta)}.$$

Also, for each pair $\theta,\theta^*\varepsilon\Theta$ and $j=1,2,\ldots$, define the random functions (r.f.'s)

$$\varphi_j(\theta,\theta^*)=\varphi(Z_j,Z_{j+1};\theta,\theta^*)=[\delta(Z_j,Z_{j+1};\theta,\theta^*)]^{\frac{1}{2}},$$

$$\psi_j(\theta,\theta^*)=\psi(Z_j,\rho_j;\theta,\theta^*)=\left\{\frac{q(Z_j;\theta^*)\exp[-\rho_j q(Z_j;\theta^*)]}{q(Z_j;\theta)\exp[-\rho_j q(Z_j;\theta)]}\right\}^{\frac{1}{2}}.$$

Next, for $T\geq 0$, define the r.v.'s $\nu(T)$ and r_T as follows:

$$\nu(T)=\begin{cases}\max\{j;\rho_1+\ldots+\rho_j<T\}\\ 0\ , \text{ if no such } j \text{ exists},\end{cases}$$

$$r_T=\begin{cases}\rho_1+\ldots+\rho_{\nu(T)}, & \text{if } \rho_1<T\\ 0 & , \text{otherwise}.\end{cases}$$

Then, it may be shown that

$$\frac{dP_{T,\theta^*}}{dP_{T,\theta}}=\delta(Z_1;\theta,\theta^*)\prod_{j=1}^{\nu(T)}[\varphi_j^2(\theta,\theta^*)\psi_j^2(\theta,\theta^*)].$$

$$\cdot\ \frac{\exp\left\{-(T-r_T)q[Z_{\nu(T)+1};\theta^*]\right\}}{\exp\left\{-(T-r_T)q[Z_{\nu(T)+1};\theta]\right\}}.$$

Some additional notation

The methodology utilized herein is based on contiguity results. This requires that for each $\theta\varepsilon\Theta$, the parameter point θ^* be of some special form, namely, θ^*, to be denoted here by θ_T, is taken to be

$$\theta_T=\theta+\frac{h_T}{[c(\theta)T]^{\frac{1}{2}}},$$

where $h_T \xrightarrow[T\to\infty]{} h\varepsilon\mathbb{R}^k$ and $0<c(\theta)=\lim_{T\to\infty}\frac{\nu(T)}{T}$ a.s. $[P_\theta]$.

Also, one of the main assumptions made is that of differentiability in q.m. of the r.f.'s

$$\varphi_j(\theta,.),\ \ \psi_j(\theta,.),\ j\geq 1.$$

Let $\dot{\varphi}_j(\theta)$ and $\dot{\psi}_j(\theta)$ be the derivatives in q.m. involved and set

$$\Delta_T(\theta)= \frac{2}{[\nu(T)]^{\frac{1}{2}}} \sum_{j=1}^{\nu(T)} [\dot{\varphi}_j(\theta)+\dot{\psi}_j(\theta)],$$

$$\Gamma_1(\theta)=4\mathcal{E}_\theta[\dot{\varphi}_1(\theta)\dot{\varphi}_1'(\theta)],$$

$$\Gamma_2(\theta)=4\mathcal{E}_\theta[\dot{\psi}_1(\theta)\dot{\psi}_1'(\theta)],$$

$$\Gamma(\theta)=\Gamma_1(\theta)+\Gamma_2(\theta).$$

It is assumed that the covariance functions $\Gamma_1(\theta)$ and $\Gamma_2(\theta)$ are positive definite on Θ.
Finally, set

$$\Lambda_T(\theta)=\log \frac{dP_{T,\theta_T}}{dP_{T,\theta}}.$$

Main results

The main results to be stated below refer to the asymptotic behavior of the log-likelihood function $\Lambda_T(\theta)$, as well as to the asymptotic distribution of $\Lambda_T(\theta)$ and that of $\Delta_T(\theta)$. More precisely, the following results hold true under suitable regularity conditions.

Theorem 1: For each $\theta\varepsilon\Theta$ and as $T\to\infty$,

$$\Lambda_T(\theta)-h'\Delta_T(\theta) \to -\frac{1}{2} h'\Gamma(\theta)h \quad \text{in } P_{T,\theta}\text{-probability.}$$

(Thus, for sufficiently large T and in the neighborhood of θ,

$$\frac{dP_{T,\theta_T}}{dP_{T,\theta}} \simeq e^{-\frac{1}{2}h'\Gamma(\theta)h} e^{h'\Delta_T(\theta)} .)$$

Theorem 2: For each $\theta\varepsilon\Theta$ and as $T\to\infty$,

$$\mathcal{L}[\Delta_T(\theta)|P_{T,\theta}] \Longrightarrow N(0,\Gamma(\theta)).$$

From Theorems 1 and 2 and standard Slutsky type results, it follows

Theorem 3: For each $\theta\varepsilon\Theta$ and as $T\to\infty$,

$$\mathcal{L}[\Lambda_T(\theta)|P_{T,\theta}] \Longrightarrow N(-\frac{1}{2} h'\Gamma(\theta)h,\ h'\Gamma(\theta)h).$$

Results analogous to those provided by Theorems 1,2 and 3 are also required under the (moving) probability measure P_{T,θ_T}.

These results are needed, for instance, for the purpose of determining power of various statistical tests.

To this end, there holds.

Theorem 4: For each $\theta\varepsilon\Theta$ and as $T\to\infty$,

(i) $\Lambda_T(\theta)-h'\Delta_T(\theta) \to -\frac{1}{2}h'\Gamma(\theta)h$ in P_{T,θ_T}-probability,

(ii) $\mathcal{L}[\Lambda_T(\theta)|P_{T,\theta_T}] \Longrightarrow N(\frac{1}{2}h'\Gamma(\theta)h,\ h'\Gamma(\theta)h)$,

(iii) $\mathcal{L}[\Delta_T(\theta)|P_{T,\theta_T}] \Longrightarrow N(\Gamma(\theta)h,\ \Gamma(\theta))$.

Some statistical applications

For the purpose of illustrating the applicability of the main results obtained above in making statistical inference about the (unknown) parameter θ, consider the following examples.

(1) Let $k=1$ and consider the problem of testing the hypothesis $H:\theta=\theta_0$ against the alternative $A:\theta>\theta_0$ at level of significance α.

Reject the hypothesis H when $\Delta_T(\theta_0)>C_T$, where the constant C_T is determined asymptotically by means of the normal tables (because of Theorem 2) and the level of significance α. Actually, $C_T=\xi_\alpha$, where ξ_α is the upper α-th quantile of the $N(0,\Gamma(\theta_0))$ distribution.

This test is asymptotically locally most powerful. It becomes asymptotically globally most powerful under a certain further assumption.

Theorem 4(iii) is utilized for determining asymptotically the power function of said test.

(2) Let again $k=1$ and consider now the problem of testing the hypothesis $H:\theta=\theta_0$ against the alternative $A:\theta\neq\theta_0$ at level of significance α.

Reject the hypothesis H when $\Delta_T(\theta_0)<a_T$ or $\Delta_T(\theta_0) > b_T$, where the constants a_T and b_T are determined again asymptotically by means of the normal tables (because of Theorem 2) and the level of significance α. More precisely, $a_T=-\xi_{\alpha/2}$, $b_T=\xi_{\alpha/2}$.

This test is as above asymptotically locally most powerful which becomes asymptotically globally most powerful under a suitable further assumption.

Again, Theorem 4(iii) is utilized for determining asymptotically the power function of said test.

(3) Let now $k\geq 2$ and consider again the hypothesis $H:\theta=\theta_0$ and the alternative $A:\theta\neq\theta_0$. Then one may construct a test based on $\Delta_T(\theta_0)$ which enjoys a number of asymptotically optimal properties. More precisely, this test, asymptotically and over certain ellipsoids, has maximum average power, is most stringent and is most

powerful among those tests whose power is asymptotically constant on the above mentioned ellipsoids.

The above mentioned properties are shown by means of Theorems 3 and 4 (iii).

(4) The same main results may be suitably used for studying asymptotic efficiency of estimators in the sense of Weiss-Wolfowitz.

An outline of the proofs of main results

Assumptions under which the main results of this paper hold true may not be explicitly presented here for lack of space. Also, complete proofs of the theorems will have to be omitted for the same reasons. They may be both found, however, in a Technical Report.

On the proof of Theorem 1: By considering sequences $\{T_n\}$, $n \geq 1$, in $[0,\infty)$ with $T_n \xrightarrow[n\to\infty]{} \infty$, one discretizes the problem and then employes suitably known results from the discrete case.

On the proof of Theorem 2: As above.

On the proof of Theorem 3: It follows from Theorems 1 and 2.

On the proof of Theorem 4: It follows from Theorems 1-3 and the contiguity of the sequences, $\{P_{T_n,\theta_{T_n}}\}$.

An Example

For the sake of illustration, an example is presented below, where the basic differentiability in q.m. assumption is satisfied. To this end, let S=R and take

$$q(\xi,A;\theta)=\int_A \frac{1}{2} e^{-|x-\theta\xi|}dx, \quad q(\xi,\theta)=1.$$

Then it may be shown that $\pi(\xi,A;\theta)=q(\xi,A;\theta)$, and consequently

$$\varphi_1(\theta,\theta^*)=\exp(\tfrac{1}{2}|Z_2-\theta Z_1| - \tfrac{1}{2}|Z_2-\theta^* Z_1|),$$

$$\psi_1(\theta,\theta^*)=1.$$

Then the classical type pointwise differentiability assumption of the r.f. $\varphi_1(\theta,.)$ is <u>not</u> satisfied whereas differentiability in q.m. <u>is</u> satisfied.

Appendix

On the concept of contiguity

Basic tools in this paper are contiguity results. For a complete account the interested reader is referred, for example, to Roussas (1972). Here the definition of contiguity is presented along with two of its possible characterizations.

Contiguity is a measure of closeness of sequences of probability measures. More precisely, for n=1,2,..., consider the probability spaces $(\Omega_n, \mathcal{A}_n)$, and let P_n and Q_n be probability measures on $\mathcal{A}_n$. Then

Definition: The sequences $\{P_n\}$ and $\{Q_n\}$ are said to be contiguous, it for any $A_n \varepsilon \mathcal{A}_n, P_n(A_n) \xrightarrow[n\to\infty]{} 0$ implies $Q_n(A_n) \xrightarrow[n\to\infty]{} 0$, and conversely.

Let $P_n \approx Q_n$, $n \geq 1$, and set

$$L_n = \frac{dQ_n}{dP_n} \text{ and } \Lambda_n = \log L_n.$$

Also, let

$$\mathcal{L}_n = \mathcal{L}(\Lambda_n | P_n), \; \mathcal{L}_n' = \mathcal{L}(\Lambda_n | Q_n).$$

Then

Theorem A: The sequences $\{P_n\}$ and $\{Q_n\}$ are contiguous iff the sequences $\{\mathcal{L}_n\}$ and $\{\mathcal{L}_n'\}$ are relatively compact.

Also,

Theorem B: The sequences $\{P_n\}$ and $\{Q_n\}$ are contiguous iff the sequence $\{\mathcal{L}_n\}$ is relatively compact and for any $\{\mathcal{L}_m\} \subseteq \{\mathcal{L}_n\}$ such that $\mathcal{L}_m \xRightarrow[m\to\infty]{} \mathcal{L}$, a probability measure, then $\int_{\mathbb{R}} e^{\lambda} \, d\mathcal{L} = 1$.

Acknowledgement

This research was supported by NSF grant No. MCS76-00951.

REFERENCES

1. Akritas, M.G. (1978). "Contiguity of probability measures associated with continuous time stochastic processes". Ph.D. dissertation, Univ. of Wisconsin, Madison.

2. Akritas, M.G. and Roussas, G.G. (1979)."Asymptotic expansion of the log-likelihood function based on stopping times defined on a Markov process". Ann. Inst. Statist. Math.

3. Billingsley, P. (1961). "Statistical Inference for Markov Processes". The University of Chicago Press.

4. Doob, J.L. (1953). "Stochastic Processes". John Wiley and Sons, New York.

5. Hājek, J. (1970). "A characterization of limiting distributions of regular estimates". Z. Wahrscheinlichkeitstheorie und Verw. Gebiete 14 323-30.

6. Inagaki, Nobuo (1970). "On the limiting distribution of a sequence of estimators with uniformity property." Ann. Inst. Statist. Math. 22 1-13.

7. Johnson, R.A. and Roussas, G.G. (1969). "Asymptotically most powerful tests in Markov processes". Ann. Math. Statist. 40 1207-1215.

8. Johnson, R.A. and Roussas, G.G. (1970). "Asymptotically optimal tests in Markov processes." Ann. Math. Statist. 41 918-938.

9. Johnson, R.A. and Roussas, G.G. (1972). "Applications of contiguity to multiparameter hypothesis testing". Proc. 6th Berkeley Symp. of Math. Statist. and Prob. 1 195-226.

10. Lecam, L. (1960). "Locally asymptotically normal families of distributions". Univ. California Publ. Statist. 3 37-98.

11. Roussas, G.G. (1965). "Asymptotic inference in Markov processes". Ann. Math. Statist. 36 978-992.

12. Roussas, G.G. (1972). "Contiguity of Probability Measures: Some Applications in Statistics". Cambridge University Press.

13. Roussas, G.G. (1978). "Asymptotic distribution of the log-likelihood function for stochastic processes". Z. Wahrscheinlichkeitstheorie und Verw. Gebiete 47 31-46.

TRANSMULTIPLEXING: FACTORS INFLUENCING THE DESIGN OF TDM TO FDM CONVERSION

A.G. Constantinides, B.Sc.(Eng.),Ph.D.,C.Eng.,M.I.E.E.

R.A. Valenzuela, B.Sc.(Eng.).

ABSTRACT:

As a result of the increased usage of Digital Systems in hitherto analogue enviroments, such as Telephone Networks, Frequency Division Multiplex (FDM) and Time Division Multiplex (TDM) are now of common use.

Several authors have so far considered the problems related to the design of an all digital FDM-TDM Translator. A review of the most representative schemes suggested in the literature is presented in this paper together with a discussion of new techniques for Transmultiplexer implementation.

1. INTRODUCTION.

Although Digital Systems are generally more complex than their analogue couterparts in the sense that a larger number of active devices is required, the ever decrasing price as well as accuracy and reproducibility of digital techniques has generated a trend towards the increased use of digital systems.

Analogue and Digital techniques are already coexisting in very large networks where a complete switch over to digital equipment is impracticable due to the massive scale of the capital investment involved. In the Thelephone Networks for instance the traditional multiplexing format has been Frequency Division Multiplex (FDM). Time Division Multiplexing (TDM) results in a much more natural choice when digital signals are

D. G. Lainiotis and N. S. Tzannes (eds.), Advances in Communications, 218–225.

available. Consequently FDM and TDM multiplexing formats are already coexisting in present Telephone Networks and will continue to do so for some time to come.

The inter connection betewen FDM and TDM multiplexing formats is usually done by tandem connection of an A-Type analogue channel bank and a D-Type digital channel bank, in other words the multiplexed channels are demultiplexed down to voicebands first and then multiplexed again into the desired format. It is clear that this process is highly inefficient. Furthermore substantial savings can be achieved if FDM-TDM conversion is implemented via an all digital processor, working directly at the multiplex level.

2. STATEMENT OF THE PROBLEM.

In the FDM multiplexing format each channel is allocated a 4 KHz slot from the available bandwidth, while in TDM systems each channel may occupy the whole bandwidth, but exists in a given time slot only.

Multiplexing and Demultiplexing of PCM signals into the TDM format is a straightforward process. Therefore the most important function of a TDM-FDM converter is by far the PCM-FDM conversion.

The typical spectrum of one PCM channel is illustrated in Figure 1 while Figure 2 shows the spectrum of a typical FDM signal.

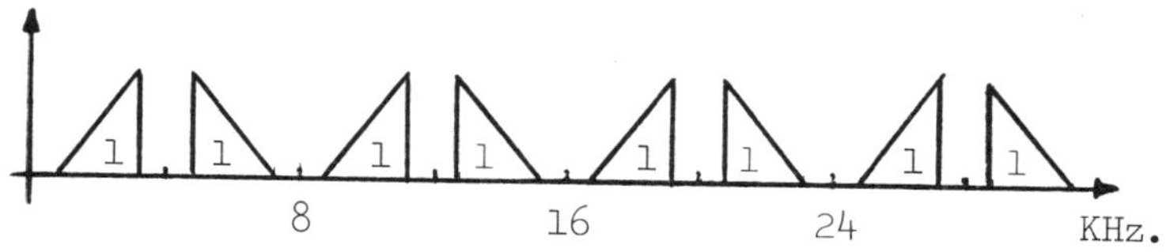

Fig. 1. Typical PCM signal Spectrum.

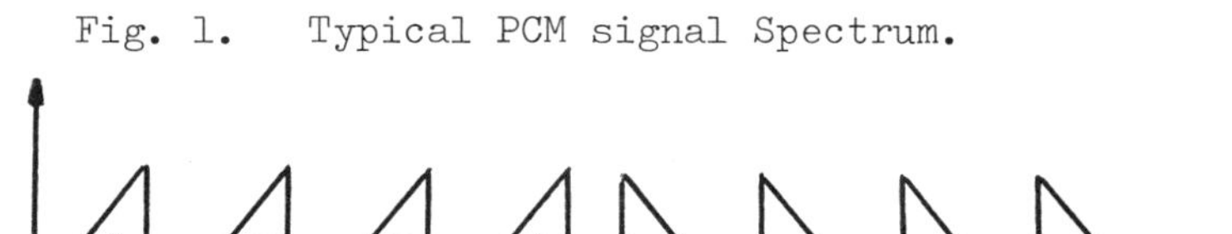

Fig. 2. Typical FDM signal Spectrum (4 Channels).

It can be seen that the Transmultiplexing function consist of building the FDM signal from a set of (60 or 12) PCM channels, in the PCM-FDM sense, and the extraction of each PCM Channel from the FDM signal, in the opposite sense.

3. REVIEW OF SYSTEMS SUGGESTED IN THE LITERATURE.

Several authors have considered the Transmultiplexing problem and it is in fact possible to classify these propossed schemes ino three broad groups according to the architecture or basic building block they are organized upon. Thus we have:

- -Weaver Modulators
- -FFT processing
- -Multistage Systems.

3.1. Weaver Modulator Based Transmultiplexors.

An early attempt to perform an all digital TDM-FDM Translator has been reported by S.L. Freeny et al. in 1971 (1,2, 3). The approach chosen was based on ideas suggested by Darlington (4) for the implementation of a digital SSB modulator of the Weaver (5) type.

In the Weaver type of SSB modulator, the signal is split into two paths then a sequence of modulation, sample rate increase and filtering is performed in each path in such a way that the unwanted sideband cancels out. This is illustrated in Figure 3. where the overlapped frequency components are shown ditinct for clarity.

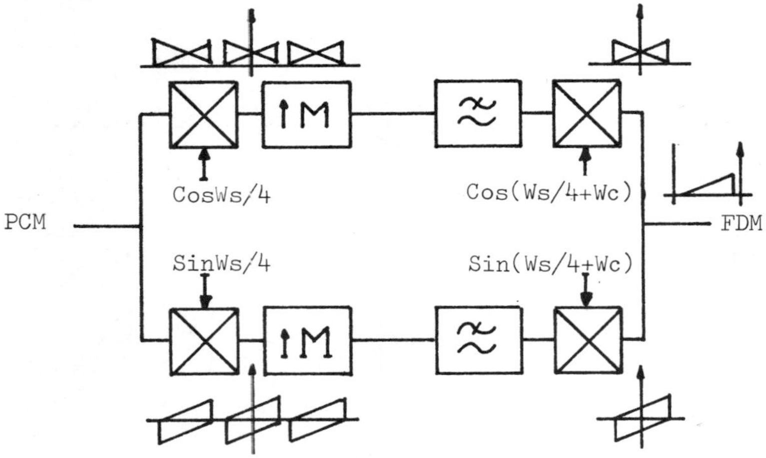

Fig. 3 Weaver Modulator.

The system based on the above ideas requires a multiplication rate of the order of one million multiplications per channel per second, for a 12 channel system.

A. peled and S. Winograd (6) in 1978 suggested a modified version of the basic bank of Weaver modulators scheme requiring reduced computation complexity and in fact the propossed algorithm requires approximately 300 thousand multiplications per channel per second. This reduction is achieved by exploiting some results from the theory of Computational Complexity.

3.2 FFT Processor Based Transmultiplexors.

The use of FFT techniques for Transmultiplexing purposes was first suggested by Darlington (4) as a mean of reducing the complexity associated to the basic bank of Weaver modulators scheme to approximately 470 Thousand multiplications per channel per second by organizing the multiplications required for the filtering and modulation steps in a way similar to the Fast Fourier Transform. The reduction in multiplier hardware is somewhat offset however by an increase in memory requirements.

P.M. Terrel and P.J.W. Rayner (7) suggested a different approach based on FFT processing featuring relatively attractive coefficient sensitivity properties but unfortunately, requiring about twice the multiplication rate of the non FFT scheme previously reported by S.L. Freeny et al. in (1).

M. Bellanger et al. (8) in 1974 developed a new FFT based scheme, representing the first significant departure from the original ideas of Darlington, opening the way to a new class of FFT based Transmultiplexors. This approach can be descrived as the combination of a bank of phase shifters (constituting a Digital Polyphase Network) with a FFT processor. In this scheme the low sampling rate is preserved throughout and the high sampling rate output is obtained by mere interleaving of the output sequences from the Polyphase Network. Consequently a very low multiplication rate is achieved. Indeed 646.4 Thousand multiplications per channel per second are required for a 60 channel system. Further development of this idea by G. Bonnerot, M. Courdeuse and M. Bellanger (9) led to a 64 channels system requiring only 170 Thousand multiplications per second, this being achieved by using a particular type of Dicrete Fourier Transform which has been called Odd-Time Odd-Frequency DFT (9) as well as combining branches of the Polyphase network involving similar coefficients.

F. Takahata et al.(11) investigated a Real input FFT based Transmultiplexor in which the Polyphase network is replaced by a bank of sub-filters operating at the low sampling rate, the output signal being determined by interleaving the sub-filter outputs. The multiplication rate acieved in this way is found to be 305 Thousand multiplications per channel per second for a 24 channel Transmultiplexor.

Another FFT based Transmultiplexing algorithm has been reported by R. Maruta and A. Tomozawa (12) in which the base-band signals are applied directly to the FFT processor. Its outputs are then applied to a set of slow complex bandpass filters. Interleaving the real outputs of the complex bandpass filters leads to the required FDM signal. The multiplication rate required in this case amounts to 571.7 Thousand multi-plications per channel per second

3.3 Multistage Systems.

T. Tsuda et al. (13) and A. Fettweiss (14,15) in 1978 have suggested Transmultiplexing algorithms that can be descrived as multistage in the sense that two channels only are multiplexed at a time, giving rise to a tree architecture of the type shown in Figure 4.

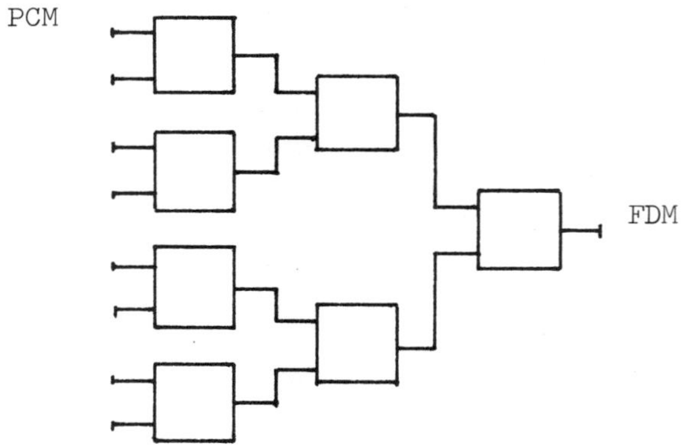

Fig. 4 Multistage Struture.

In the scheme given in (13) the building block consits of complex modulation by one quarter or three quarters of the sampling frequency and filtering, resulting in a multiplication rate of 380.8 Thousand multiplications per channel per second for a 60 channel converter.

A. Fettweiss (15) has suggested a system in which the basic building block can be seen as a sequence of real modulation and filtering operations, the choice of modulating frequencies is increased to Fs/2, Fs/3, Fs/4 and Fs/6 all of which are shown to require multiplications by 1, -1 or 0 only. The complexity of this approach is determined by the different types of filters required.

4. NEW ALTERNATIVES.

As shown in (16) the Transmultiplexing schemes base upon FFT processor followed by a bank of filters is a particular case of a very large class of systems characterized by the cascade of a processor and a bank of filters. In fact for any given processor, it is possible to determine the required bank of filters or, conversely, it is possible to determine the processor required to match a given bank of filters (16).

An endless series of Transmultiplexing algorithms can be generated in this way and suggestions have already been made in (16) with respect to the use of multiplier free processors such as Hadamard, for instance. It is not known, however, whether the simplicity gained in the processor is offset or not by the requirements on the resulting bank of filters.

Approaching the above from another point of view it appears that there is no reason for the processor to be orthogonal and therefore a large variety of systems can be envisaged and a procedure for directing the search towards simpler implementations is consequently needed.

In addition a completly different line of researh is being followed by the authors, in order to establish the effectivness of frequency domain techniques in the implementation of Transmultiplexors. Due to the fact that FFT processing is being used in such a wide range of applications, much effort is being devoted to new algorithms faturing reduced multiplication requirements. Developments are also to be expected in dedicated hardware for FFT processing. Therefore a system based almost solely upon FFT processing will be very attractive provided that the level of noise resulting from short-time spectral estimation can be kept under an acceptable limit. Reults obtained so far are very encouraging indeed.

A final remark has to be made with respect to the usage of multiplication rate in the comparison of Transmultiplexing schemes. Due to the complexity of the techniques under consideration, storage accuracy and number of data movements required may contribute significantly towards the overall hardware requirements. Therefore the multiplication rate should not be considered as the only mesure of hardware complexity related to the implementation of a given scheme.

REFERENCES:

(1) Freeny, S.L. et al.,"System Analysis of TDM-FDM Translator/ Digital A-Type Channel Bank",IEEE Trans. on Comm. Technology, Vol. COM-19,6,Dec. 1971,pp. 1050-1059.
(2) Kieburtz,R.B. and Tewksbury,S.K.,"Experimental Test of a Digital A-Type Channel bank",Proc. of International Comm. Conference ICC 1973,pp. 37-12 to 37-17.
(3) Freeny, S.L. et al.,"Design of Digital Filters for an all Digital Frequency Division Multiplex Time Division Multiplex Translator",IEEE Trans.on Circuit Theory,Vol. CT-18,Nov. 1971, pp. 702-711.
(4) Darlington, S.,"On Digital Single-Sideband Modulators", Proc. 1969 IEEE Int. Sym. Circuit Theory, pp. 14-15.
(5) Weaver, D.K.Jr.,"A Third Method of Generation and Detection of Single-Sideband Signals",Proc. IRE,Vol.44,Dec.1956, pp. 1703-1705.
(6) Peled, A. and Winograd, S.,"TDM-FDM Conversion Requiring Reduced Computation Complexity",IEEE Trans. on Comm.,Vol.26 No. 5,May 1978.
(7) Terrell, P.M. and Rayner, P.J.W.,"A Digital Block Processor For SSB-FDM Modulation and Demodulation",IEEE Trans. on Comm., Vol. COM-23,No. 2,Feb. 1975, pp. 282-286.
(8) Bellanger, M.G. et al.,"TDM-FDM Transmultiplexer:Digital Polyphase and FFT",IEEE Trans. on Comm.,Vol.COM-22,No 9, Sept. 1974, pp.1199-1205.
(9) Bonnerot, G. et al.,"Digital Processing Techniques in The 60 Channel Transmultiplexer",IEEE Trans. on Comm.,Vol. COM-26, No 5,May 1978., pp. 698-706.
(10) Bonnerot, G. et al.,"Odd-Time Odd-Frequency Discrete Fourier Transform for Symetric Real Valued Series",Proc. of the IEEE,Vol. 64,No 3,March 1976, pp.392-393.
(11) Takahata, F. et al.,"Develpment of a TDM/FDM Transmultiplexer", IEEE Trans. on Comm.,Vol.COM-26,No 5,May 1978,pp. 726-733.
(12) Maruta, R. et al.,"An improved Method For Digital SSB-FDM Modulation and Demodulation",IEEE Trans. on Comm.,Vol COM-26, No 5,May 1976, pp. 720-725.
(13) Tsuda, T. et al.,"Digital TDM-FDM Translator with Multistage Structure",IEEE Trans. on Comm.,Vol COM-26,No 5,May 1978.
(14) Fettweiss, A.,"Principles for Multiplier-free Single-way PCM/FDM and Audio/FDM Conversion",AEU 32,1978,pp. 441-446.

(15) Fettweiss, A.,"Multiplier Free Modulation Schemes for PCM/FDM and Audio/FDM Conversion",AEU 32,1978,pp. 477-485.
(16) Claasen, T.A.C.M. et al.,"A Generalized Scheme For An All Digital Time Division Multiplex to Frequency Division Multiplex Translator",IEEE Trans. on Circuit and Systems, Vol. CAS-25,May 1978, pp. 252-259.

REALISABLE DATA TRANSMISSION FILTERS WITH OPTIMAL PROPERTIES

L. F. Lind

Department of Electrical Engineering Science,
University of Essex, Wivenhoe Park, Colchester,
Essex, England.

1. INTRODUCTION

This paper gives a descriptive account of recent work in the field of filters for data transmission, and includes a summary of the main results.

An ideal band limited data transmission filter[1] would have an impulse response as shown in figure 1.

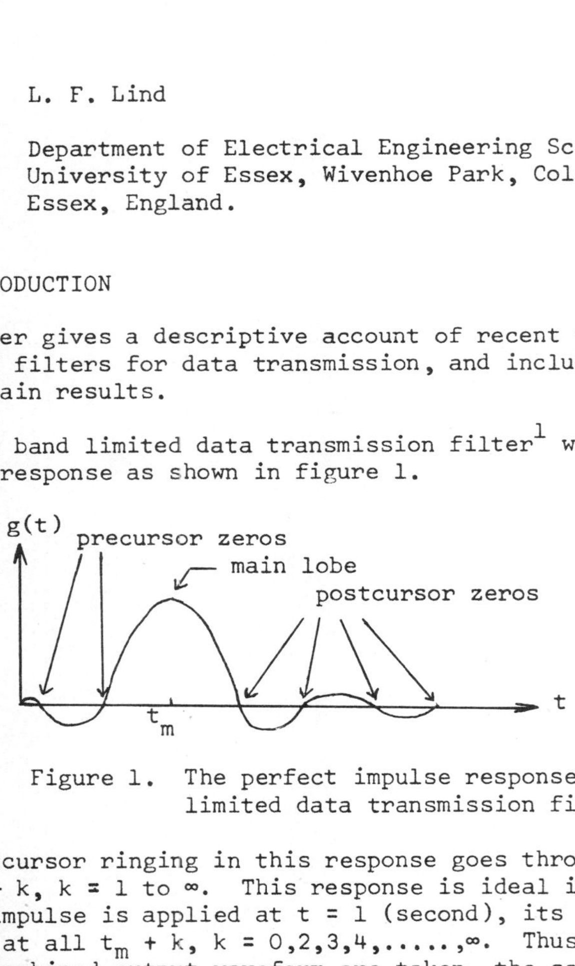

Figure 1. The perfect impulse response for a band limited data transmission filter.

The postcursor ringing in this response goes through zero at all $t_m + k$, $k = 1$ to ∞. This response is ideal in that, if a second impulse is applied at $t = 1$ (second), its response will be zero at all $t_m + k$, $k = 0,2,3,4,\ldots\ldots,\infty$. Thus if samples of the combined output waveform are taken, the sample at t_m relates to only the first impulse input, the sampled $t_m + 1$ relates to only the second input, and so on. These samples are decoupled, and so no intersymbol interference (isi) is present.

D. G. Lainiotis and N. S. Tzannes (eds.), Advances in Communications, 226–231.

The other post-cursor zeros are added to ensure zero isi between any two impulses of an isochronous data stream. The response of Figure 1 is ideal in that the postcursor ringing is constrained to be zero at an infinite number of points. Any realisable filter will have a finite number of elements (or degrees of freedom), and so cannot give the exact performance of Figure 1. However, we can attempt to approximate the response of a realisable filter to that of Figure 1 in some optimal manner. This approximation is the concern of this paper.

In the past the approximation problem has been tackled in the frequency domain. There are good reasons for this approach. Firstly, the approximation need only be made over a finite interval, due to the bandlimited nature of the frequency response. Secondly, ideal target functions, such as the raised cosine response, are readily available. Contrast this to the approximation problem in the time domain, where an infinite point set (the precursor and postcursor zeros) is involved. Actually, we will see that this apparent difficulty can be overcome.

With the frequency domain approach, there are two main difficulties. The first is the selection of a suitable target function. Nyquist showed[2] that any amplitude response with vestigal symmetry about the cut-off point will do. Therefore, for a given transition bandwidth, there are an infinity of target functions, all of which give zero isi. Which should be used? The second problem is concerned with the rational function approximation to the target function. What is the best shape for the error curve? Should it be equiripple, ripply, maximally flat, or some other shape? Could we improve the error curve by using another target function? These questions are never resolved in an optimal manner. Of course, it is true that as we increase the degree of the filter (and hence the degree of the rational transfer function), we can steadily reduce the error, and therefore improve the time domain performance. On an iterative basis, the degree can be increased until the performance is satisfactory. It is not certain however, that this is the minimal degree which will meet the specification.

The view taken in this paper is: for a given degree of filter, and a given isi, find a realisable filter that achieves this isi at a global minimum. Then the resulting filter should give the best frequency selectivity performance for all filters of that degree (which has been verified by numerical experiment). The degree is then increased, until the selectivity is satisfactory. Optimisation is therefore performed in the time domain. The problem of dealing with an infinite point set is resolved by noting that a realisable filter has an impulse response given by a sum of exponentials. The sum over an equally spaced infinite point set can be performed by using a geometric series. Putting

the specified isi at a global minimum also has the desired feature of giving the filter low sensitivity. In fact, the differential sensitivity is exactly zero at the solution point.

2. SUMMARY OF MAIN EQUATIONS

A realisable filter will have a tranfer function F(s) in partial fraction form as follows

$$F(s) = \sum_{n=1}^{N} \frac{r_n}{s-p_n}$$

where the r_n are the residues and p_n the pole locations. The corresponding impulse response is

$$f(t) = \sum_{n=1}^{N} r_n e^{p_n t}$$

The criterion to be minimised is the total normalised squared error at the postcursor sample points $t_m + 1$, $t_m + 2$,...... Thus

$$E \triangleq \frac{\sum_{k=1}^{\infty} f^2(t_m + k)}{f^2(t_m)}$$

The denominator term is necessary. If it were not included, then a feasible solution is to set all the r_n to zero. A functional y, containing a Lagrange multiplier, is formed from E. This y is differentiated with respect to each r_n in turn, in order to find a global optimum. The resulting equation set is finite (by using a geometric summation formula), linear, and has the closed form solution[3,4]

$$r_n = \frac{f(t_m)e^{-p_n t_m}}{\sinh(\sum_{i=1}^{N} p_i)} \left[\sinh p_n\right] \prod_{\substack{i=1 \\ i\neq n}}^{N} \frac{\sinh(p_n + p_i)/2}{\sinh(p_n - p_i)/2} \qquad (1)$$

$$n = 1,2,.....,N.$$

These equations are called the optimal residue equations. It can also be shown[3] that there is a simple relation between the pole locations and E, namely,

$$E = 1/\left[\exp(-2 \sum_{n=1}^{N} p_n) - 1\right]$$

Since the poles are in complex conjugate pairs, the error is related to the sum of the real parts of the pole locations.

There is another set of residue equations, which are simply due to partial fraction expansion. Let

$$F(s) = \frac{N(s)}{\prod_{n=1}^{N} (s - p_n)} \qquad (2)$$

where N(s) is a fixed polynomial. Then

$$r_i = \frac{N(p_i)}{\prod_{\substack{n=1 \\ n \neq i}}^{N} (p_i - p_n)}, \quad i = 1,2,\ldots,N \qquad (3)$$

is the set of normal residue equations. The normal and optimal residue equations can be equated to each other, to give a set of non-linear equations for the pole locations. This pole equation set (for a fixed N(s)) has been solved with a computer program. It was found that there are many possible solution sets, but only one set contains all poles in the strip $-\omega_N < I_m \{s\} < \omega_N$, where ω_N is the Nyquist frequency. This set is chosen, as it gives the best selectivity in the s-plane.

The role of N(s) is to control the precursor isi. An all-pole filter (N(s) = 1) is used at the start, which normally has $t_m > 1$. Then, N(s) is increased in degree so as to eliminate the precursor isi. Next, the pole equations are solved again, and so on, until both the pole equations are satisfied and the precursor isi has been eliminated. In practice, this iteration has been found to be rapidly convergent.

Overall, the inputs to the design procedure are the degree N of the filter and the required isi E. A computer program then finds the poles and zeros, as well as t_m, the main sample point. This point is constrained to lie at the peak of the main lobe[3]. Analysis programs can be used to plot amplitude and group delay responses, as well as the impulse response.

3. RESULTS

The poles have been found to lie approximately on three sides of a rectangle, whose top and bottom sides are at $\pm\omega_N$. The pole on the top side is at the middle of this side. The poles on the vertical locus are approximately equally spaced from each other.

Turning to the transmission zeros, it is found that they occur in both the left and right-half of the s-plane. Hence this class of filters is non-minimum phase. The right-half plane zeros can lie on the real axis, out in the complex plane, or both. In many ways these filters are similar to the linear phase filters of Rhodes[5], but of course have been optimised for a different criterion. Two interesting properties to consider in the frequency domain are the folded summations of the real and imaginary parts of the transfer function. Ideally they should be equal to 1 and 0, for all frequencies. The errors turn out to be equiripple (of small amplitude), and interlaced[3]. This performance, although not designed for, is indicative of the excellent freuquency domain properties of these filters.

In the time domain the impulse response has the desired zero-crossing performance. It has also been found that the ringing peak amplitudes are close to the ideal, and that the main lobe is approximately symmetrical about the main sample point[3]. The eye diagram is diamond shaped[3], and is fully open at the main sample point.

Simple methods have been developed to extend this design method to handle non-impulsive inputs[6], and non-ideal channels[3]. It has been found that only small changes in the pole and zero locations are necessary to handle these extensions. Finally, the error surface for changes in pole and zero locations has been plotted. It has been found that a great many local minima are clustered about the global one, indicating the difficulty of finding the global minimums by computer-aided-design search programs.

REFERENCES

1. R. W. Lucky, et al, "Principles of Data Communication". McGraw-Hill, 1968, Chapter 4.

2. H. Nyquist, "Certain Topics in Telegraph Transmission Theory," AIEE Trans., Vol. 47, pp 617-644, April 1928.

3. S. E. Nader and L. F. Lind, "Optimal Data Transmission Filters", IEEE Trans. on Circuits and Systems, Vol. CAS-26, No. 1, January 1979, pp 36-45.

4. L. F. Lind and S. E. Nader, "Optimal Analog Filters for Data Transmission", Proc. IEEE Int. Symp. on Circuits and Systems, (Arizona, NM), pp 344-347, April 1977.

5. J. D. Rhodes, "A Low-Pass Prototype Network for Microwave Linear Phase Filters", IEEE Trans. on Microwave Theory and Techniques, Vol. MTT-18, pp 290-301, June 1970.

6. L. F. Lind and S. E. Nader, "Realisable Filters that Minimise Intersymbol Interference", Radio and Electronic Engineer, Vol. 48, No. 12, pp 612-618, December 1978.

SYNCHRONIZATION STRATEGIES IN DIGITAL DATA COMMUNICATION

L. E. Franks

Department of Electrical and Computer Engineering
University of Massachusetts, Amherst, MA 01003

The problem of joint recovery of carrier phase and symbol timing from measurements on a suppressed-carrier digital data signal is considered. Strategies based on maximum-likelihood estimation of these parameters are described and a simple method for evaluation of estimation errors is presented.

INTRODUCTION

In digital communication systems using a modulated carrier, it is necessary to obtain accurate estimates of both the carrier phase and symbol timing phase in order to successfully recover the data sequence from the received signal. A specific problem of this type is to estimate the parameters θ and τ based on measurements performed on the suppressed-carrier signal

$$z(t) = \mathrm{Re}[\beta(t-\tau)\exp(j2\pi f_o t + j\theta)] + u(t) \tag{1}$$

where $u(t)$ is white Gaussian noise with spectral density N_o. For the case of one-dimensional pulse amplitude modulation formats; e.g., DSB/PAM and VSB/PAM, we have

$$\beta(t) = \sum_k a_k \gamma(t-kT) \tag{2}$$

where $\{a_k\}$ is the data sequence and $\gamma(t)$ is the complex envelope of a single data pulse. For DSB we can use a real $\gamma(t)$, but for VSB or SSB, $\gamma(t)$ will be a complex-valued pulse with real and imaginary parts related through convolution integrals. For two-dimensional modulation schemes; e.g. QAM, there are two data sequences and

D. G. Lainiotis and N. S. Tzannes (eds.), Advances in Communications, 232–237.

$$\beta(t) = \sum_k a_k g(t-kT) + jb_k h(t-kT) \tag{3}$$

where g(t) and h(t) are normally real data pulses.

It is helpful to have a standard estimation strategy, preferably optimal in some sense, to study the effectiveness of various suboptimal recovery schemes and to evaluate the dependence of estimator performance on various system parameters, such as signal bandwidth, signal-to-noise ratio (SNR), pulse shape, modulation scheme, etc. A natural choice for this standard estimator is the maximum-likelihood (ML) estimator, where estimates are based on observations of z(t) over a prescribed time interval T_0. The observation interval T_0 is itself an important system parameter as it is desired to keep T_0 small for rapid acquisition of the timing parameters, but estimation error decreases as T_0 increases.

The ML estimator chooses as estimates of θ and τ the values of $\hat{\theta}$ and $\hat{\tau}$ that maximize the likelihood function

$$L(\hat{\theta},\hat{\tau}) = \exp[-\frac{1}{4N_0}\int_{T_0} |\alpha(t) - \beta(t-\hat{\tau})\exp(j\hat{\theta})|^2 dt] \tag{4}$$

where $\alpha(t)$ is the complex envelope of the received signal z(t) in (1). When T_0 is relatively long, compared to a symbol period T, then only the cross-product term in the integrand in (4) has a significant dependence on $\hat{\theta}$ and $\hat{\tau}$. Hence we can use a simplified log-likelihood function in the maximization process given by

$$\Lambda(\hat{\theta},\hat{\tau}) = \frac{1}{2N_0}\sum_k a_k\, q_k(\hat{\theta},\hat{\tau}) \tag{5a}$$

where

$$q_k(\hat{\theta},\hat{\tau}) = \mathrm{Re}[\exp(-j\hat{\theta})\int_{T_0} \alpha(t)\gamma^*(t-kT-\hat{\tau})dt] \tag{5b}$$

for the one-dimensional modulation case. The two-dimensional case develops along completely parallel lines and will not be discussed further here. Performance evaluation of the ML estimator, and various suboptimal approximations to it, for specific signal formats is presented in References [1]-[5]. This paper discusses, in somewhat more detail, the basic procedures and viewpoints that were found helpful in obtaining the results presented.

ANALYSIS OF ESTIMATOR PERFORMANCE

One approach to an ML strategy would be to try a prescribed set of $\hat{\theta}$ and $\hat{\tau}$ values in (5) and to choose the maximizing set. This would clearly be an inefficient search procedure even if only a moderate precision in estimation is required. A better approach is to use an iterative search procedure, giving refined estimates of the parameters at each stage. One could subdivide the T_o-second interval so that sufficiently accurate estimates result when receiver measurements from the last interval are used. A notationally more convenient approach is to leave T_o fixed, then specify a permissible number of iteration steps. One of the several types of gradient search procedures is considered appropriate here. The simplest type is the "steepest ascent" procedure wherein the estimates are incremented by amounts proportional to the gradient components, $\partial\Lambda/\partial\hat{\theta}$ and $\partial\Lambda/\partial\hat{\tau}$. The problem here is that a suitable proportionality constant (step size) must be determined and typically an excessive number of iterations are required. A much faster convergence, in the vicinity of a local maximum, is obtained with a Newton-Raphson procedure where the gradient vector is multiplied by a 2x2 matrix to determine the increments. At the m-th iteration,

$$\begin{pmatrix}\hat{\theta}_m \\ \hat{\tau}_m\end{pmatrix} = \begin{pmatrix}\hat{\theta}_{m-1} \\ \hat{\tau}_{m-1}\end{pmatrix} - \begin{pmatrix}\Lambda_{\theta\theta} & \Lambda_{\tau\theta} \\ \Lambda_{\tau\theta} & \Lambda_{\tau\tau}\end{pmatrix}_{m-1}^{-1} \begin{pmatrix}\Lambda_{\theta} \\ \Lambda_{\tau}\end{pmatrix}_{m-1} \tag{6}$$

where the subscripts on the Λ variables denote partial derivatives. The difficulty with this (from both an analysis and implementation standpoint) is in the evaluation and inversion of the matrix at each step. This difficulty is circumvented by replacing the 2x2 matrix with its expected value so that the matrix in (6) is fixed for all iteration steps. Then the estimation strategy is

$$\begin{pmatrix}\hat{\theta}_m \\ \hat{\tau}_m\end{pmatrix} = \begin{pmatrix}\hat{\theta}_{m-1} \\ \hat{\tau}_{m-1}\end{pmatrix} - \begin{pmatrix}E(\Lambda_{\theta\theta}) & E(\Lambda_{\tau\theta}) \\ E(\Lambda_{\tau\theta}) & E(\Lambda_{\tau\tau})\end{pmatrix}^{-1} \begin{pmatrix}\Lambda_{\theta} \\ \Lambda_{\tau}\end{pmatrix}_{m-1} \tag{7}$$

and this modified strategy is accurate at moderately large values of SNR and T_o.

A relatively simple expression for estimator performance can be obtained from the variance of the steady-state ($m\to\infty$) errors called phase and timing jitter, obtained by setting $\hat{\theta}_{m-1} = \theta$ and $\hat{\tau}_{m-1} = \tau$ in (7). Thus only the 2nd-order moments of Λ_θ and Λ_τ and the 1st-order moments of $\Lambda_{\theta\theta}$, $\Lambda_{\tau\tau}$, and $\Lambda_{\tau\theta}$ are needed to evaluate performance. The resulting jitter variance expressions are known to be equal to the Cramer-Rao bounds [3] except when the log-likelihood function is modified to represent a more practical receiver implementation.

IMPLEMENTATION CONSIDERATIONS

Receiver implementation is greatly simplified by making q_k in (5b) a convolution integral by changing the integration interval to $(-\infty,\infty)$. Then q_k can be implemented as the sampled output of a coherent demodulation whose input is the received signal passed through a bandpass filter matched to the carrier data pulse. The constraint of a finite observation interval is maintained by changing the infinite sum in (5a) to a K-term sum with an equivalent $T_o = KT$. Then the gradient components in (7) are given by

$$\Lambda_\theta = \sum_{k=0}^{K-1} a_k \mathrm{Re}[-j\exp(-j\hat{\theta})\int_{-\infty}^{\infty} \alpha(t)\gamma^*(t-kT-\hat{\tau})dt] \tag{8a}$$

$$\Lambda_\tau = \sum_{k=0}^{K-1} a_k \mathrm{Re}[-\exp(-j\hat{\theta})\int_{-\infty}^{\infty} \alpha(t)\dot{\gamma}^*(t-kT-\hat{\tau})dt] \tag{8b}$$

which, relative to evaluation of Λ, requires a $\pi/2$ phase shift in the local oscillator for (8a) and a time differentiation for (8b).

The effect of this approximate implementation is to introduce an additive, noise-independent term in the jitter variances. The true ML receiver has jitter variances inversely proportional to SNR and K. The additive term for the suboptimal receiver varies approximately as K^{-2}. At high SNR and small K, this term can dominate and performance will be substantially poorer than for the ML receiver. This is often called pattern-dependent jitter, resulting from the random data sequence, and for a given modulation scheme, this component of jitter can be reduced by proper pulse shaping, in a manner similar to the control of intersymbol interference by means of suitable shaping of the data pulses.

We note from (5a) that the estimation procedure requires knowledge of the data sequence. In practice, this information is provided in either of two ways: (1) Synchronization may be accomplished during a training period where a known sequence of K symbols is transmitted and this sequence is stored at the receiver. (2) Another mode of operation uses the detected symbols at the receiver in place of the true values of a_k, resulting in a decision-directed synchronization strategy. We refer to both modes of operation as "data-aided" timing recovery. For simplicity of implementation, a "non-data-aided" approach can be used wherein the likelihood function in (4) is averaged over the data sequence. If certain constraints on pulse shape are imposed [2], then the averaging is readily performed and modified log-likelihood functions are used in (7). Analysis of jitter variances for non-data-aided cases shows that the results are strongly dependent upon signal bandwidth. The approach generally fails when the bandwidth is close to the Nyquist bandwidth (1/2T) for the symbol rate used. However, for bandwidths of 50% or more in excess of the Nyquist bandwidth, the performance is not significantly poorer than for the data-aided strategy.

In many cases, the preferred receiver implementation takes the form of a tracking loop where the local oscillator has a voltage-controlled frequency (VCO) and the sampler is also operated by a voltage-controlled clock (VCC). The strategy that we propose in this case is to update the VCO and VCC frequencies at the symbol rate (1/T). The dynamic behavior of the interconnected tracking loops is then characterized by (7) with K=1 used for evaluating the gradient components in (8) and the coupling matrix. With this strategy, the loop does not use a summer for the $a_k q_k$ samples and relies instead upon the recursive nature of (7) to average over past measurements. The result is a pair of first-order loop equations which can be linearized to evaluate the "small-jitter" steady-state performance of the combined loops. From an analysis of this type, we can determine the loop bandwidth that produces a steady-state jitter variance equal to that of a "one-shot" estimator based on a K-term sum. This allows us to evaluate tracking loop performance by the previously described method, with a loop bandwidth parameter inversely proportional to the T_o parameter.

As a final comment, we note that the VCO and VCC control voltages in the interconnected loops are formed as linear combinations of the gradient components if the coupling matrix in (7) is not diagonal. In the VSB or SSB case, this matrix has significant off-diagonal terms [1] and coupling of the gradient components is essential for rapid loop convergence. This aspect has been reported in other studies [6].

REFERENCES

1. Franks, L.E., "Timing recovery problems in data communication," Communication Systems and Random Process Theory, pp. 111-127, Sijthoff and Noordhoff, 1978.

2. Franks, L.E. and Meyers, M.H., "Synchronization performance of PAM carrier systems," ICC'78, pp. 42.5.1-42.5.5.

3. Meyers, M.H., "Timing recovery in data communications," Ph.D. dissertation, University of Massachusetts, Amherst, 1978.

4. Meyers, M.H. and Franks, L.E., "New results in timing recovery for data communication," NTC'78, pp. 27.5.1-27.5.4.

5. Meyers, M.H. and Franks, L.E., "Carrier phase and symbol timing recovery for PAM systems," submitted to IEEE Trans. on Communications, 1979.

6. Manicianti, M., Mengali, U., and Reggiannini, R., "A fast start-up algorithm for channel parameter acquisition in SSB-AM data transmission," ICC'79, pp. 25.1.1-25.1.5.

ON SOME POLLING STRATEGIES*

Einar P. Gudjohnsen, Donald F. Towsley and Jack K. Wolf

University of Massachusetts, Amherst, MA USA

ABSTRACT

Various techniques for polling, especially as applied to a computer-communication network, are discussed. Some modifications are suggested for improving the performance of an adaptive scheme previously described in the literature. A new and more efficient polling technique is described which is based upon a conflict resolution scheme for multiple access communications.

INTRODUCTION

Consider an ordered collection of n objects, each of which is identified by a unique identifier such as an address or a name. Each object is assumed to be in one of two states (say 0 or 1). We assume that each object is in state 1 with probability p, where p is a real number in the range $0 \leq p \leq 1$. We further assume that the random variables associated with the n objects being in state 0 or 1 are statistically independent.

A central authority desires to <u>poll</u> these n objects to determine their states. The form of the polling is a series of questions asked of the n objects. The types of responses that can be obtained from these n objects vary from system to system and will be discussed later in more detail. In general, however, we would like to minimize the number of questions (and responses) required in order for the central authority to determine the states of the n objects.

As an example of such a scheme, consider the central authority to be a teacher in a class and let the n objects be n students en-

D. G. Lainiotis and N. S. Tzannes (eds.), Advances in Communications, 238–244.

rolled in the class (but who are not all always present). Let p be the probability that any given student is present. Polling now consists of taking attendance in the class. Normally, the teacher would call the name of each student and the response obtained would either be "silence" if the student were absent or some sort of verbal response (e.g., the word "here") if the student were present. Such a scheme requires asking n questions. Assume for the moment that p was much less than 1. There would be many queries for which "silence" was the answer—clearly an embarrassing and inefficient method of taking attendance. Faced with such a dilemma, a teacher might consider breaking the students up into groups and asking questions like: "Is anyone here from group 1?" Only if the teacher received a positive response would he or she need to ask any further questions of group 1.

Another example, which is closer to the theme of this paper, is where the central authority is a central computer and where the objects are remote terminals. The state of the terminal would be a 1 if and only if it had a message to send to the central computer. The central computer would then poll the terminals to determine which of them had messages to send.

Two classic approaches to polling are <u>roll-call polling</u> and <u>hub polling</u> [1]. In roll-call polling the central authority queries each object in order, waiting for a response (or absence thereof) from the ith object before proceeding to query the (i+1)st object. In hub polling, the central authority merely starts the process by interrogating the first object, the second object responding immediately after the response (or lack thereof) from the first object without the need for a further question. Since both systems essentially interrogate every object, n algorithmic steps are required to poll n objects. Our aim will be to find polling strategies which require fewer (on the average) than n algorithmic steps.

<u>Probing</u> [2] is a technique whereby a group of objects are queried simultaneously. Only if there is any positive response from the group are further questions asked of members of that group. We will always assume that the form of question used in probing a group is: "Is any member of the group in state "i"?" (Note that we have left some choice in the question as to whether "i" should be 0 or 1. Normally we will take i equal to 1 but later we shall see that if p is close to 1, it might be useful to have "i" equal to 0.)

We will assume two different forms for the response received by the central authority to such a probing question. These correspond to two different models which we will refer to simply as model 1 and model 2 and which are described below.

Model 1 In model 1, each object in the group that was probed returns an answer "yes" or "no" to the question. The central authority receives a "yes" if and only if one or more object responds with a "yes". Otherwise it receives a "no". Thus, if the central authority receives a "yes" it knows that one or more objects in the group is in state "i" while if it receives a "no" it knows that no member of the group is in state "i".

Model 2 In model 2, any object that is in state "i", responds with its address. If one and only one object in the group responds, the central authority is assumed to receive the address of that object. On the other hand, if two or more objects in the group give a positive response (by sending their addresses), the resulting signal received by the central authority is distorted to the point where the addresses are unreadable. However, upon receiving such a distorted signal, the central authority knows that two or more objects are in state "i". If no objects respond, the central authority knows that no object in the group is in state "i". In summary, three types of signals are received by the central authority in this model. These signals indicate that:

(a) no object in the group is in state "i".
(b) one object in the group is in state "i" (and the signal gives the address of that object).
(c) two or more objects in the group are in state "i" (but the addresses of these objects are not known except for the knowledge that these objects were in the group queried).

PROBING USING MODEL 1

Hayes [2] described a probing scheme which utilized model 1 for the response to the questions. Since in model 1, a positive response does not give the address of the respondent, smaller and smaller groups need be probed until questions are asked of individual objects. His technique is best illustrated by an example.

Example

Consider that 8 objects denoted by the integers 0, 1, 2, 3, 4, 5, 6 and 7 are arranged so as to be the leaves of a binary tree as shown below in Figure 1. The other (internal) nodes in the tree are denoted by the integers 8, 9, 10,..., 14. The form of the tree is to suggest how objects are formed into groups for probing. If for example, we probe the subtree which grows from node 12, we are probing the objects 0, 1, 2 and 3 while if we probe the subtree which grows from node 10, we are probing the two objects 4 and 5. Node 14 is called the root node and a probe at this node, queries all eight objects. A probe at a leaf node (e.g., node 2) probes that single object. We say that the root

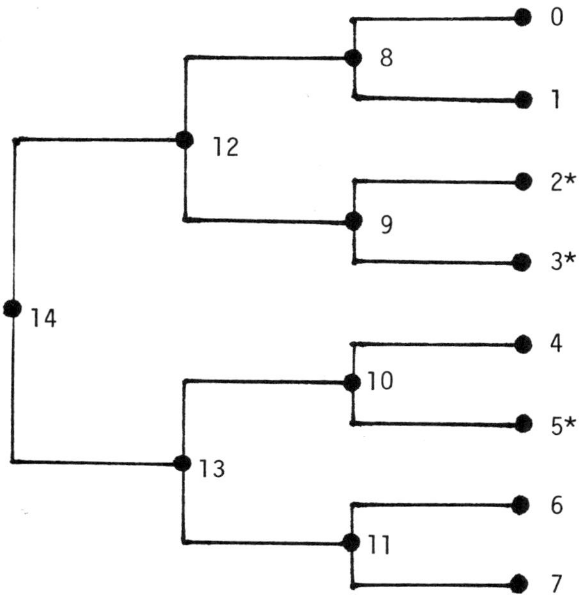

Figure 1. Binary Tree for Eight Objects

node, node 14, is at depth 0 in the tree, nodes 12 and 13 are at depth 1 in the tree, nodes 8, 9, 10 and 11 are at depth 2 in the tree and the leaf nodes (0, 1, ..., 7) are at depth 3 in the tree.

For this example assume that objects 2, 3 and 5 are in state 1 (indicated by a star) while objects 0, 1, 4, 6 and 7 are in state 0. Furthermore, we will assume that we initially start our probing at the root node although as will be discussed later this is not necessarily the best thing to do. Furthermore we will assume that in probing any group, the central authority asked the question: "Is any member of this group in state 1?"

For this example, the dialogue would proceed as follows: Comments as to who did or did not answer are shown in parentheses.

<u>Central Authority</u>: Is any member of the subtree growing from mode 14 in the 1 state?
<u>Answer</u>: Yes. (Answers from 2, 3 and 5.)
<u>C.A.</u>: ... subtree growing from node 12 ...?
<u>Answer</u>: Yes. (Answers from 2 and 3.)
<u>C.A.</u>: ... subtree growing from node 8 ...?
<u>Answer</u>: No. (Objects 0 and 1 are in 0 state.)
<u>C.A.</u>: ... subtree growing from node 9 ...?
<u>Answer</u>: Yes. (Again answers from 2 and 3.)

C.A.: ... subtree growing from node 2 ...?
Answer: Yes. (Only 2 is in this subtree and it answers.)
C.A.: ... subtree growing from node 3 ...?
Answer: Yes. (Only 3 is in this subtree and it answers.)
C.A.: ... subtree growing from node 13 ...?
Answer: (Answer from 5.)
C.A.: ... subtree growing from node 10 ...?
Answer: Yes. (Answer from 5.)
C.A.: ... subtree growing from node 4 ...?
Answer: No. (Only 4 is in this subtree and it is in the 0 state.)
C.A.: ... subtree growing from node 5 ...?
Answer: Yes. (Only 5 is in this subtree and it answers.)
C.A.: ... subtree growing from node 11 ...?
Answer: No. (Both 6 and 7 are in the 0 state.)

This completes the dialogue for this example.

Hayes was able to analyze his scheme to find the average number of queries for a given value of p and n. For small values of p, the average is much less than n but as p approaches 1, the average approaches 2n-1 which is the number of nodes in the entire tree. Hayes then commented that one need not start the probing at the root node (at depth 0) but rather one could start at any depth in the tree. Thus as p increases, one would increase the depth for the initial probe until for large p one would be probing only the leaf nodes which would take n queries.

The reader may have observed that certain questions in the previous example were redundant. For example, after the central authority received a "yes" after probing the subtree which grew from node 12 and then received a "no" when probing the subtree which grew from node 8, it would automatically know that it must receive a "yes" from probing the subtree which grew from node 9. Thus this question could be eliminated from the algorithm. Similarly, when the central authority received a "yes" from probing the subtree at node 10 and then received a "no" from probing node 4, it knows that node 5 will respond with a "yes" and this query can be eliminated.

A further improvement on Hayes' scheme is possible if $p>1/2$. Now it would be better to query groups with the question: "Is any object in this group in state 0?" Thus $p = 1/2$ rather than $p = 1$ will result in the largest number of questions.

PROBING USING MODEL 2

In this section a new probing technique is described which is based upon the answers allowable in model 2. Since more information is conveyed in the response to the question, fewer questions need be asked. The basic difference between this scheme and the previously described scheme is that if the group queried contains a single object which responds positively, no further questions need be asked of objects in the group since the address of the respondent is conveyed in the answers.

We illustrate the technique by considering the same example as in the previous section.

Example. Consider the same tree as shown in Figure 1 with objects 2, 3 and 5 in state 1. The dialogue between the central authority and the objects is as follows:

Central Authority: Is any member of the subtree growing from node 14 in the 1 state?
Answer: A garbled response. (Objects 2, 3 and 5 all send their addresses so response is garbled.)
C.A.: ... subtree growing from node 12 ...?
Answer: A garbled response. (Objects 2, and 3 respond.)
C.A.: ... subtree growing from node 8 ...?
Answer: No. (Neither objects 0 or 1 respond with an address.)

At this point the central authority knows that objects 2 and 3 are in state 1.

C.A.: ... subtree growing from node 13 ...?
Answer: Address of object 5.

This ends the polling cycle.

As in Hayes' scheme one need not start the probing process at the root node of the tree but could rather initially probe smaller subgroups. Furthermore, if $p>1/2$, one should ask questions regarding elements which are in state 0 rather than in state 1. It should be noted that an absolute upper bound to the number of questions required is $\frac{M}{2}$. This follows since there are $\frac{M}{2}$ subgroups containing 2 objects each, and probing each such subgroup requires only one question and answer. An analysis of this scheme is given in [3]. The idea for the scheme came from Capetanakis's scheme for resolving conflicts in a multiple-access communications system [4].

REFERENCES

(1) M. Schwartz, Computer Communication Network Design and Analysis, Prentice Hall, Inc., 1977, Chapters 12-13.

(2) J. F. Hayes, "An Adaptive Technique for Local Distribution," IEEE Trans. on Comm., Vol. COM-26, No. 8, August 1978.

(3) E. P. Gudjohnsen, "Adaptive Polling in Computer Communication Networks Using Random Access Techniques," M.S. Dissertation, University of Massachusetts, September 1979.

(4) J. I. Capetanakis, "The Multiple Access Broadcast Channel: Protocol and Capacity Considerations," Ph.D. Dissertation, M.I.T., August 1977.

A DIGITAL TRANSMITTER AND RECEIVER FOR AUTOMATIC MONITORING OF CHANGES IN THE STATE OF A GROUP OF SWITCHES

N.C. Voulgaris, L.P. Lambrianides and C.C. Halkias

National Technical University of Athens
Chair of Electronics

Digital terminal units have been developed capable of scanning continuously the state of a group of switches of a system and transmitting information only when changes in their state have occurred.

Since the digital information can be transmitted via FSK modems on regular telephone lines, these units offer the capability of supervising the status of remotely-located switches from a control center.

In applications where the switch status changes are infrequent, many of these units can use the same line, in a time-sharing scheme, resulting in considerable saving in telephone channel use time.

System considerations and general design principles are given in some detail. A design example applying these principles is described for a group of eight switches. The transmitted digital word contains error check bits for single error detection and correction (Hamming code), but other codes can be used for higher transmission reliability.

1. INTRODUCTION

A previous work (1) described the design and development of terminal units capable of encoding, transmitting, receiving and decoding the status of a group of switches.
The digital units described in this paper are capable of scanning continuously the state of a group of switches and transmitting

D. G. Lainiotis and N. S. Tzannes (eds.), Advances in Communications, 245–252.

information to a control center only when changes in their status have occured. In typical applications, such as monitoring the status of the automatic switches of an electric power station, or the remote supervision of doors and windows in a factory, bank etc.as part of a security alarm system, switch status changes are infrequent and, consequently, the proposed units can result in considerable saving in telephone channel use time. Thus, in a time-sharing scheme, the same telephone channel can be used more effectively.

This paper gives the system considerations and general design principles for a pair of digital terminal units consisting of a transmitter and a receiver, and describes, in turn, the development of a prototype for these units capable of monitoring the state of eight switches. The design can be extended to groups of 10, 16, 20 etc. switches, the upper limit determined by the required sampling speed of each switch and the bandwidth of the available transmission channel.
The system considerations are stated in section 2 while the system design principles, a description of operation and the development of an experimental prototype for these digital units are given in section 3. The conclusions drawn from this design are finally stated in section 4.

2. SYSTEM CONSIDERATIONS

The basic idea to transmit only changes in the state of a group of switches involves the transmission of a logical word consisting of the address of the switch and its new state. In general, the word may include check bits for detection and correction of one or more errors.

2.1 Word Format and Error Correction Code

The word format, therefore, will be, in general, as shown below:

← switch address →				state	← check bits →			
I_1	I_2	...	I_{m-1}	I_m	P_1	P_2	...	P_k

where I_1, I_2,.... I_m are information bits and P_1, P_2,... P_k are error check bits.

The number of bits required by the switch address should be sufficient to identify not only the number of the switch in the group but also the place where the group of switches is located. For the switch state only one bit is needed. The number of required check bits depends on the desired transmission reliability and the code used. In the design example given in section 3, a single error detection and correction Hamming code is used with eight switches for illustration purposes.

Thus, for m information bits, k check bits are needed, where

$$2^k \geq m + k + 1$$

and with $m = 4$, $k = 3$.

2.2 Synchronization

In order to decode the received logical word, it is necessary either to have a synchronized clock at the receiver or to encode synchronization pulses within the word at the last stage of the transmitter. The latter method is used in the design example given here and involves the following:

A synchronization pulse consisting of a transition from low to high is encoded in the second half of an information cell. The result obtained from a zero-zero-one information cell sequence is shown below:

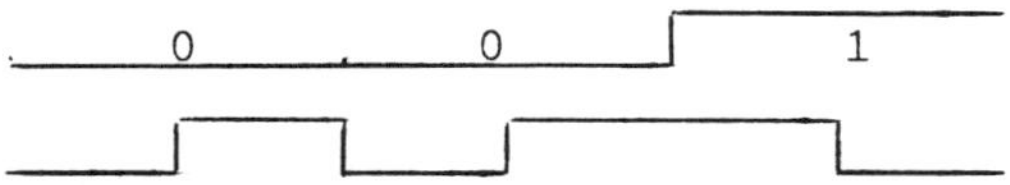

without synchronization

with encoded sync pulses

This encoding format insures that there will be a transition in each data cell time. These transitions can be sensed at the receiver to regenerate the required clock pulses. It should be noted that as a result of the above clock pulse encoding and decoding, the first bit of the received word is lost, and therefore, provisions must be made in the transmitter and receiver to take this situation into consideration.

3. SYSTEM DESIGN PRINCIPLES

3.1 General Design Criteria

Figure 1 shows a block diagram indicating a generalized configuration of a pair of terminal units interconnected through a telephone line and FSK modems. The following design criteria were taken into consideration:

- Whenever any of the switches S_1, S_2, ... S_N changes its state, the transmitter detects the change and forms a logical word, consisting of the switch address and new state, as well as the appropriate number of check bits for error detection and correction, and sends that word to the receiver.

- The receiver decodes the received signal, extracts the necessary synchronizing clock pulses, checks and corrects errors that may have occurred during signal transmission and finally stores in memory and displays the received change.

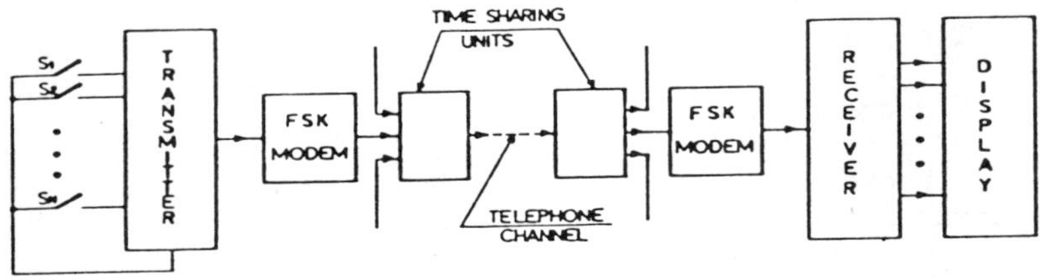

Fig. 1: Block Diagram of a Pair of Terminal Units in a Generalized Configuration

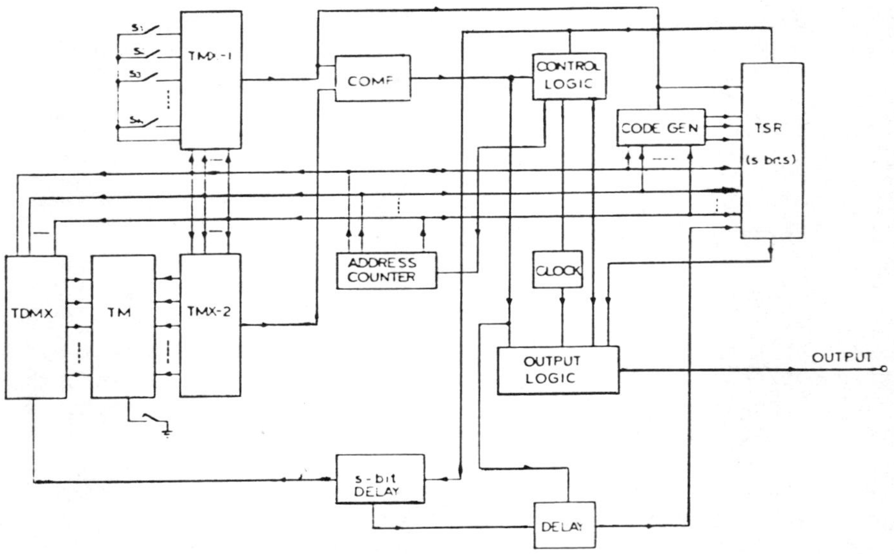

Fig. 2: Transmitter Block Diagram

The transmitter is also equipped with a pair of contacts which may be closed by a telecontrol signal or manually by an operator, in a test mode, to send the state of all switches to the receiver.

The transmitter encodes within the logical-word-bits synchronization pulses which can be sensed at the receiver to regenerate clock pulses.

3.2. Digital Transmitter

A more detailed block diagram of the transmitter is shown in Fig. 2. The present state of a switch S_i is checked by multiplexer TMX-1, while its previous state, which is stored in memory TM, is checked by multiplexer TMX-2. The outputs of TMX-1 and TMX-2 are compared in the digital comparator, COMP, and the result is delivered to the CONTROL LOGIC. If the comparison shows that the two outputs are identical, then the multiplexers are allowed to check the next switch. However, if there is a difference, which means that this switch has changed state, the CONTROL LOGIC stops the multiplexers and the following information is loaded in the parallel-in/serial-out shift register, TSR : The address of the switch that changed state, the new state and the appropriate check bits according to the chosen error code. On completion of the loading process, shift pulses are applied to the shift register and the resulting serial word is then delivered to the OUTPUT LOGIC stage. A sync pulse is encoded here between the bits of the logical word in order to insure the required synchronization at the receiver. After the logical word is sent out, the new state of the switch is stored in the appropriate memory position, and the multiplexers are released to resume the scanning process, until a change is found when the above-described procedure is repeated.

3.3. Digital Receiver

As shown in the detailed receiver block diagram of Fig. 3, the received logical word is applied to a D-type flip-flop, FF, and to a clock pulse regeneration circuit. The latter circuit extracts the clock pulses encoded within the information bits and acts then as the effective clock for the receiver. The output of the D-F/F consists of only the information bits which then enter into the serial-in/parallel-out shift register RSR. Error check and correction, if necessary, is performed in the ERROR CHECK & CORRECT circuit and the BCD address of the switch, whose state has changed, is applied to the demultiplexer, RDMX. The output of the RDMX writes in the memory, RM, the new switch state and the result then is either recorded or shown on a display.

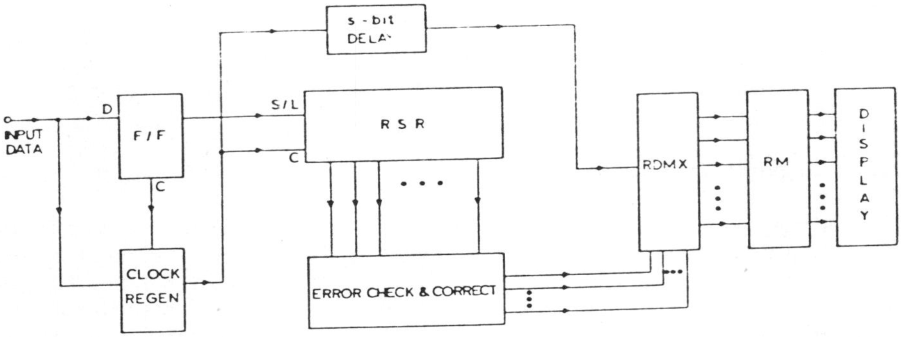

Fig. 3: Receiver Block Diagram

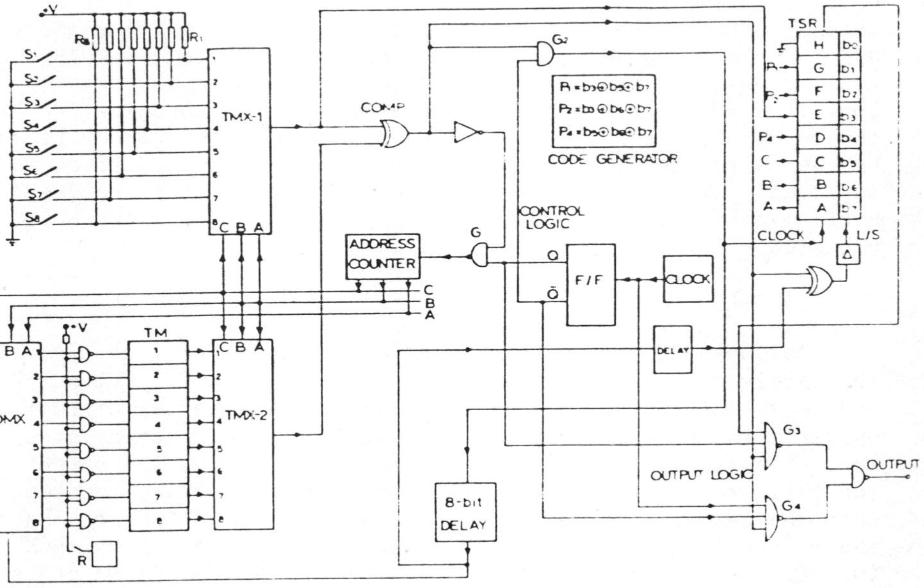

Fig. 4: Transmitter Diagram.

3.4 Development of a Prototype

A prototype based on the previous described general design principles has been developed. This development is part of a project designed to monitor the state of the automatic switches of an electric power generating station to a control center.
In this design, the state of eight switches is scanned at a rate of 25 Hz. The transmitter output goes through an FSK modem and through a time-sharing control unit to a telephone channel.At the receiving end, the signal goes through similar units and is finally decoded and the appropriate switch indication is displayed. A circuit diagram of the transmitter is shown in Fig. 4 and that of the receiver is shown in Fig. 5.

In the design example the Hamming code for single error detection and correction is used [2-5]. Since according to a CCITT study [3], up to 60% of all bit errors in a similar type of transmission are isolated single-bit errors, it follows that this code is quite adequate.

4. CONCLUSIONS

Digital units, consisting of a transmitter and a receiver , were described in this paper. These units are capable of scanning continously the state of a group of switches and transmitting information to a remotely-located control center only when changes in the switch status have occured. In conjuction with FSK modems, the transmission can be made on regular telephone lines. If switch status changes are infrequent, then, these units can result in considerable saving in telephone channel use time, and in a time-sharing scheme, a telephone channel can be used more effectively.General design principles are described and an example is given for a design of a transmitter and receiver capable of monitoring the state of 8 switches. This design was inplemented using mostly TTL ICs and the test results are completely satisfactory. The transmitted word contains error check bits for single error detection and correction, but other codes can be used for higher reliability. The design requirements specify a 50-baud transmission speed, but higher speeds can easily be accomplished by changing the appropriate time constants. The design can be extended also to groups of more switches according to the required sampling speed of each switch and the transmission channel bandwidth.

A comparison between a system in which information on the state of 8 switches is transmitted continuously and a system in which only switch state changes are sent reveals that the latter method is more efficient when the rate of change of states is less or equal on the average to two switch changes per scanning period, i.e. 16T where T is the clock period. The minimum time a switch has to remain in its new state in order for the change to be trans-

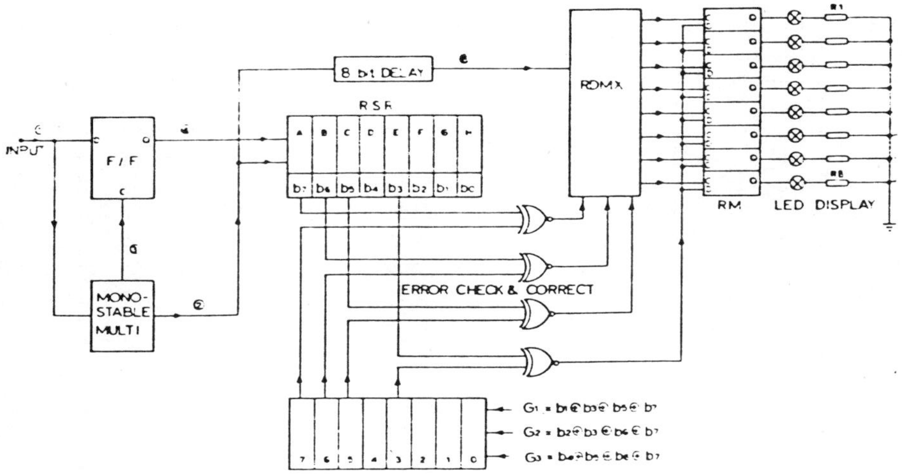

Fig. 5: Receiver Diagram

mitted depends on the scanning speed. The worst case occurs when all switches change states simultaneously. With the continuous system, the switch change will be noted if the switch remains in its new state for more than 16T. With the second system, the switch is required to remain in its new state for at lease 64T.

REFERENCES

(1) N.C. Voulgaris: "Digital terminal units for continuously monitoring the state of a group of switches", Greek Public Power Corp. internal report, July 1974.

(2) T. Sifferlen and V. Vartanian: "Digital Electronics with Engineering Applications", Prentice-Hall, Inc., Englewood Cliffs, N.J.

(3) J. Martin: "Security, Accuracy and Privacy in Computer Systems", Prentice-Hall, Inc., Englewood Cliffs, N.J. 1973,p.97.

(4) J. Martin: "Systems Analysis for Data Transmission", Prentice-Hall, Inc., Englewood Cliffs, N.J. 1972.

(5) E. Jorgensen: "Communications Alternatives for Monitoring and Control Systems", IEEE Conference on Computer Monitoring of Power Systems, Bell Telephone Laboratories, 1972.

SWITCHED POLARITY DPLL's FOR PSK DEMODULATION

N.A.D'Andrea and F.Russo

CSMDR - Consiglio Nazionale delle Ricerche
Via Diotisalvi, 2 - Pisa - Italy -

A class of digital phase-locked loops is analyzed which are conceived for PSK carrier recovery and for digital data stream detection.
Design constraints and a closed-form expression of the acquisition time are derived in the absence of noise for a binary quantized loop.

1. INTRODUCTION

In this paper we suggest some changes in the basic scheme of a non-uniform sampling digital phase-locked loop (DPLL) /1-6/.
These implementations are useful for PSK carrier recovery since they exhibit the usual π radians phase ambiguity inherent in all systems that attempt to recover the carrier phase from a biphase modulated signal. This is because changing the sign of the received signal leaves the sign of the recovered carried unaltered.
Moreover, the proposed circuits also provide the detection of the transmitted digital data stream, so that in this sense they represent a digitalized implementation of a Costas loop /7/.
The demodulation properties and the acquisition behavior of the loops are analyzed in the absence of noise.
Design constraints and closed-form expression of the acquisition time are also derived in the case of phase- and frequency-step inputs, when signum quantizers are employed.
This class of loops enjoys the usual advantages of digital systems, i.e. extreme stability and component economy, as ecompared with conventional analog loops; in addition, the quiescent clock period can be easily modified, so that one loop might be time-shared to synchronize several inputs.

D. G. Lainiotis and N. S. Tzannes (eds.), Advances in Communications, 253–260.

2. LOOP ANALYSIS

The loop configuration, proposed for PSK demodulation is sketched in fig. 1.

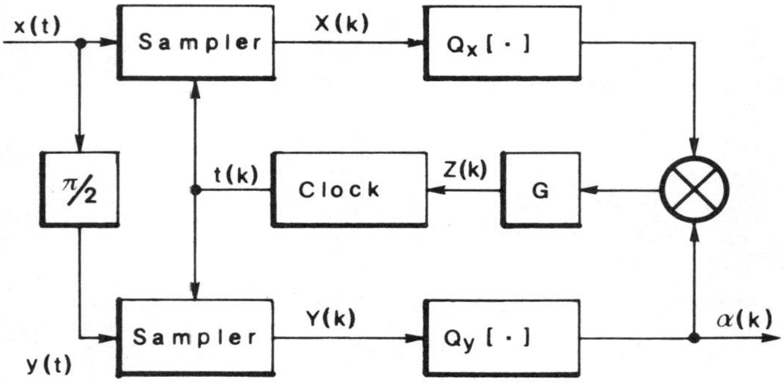

Fig. 1: Block diagram of proposed loop

The received signal is

$$x(t)=V\sin\left[\omega_o t+\theta(t)\right] \tag{1}$$

with

$$\theta(t)\triangleq \omega_o(\xi-1)t+\alpha(t)\,\pi/2+\theta_o; \tag{2}$$

here ξ represents the input frequency normalized with respect to the clock nominal frequency ω_o, $\alpha(t)$ may assume the value +1 or -1 in each symbol period T and θ_o is the input phase step ($|\theta_o|\leq\pi$). The input signal x(t) and its in-quadrature version y(t) are sampled at time produced by the clock according to the rule:

$$t(k)=t(k-1)+\frac{2\pi}{\omega_o}-Z(k-1)=\frac{2k\pi}{\omega_o}-\sum_{i=0}^{k-1}Z(i) \tag{3}$$

In (3), t(0) is assumed equal to zero and Z(k) is the output of the digital filter which in a first-order loop is modeled as an amplifier of gain G. By using (3), the sampler outputs result in

$$x[t(k)]\triangleq X(k)=V\sin\varphi(k)$$
$$y[t(k)]\triangleq Y(k)=V\cos\varphi(k) \tag{4}$$

where:

$$\varphi(k)\triangleq\theta(k)-\theta_c(k) \tag{5}$$

is the phase error between the incoming signal and the clock reference

$$\theta_c(k) \triangleq \omega_o \sum_{i=0}^{k-1} Z(i) \tag{6}$$

at the sampling instant t(k).
In this way the clock control signal becomes

$$Z(k)=GQ_x[V\sin\varphi(k)]\,Q_y[V\cos\varphi(k)] \tag{7}$$

$Q_x[\cdot]$ and $Q_y[\cdot]$ being the outputs of the quantizers (fig. 1), which are supposed to be odd functions of the argument.
From (6) we obtain the difference equation of the clock error

$$\theta_c(k+1)=\theta_c(k)+\omega_o Z(k) \tag{8}$$

whereas, by utilizing (2), (3) and (8), from (5) we have:

$$\varphi(k+1)=\varphi(k)-\omega_o\xi Z(k)+2\pi(\xi-1)+[\alpha(k+1)-\alpha(k)]\pi/2=$$

$$=\tilde{\varphi}(k+1)+[\alpha(k+1)-\alpha(k)]\pi/2 \tag{9}$$

which is the loop phase error difference equation.
In (9) $\tilde{\varphi}(k+1)$ represents the value the phase error should assume if $\alpha(k+1)=\alpha(k)$.
The initial conditions are $\varphi(0)\triangleq\varphi_o=\theta_o+\alpha(0)\pi/2$ and $\theta_c(0)=0$.
Finally, from (3) and (4), the time the phase error takes to pass from φ_o to $\varphi(m)$ in m samplings is expressed by:

$$t(m) = \frac{\varphi(m)-\varphi_o+2m\pi}{\xi\omega_o} \quad . \tag{10}$$

Now, to place into evidence the effect of the data modulation on the phase error $\{\varphi(k)\}$, we suppose that a change in the bit data stream occurs between the h-th and (h+1)-th sampling:

$$\alpha(h+1)\neq\alpha(h) \tag{11}$$

Making use of (7) and (9), we have:

$$Z(h+1)=GQ_x[V\sin\tilde{\varphi}(h+1)]\,Q_y[V\cos\tilde{\varphi}(h+1)] \quad . \tag{12}$$

This entails

$$Z(h+1)=\tilde{Z}(h+1) \tag{13}$$

i.e. $\{Z(k)\}$ is independent of the data modulation. Hence $\varphi(h+2)=$ $=\tilde{\varphi}(h+2)+[\alpha(h+1)-\alpha(h)]\pi/2$ and so on, as long as a new change in

the bit stream occurs.
In conclusion, if the loop were tracking the positive going zero-crossing, after the bit transition (11) it would track the negative going ones and viceversa. In other words, if the phase error sequence fluctuates around $\varphi=0$ when $k\leq h$, it will fluctuate around π radians when $k\geq h+1$, without any phase transient in the loop; in fact from (8) and (13) it seen that the sequence $\{\theta_c(k)\}$ coincides with the sequence $\{\tilde{\theta}_c(k)\}$.
From (7) and (9) it is evident that, to gain a right loop operation, it is sufficient for $Q_y[\cdot]$ to recognize if each sampling instant is nearer to a positive or negative going zero crossing of x(t) and that $Q_x[\cdot]$ provides the proper correction. With this aim we choose a signum quantizer for $Q_y[\cdot]$, while, as in usual DPLL's, the functional characteristic of $Q_x[\cdot]$ determines the loop performance /3,8/.

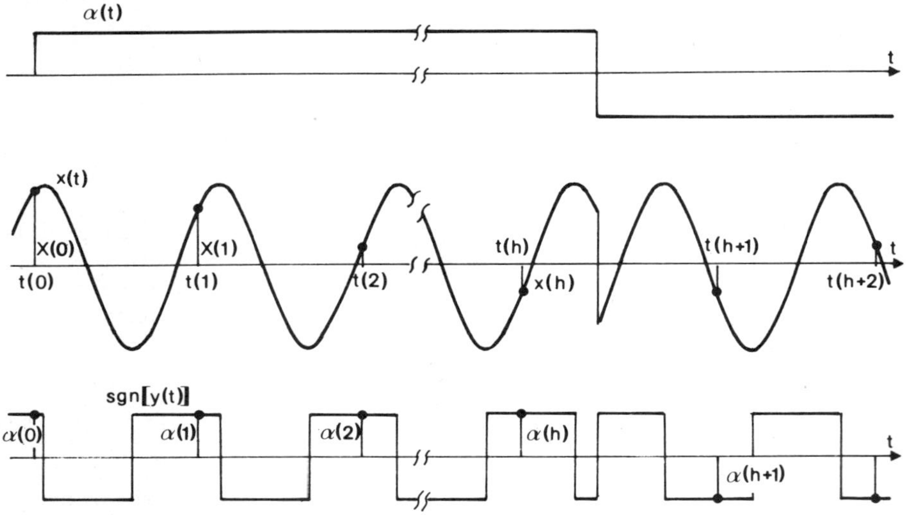

Fig. 2: Loop operation

Fig. 2, where the loop operation is visualized, shows that $Q_y[V\cos\varphi(k)]=\mathrm{sgn}[\cos\varphi(k)]>0$ when the loop is tracking the positive going zero-crossing while $Q_y[.]<0$ when the loop is tracking the negative ones. In this way $\{Q_y[V\cos\varphi(k)]\}$ represents a sampled version of $\alpha(t)$ or its inverse. This ambiguity and the symbol synchronization may be resolved by the usual methods /7,9/. In the sequel, to simplify things, we assume both $Q_x[.]$ and $Q_y[.]$ to be implemented as signum quantizers, so that we have (7):

$$Z(k)=G\mathrm{sgn}[\sin\varphi(k)]\,\mathrm{sgn}[\cos\varphi(k)]=G\mathrm{sgn}[\mathrm{sen}2\varphi(k)] \tag{14}$$

and (9):

$$\varphi(k+1)=\varphi(k)-\lambda\,\xi\,\mathrm{sgn}[\mathrm{sen}2\,\varphi(k)]+2\pi(\xi-1)+$$

$$+[\alpha(k+1)-\alpha(k)]\;\pi/2 \qquad (15)$$

where $\lambda \triangleq \omega_o G$ represents the model gain.

3. ACQUISITION BEHAVIOR

A useful tool to understand the acquisition behavior of the loop is the graphical approach proposed in /3,4/.
Let us consider the following functions (fig. 3):

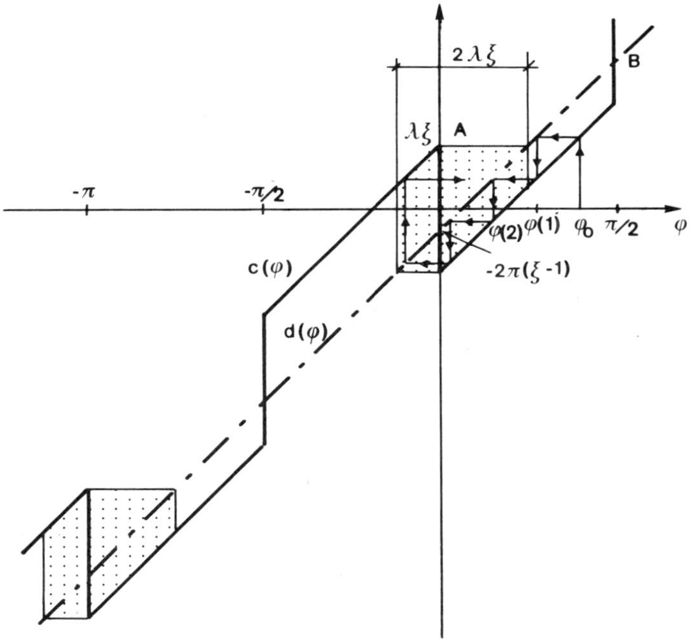

Fig. 3: Graphical analysis of the loop behavior

$$c(\varphi)=\varphi-\lambda\,\xi\,\mathrm{sgn}(\mathrm{sen}2\,\varphi)$$

$$d(\varphi)=\varphi-2\pi(\xi-1)$$

It is a simple matter to verify that, when no symbol transition occurs, the phase error sequence φ_o, $\varphi(1)$, $\varphi(2)\ldots$, obtained

through the graphical construction explained in fig. 3, is the same as the sequence given in (15). Whereas, when a transition (11) occurs between the h-th and the (h+1)-th sampling the phase error sequence $\varphi(h+1)$, $\varphi(h+2)$... may be obtained by the same construction on condition that we substitute $\varphi(h)$ with $\varphi(h) \pm \pi$, according to the results in the previous section.
With the same reasoning as in /4/, we deduce that, in the absence of noise, no cycle slipping occurs if $d(\varphi)$ intersects $c(\varphi)$:

$$\frac{2\pi}{2\pi+\lambda} \leq \xi \leq \frac{2\pi}{2\pi-\lambda} \tag{16}$$

and if the ordinate of point A is less than that of point B (fig. 3):

$$\lambda\xi \leq \pi/2-2\pi(\xi-1), \tag{17}$$

From (16) and (17) we find that the model gain must be

$$\lambda \leq 2\pi/9 \quad , \tag{18}$$

Substituting $\lambda=2\pi/9$ in (16) we find the largest normalized frequency range for which the loop will phase lock regardless of the initial condition and without cycle slippings:

$$9/10 \leq \xi \leq 9/8 \tag{19}$$

Moreover, as in /4/, we define the acquisition to be accomplished as soon as the phase error reaches one of the intervals (dotted regions in fig. 3):

$$2\pi(\xi-1)-\lambda\xi+n\pi \leq \varphi \leq 2\pi(\xi-1)+\lambda\xi+n\pi \ , \quad (n=0,\pm1,\ldots). \tag{20}$$

Therefore denoting by 2ϵ the extent of each region in fig. 3, when ξ assumes the maximum value consistent with (16) (note that 2ϵ also represents the maximum final phase error range), we get:

$$\epsilon=2\pi\lambda/(2\pi-\lambda) \leq \pi/4 \tag{21}$$

or

$$\lambda=2\pi\epsilon/(2\pi+\epsilon) \leq 2\pi/9 \tag{22}$$

and

$$9/10 \leq \frac{2\pi+\epsilon}{2\pi+2\epsilon} \leq \xi \leq \frac{2\pi+\epsilon}{2\pi} \leq 9/8 \ . \tag{23}$$

In this way, choosing ϵ according to the specific application, by (22) and (23) we get the model gain λ and the relevant pull-in range

$$B=\epsilon f_o \frac{2\pi+\epsilon}{2\pi(\pi+\epsilon)} \leq \frac{9\pi f_o}{40} \quad . \tag{24}$$

Let us note that the trade-off between B and ϵ in (24), if too severe, may be overcome employing a multilevel quantizer /3/ instead of a signum one in $Q_x[\cdot]$.

There remains now to evaluate the acquisition time. With reference to fig. 3, we note that, when $|\varphi_o|<\pi/2$, the acquisition is achieved (see (20)) as soon as the phase error reaches the value: $\epsilon\,\mathrm{sgn}(\varphi_o)+2\pi(\xi-1)$ and that each correction is: $\lambda\xi\,\mathrm{sgn}(\varphi_o)-2\pi(\xi-1)$, so that the number of samplings to acquire is expressed by

$$\mathcal{N}(\varphi_o)=\left.\frac{\varphi_o-[2\pi(\xi-1)+\epsilon\,\mathrm{sgn}(\varphi_o)]}{\lambda\xi\,\mathrm{sgn}(\varphi_o)-2\pi(\xi-1)}\right|_{u.i.} \tag{25}$$

In (25) the subscript u.i. stands for "upper integer" since the number of samplings is obviously an integer. Finally, from (10) we get the acquisition time:

$$t(\mathcal{N})=\frac{2\pi\mathcal{N}(\varphi_o)}{\omega_o}\left[1-\frac{\lambda}{2\pi}\,\mathrm{sgn}(\varphi_o)\right] \quad . \tag{26}$$

If $|\varphi_o|>\pi/2$, the acquisition is achieved when the phase error reaches the region $[2\pi(\xi-1)+\pi\,\mathrm{sgn}(\varphi_o)-\epsilon, 2\pi(\xi-1)+\pi\,\mathrm{sgn}(\varphi_o)+\epsilon]$, so that the number of samplings and the time to acquire may be obtained by (25) and (26) replacing φ_o with $\varphi_o-\pi\,\mathrm{sgn}(\varphi_o)$. In fact, being in this case $|\varphi_o-\pi\,\mathrm{sgn}(\varphi_o)|<\pi/2$, (25) applies, owing to π radians periodicity of $c(\varphi)$.

To fully specify the loop transient behavior, it seems reasonable to assume the product $\mathcal{P}=Bt(\mathcal{N})$ as a performance factor. In fig. 4 $\mathcal{P}$ is represented versus ϵ/π for various φ_o and $\xi=1$. To get on easy representation, fig. 4 has been drawn disregarding the rounding to the upper integer in (25).

4. CONCLUSIONS

A digital loop has been proposed, which recovers the PSK carrier and provides the detection of the transmitted data stream. The acquisition behavior and demodulation properties have been investigated in the absence of noise. Moreover a closed-form expression of the acquisition time has been derived when signum quantizers are employed. In the presence of noise is believed that the previous analysis is accurate for large enough signal-to-noise ratios. In fact, the analysis in /4/, relevant to a conventional DPLL with a binary quantizer , shows that the acquisition time is practically independent of the signal-to-noise ratio when this is greater than 10 dB. Finally we note that the acquisition time may be reduced and the trade-off between the final phase error range 2ϵ and the pull-in range B may be removed by properly shaping the quantizer $Q_x[\cdot]$/3/.

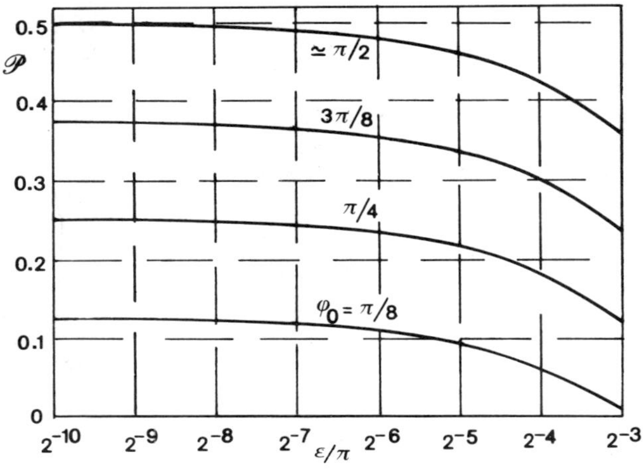

Fig. 4: Product pull-in range x acquisition time -

- REFERENCES -
/1/ C.P.Reddy,S.C.Gupta:"A Class of All Digital Phase Locked Loops: Modeling and Analysis", IEEE Trans., vol.IECI, November 1973.
/2/ A.Weinberg, B.Liu:"Discrete Time Analysis of Nonuniform Sampling First-and Second-Order Digital Phase-Lock Loops", IEEE Trans. vol.COM-22, February 1974.
/3/ F.Russo, N.A.D'Andrea:"On a Multilevel Quantized Digital Phase-Locked Loop", Proc.of XV Annual Allerton Conf.on Communication, Control and Computing, September 1977, Monticello, Illinois.
/4/ N.A.D'Andrea, F.Russo:"A Binary Quantized Digital Phase-Locked Loop: A Graphical Analysis", IEEE Trans., vol. COM-26, Sept. 1978.
/5/ W.C.Lindsey, C.M.Chie:"Acquisition Behavior of a First-Order Digital Phase Locked Loop", IEEE Trans.,vol.COM-26, Sept. 1978.
/6/ F.Russo:"Graphical Analysis of a Digital Phase-Locked Loop", IEEE Trans. on AES-15, January 1979.
/7/ W.C.Lindsey: "Synchronization Systems in Communication and Control", Prentice Hall, N.J., 1972.
/8/ N.A.D'Andrea, F.Russo:"Analysis of a Multilevel Quantized DPLL with Phase Step and Input Noise", Internal Report of CSMDR, CNR, Pisa, February 1979.
/9/ W.C.Lindsey, M.K.Simon: "Telecommuniation Systems Engineering", Prentice Hall, N.J., 1973.

COMBINATION OF THE DUOBINARY TECHNIQUE WITH FM FOR ACHIEVING HIGH BIT TRANSMISSION RATES

C. Dagakis* , G. Papadopoulos**, Ch. Mantakas*

* Nuclear Research Center Demokritos, Athens, Greece
** Applied Electronics Laboratory, University of Patras, Greece

Abstract: In this paper a new technique is presented for transmitting binary data over a telephone line at rates considerably higher than what is achieved by the straight transmission of the duobinary data. In the proposed method, the duobinary signal modulates a VCO that has a free-run frequency half the bit rate, in contrast to the existing frequency shift-keying techniques where the carrier frequency is much higher than the bit rate. Thus, for each level of the duobinary signal one half of the cycle is transmitted, having a duration that corresponds to the particular level. The receiver measures this duration between zero crossings and converts it to a voltage level, thus, reconstructing the duobinary signal. The system was used in a city telephone link using an equalizer at the receiver. Bit rates as high as 10 KHz were achieved.

I. INTRODUCTION

The problems encountered in the design of modems for the transmission of information through the already existing channels are twofold. First, the power spectrum of the signal to be transmitted must be reduced. Second, the most efficient use of the available bandwidth must be made using the appropriate modulation. There is a great variety of methods for handling these two problems. For the bandwidth reduction, the duobinary technique has been widely used. This technique forms the basis of the approach described in this paper. Regarding the problem of modulation, a large variety of modems has been developed. These use various methods of modulation to achieve the desired efficient transmission through a telephone line. Bit rates up to 9600 bits per sec have

D. G. Lainiotis and N. S. Tzannes (eds.), Advances in Communications, 261–268.

been attained. For these high bit rates the resulting circuitry is quite complex.

The technique proposed here is based on a simple idea and can be easily implemented. It offers the capability of voice and data transmission up to 10 kbits per sec. In the following section we present the description of the proposed method.

II. DESCRIPTION OF THE PROPOSED TECHNIQUE

As mentioned in the introduction, the duobinary technique forms the basis of the method proposed in this paper. With this technique a controlled amount of intersymbol intereference is introduced, which results in a redistribution of the signal power (1), (2).

The spectrum for the duobinary signal has the form:

$$|H(f)| = 2\cos \pi f/R \qquad |f| \leqslant R/2 \qquad (1)$$

where R is the bit rate of the binary sequence.

There is also the so called "modified" duobinary in which there is no D.C. component. The spectrum that results in this case is,

$$|H(f)| = 2\sin \frac{2\pi f}{R} \qquad |f| \leqslant R/2 \qquad (2)$$

The system proposed in this paper is shown in Fig. 1. The raw binary data are synchronized with the system clock generated by means of a VCO. The synchronized data are next used to generate the three-level duobinary signal which in turn is applied to the control input of the VCO, thus modulating the VCO frequency. The output of the VCO is the signal that is transmitted in the channel, after it has gone through a pre-equalizer. The novel aspects of this approach are:

a) The bit-rate is defined by the VCO and the zero-crossing detector. Thus the duration of each bit corresponds to one half of the VCO period.
b) The system clock, that is the VCO, is modulated by the duobinary signal and thus forms the signal that is transmitted in the channel. Each half period of the VCO is now modulated by the corresponding level of the duobinary signal.

The above two considerations result in a simple transmitter and a simple receiver.

As can be seen from the above discussion, the approach described here is, essentially, an FM modulation method which changes the frequency deviation of the FM signal according to the duobinary level. It must be also stressed that the duobinary level corresponding to each bit interval modulates half of the VCO period, thus making efficient use the bandwidth.

The receiver is shown in Fig. 2. It consists of an equalizer, a zero-crossing detector, a half-period duration estimator and a D/A converter. The equalizer compensates for the distortion caused by the channel. The zero-crossing detector is the same as the one in the transmitter and generates a clock every half period of the VCO signal. This clock triggers the "half-period duration estimator" circuit (a counter) which counts in the time interval between two successive clock pulses. Each counting is converted, by means of the D/A converter, to a level which corresponds to the duobinary level that caused this deviation. By passing this through a full-wave rectifier we get the binary sequence.

The waveforms obtained at the output of the VCO are shown in the upper part of Fig. 3. In our implementation the VCO had a free-run frequency f_o=2.5 KHz and was modulated by the duobinary signal seen in the bottom of Fig. 3. The frequency deviation was Δf=500 Hz. From this figure it can seen that three sinusoids appear in the first semi-period. Each one of them corresponds to one of the three duobinary levels (-1, 0, +1).

III. FREQUENCY SPECTRUM EVALUATION OF THE MODULATED SIGNAL

An FM signal has the form,

$$\upsilon(t) = A\cos\left[\omega_o t + \Delta\omega \int^{t} f(\vartheta)d\vartheta + \vartheta_o\right] \tag{3}$$

where $\Delta\omega$ is the frequency deviation. It can be shown that the bandwidth of $\upsilon(t)$ obtained with $f(t) = \cos\omega_m t$ provides an estimate for the maximum bandwidth needed when $\upsilon(t)$ is modulated by a signal $f(t)$ band-limited to ω_n. With $f(t)=\cos\omega_m t$ we obtain,

$$\begin{aligned}\upsilon(t) = A\Big[J_o(\beta)\cos\omega_o t &+ J_1(\beta)[\cos(\omega_o+\omega_m)t - \cos(\omega_o-\omega_m)t] \\ &+ J_2(\beta)[\cos(\omega_o+2\omega_m)t + \cos(\omega_o-2\omega_m)t] \\ &+ J_3(\beta)[\cos(\omega_o+3\omega_m)t - \cos(\omega_o-3\omega_m)t] + \Big]\end{aligned} \tag{4}$$

where $\beta=\Delta\omega/\omega_m$.

For the system presented here, an optimum value of β was found between 0.1 to 0.2. The worst case, that is, the one that requires the greatest bandwidth occurs for β=0.2. For this value of β, $J_o(\beta)\simeq 1$, $J_1(\beta)$=0.2 and $J_2(\beta)\simeq 0$. Since $\omega_m=\omega_o$, the spectrum of $\upsilon(t)$ is bounded between 0 and $2\omega_o$, having a sharp concetration maximum around ω_o. This agrees with the experimentally obtained spectra. In Fig. 4 the spectrum of the VCO output is depicted, having a free-run frequency f_o=2.5 KHz and being modulated by a duobinary signal with β=0.2. In Fig. 5 the modulating input to the VCO was a "modified" duobinary signal again with a modulation index β=0.2. The difference between the two spectra is due to the difference in the spectra of the modulating signals used for the two cases.

IV. PERFORMANCE CHARACTERISTICS OF THE SYSTEM

The modem described was designed to be used mainly for the transmission of digitally ciphered voice data through a narrow-band voice channel like a telephone line. Of course, there are some very sophisticated systems available today. Our objective was to find a simpler and more economic solution to this problem. It is assumed that voice can be converted into digital data with bit-rates as low as 9.6 kbits/sec, by using the Continuously Variable Slope Delta Modulation method (C.V.S.D.). This technique in conjunction with the proposed system can provide the simple and economic solution that is desired.

The key to the simplicity of our method is the use of the frequency modulated VCO output as the system clock, which at the same time defines the data rate. Of course, this causes the duration of the received bits to vary. At first view, this may seem to cause distortion. However, when our system was used in conjunction with the DM technique mentioned above, no distortion was observed, especially in the useful voice bandwidth range. In addition, there was no problem in recovering the received digital data for further digital processing like ciphering and deciphering.

For the case of data transmission, the receiver needs a fixed clock for data recovering, retiming and processing. If the modulating signal is the duobinary, there is a possibility to have a number of successive symbols with the same nonzero level (±1). This will cause some bits to move beyond the expected duration and data retiming will give some errors. This problem does not appear when the "modified" duobinary technique is used to generate the modulating signal.

Next, the system performance was examined by transmitting through a simulated channel having a bandwidth similar to that of a telephone line. The channel was simulated by two band-pass filters, the first one of band-width 330-1800 Hz and the second one of bandwidth 650-3100 Hz. This is shown in Fig. 6.

In Fig. 7, at the top, the VCO output (f_o = 2.5KHz) is shown, after going through the first filter. The distortion was small so that no equalization was necessary. The frequency deviation, Δf,

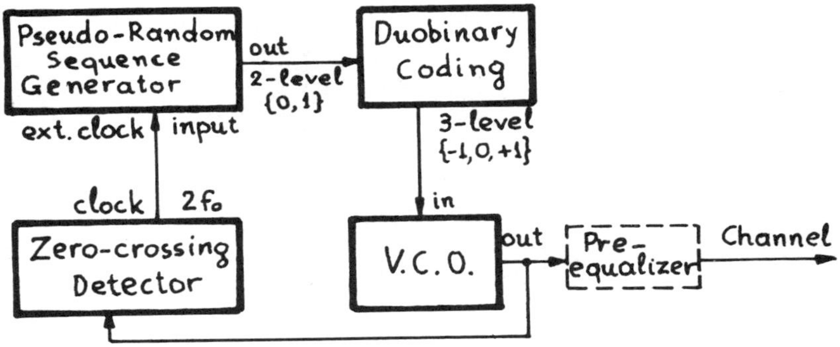

Fig.1: Transmitter block diagram

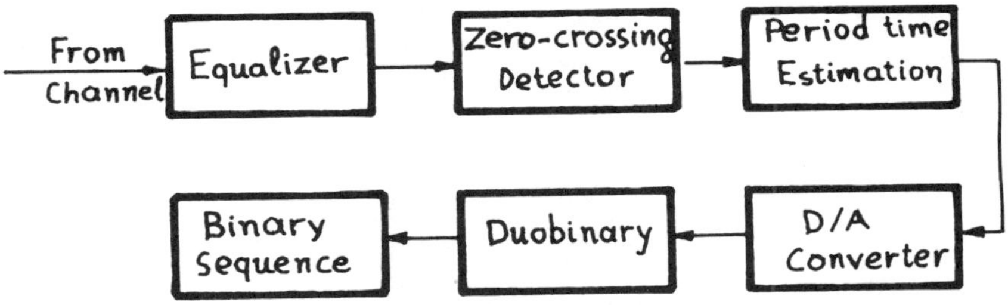

Fig.2: Receiver block diagram

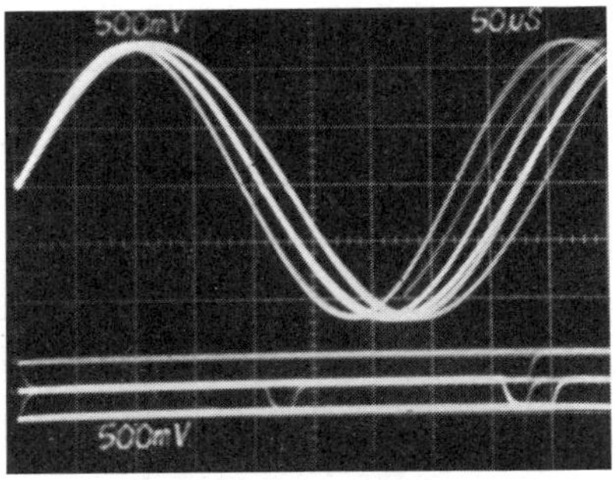

Fig.3: V.C.O. output modulated by the duobinary signal

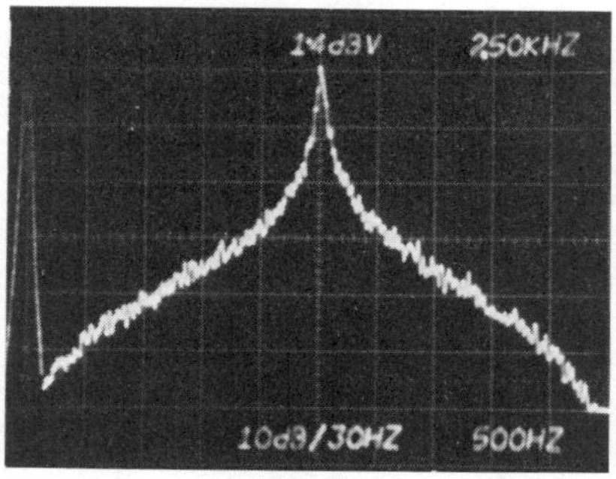

Fig.4: Spectrum for the V.C.O. output of fig. 3

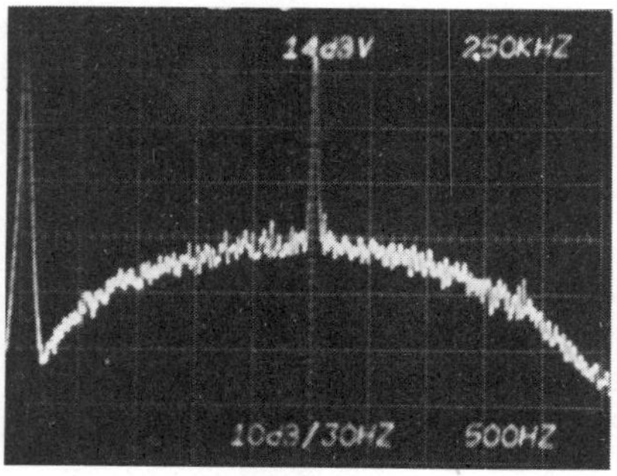

Fig.5: Spectrum of the V.C.O. output modulated by the modified duobinary signal

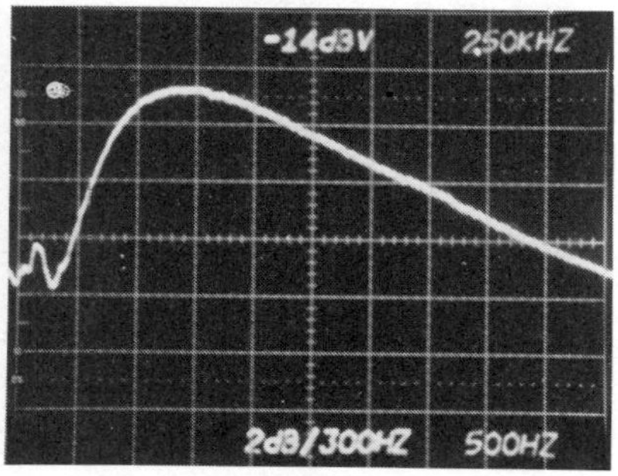

Fig.6: Filter response

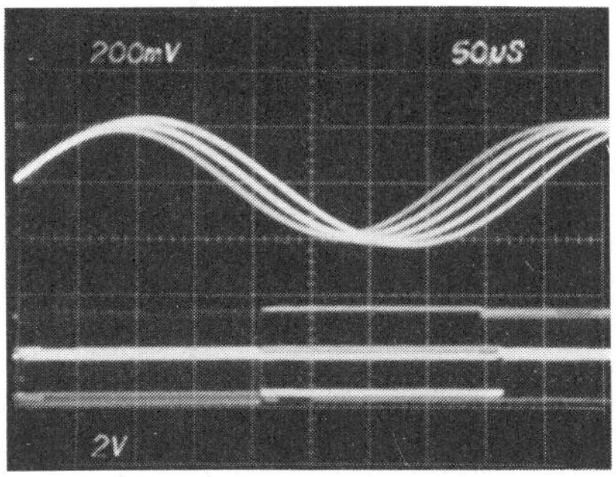

Fig.7: V.C.O. output after going through the first filter

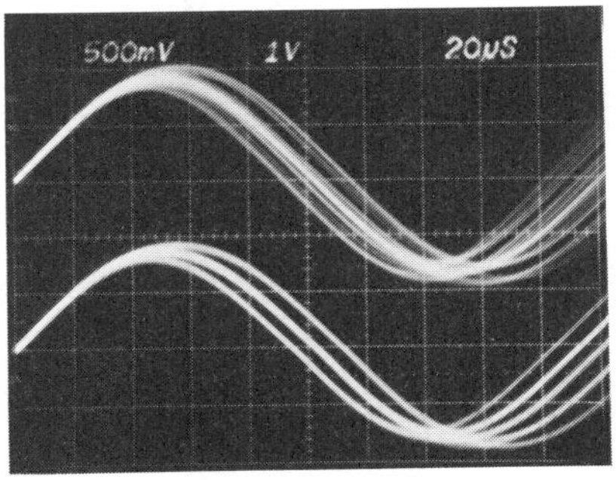

Fig.8: V.C.O. output with the second filter

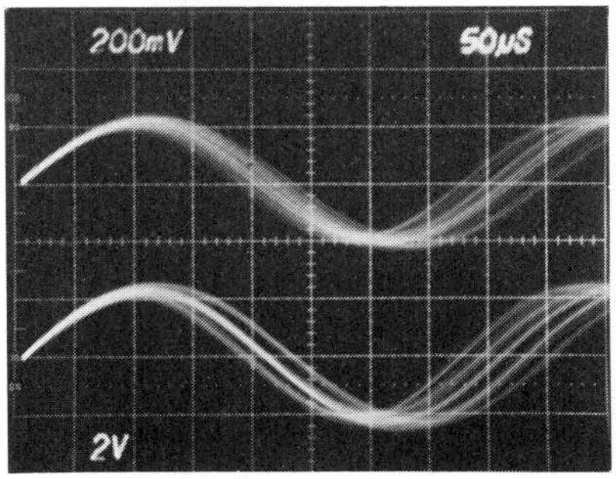

Fig.9: Received signal after going through a telephone line

is approximately 400 Hz. At the bottom of Fig. 7 the reconstructed duobinary signal is shown. The same results were obtained when the VCO output was passed through the second filter for the same f_o and Δf.

With the VCO frequency f_o=5 KHz, the output of the filters presents a distortion as can be seen at the top of Fig. 8, for which the second filter was used to simulate the channel. The signal obtained after equalization is seen at the bottom of the same figure. The equalizer used was a simple one, manually adjusted, having only seven taps. The deviation was about 500 Hz.

The VCO output was next transmitted through a telephone line of the switched network. The signal passed through the internal telephone center of N.R.C "Demokritos", through the telephone center of "OTE" and all the way back to Demokritos. The received signal without equalization is shown at the top of Fig. 9, for f=2.5 KHz and Δf=500 Hz. At the bottom of the same figure, the received signal is shown after equalization. Similar results were obtained for f_o=5 KHz and Δf=500 Hz.

V. CONCLUSIONS

A new technique has been presented for digital voice and data transmission through a telephone line. This technique was tested with filters simulating the telephone line and gave excellent results at 5 and 10 KHz bit rates. When used with an actual telephone line the results were of lower quality, mainly due to the lack of proper equalization.

For digital voice data coded at a bit rate of 10 KHz the required VCO frequency that is transmitted is 5 KHz. At this frequency the telephone line tested presents a loss of approximately 5 dB in reference to the peak at 800 Hz. This loss could increase when the connection goes through more than one telephone center. For this situation more experimental testing is required. However, this problem would not appear in leased lines.

REFERENCES

(1). S. Pasupathy, "Correlative coding. A bandwidth efficient signaling scheme", IEEE Comm. Society Magazine, pp. 4-11, July 1977.

(2). E.R. Kretzmer, "Generalization of a technique for binary data communication", IEEE Trans. Comm. Techn., Vol. COM-14, pp. 67-68, Feb. 1966.

SIGNAL PARAMETER OPTIMIZATION FOR THE IMPULSIVE NOISE CHANNEL

J.C. Vanelli
N.M. Shehadeh

Control Applications Inc., 17250 El Camino Real, Houston,TX
Elec. Eng. Dept., Univ. of Houston, Houston, TX

ABSTRACT

A common assumption in the analysis of impulsive noise channels is that at most one noise impulse occurs to affect any given decision. In practice this assumption is not always valid. However, in this work it is shown that optimum signal parameter values can be determined subject to the single noise impulse assumption. The analysis is based on a linear channel model and the optimality criterion used throughout is minimization of the decision error probability. Optimality tests based on the characteristic function of the filtered impulsive noise component are also derived. Numerical results are presented to substantiate the analysis.

INTRODUCTION

Data communication systems often encounter impulsive interference characterized by strong noise pulses occuring more frequently than predicted by Gaussian noise statistics. This impulsive noise is typically introduced in the transmission medium and is not related to thermal noise generated within the receiver itself. Consequently, a reasonable representation of such communication channels might involve two independent noise sources, one Gaussian and the other impulsive. For simplicity, such channels are herein referred to merely as impulsive noise channels, and the presence of Gaussian noise is understood, unless otherwise noted.

D. G. Lainiotis and N. S. Tzannes (eds.), Advances in Communications, 269–281.

Various signal designs and receivers have been proposed to combat impulsive noise. As described by Wainwright [1], the smear-desmear technique involves the introduction of a nonlinear phase component in the transmitted signal. A complimentary nonlinear phase component is employed in the receiver response. Richter and Smiths [2] have developed a method based on the calculus of variations to determine signals of this type that are optimum under certain conditions. The signals thus determined are optimum in a wide sense and do not necessarily minimize the probability of error. More recently, Miller and Thomas [3], have derived nonlinear detectors that are optimum in the sense of maximizing asymptotic relative efficiency. Finally, Spaulding and Middleton [4] have employed a canonical, statistical-physical model of the impulsive noise channel to obtain likelihood ratios for binary antipodal, orthogonal and ON-OFF keyed signals. The likelihood ratios thus obtained are cumbersome and generally not economically realizable. However, the authors show that a small-signal assumption can be used to greatly reduce this complexity and yield realizable, locally optimum receivers.

Now consider the case in which certain signal parameters are fixed or otherwise constrained. The designer must then obtain the best possible system performance given only limited control of the signal design. Such cases are dealt with in this paper. Parameter optimization techniques are used to establish fundamental tests for optimality of signals and linear receivers operating over an impulsive noise channel. The optimality criterion used throughout is minimization of the decision error probability $P(\varepsilon)$. It is first noted that in order to obtain optimum system performance the signal parameter optimization must be done subject to the constraint of a receiver response that is matched to the signal in the customary sense. It is then shown that in such cases, signal parameter values that are optimum in a particular strict sense can be selected without regard to the Gaussian noise component and on the assumption that exactly one significant noise impulse occurs to affect any given decision. These conditions for signal parameter optimization are further reduced to equivalent tests based on the characteristic function of the filtered impulsive noise component affecting the decision.

CHANNEL MODEL

In order to formulate the problem as concisely as possible the baseband channel model of Figure 1 is used. The binary input B is selected with equal probability from the set $\{-1,1\}$. The basic signal waveform is $s(t)$ and the linear receiver impulsive response is $h(t)$. The impulsive noise model incorporates the usual Poisson process, generating randomly weighted noise impulses at an average rate ν. The resultant interference process is:

$$y(t) = \sum_{k=1}^{\infty} A_n \, q(t-\tau_n) \qquad (1)$$

Here A_n and τ_n are the random weight and occurrence time, respectively, of the nth noise impulse and q(t) denotes the response of the receiver to the characteristic noise impulse waveform. It is assumed that the $\{A_n\}$ are independent, zero-mean Gaussian random variables of variance σ^2. In addition to the impulsive noise, additive white Gaussian noise of two-sided power spectral density $N_0/2$ is also introduced at the receiver input.

The binary decision $\hat{B}$ is determined by the polarity of a sample of the receiver response taken at time t=0. This sample consists of three components:

$$R = V + X + Y \qquad (2)$$

where the random variables X and Y represent the contributions of Gaussian and impulsive noise, respectively, and V is the component due to the signal. There is no loss of generality, in selecting B such that V is positive. Thus, in terms of the distribution function of the sum X + Y:

$$P(\varepsilon) = 1 - F_{X+Y}(V) \qquad (3)$$

Since this is a monotonic, nonincreasing function of V, the optimum signal of energy E for a specified linear receiver response must be matched in the usual sense, i.e.

$$s(t) = \sqrt{E}\, h(-t) \qquad (4)$$

However, the impulsive noise channel differs from the channel involving only white Gaussian noise in that not all such matched filter systems are equivalent. If the system designer is at liberty to specify both s(t) and h(t) then (4) establishes a necessary but not sufficient condition for minimizing $P(\varepsilon)$.

It is convenient at this point to separate $P(\varepsilon)$ into two distinct components:

$$P(\varepsilon) = P(\varepsilon/X) + \Delta P(\varepsilon/Y) \qquad (5)$$

Here $P(\varepsilon/X)$ is the probability of error due solely to the Gaussian noise and [5]:

$$\Delta P(\varepsilon/Y) = \frac{1}{\pi} \int_0^{\infty} \frac{\sin(\lambda V)}{\lambda} M_X(\lambda)[1-M_Y(\lambda)]d\lambda \qquad (6)$$

is the incremental probability of error caused by the impulsive noise. $M_X(\lambda)$ and $M_Y(\lambda)$ are the characteristic functions of X and Y, respectively.

The characteristic function of the filtered impulsive noise component can be derived in straightfoward fashion to obtain:

$$M_Y(\lambda) = \exp\{\nu \int_{-\infty}^{\infty}[\exp(-\lambda^2\sigma^2 q^2(\tau)/2)-1]d\tau\} \tag{7}$$

for the model defined herein. If the sample component Y is caused by exactly one noise impulse occurring at a random point in time, with uniform distribution over the interval [-D/2,D/2], this reduces to:

$$M_Y(\lambda/1) = \frac{1}{D} \int_{-D/2}^{D/2} \exp(-\lambda^2\sigma^2 q^2(\tau)/2)d\tau \tag{8}$$

The corresponding conditional density function can be found as the Fourier transform of $M_Y(\lambda/1)$:

$$p_Y(\beta/1) = \frac{1}{\sigma D\sqrt{2\pi}} \int_{-D/2}^{D/2} \frac{\exp(-\beta^2/2\sigma^2 q^2(\tau))}{|q(\tau)|} d \tag{9}$$

If q(t) is such that noise impulses occurring outside the interval [-D/2,D/2] cannot affect the sample taken at time t=0, then Y is a sum of some finite number of independent random variables, each having density function $P_Y(\beta/1)$. Thus the density function of Y is:

$$p_Y(\beta) = P(0)\delta(\beta) \sum_{N=1}^{\infty} P(N)[\underbrace{p_Y(\beta/1)*p_Y(\beta/1)*\ldots*p_Y(\beta/1)}_{N \text{ terms}}] \tag{10}$$

where P(N) denotes the probability that exactly N impulses occur within the specified time interval.

SIGNAL PARAMETER OPTIMIZATION

Now suppose the signal depends in some way on a parameter, ρ. Then for a matched filter system:

$$q(t,\rho) = g(t) * s(-t,\rho)/\sqrt{E} \tag{11}$$

Thus Y and $P(\varepsilon)$ are also dependent on ρ. If a particular value of ρ minimizes $P(\varepsilon)$ for all values of σ, ν and E, then it is said to be optimum in the strict sense.

For any matched filter system, $P(\varepsilon/X)$ is only a function of E/N_0, without regard to the value of ρ. This implies that optimum parameter values can be selected on the condition that $N_0=0$, by minimizing the probability of error due to the impulsive noise acting alone. This can be proven for the particular channel model under consideration here. The following lemma is essential to the proof.

Lemma 1

Let Z_0, Z_1 and W be independent random variables with continuous density functions $p_{Z_0}(\beta)$, $p_{Z_1}(\beta)$ and $p_W(\beta)$ that are unimodal and even-symmetric about the origin. If the distribution functions of Z_0 and Z_1 satisfy the condition:

$$F_{Z_0}(\alpha) \geq F_{Z_1}(\alpha) \qquad \text{for } \alpha > 0 \tag{12}$$

then the distribution functions of the sums Z_0+W and Z_1+W satisfy a similar condition:

$$F_{Z_0+W}(\alpha) \geq F_{Z_1+W}(\alpha) \qquad \text{for } \alpha > 0 \tag{13}$$

Furthermore, (13) is satisfied with equality if and only if the equality also holds in (12).

The proof of this lemma is straightforward and is omitted here.

It is now shown that optimum signal parameter values can be selected on the assumption that no Gaussian noise is present.

Theorem 1

Suppose that a matched filter system with signal $s(t,\rho)$ of energy E operates over the impulsive noise channel with arbitrary noise parameters σ, ν and N_0. Let $P(\varepsilon/Y)$ denote the probability of error that would result due to the impulsive noise acting alone. Then those values of ρ that minimize $P(\varepsilon/Y)$ are optimum in the strict sense.

Proof

By definition $P(\varepsilon/Y)$ is independent of N_0 and can be expressed in terms of the distribution function of Y:

$$P(\varepsilon/Y) = 1-F_Y(\sqrt{E}) \tag{14}$$

Let $P(\varepsilon/Y,\rho_0)$ and $P(\varepsilon/Y,\rho_1)$ denote forms of $P(\varepsilon/Y)$ corresponding to particular signal parameter values ρ_0 and ρ_1. Suppose that for all possible values of ν and E:

$$P(\varepsilon/Y,\rho_0) < P(\varepsilon/Y,\rho_1) \tag{15}$$

Then the corresponding conditional distribution functions must satisfy the condition:

$$F_Y(\sqrt{E}/\rho_0) > F_Y(\sqrt{E}/\rho_1) \tag{16}$$

Now let $p_Y(\beta/\rho_0)$ and $p_Y(\beta/\rho_1)$ denote the conditional density functions of Y. It is apparent from (9) and (10) that these functions are unimodal and even-symmetric about the origin. Of course $p_X(\beta)$, the density function of X, is also unimodal and even-symmetric. Furthermore, the distribution function $F_X(\sqrt{E})$ does not depend on ρ. But the total distortion at the matched filter output is the sum of the independent components X and Y. Thus Lemma 1 can be invoked to obtain the following condition on the conditional distribution functions of the sum:

$$F_{X+Y}(\sqrt{E}/\rho_0) > F_{X+Y}(\sqrt{E}/\rho_1) \tag{17}$$

It follows from (3) that

$$P(\varepsilon/\rho_0) < P(\varepsilon/\rho_1) \tag{18}$$

for all possible values of σ, ν and E. This is a direct consequence of the assumption stated in (15). Obviously the converse is also true. Q.E.D.

It is important to note that this result applies only insofar as X is in fact the response of the matched filter to a white Gaussian noise process. Otherwise, $F_X(\sqrt{E})$ is not independent of ρ and the proof cannot be made.

SINGLE IMPULSE TEST

For any real system it is possible to select some value of D sufficiently large so that $q(t,\rho)$ is essentially zero for $|t|>D/2$. It is then convenient to express $P(\varepsilon/Y)$ as an expectation over the number of noise impulses that might occur in the interval [-D/2,D/2]:

$$P(\varepsilon/Y) = \sum_{N=1}^{\infty} P(N)P(\varepsilon/Y,N) \tag{19}$$

Here $P(\varepsilon/Y,N)$ denotes the probability of error due solely to the impulsive noise and subject to the condition that exactly N interfering noise impulses occur in the interval. For the subject impulsive noise channel it can be shown that optimum signal parameter values obtain coincident minima of $P(\varepsilon)$ and $P(\varepsilon/Y,1)$. The following lemma is used in the proof.

Lemma 2

Let Z_0, Z_1, W_0, and W_1 be independent random variables with continuous density functions $p_{Z_0}(\beta)$, $p_{Z_1}(\beta)$, $p_{W_0}(\beta)$ and $p_{W_1}(\beta)$ that are unimodal and even-symmetric about the origin. If the distribution functions satisfy the conditions:

$$F_{Z_0}(\alpha) \geq F_{Z_1}(\alpha) \qquad \text{for } \alpha > 0 \qquad (20)$$

$$F_{W_0}(\alpha) \geq F_{W_1}(\alpha) \qquad \text{for } \alpha > 0 \qquad (21)$$

then the distribution functions of the sums Z_0+W_0 and Z_1+W_1 satisfy a similar condition:

$$F_{Z_0+W_0}(\alpha) \geq F_{Z_1+W_1}(\alpha) \qquad \text{for } \alpha > 0 \qquad (22)$$

Furthermore, (22) is satisfied with equality if and only if the equality also holds in (21) and (22).

The proof of this lemma follows immediately from Lemma 1 and is omitted here.

It is now shown that optimum signal parameter values can be selected on the assumption that only one noise impulse occurs to interfere with the decision made by the receiver.

Theorem 2

Suppose that a matched filter system with signal $s(t,\rho)$ of energy E operates over the impulsive noise channel with arbitrary noise parameters σ, ν and N_0. Let $q(t,\rho)$ be such that noise impulses occurring outside the interval $[-D/2,D/2]$ do not affect the sample of the matched filter output taken at t=0. Let $P(\varepsilon/Y,1)$ denote the probability of error that would result due to the impulsive noise acting alone and subject to the condition that exactly one noise impulse occurs within the interval. Then those values of ρ that minimize $P(\varepsilon/Y,1)$ are optimum in the strict sense.

Proof

By definition, the $\{P(\varepsilon/Y,N)\}$ are independent of N_0 and can be expressed in terms of a corresponding set of conditional distribution functions of Y:

$$P(\varepsilon/Y,N) = 1-F_Y(\sqrt{E},N) \qquad \text{for } N = 1,2,3,\ldots \qquad (23)$$

Let $P(\varepsilon/Y,1,\rho_0)$ and $P(\varepsilon/Y,1,\rho_1)$ denote forms of $P(\varepsilon/Y,1)$ corresponding to particular signal parameter values ρ_0 and ρ_1. Suppose that for all possible values of E:

$$P(\varepsilon/Y,1,\rho_0) < P(\varepsilon/Y,1,\rho_1) \tag{24}$$

Then the corresponding conditional distribution functions must satisfy the condition:

$$F_Y(\sqrt{E}/1,\rho_0) < F_Y(\sqrt{E}/1,\rho_1) \tag{25}$$

Now let $p_Y(\beta/N,\rho_0)$ and $p_Y(\beta/N,\rho_1)$ denote the conditional density functions of Y given that exactly N noise impulses occur in the interval. It is apparent from (9) and (10) that these functions are unimodal and even-symmetric about the origin. Furthermore, the corresponding random variable Y is a sum of N independent and identically distributed random variables. Consequently, Lemma 2 can be applied successively in conuunction with (25) to obtain a set of inequalities involving the conditional distribution functions of Y:

$$F_Y(\sqrt{E}/N,\rho_0) > F_Y(\sqrt{E}/N,\rho_1) \quad \text{for } N = 1,2,3,... \tag{26}$$

It follows immediately from (19) and (23) that:

$$P(\varepsilon/Y,\rho_0) < P(\varepsilon/Y,\rho_1) \tag{27}$$

and therefore, according to Theorem 1:

$$P(\varepsilon/\rho_0) < P(\varepsilon/\rho_1) \tag{28}$$

for all possible values of σ, ν and E. This is a direct consequence of the assumption stated in (24). Obviously the converse is also true. Q.E.D.

Thus, if a matched filter system is not designed to adapt to variations in noise parameters, and if it is necessary to select a signal design which is optimum in the strict sense, then this selection can be made subject to the assumption of a single interfering noise impulse. Furthermore, although a Gaussian distribution of noise impulse amplitudes has been assumed, it is in fact only necessary that $p_y(\beta)$ be unimodal and even-symmetric.

CHARACTERISTIC FUNCTION TESTS

In this section the conditions for signal parameter optimization are reduced to equivalent and generally more useful tests based on characteristic functions of the sample component due to impulsive noise. This analysis is formulated in terms of the normalized random variable

$$U = Y/\sigma \tag{29}$$

and its characteristic function. It is first shown that this characteristic function is stationary with respect to variation of the parameter about any optimum value.

Theorem 3

Suppose that a matched filter system with signal $s(t,\rho)$ of energy E operates over the impulsive noise channel with arbitrary noise parameters σ, ν and N_0. Let $M_U(\lambda)$ denote the characteristic function of the normalized random variable U. If $M_U(\lambda)$ is a well-behaved function of ρ, then all values of the parameter that are locally optimum in the strict sense satisfy the necessary condition:

$$\frac{\partial M_U(\lambda)}{\partial \rho} = 0 \qquad (30)$$

for finite, non-zero λ and all possible values of ν.

Proof

It follows from Theorem 1 that a necessary condition for optimality of a particular parameter value is that the first partial derivative of $P(\varepsilon/Y)$ with respect to the parameter must equal zero. In terms of $M_U(\lambda)$ this condition becomes:

$$\frac{\partial P(\varepsilon/Y)}{\partial \rho} = \frac{1}{\pi} \int_0^\infty \frac{\sin(\lambda\sqrt{E/\sigma^2})}{\lambda} \frac{\partial M_U(\lambda)}{\partial \rho} d\lambda = 0 \qquad (31)$$

The arbitrary ratio E/σ^2 appears explicitly in the integrand of (31). However, the function:

$$M_U(\lambda) = \exp\{\nu \int_{-\infty}^{\infty} [\exp(-\lambda^2 q^2(\tau,\rho)/2)-1]d\tau\} \qquad (32)$$

as determined by (7) and (29) has no such dependence. Therefore, if (31) is to hold for all possible values of E and σ then (30) must hold for finite, non-zero λ and all possible values of ν. Q.E.D.

Corollary

The rate-normalized second characteristic function:

$$L_U(\lambda) = \frac{1}{\nu} \ln M_U(\lambda) = \frac{1}{\nu} \ln M_Y(\lambda/\sigma) \qquad (33)$$

is a monotonic function of $M_U(\lambda)$. Thus if $L_U(\lambda)$ is a well-behaved function of ρ, then all values of the parameter that are locally optimum in the strict sense satisfy the necessary condition:

$$\frac{\partial L_U(\lambda)}{\partial \rho} = 0 \qquad (34)$$

for finite, non-zero λ and all possible values of ν.

Yet another optimality test based on the single noise impulse model is developed in the following theorem.

Theorem 4

Suppose that a matched filter system with signal $s(t,\rho)$ of energy E operates over the impulsive noise channel with arbitrary noise parameters ν, σ^2 and $N_0/2$. Let $q(t,\rho)$ be such that noise impulses occurring outside the interval $[-D/2,D/2]$ do not affect the sample of the matched filter output taken at t=0. Let $M_U(\lambda/1)$ denote the characteristic function of U subject to the condition that exactly one noise impulse occurs within the interval. If $M_U(\lambda/1)$ is a well-behaved function of ρ, then all values of the parameter that are locally optimum in the strict sense satisfy the necessary condition:

$$\frac{\partial M_U(\lambda/1)}{\partial \rho} = 0 \tag{35}$$

for finite, non-zero λ.

Proof

It follows from Theorem 2 that a necessary condition for optimality of a particular parameter value is that the first partial derivative of $P(\varepsilon/Y,1)$ with respect to the parameter must equal-zero. In terms of $M_U(\lambda/1)$ this condition becomes:

$$\frac{\partial P(\varepsilon/Y,1)}{\partial \rho} = -\frac{1}{\pi}\int_0^{\infty}\frac{\sin(\lambda\sqrt{E/\sigma^2})}{\lambda}\,\frac{\partial M_U(\lambda/1)}{\partial \rho}d\lambda = 0 \tag{36}$$

The arbitrary ratio E/σ^2 appears explicitly in the integrand of (36). However, the function:

$$M_U(\lambda/1) = \frac{1}{D}\int_{-D/2}^{D/2} \exp(-\lambda^2 q^2(\tau,\rho)/2)d\tau \tag{37}$$

as determined by (8) and (29) has no such dependence. Therefore, if (36) is to hold for all possible values of E and σ then (35) must hold for finite, non-zero λ. Q.E.D.

The conditions for strict sense optimality established in these last two theorems and corollary are necessary but not sufficient conditions. Consequently, other devices must be employed to determine if a particular parameter value satisfying these conditions corresponds to a maximum or minimum value of $P(\varepsilon)$. Often this can be done by inspection, but in some cases the desired solutions must be determined by numerical calculation of $P(\varepsilon)$ or $P(\varepsilon/Y)$ for all parameter values satisfying the conditions.

Another point worthy of emphasis is that the results stated by the Corollary and Theorem 4 are not dependent on the impulsive noise parameters σ and ν because the functions $L_U(\lambda)$ and $M_U(\lambda/1)$ have no such dependence.

NUMERICAL EXAMPLE

The application considered in this section serves to illustrate the utility of the above-described optimality tests. This example involves a split-phase signal of energy E and duration T:

$$s(t,\rho)= \frac{\sqrt{E}}{T} [-U(t+T/2)+2U(t-\rho T+T/2)-U(t-T/2)] \text{ for } 0\leq\rho\leq 1 \quad (38)$$

The corresponding ideal matched filter response is employed at the receiver. The noise impulses are assumed to cause rectangular interferences pulses of duration T:

$$g(t) = \frac{\sqrt{T}}{T} [U(t+T/2-U(t-T/2)] \quad (39)$$

and therefore:

$$q(t,\rho) = \frac{t}{T} [U(t+T)-2U(t+\rho T)+2U(t+\rho T-T)-U(t-T)] \text{ for } 0\leq\rho\leq 1 \quad (40)$$

Expressing $L_U(\lambda)$ according to (7) and (33) yields:

$$L_U(\lambda) = \frac{T\sqrt{2\pi}}{\lambda} [\mathrm{erf}(\rho\lambda/\sqrt{2})+\mathrm{erf}((1-\rho)\lambda/\sqrt{2})]-2T \quad (41)$$

Applying (35) and solving for ρ then provides the desired result:

$$\frac{\partial L_U(\lambda)}{\delta\rho} = T\sqrt{\pi} [\exp(-\rho^2\lambda^2/2)-\exp(-(1-\rho)^2\lambda^2/2)]=0 \quad \rho=0.5 \quad (42)$$

Clearly, this is the only possible solution that satisfies (35) for all values of λ.

More conclusive evidence of the optimality of the parameter value $\rho=0.5$ is shown in Figures 2 and 3 where $P(\varepsilon)$ is plotted as a function of ρ for various signal levels and combinations of the impulsive noise parameters. In interpreting these numerical results it should be noted that the quantity νT denotes the average number of noise impulses occurring over the duration of the signal. Because of the ideal nature of the channel and signal this is the average number of noise impulses affecting each decision made by the receiver. Further note that in all the cases shown, $N_0/2 = 2.5\ \rho\sigma^2$; i.e. the spectral density of the white Gaussian noise is 2.5 times the spectral density of the impulsive noise. Each set of curves was generated by evaluating Equation (6) using numerical integration techniques as described in [5].

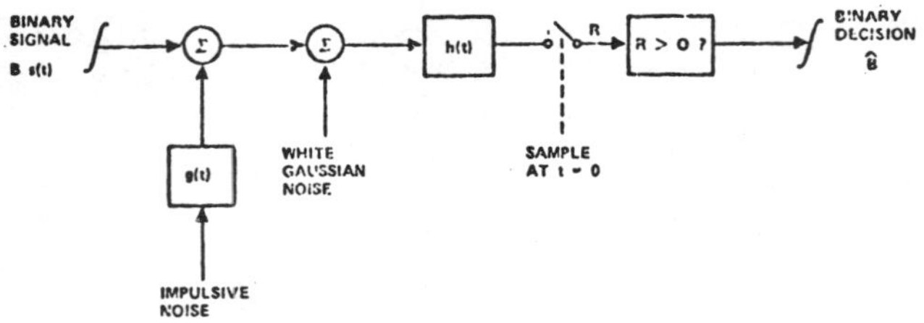

Figure 1. Baseband model of the impulsive noise channel

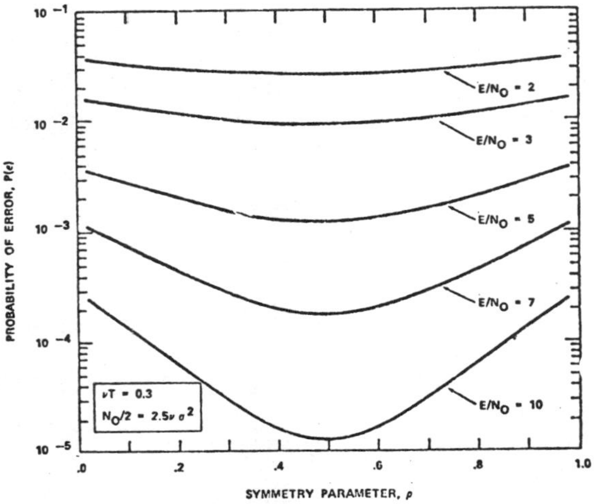

Figure 2. Ideal split-phase performance, $\nu T = 0.3$

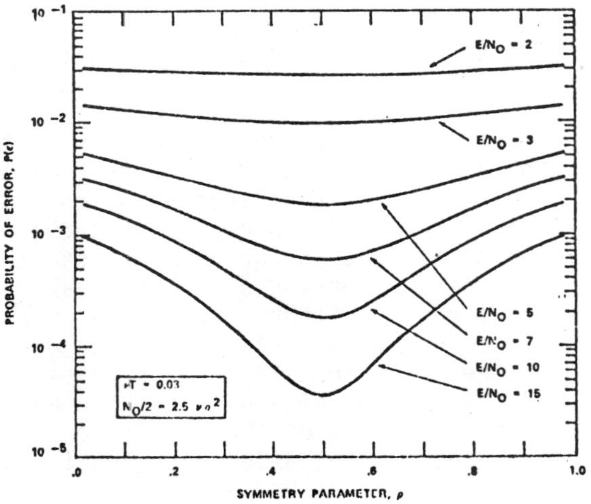

Figure 3. Ideal split-phase performance, $\nu T = 0.03$

CONCLUSION

It has been shown that the optimum linear communication system for the impulsive noise channel modeled herein is a matched filter system with appropriately selected signal parameter values. Furthermore, these strict sense optimum parameter values can be selected without regard to the white Gaussian noise affecting the channel and subject to the assumption of a single interfering noise impulse. Finally, it is shown that the characteristic function of the filtered impulsive noise component Y is stationary with respect to signal parameter variation about any strict sense optimum value. Although a Gaussian distribution of noise impulse amplitudes was assumed for this work, it is only necessary that the density function of the filtered impulsive noise component be unimodal and even-symmetric.

It is important not to overstate these results. The strict sense optimality condition precludes consideration of systems capable of adapting to variations in the noise parameters of the channel. Nonetheless, utilization of the optimality tests established herein can result in worthwhile simplification of the signal design problem for non-adaptive matched filter systems operating over a broad class of impulsive noise channels.

REFERENCES

[1] R.A. Wainwright, "On the Potential Advantage of a Smearing-Desmearing Filter Technique in Overcoming Impulse Noise Problems in Data Systems," IRE Transactions on Communication Systems, Vol. CS-9, pp. 362-366, December 1961.

[2] W.J. Richter and T.I. Smits, "Signal Design and Error Rate for an Impulse Noise Channel," IEEE Transactions on Communication Technology, Vol. COM-19, pp. 446-458, August 1971.

[3] J.H. Miller and J.B. Thomas, "Robust Detectors for Signals in Non-Gaussian Noise," IEEE Transactions on Communications, Vol. COM-25, pp. 686-690, July 1977.

[4] A.D. Spaulding and D. Middleton, "Optimum Reception in an Impulsive Interference Environment-Part II: Incoherent Reception," IEEE Transactions on Communications, Vol. COM-25, No. 9, September 1977. pp. 924-934.

[5] J.C. Vanelli and N.M. Shehadeh, "Computation of Bit-Error Probability Using the Trapezoidal Integration Rule," IEEE Transactions on Communication. Vol. COM-22, pp. 331-334, March 1974.

HYBRID CORRELATOR BASED ON EXPONENTIAL DELTA SIGMA MODULATION

D. Lagoyannis
Electronics Department,
N.R.C. "DEMOKRITOS"
Athens, Greece.

ABSTRACT

In this paper a simple short time hybrid correlator is being described which makes use of an Exponential Delta Sigma Modulator (EDSM) as ADC of one of the two input signal to be correlated. The use of EDSM permits the development of a hybrid multiplier of high simplicity. The integration of the product is done by an RC integrating network. The simplicity of the integration and multiplication circuits permits the development of a parallel type correlator used in real time determination of correlation functions with application to signal analysis.

1. INTRODUCTION

Many times, in the field of voice analysis and other acoustical applications a short time correlation technique is necessary. In this particular form of correlation the necessary short integration time can be achieved by simple RC networks. Most of the complexity of such a correlator remains thus in the multiplying circuits. If therefore a simple multiplication circuit could be achieved the development of a simple parallel hybrid correlator becomes possible (1). Such a correlator has been actually developed based on an EDSM (2).

As it is known, the correlation function of two signals x(t) and y(t) is given by

D. G. Lainiotis and N. S. Tzannes (eds.), Advances in Communications, 282–288.

$$C_{xy}(\tau) = \lim_{T} \frac{1}{T}\int_{T}^{T} x(\tau)\,y(t-\tau)\,dt \qquad (1)$$

For short time correlation the above expression becomes

$$C_{xy}(t_0,T,\tau) = \frac{1}{T}\int_{t_0}^{t_0+T} x(t)\,y(t-\tau)\,dt \qquad (2)$$

The necessary operations involved in this formula are: Delay, multiplication and integration. These operations can simply realized if one of the two signals, say y(t), is digitized by means of an EDSM as follows.

2. THE DELAY

The signal y(t) is digitized by an EDSM in the output of which is obtained a binary sequence $Y_n = Y(nT_c)$ where $T_c = 1/f_c$ and f_c is the clock frequency. This output sequence drives an m-bit shift register whose output is a new sequence $Y_m = Y\ (n-m)T_c$ corresponding to the delayed analog signal $y(t-\tau)$, where $\tau = mT_c$. Since we are interested in the simultaneous determination of $C_{xy}(\tau)$ for N values of τ, it becomes necessary to use a shift register of (n-1) stages of m-bit each (fig. 1).

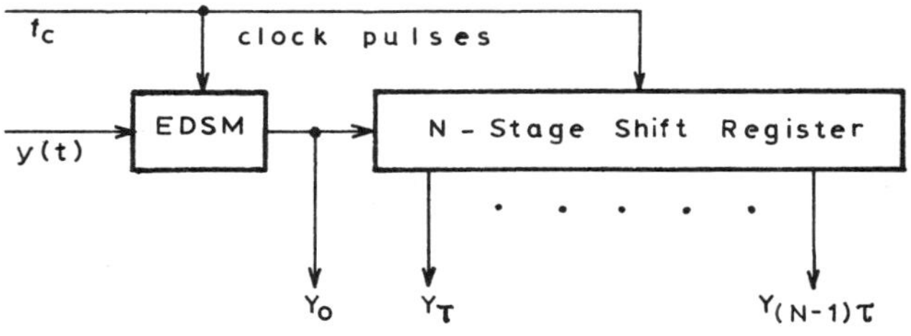

Figure 1. Diagram of the delay section.

3. THE MULTIPLIER

The proposal multiplier is based on the variable area analog multiplier (3). Let the time interval $[t_0-T/2,\ t_0+T/2]$ be such that $T=KT_c$ with K a positive integer. Also let function y(t) be constant and equal to $y(t_0)$ within this interval. If p and n are respectively the numbers of positive and negative pulses of the sequence Y_n in the time interval T, then (4)

$$y(t_0) = k\,\frac{(p-n)T_c}{T} \qquad (3)$$

where k is a constant.

From the equation (3) it becomes evident that y(t) is proportional to the average area of the pulse sequence Y_n in the interval T. If the pulse sequence Y_n is amplitude modulated by the second signal x(t) and filtered by a low-pass filter (LPF) then, as it is known, at the output of the filter a signal proportional to the product x(t)y(t) is obtained. Such a multiplier can be very easily realized by the use of CMOS analog gates. In figure 2 the schematic diagram of the multiplier is given. The transfer characteristic of the multiplier is given in figure 3 for clock frequency f_c 1MHz and a LPF with cut-off frequency 10 KHz.

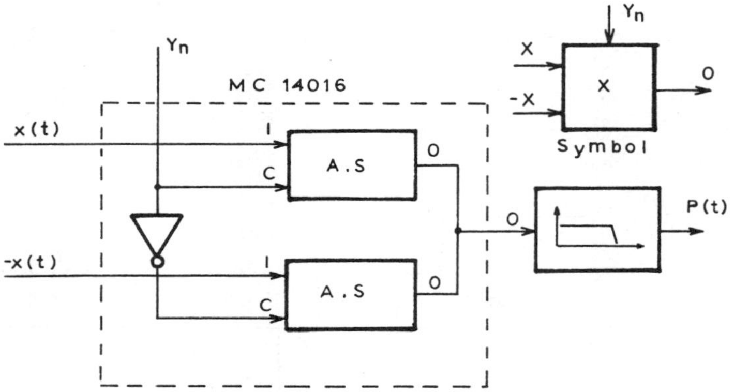

Figure 2. Implemantation of the multiplier.

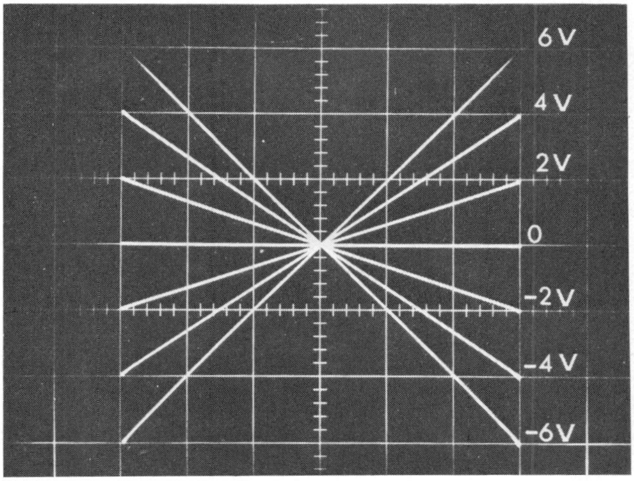

Figure 3. The product of a ramp voltage multiplied by different constants.
Horizontal: 2V/div., Vertical: 2V/div.

igure 4 the output of the multiplier is given re-
enting the product (trace b) of two sinusoids of
uencies 500 Hz (trace a) and 8 KHz.

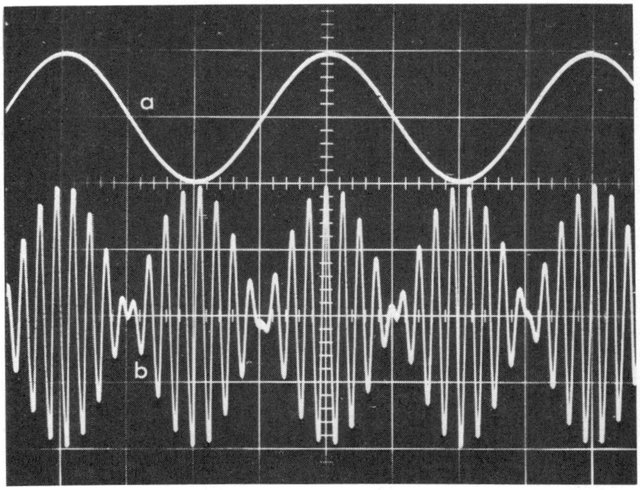

Figure 4. The product of two sinusoids.
Trace a) Hor.: 0.5 msec/div., Vert.: 5V/div.
Trace b) Hor.: 0.5 msec/div., Vert.: 2V/div.,

4. THE INTEGRATOR

The integrator is a simple RC network that replaces the LPF at the output of the multiplier. The effective noise bandwidth W_e of this circuit is given by

$$W_e = \frac{\pi}{2} \omega_o \qquad (4)$$

where ω_o is the cutoff angular frequency. From (4) the effective integration time is derived

$$T_e = \frac{2\pi}{W_e} = 4RC \qquad (5)$$

This effective Integration time is determined by the lowest frequency of the signal to be correlated.

5. THE COMPLETE CORRELATOR

After the simple method of realizing the three basic operations the develompent of the complete short time parallel correlator is straightforward. The block diagram of the correlator is given in figure 5. At the output of the correlator the value of the $\varphi_{xy}(\tau)$ will be given simultaneously for 64 different values of the variable τ. The values of $\varphi_{xy}(\tau)$ are displayed on a CRT by means of a multiplexer.

The above correlator has been constracted on a rack and tested as an autocorrelator and the results are given below.

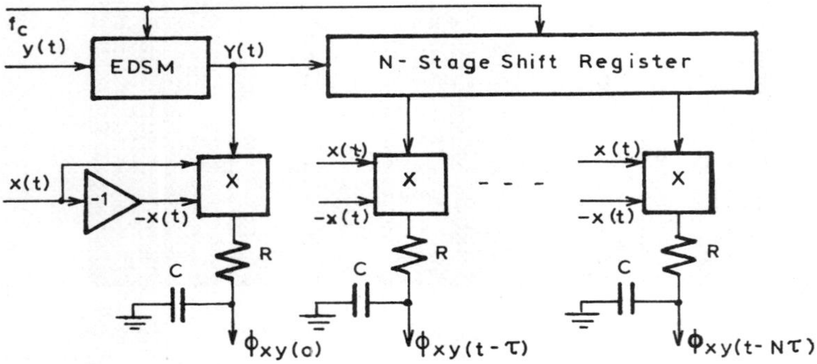

Figure 5. The block diagram of the correlator.

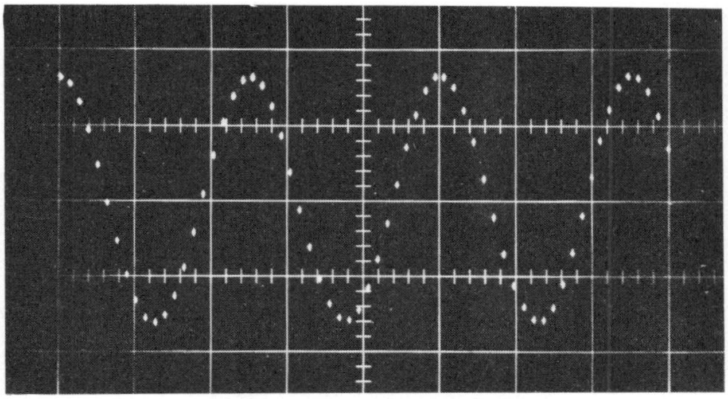

Figure 6. The autocorrelation function of a sinusoid against τ.
Vert.: 2V/div., Hor.: .2msec/div.,

In figure 6 is shown the autocorrelation function of a sinusoid $x(t) = A\sin\omega t$ which is given by

$$\varphi_{xx}(\tau) = \frac{A^2}{2}\cos\omega\tau \qquad (6)$$

The frequency of the signal is 2KHz and the clock frequency f_c=640 KHz. The distance between the points is 25µsec.

Another application of the correlator is to recover a sinusoidal signal of 2,5 V amplitude and 1KHz frequency, masked by a low-pass band limited white noise of 10 KHz bandwidth and 3,6 volts rms. (figure 7a). The autocorrelation function (figure 7b) of the band limited white noise is given by

$$\varphi_{xx}(\tau) = \frac{G}{2\pi\tau}\sin 2\pi f_h\tau \qquad (7)$$

where G is the power density spectrum of the noise and f_h the bandwidth. From (7) it can be seen that $\varphi_{xx}(\tau)$ approaches a very narrow pulse which can be seen in figure 7b (encircled point). The autocorrelation function of the masked signal is the sum of the autocorrelation function of the two signals which also can be seen in figure 7c.

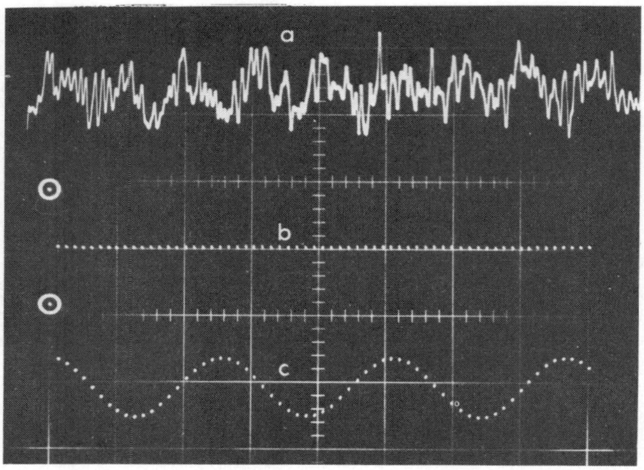

Figure 7. Trace a) Sinusoid masked by white noise. Vert.: 10V/div., Hor.: 1msec/div.
Trace b) $\varphi_{xx}(\tau)$ of white noise against τ.
Vert.: 1V/div., Hor.: 50 μsec/dot.
Trace c) The recovered signal against τ.
Vert.: 1V/div., Hor.: 50 μsec/dot.

CONCLUSIONS

A parallel type hybrid correlator has been developed which employes an EDSM as the A-D converter. Using the EDSM the operations implied by the correlation function are carried out with simplicity from the circuitry point of view.

REFERENCES

1. Max, J., "Traitement du signal" (MASSON et Cie, Paris 1972).
2. Steel, R., "Delta Modulation Systems", (Pentech Press London 1975).
3. Evangelsti, A., "A Transistorized variable area analog multiplier", (Application report AR 126, SGS-FAIRCHILD, January 1965).
4. Lagoyannis, D., "Multiplier for Delta-Modulated signals", (Electronics Letters, Vol. 14, No. 19, Sept. 1978, pp. 615-616).

THE DESIGN OF A BROADBAND DELTA MODULATOR WITH OPTIMUM PERFORMANCE

N.C.David and V.Makios

Laboratory of Electromagnetics, School of Engineering, University of Patras, Patras, Greece.

ABSTRACT

Recently we introduced (1) an algorithm for an adaptive DM system and presented a detailed theoretical analysis of the model which led to the derivation of closed-form equations for the performance of the modulator in terms of S/N and DR.

In this contribution we intend to give more emphasis on the experimental results in order to confirm the validity of our theory. For this reason we describe in some extend the design of the proposed DM system and present experimental results.

I. INTRODUCTION

In previous papers (1),(2) we presented a general theory referring to the optimum performance of an Adaptive Delta Modulation (ADM) system. We examined the variations of the performance of the system in terms of its complexity, signal-to-noise-ratio and dynamic range and we found the parameters for which the performance of the system was optimum.

In this contribution we are intended to give more emphasis on the experimental results and thus confirm the validity of our theoretical approximations. With this aspect in mind we describe in some extend the design of the proposed delta modulation system for two different sampling rates of interest. We present the block diagram of the modulation system and explain step by step the overall function (operation) of the system. Afterwards we give experimental results for bandlimited Gaussian input signals (2MHz) and

D. G. Lainiotis and N. S. Tzannes (eds.), Advances in Communications, 289–296.

attempt a comparison between the experimental and theoretical results as well as a justification of their differences (wherever these differences exist).

II. SUMMARY OF THEORY

The block diagram of the proposed DM system is shown in fig. 1. Fig.2 shows the response of an ideal (optimum) delta modulator.

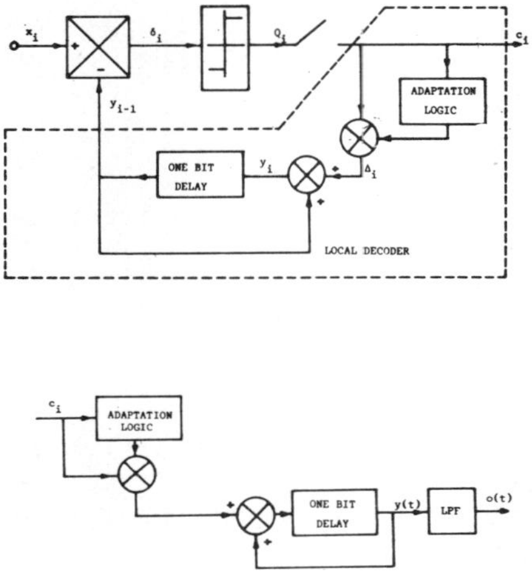

Fig.1. Block diagram of a DM system

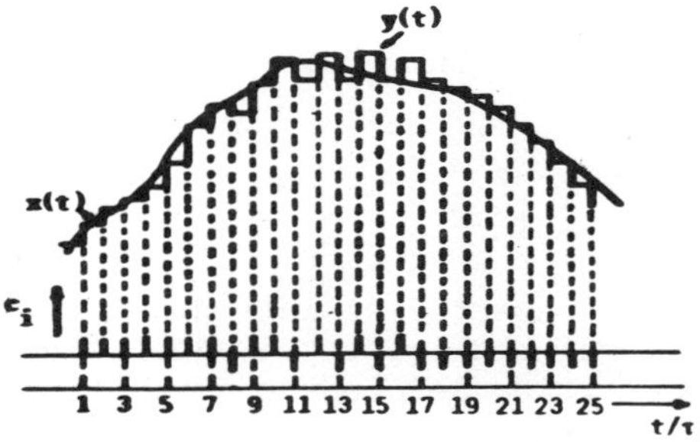

Fig.2. Response of an ideal delta modulator

Although such a response is not encountered in actual DM systems, fig.2 may be used as an assistant to the clarification of some main assumptions as follows:

$$P(c_i=1) \simeq P(\xi_i>0)$$

$$P(c_i=-1) \simeq P(\xi_i \leq 0)$$

where $\xi_i = \tau dx_i/dt$

is the first order difference of x_i.

The adaptation logic is such that,

$$\Delta_i = \sum_{k=0}^{n-1} a_{n-k} c_{i-k} \tag{1}$$

where n is the bit delay of the output sequence c_i (memory of the system).

The signal-to-noise ratio of the modulator is defined as,

$$S/N \triangleq \frac{E(x_i^2)}{E(e_i^2)} \tag{2}$$

where e_i is the quantizing noise (error) defined as,

$$e_i \triangleq x_i - y_i \tag{3}$$

We also define the dynamic range as,

$$DR = 20\log(x_1/x_2) \tag{4}$$

where x_1 and x_2 are the rms values of the input signal for which the S/N falls by 3 DB relatively to the peak value.We relate the parameters a_{n-k}, k=0,1,...,n-1 of eq.1 as follows:

$$a_{n-k} = aQ^{k^p}$$

where $a > 0$ and $0 < Q \leq 1$.

Now assuming that the input signal is a normal distribution function with flat spectrum up to f_b and zero beyond f_b, theory led (1), after long mathematical manipulations, to the following equations for the signal-to-noise ratio at the output of the demodulator: (for n=4).

1. $f_s/f_b=8$

(where f_s is the sampling frequency, and f_b is the bandwidth of the input signal x(t).

$$(S/N)^8_{LP} = \frac{4(x/a)^2}{.206(x/a)^2-1.0278(x/a)+4.99} \tag{5}$$

where a is a constant.

2. $f_s/f_b=12$

$$(S/N)^{12}_{LP} = \frac{6(x/a)^2}{.0914(x/a)^2-.8(x/a)+3.08} \tag{6}$$

3. $f_s/f_b=16$

$$(S/N)^{16}_{LP} = \frac{8(x/a)^2}{.05139(x/a)^2-.65(x/a)+3.2} \qquad (7)$$

The plots of eqs. (5),(6) and (7) are shown in fig.3.

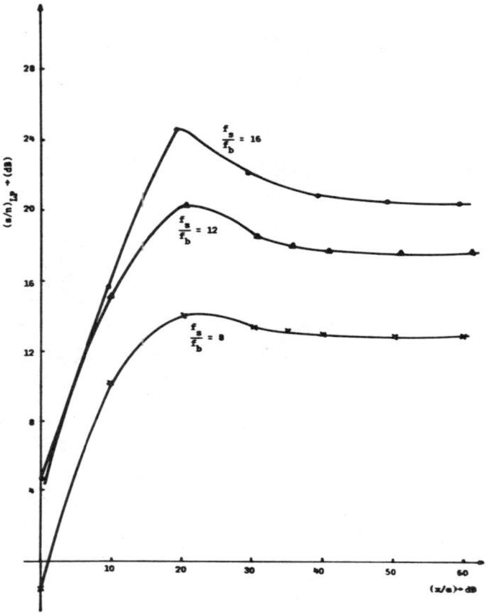

Fig.3. Plots of eqs. (5),(6) and (7).

III. IMPLEMENTATION OF THE PROPOSED DM SYSTEM

Fig.4 shows a detailed block diagram of the circuit implementation. The input signal x(t) is compared to the feedback signal y(t); the output of the comparator is passed through a D-type flip-flop which acts as a sample and hold circuit. The two complementary outputs of the flip-flop are fed to two shift registers to form the binary signals C_{i-1}, C_{i-2}, C_{i-3} and their complements. These binary signals are then used to implement the algorithm described in the previous section. According to this algorithm the step size at every instant should obey the following equation,

$$\Delta_i/a = \sum_{k=0}^{3} C_{i-k} Q^{k^p}$$

The values of Q and p for an optimum S/N and DR=∞ have been found to be equal to .43 and .45 respectively.

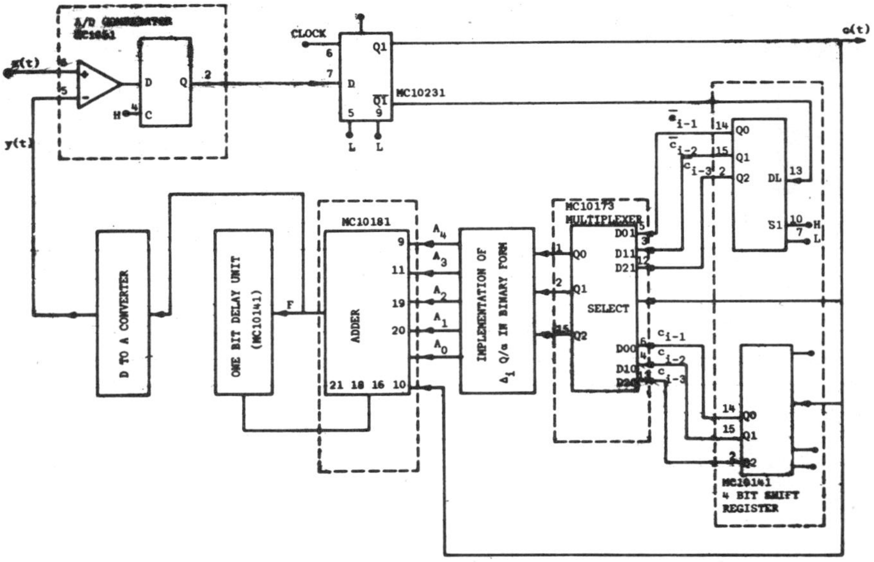

Fig.4. Detailed block diagram of the proposed delta modulator.

The values that Δ_i/a takes on, when $C_i=1$ are shown in table 1 for the various states of C_{i-1}, C_{i-2}, C_{i-3}. The parameter Δ_i^Q/a

Table 1: Values of Δ_i/a for the various states of C_{i-1}, C_{i-2}, C_{i-3} and $C_i=1$.

c_{i-3}	c_{i-2}	c_{i-1}	Δ_i/a	Δ_i^Q/a	A_0	A_1	A_2	A_3	A_4
1	1	1	2.	20x.1	0	0	1	0	1
0	1	1	1.476	15x.1	1	1	1	1	0
1	0	1	1,368	14x.1	0	1	1	1	0
0	0	1	,844	8x.1	0	0	0	1	0
1	1	0	1,140	10x.1	0	1	0	1	0
0	1	0	,616	6x,1	0	1	1	0	0
1	0	0	.508	5x,1	1	0	1	0	0
0	0	0	0	0x.1	0	0	0	0	0

is the normalized quantized step with base .1 while A_4, A_3, A_2, A_1 and A represent the values of Δ_i^Q/a in binary form.

The signals A_4, A_3, A_2, A_1 and A_0 pass through the binary adder (subtractor) causing the step to be added to (or subtracted form) the previous value of y. The output F's of the adder represent in binary form the tracking signal y. This binary signal is then converted to an analogue signal and fed to the comparator. It is noted that all the digital circuits used in this implementation belong to the ECL family.

IV. EXPERIMENTAL RESULTS

The input to the coder was a Gaussian noise source with flat spectrum as shown in fig.5a. Fig.5b shows the spectrum of the de-

Fig.5a. Gaussian input to the modulator

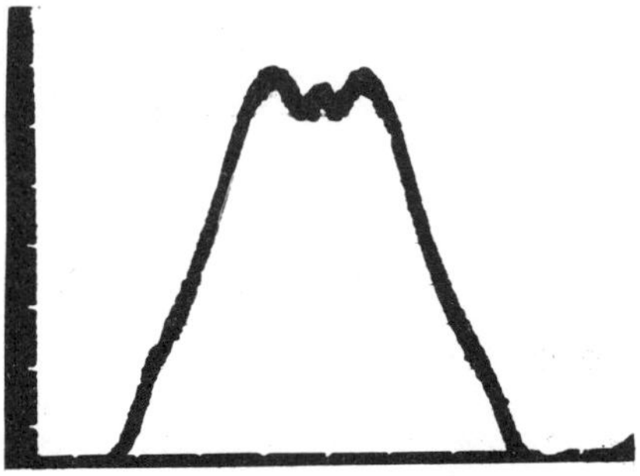

Fig.5b. Spectrum of the demodulated signal

modulated signal (after the LPF of the demodulator).

The bandwidth of the signal was 2MHz (f_b=2MHz). The resultant output power was measured in selected 5KHz bands with the aid of a wave analyzer. This power comprises a representation of the input power with additional noise generated in the modulator itself. A band rejection filter was then inserted before the modulator to block the applied signal in the frequency band where the measurement is made. The measured power is therefore the noise generated in the coder alone. A signal-to-noise ratio can be determined from these two measurements for the particular band of frequency chosen.

In fig.6 the experimentally measured $(S/N)_{LP}$ results are shown for f_s=16MHz. In fig.7 the experimentally measured results are shown for f_s=24MHz. Comparing the theoretical results with the experimental results (i.e. fig.3 $(f_s/f_b)=8$ with fig.6 and fig.3 $(f_s/f_b)=12$ with fig.7) the following conclusions may be deduced.
1. The DR is about 40db. This is not in agreement with the theoretical results where the DR was predicted to be infinite. This dramatic reduction of the S/N (and consequently of the DR) which occurs for high values of (x/a) is due to the fact that at high

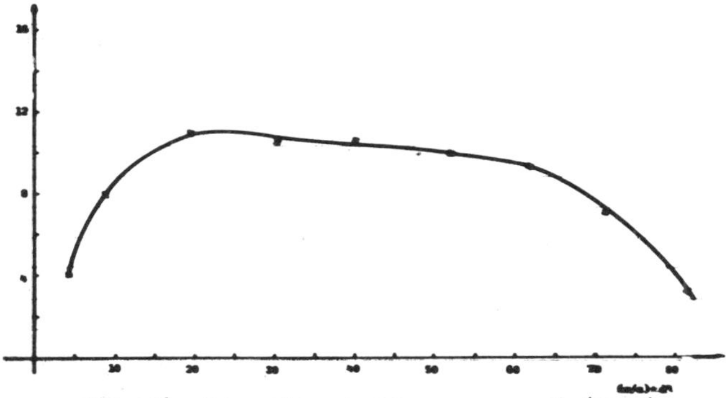

Fig.6. Experimentally measured $(S/N)_{LP}$ for f_S=16MHz.

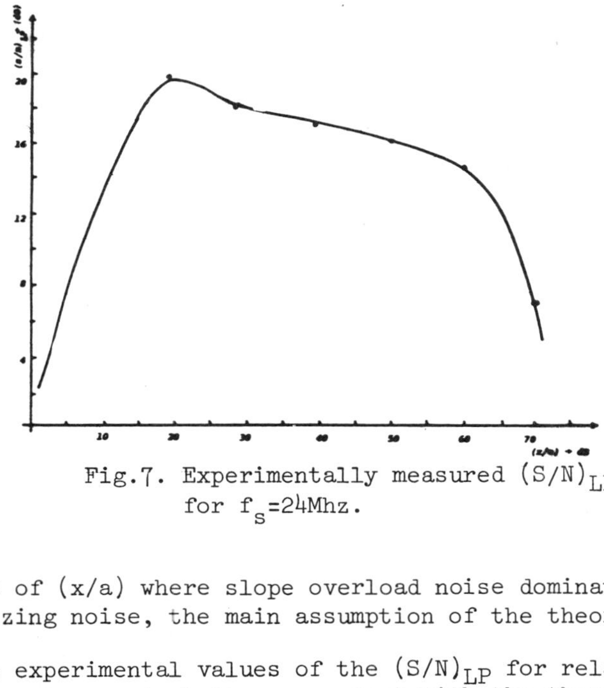

Fig.7. Experimentally measured $(S/N)_{LP}$ for f_s=24Mhz.

values of (x/a) where slope overload noise dominates rather than quantizing noise, the main assumption of the theory becomes invalid.

2. The experimental values of the $(S/N)_{LP}$ for relatively small input signals are in better agreement with the theoretical results at higher sampling rates. This was expected since the larger the sampling rate the betteris the tracking, and consequently the more valid are the main assumptions of the theory.

V. CONCLUSIONS

The experimental results seem to be in good agreement with the theoretical ones for small values of the input signal. The agreement becomes better as we increase the sampling rates.

REFERENCES

1. David N.C., "Analysis and Design of a broadband Adaptive Delta Modulation System" Master Thesis, Carleton University, Ottawa Canada, 1978.
2. David N.C., Chudobiak W.J. and Makios V. "Analysis and Design of a broadband Delta Modulation System". A Manuscript submitted for possible publication in the Proceeding of IEE.
3. Steele R., "Delta Modulation Systems" Pentech Pres. London, 1975.
4. Basworth, R.H. and Candy J.C., "A companded one-bit coder for television transmission", Bell System Techn.J.,48, 1459-1479, July (1969).

A NEW PROTOCOL FOR A DISTRIBUTED LOOP COMMUNICATION NETWORK

G. Papadopoulos, S. Leventis, S. Koubias

Applied Electronics Lab. University of Patras, Greece.

ABSTRACT: This paper presents a new loop structure and its associated protocol for a mini-micro computer communication network. The proposed structure consists of two loops, both of which are used for control and data transmission one in the clockwise direction and the other in the counter-clockwise direction. This topology has all the advantages of ring structures such as simplicity, cost modularity, space modularity, concurrent service and high reliability. In addition, it provides better line utilization and traffic regulation than previously proposed loops because a) it has a distributed control mechanism and b) it allows two stations to communicate for long periods of time without blocking completely communication between other stations. Furthermore, it allows limited communication in case that a break occurs in the line. The loop protocol is based on the SDLC frame format. This has made possible the use of the latest LSI technology that disasembles and reassembles the character of this frame.

I. INTRODUCTION

The objective of this work is to develop a network that is efficient, inexpensive and reliable and that will provide flexible service to a localized groop of users. Such a network must have the following properties:

- Cost modularity
- Space modularity
- Reliability
- Dynamic reconfiguration
- Universality

D. G. Lainiotis and N. S. Tzannes (eds.), Advances in Communications, 297–304.

. High throughput
. Small installation cost

There are many networks in use today, that have different topological structure. The most widely used structures are: a) Complete interconnection requiring N(N-1)/2 links for N nodes, b) Star configuration requiring N interconnections, c) Loop configuration, d) Bus connection. These structures have been examined by many researchers. The main conclusions with respect to the properties mentioned above are summurized in Table 1.

TABLE 1

	INTERCONNECTION			
	FULL	STAR	LOOP	BUS
Cost Modularity	No	No	Yes	Yes
Space Modularity	No	*	Yes	*
Installation Cost	High	*	Low	*
Reliability				
Distributed Control	Yes	No	Possible	No
Dynamic Reconfiguration	Yes	No	Possible	No
Logical Complexity	No	Yes	Yes	Yes*
Throughput	Good	?	?	No Good

* trade-off

From this comparison it is clear that the full connected network is too costly. The star network is less costly, has very good throughput but lacks in reliability, since, as soon as the central computer exhibits faulty operation, the whole system breaks down.

The loop configuration seems to satisfy most of the desired properties and this is the reason we have addressed our attention to the design and construction of such a network. In addition, loop system are attractive for mini and microcomputer networks because of their simplicity and their high line utilization. Therefore, it was not surprising that loop systems have attracted the attention of many reasearchers who have designed a variety of loop networks (1), (2), (3), (4), (5), (6).

The first loop system was suggested by Farmer and Newhall (1). In this system a round-robin control passing mechanism circulates around the loop and allows only one interface at a time to transmit one or more messages through the loop. During this transmission by one node, the other nodes have to wait. This results in a very inefficient loop channel utilization and long message delays.

In the Pierce loop (3) a time multiplexing mechanism is incorporated by dividing the communication space into one or more fixed size slots. In this scheme the messages are divided into packets, so that each packet will occupy one slot in the loop. This scheme improves the channel utilization, but wastes the communication space when a variable-length message is divided into fixed-size packets. Another disadvantage of this system is the extra hardware that must be designed for message disassembly and reassembly.

It is apparent that neither the Farmer-Newhall or the Pierce loops make very efficient use of loop topology. Reames and Liu. (5) introduced a new message transmission mechanism called DLCN (Distributed Loop Computer Network) which combines the advantages of the Newhall and the Pierce loops. That is it allows multiple messages of variable length to circulate in the loop. This is done by inserting a variable length shift register before each node's transmitter. A condition that must hold is that the length of the locally generated messages cannot exceed the size of the delay buffer.

Although the DLCN combines the advantages of the Farmer-Newhall and Pierce loops and also provides automatic traffic regulation (by means of the size of the delay buffer), it favors infrequent requests while delaying more frequent requests for network service. Other disadvantages of the DLCN are:

a) Interface complexity and, therefore, cost.
b) Lower reliability due to the insertion of the variable shift register.
c) The queuing time increases drastically with the number of nodes inserted in the loop.

In addition to the disadvantages discussed above, the previous loops have some additional common shortcomings. The most significant ones are:

a) The stream of data is in one direction. Therefore, when a node communicates with the node just behind it in the loop, the transmission of data takes place through the rest of the nodes. This results, in more delay and lower reliability.
b) A node can block out all other transmissions by sending a stream of messages to another node. This means that the above loops are sensitive to local demands which results in an over all decrease of the network performance.

With the motivation to the eliminate the above mentioned shortcomings a new loop structure was proposed by Jafari and Lewis (6). The main innovation in this loop is to distribute data and control into two different loops, a data loop and a controller loop. The data loop consists of bi-directional segments connecting the nodes. Each node is interfaced to the loop by a zero-delay switch that may be turned "on" or "off". The operation of the control loop is based on a simple arbiter, which selects the minimum route.

The structure of this loop permits concurrent data transmission between pairs of nodes connected by non-interfering segments. The zero delay relay nodes together with the capability of bidirectional message transfer between more than one pair of nodes improve considerably the throughput and the line utilization of this loop compared to the ones described earlier. However, the Jafari loop has a few disavantages of its own:

a) The control loop is under-utilized, since it remains idle for most of the time.
b) Concurrent message transfer is allowed only between non-interfering nodes.
c) The control mechanism is centralized.

In the loop structure to be proposed here, the two-loop concept is maintained. However, both loops are used for message transfer. One loop allows transmission only in the clockwise direction (called forward), while the other only in the counter-clockwise direction (called backward). Thus, we can have messages of variable length transfered simultaneously between interfering node pairs. As far as the control operation is concerned, we have returned to the distributed control concept employed in the earlier loops.

The distributed control operation together with the simultaneous bi-directional message transfer provide also automatic traffic regulation. This can be asserted on the basis of the following properties of the new loop:

a) There is no central supervisor
b) The previous individual accesses to the loop affect in an automatic way the passing of the control.

II. DISCRIPTION OF THE NEW LOOP STRUCTURE AND PROTOCOL

As mentioned above the proposed network consists of two loops a forward and a backward. As shown in Fig. 1 the outside loop carries the forward traffic (clockwise), while the inside carries the backward traffic (counter-clockwise). Every node can become the controllerof the network as long as it has a message to transmit.

Once a node gains control of the loop it establishes a forword communication link with the node to which it wants to transmit. The intervening nodes act as relays and their design is based on the SDLC chip (7), (8), which acts like a relay when it is programmed in the one-bit delay mode. Two SDLC chips (one for each

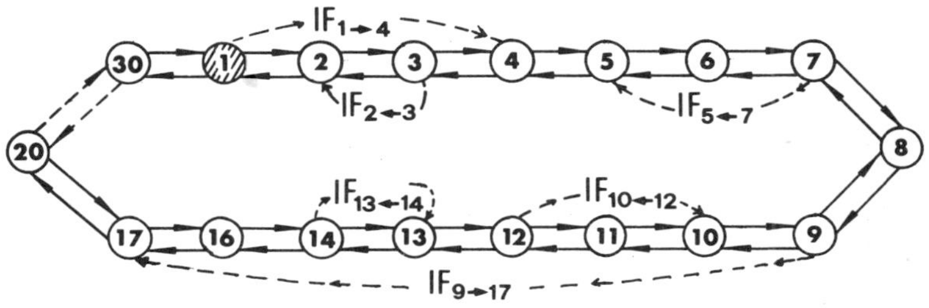

(i) node (i) controller
$IF_{i \to j}$ forward transmission
$IF_{j \leftarrow i}$ backward "

FIG.1. The New Loop Structure

loop) under microprocessor control form the basic hardware for each node. The SDLC chip was found very convenient because it combines the relay, transmit and receive funtions all in one unit.

The controller node makes first its identity known to all other nodes and starts transmitting the Information Frame (IF). The receiving node becomes itself a subcontroller, at the same time that it starts to receive the message. In this capacity it sends a Forward Subcontrol Frame (FSF) and a Backward Subcontrol Frame (BSF). The FSF gives the control to any other node beyond the subcontroller which wants to communicate with some other node in the forward direction but not past the main controller or with a node in the backward direction but not past the subcontroller. The node that made use of the FSF from the subcontroller acts now like a second controller. The node it communicates with acts like a secondary subcontroller. That is, the secondary subcontroller can transmit another FSF and BSF and so on. Thus, there can be more than one pair of secondary controller-subcontroller connections.

From the above discussion it is clear that we can have simultaneous message transfer between many interfering node pairs. From

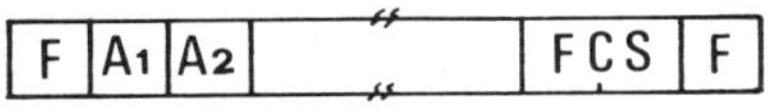

F Opening, Closing Flag, 01111110.
A_1 Destination Address
A_2 Control: Source or End Address
I Information Field (variable leng.)
FCS Frame Check Sum

FIG.2. The Frame Format

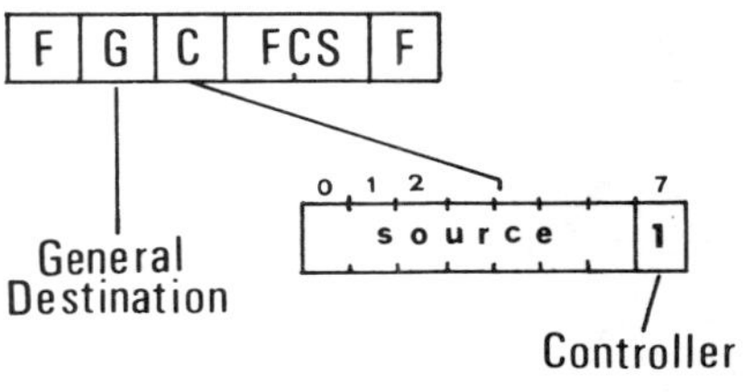

FIG.3. The Start Frame (SF)

an intuitive point of view this must increase considerably the line utilization over the Jafari loop.

The communication protocol used in the proposed structure has the form of the SDLC protocol, the frame format of which is shown in Fig. 2, (7). The existence of chips developed for the implementation of this protocol made it particularly attractive for our application.

The proposed method will best be described by an example, refering to Fig. 1. Let us assume that node 1 is the main controller and node 4 is the node that it wants to communicate with. That is node 4 is the first subcontroller. Also, let us assume that node 1 has been set in the transmit mode and all the others in the relay mode. Next, node 1 transmits a general frame that notifies all nodes about the identity of the main controller. This frame is called Start Frame (SF) and has the format shown in Fig. 3.

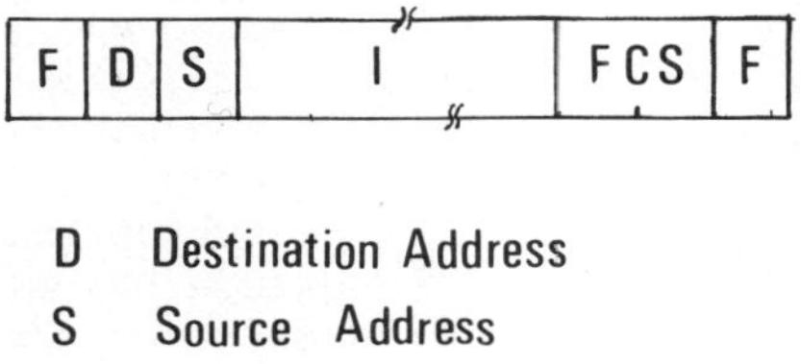

FIG. 4. The Information Frame (IF)

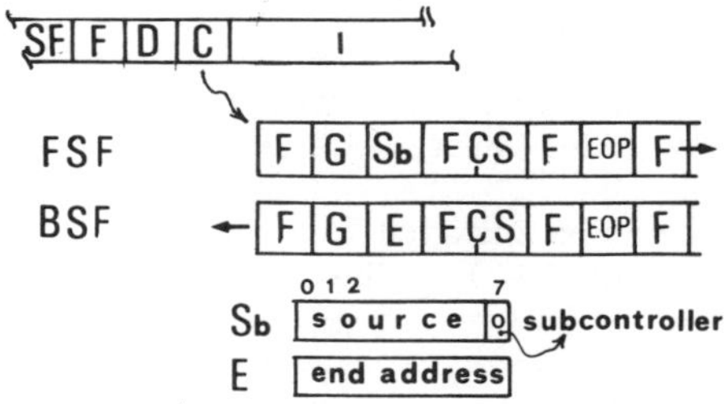

FIG. 5. The Formats of FSF and BSF

After the transmission of the SF, node 1 starts to transmit the Information frame (IF) shown in Fig. 4. As soon as the subcontroller detects its address code in the destination field, it exits from the relay mode and enters the Receive Forward (RF) mode. At the same time it becomes the subcontroller and transmits two frames, one forward (FSF) and one backward (BSF), as mentioned in the beginning of this section. The formats of the FSF and BSF are shown in Fig. 5. It is seen that the subcontrol frames are followed by an EOP (End of Polling) character and a series of flags. The EOP character gives the opportunity to any of the nodes that it goes through to transmit an IF in the loop forward up to the main controller and backward up to the previous subcontroller.

It is clear that he nodes that receive an FSF or a BSF become themselves in a sense traffic regulators. The format of the frame that is transmitted by a node k that receives an FSF and wants to transmit backward is shown in Fig. 6. It is seen that the EOP character that follows and FSF is converted a) to an opening flag for a new FSF that notifies the nodes forward of k about the current status of the network and b) to an opening flag for the IF that is transmitted backward.

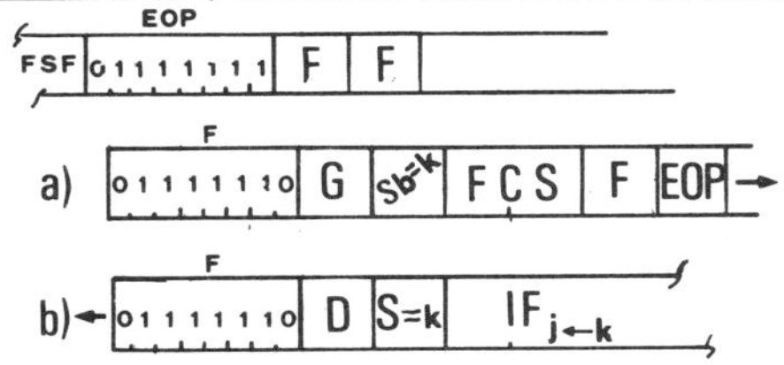

FIG. 6. The response of a node k that receives FSF, EOP, and wants to send $IF_{j \leftarrow k}$

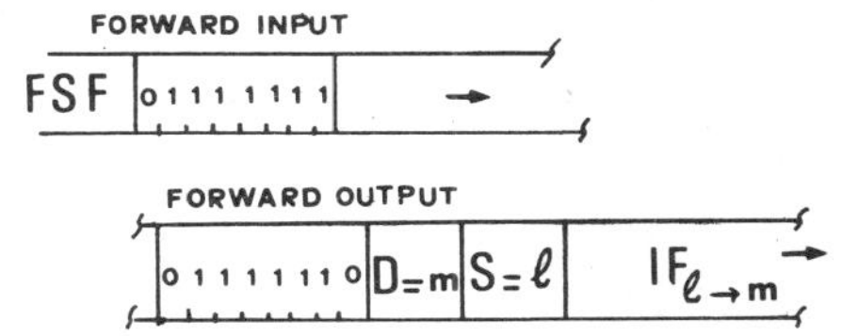

FIG. 7. The response of a node ℓ that receives FSF, EOP, and wants to send $IF_{\ell \to m}$

In Fig. 7 is depicted the response of a node ℓ that receives an FSF and wants to transmit a message forward. All the nodes forward of ℓ have received the FSF that went through node ℓ but do not find the EOP character that has been converted to an opening flag by node ℓ. Thus, all the nodes forward of ℓ up to node m that is supposed to receive the IF transmitted by ℓ cannot transmit forward. The response of node m to the frame $IF_{\ell \to m}$ is shown in Fig. 8.

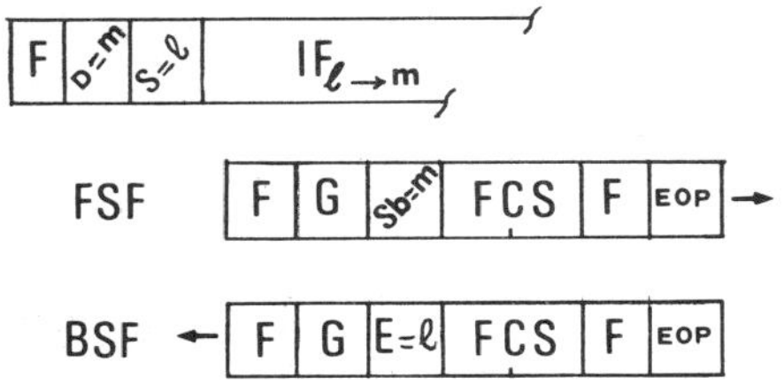

FIG. 8. The response of node m to $IF_{\ell \to m}$

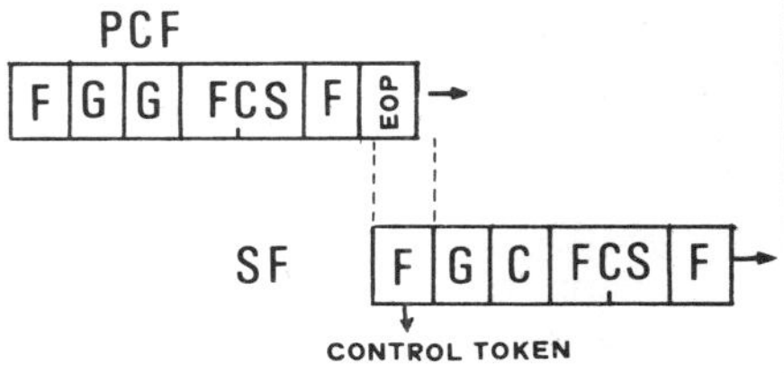

FIG. 9. The format of Pass Control Frame (PCF)

The above described process continues as long as there are available nodes in the loop. When the reception or transmission of a message at a node is completed, that node returns automatically to the one-bit delay (relay) mode, waiting for a new sequence of events to start.

The new sequence is initiated upon the completion of the message transmitted by the main controller. As soon as this occurs, the controller will transmit forward and backward ABORT characters. These characters will return to the controller as long as all the other nodes have returned to the one-bit delay mode. Then, the controller transmitts forward the so-called PASS CONTROL FRAME (PCF), followed by an EOP character. This gives the opportunity for control take-over by the first forward node that has a message to transmit. In Fig. 9 the format of the PCF is depicted.

III. CONCLUSIONS

A new loop structure has been presented. This loop consists of two loops transmitting control and information packets in the clockwise and counter-clockwise directions respectively. It has been asserted that this new loop structure provides better line utilization than previously proposed loops. It does this while retaining the property of automatic traffic regulation in the sense that control is distributed and access to the loop is based on the current system load and previous individual accesses. The node hardware is based on the SDLC chip that handles the SDLC frames automatically, thus relieving the node processor of many tasks and reducing the required hardware. Preliminary investigation by the authors has shown that the proposed scheme is feasible. Hardware realization as well as analytic and simulation studies are currently underway.

REFERENCES

(1) Farmer, W. W., Newhall, E. E., "An Experimental Distributed Switching System to Handle Bursty Computer Traffic", Proc. ACM Symposium. "Problems in the Optimization of Data Communications System", Pine Mtn. Georgia, Oct. 1969.

(2) Farber, D. J., Larson, K., "The Structure of a Distributed Computer System - The Communication System", Proc. Symp. on Computer Communications, Networks and Teletraffic, Polytechnic Institute of Brooklyn Press, 1972, pp. 21-27.

(3) Pierce, J. R., "Network for Block Switching of Data", BSTJ, Vol. 51, No. 6, 1972, pp. 1133-1145.

(4) Fraser, A. G. "Spider - A Data Communication Experiment", Computing Science Technical Report No. 23, Bell Telephone Lab.

(5) Reames, C. C., Liu, M. T., "Design and Simulation of the Distributed Loop Computer Network (DLCN)", in Proc. 3rd Annual Sympsium on Computer Architecture, Clearwater, Florida, January 1975, pp. 7-12.

(6) Jafari H., Spragins J., Lewis T., "A New Modular Loop Architecture for Distributed Computer Systems", Thrends and Applications: 1978 Distributed Processing.

(7) IBM Synchronous Data Control, Report No. GA27-3093-1

(8) Intel Peripheral Design Handbook, 1978.

FIBER OPTICS vs. CONVENTIONAL TRANSMISSION LINES IN DISTANCE COMMUNICATION SYSTEMS

C.J. Georgopoulos and N.P. Athanasiadis

University of Thrace, Xanthi, Greece

ABSTRACT

It has been clear that the low-loss high bandwidth fibers offer many significant advantages over conventional transmission lines, especially in the areas of long-distance transmission data, security data handling, EMI/RFI/EMP noise immunity and weight and volume reduction. Since more and more new data link systems are designed around fiber optics and old ones are retrofited or replaced by modern electro-optical elements, it is an urgent need to establish certain relationships between the two, taking into account those parameters that make the difference in the quantity and quality of data transmission. In this paper, the main parameters that are affecting maximum information transmission rates, attenuation and delay distortion are examined and an analogy is drawn.

1. INTRODUCTION

The optical fibers that are in common use today are of either step or graded index type. The step index optical fibers present multimode or single mode ray paths, while the graded index present only multimode ray paths. In designing a fiber optic system, the first step is to select either an LED or laser source, an appropriate fiber for the required link length and bandwidth or data rate and the type of optical detector (avalanche or p-i-n photodiode).

An LED is favored over the faster, higher power laser, when the application permits, because of its longer lifetime, better linearity and greater optical range. Moreover, a laser requires costly optical-feedback stabilization.

The lowest attenuation for lower-grade step index fiber is

D. G. Lainiotis and N. S. Tzannes (eds.), Advances in Communications, 305–312.

achieved between 0.7 and 0.9 μm, while multimode graded index fibers have their lowest attenuations in the 0.9 μm region.

Fiber optic communication systems are getting faster and larger. Although fiber optics is generally regarded communications technology for transmitting digital signals, it is also capable of handling analog signals. Moreover, the optical fibers hold losses low enough that CATV distributed networks may operate without repeaters over distances of up to 5 Km. With conventional transmission lines, repeaters may be required sometimes every 400 m.

Certain characteristics, such as immunity to electromagnetic interference, large bandwidth combined with low weight and proof against interception, give the fiber optic systems a definite advantange for selective use in some special fields. It is expected that fiber optic transmission systems will eventually be able to compete with conventional systems on economic terms, also, and will then be widely adopted. It is therefore necessary now to develop various types of fiber optic systems and make trade offs comparing them with their conventional cable counterparts.

2. FIBER OPTICS vs. CONVENTIONAL CABLES

2.1. The two technologies

Figure 1 shows the main parameters of coaxial cables and optical fibers. Cabling techniques have reached a level were optical waveguides can be processed without impairing the transmission

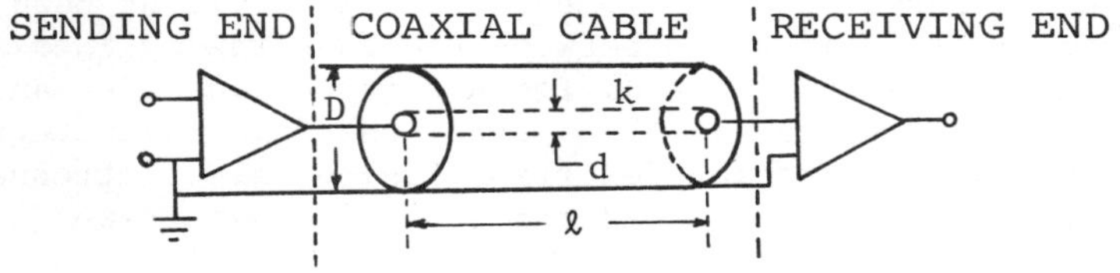

- Electrical parameters:R,L,C.
- Geometrical parameters:D,d,ℓ.
- Material parameters:k.

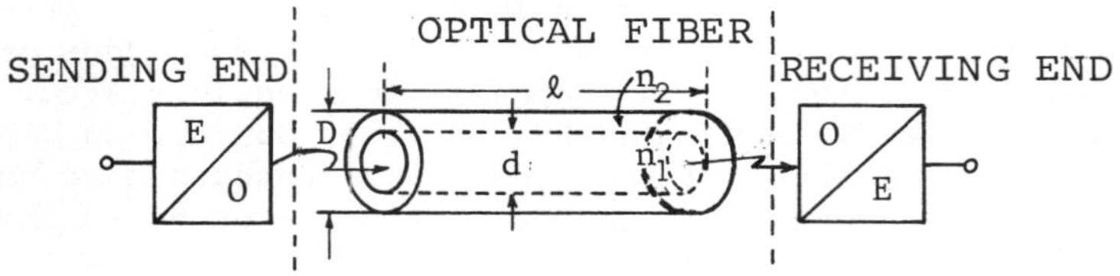

- Optical material parameters :n_1,n_2, impurities.
- Geometrical parameters : d, D-d, ℓ.

Figure 1. Main parameters of coaxial cables and optical fibers.

characteristics of the fibers during processing. It is also an important objective of the cabling process to preserve those characteristics during installation and for a long time afterwards. Protection of the fiber against mechanical destructions is generally achieved by stranding the cores with stress members and providing robust sheath structures. The stress members require materials with a high modulus of elasticity, and other materials such as the plastic KEVLAR or glass. Steel spring is also suitable for use as a stress member and provides additional reinforcement for particularly highly stress cables, such as aerial cables.

2.2. Attenuation

Coaxial cable losses can be attributed to dielectric and resistive factors as shown by the following relationships:

$$a(dB/100m) = \frac{36.32}{Z_o}\left(\frac{1}{d}+\frac{1}{D}\right)\sqrt{f} + 9.1p\sqrt{k}\,f \tag{1}$$

$$a(dB/100m) = 4.85 \times 10^6 \sqrt{T_o f}\,/\ell, \tag{2}$$

where Z_o = characteristic impedance of the line [Ω]
d = diameter of the inner conductor [mm]
D = diameter of the outer conductor [mm]
f = frequency [MHz]
k = dielectric constant
p = power factor
T_o = time for the output to reach 50% of the input magnitude [ns]
ℓ = line length [m]

The above equations can be put in the following general forms:

$$a(dB) = A_1 f^{B_1} + C_1 f, \tag{3}$$

and

$$a(dB) = A_2 f^{B_2}, \tag{4}$$

where $B_1 = B_2 = 1/2$

Below about 1 GHz the resistive loss predominates and is a function of the diameter of the center conductor. Resistive loss varies as the square root of the frequency due to skin effect. Therefore, the larger the center conductor, the lower the loss. Fig. 2 shows plots of calculated attenuation vs. frequency for a coaxial cable RG $59/U_2$ (75 Ω) for both of the above relationships. It should be mentioned here that unshielded twisted pair cables exhibit approximately the same square law characteristics as the coaxial cables, except that the overall loss is higher.

Early fibers suffered transmission losses of several decibels per meter in the visible region. Today, for the near infrared

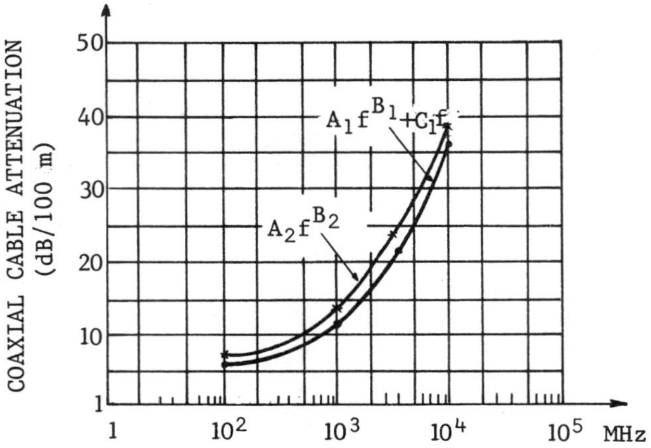

Figure 2. Coaxial cable attenuation vs. frequency.
Type of cable : RG/U_2 (75 Ω).
(A_1=0.956, B_1=1/2, C_1=6.85x10^{-3}, A_2=1.19, B_2=1/2).

region, these losses are three orders of magnitude lower.

After fitting some curves to already known data for fiber optics[1,2], it was found that those curves are similar to the form of previous equations that are valid for conventional coaxial or twisted pair cables. Figures 3 and 4 show such plots. The plot in Fig.3 represents two types of low loss fibers[2], while Fig.4 represents a medium loss fiber. The attenuation, the cutoff frequency, the rise time and the pulse spreading are closely linked with each other in a fiber optic transmission line.

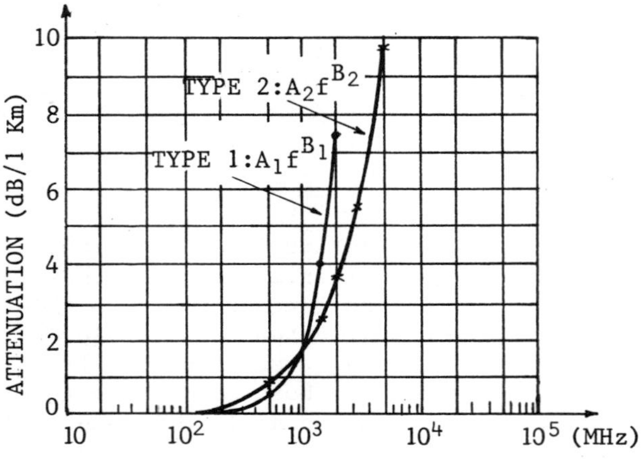

Figure 3. Low-loss optical fiber attenuation vs. frequency [Ref.2]
(For type 1:A_1=1.87, B_1=2. For type 2:A_2=1.82, B_2=1.05).

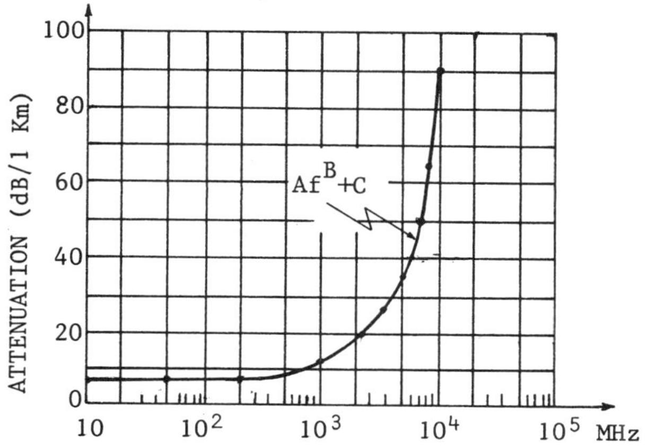

Figure 4. Medium-loss optical attenuation vs. frequency
($A=3.65x10^{-5}$, B=1.585, C=8.4)

2.3. Bit rates

The performance of any system for transmitting information becomes most significant when the system is being used at or near its response limit.

Figure 5 shows the relationship between the maximum length of

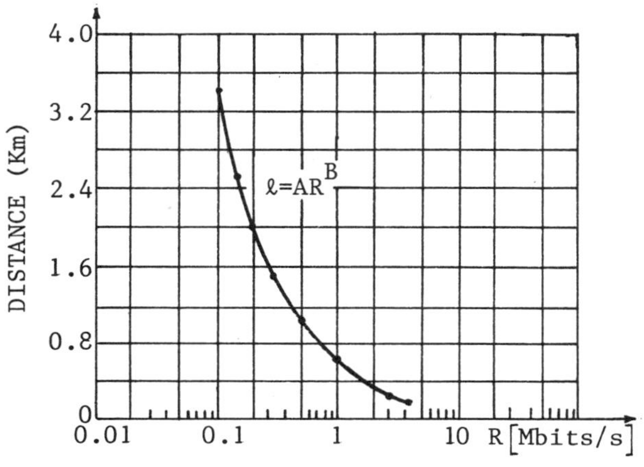

Figure 5. Data transmission rates for balanced lines.
(A=0.637, B=-0.724, Duty cycle:10%).

a balanced cable line and the bit rate (R) in Mbits/s. The fitting curve for the region shown[3] has the form:

$$\ell = AR^B \tag{5}$$

where ℓ = line length [Km]
A,B = constants
R = bit rate [Mbits/s]

In the case of optical fibers a trade off is usually made between the type of source and type of fiber optic line, depending on the bit rate constraints. For a long-range communication the best solution is the binary transmission of pulse-code modulation (PCM). From Fig.6 we can find the maximum distance before a repeater is needed for the transmission of binary signals at a bit rate (R) with either an LD or LED source. This is a fitting curve[4]. which has the following form:

$$\ell = AR^{B} \tag{6}$$

Here, the line imposes a length limit of approximately 10 Km. However, the source may impose a lower limit depending on the magnitude of the power it can pump into the optical fiber line.

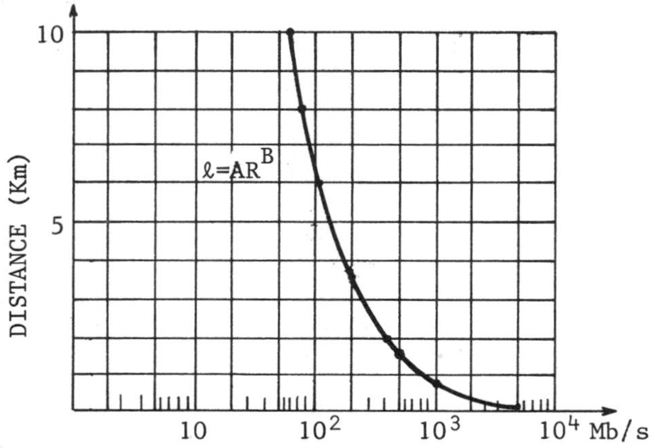

Figure 6. Transmission distance vs., bit rate (A=680, B=-1).

2.4. Time delay

For single mode fiber optic guides[5], time delay per unit length between the longest and the shortest path length is given by:

$$t_1 = \frac{1}{c} \cdot \frac{n_1}{n_2} (n_1 - n_2) = \frac{n_1}{c} \Delta \tag{7}$$

where n_1 = index of refraction of the core

n_2 = index of refraction of the cladding

$$\Delta = \frac{n_1}{n_2} - 1$$

c = the velocity of light

If n=1.5 and Δ=1%, time spread is only 50 ns/km. In this case, the bandwidth of 1 km fiber is about 50 MHz.

For graded index light waveguides[6], the index varies between n_o at the core (r=0) and $n_o(1-\Delta)$ at the clad (r=α). Among the possible index profiles, the quadratic dependance

$$n(r) = n_o(1-\Delta \frac{r^2}{\alpha^2}) \text{ for } r \leqslant \alpha \tag{8}$$

is of particular interest because of its lens-like characteristics.

The ray representetion helps to make an estimate for the delay distortion in the guide. There is no index profile which could completely eliminate delay (achieving equal transit times along all possible ray trajectories from one guide cross-section to another). However, the parabolic profile is close to optimal.

The difference between the longest and the shortest transit time (the longest transit time corresponds to an helical path and the shortest to the axial path) per unit of guide length is:

$$t_2 = \frac{n_o}{c}(\frac{1-\Delta}{\cos\theta_\rho} - 1) \simeq \frac{3}{2} \cdot \frac{n_o}{c}\Delta^2 \tag{9}$$

where θ_ρ is the pitch angle of the helical path

$$\theta_\rho = \cos^{-1}(\frac{1-3\Delta}{1-\Delta})^{\frac{1}{2}} \tag{10}$$

Considering the equations (7) and (9) we can draw an analogy between time delay in conventional and fiber optic cables. Since

$$n = \sqrt{k} = \sqrt{\varepsilon/\varepsilon_o}, \tag{11}$$

where n is the index of refraction of a dielectric medium with permittivity ε and dielectric constant k, we obtain

$$t_d(\text{per unit length}) = \begin{cases} A\sqrt{k} & \text{(cable)} \\ A'n & \text{(fiber)}, (n=n_o, n_1) \end{cases} \tag{12}$$

3. CONCLUSION

In the conventional cables the information is transmitted by analog or digital electrical signals. This form of transmission, however, has reached its limit of development. It is also uncertain whether any further significant improvements are possible in crosstalk reduction, external interference limitation, bit rate, attenuation, cable miniaturization, e.t.c.

Optical transmission offers an advantageous alternative over these electrical cable types. But unfortunately, serious standardization has not started yet for the optical components and there is very little reference to corresponding electrical parameters. This work, which is only the first step into the area of conventional vs. fiber optic cables competition, shows that definite analogies can be drawn and various relationships can be developed that will relate the parameters and the functional characteristics of the two types of cable technology.

REFERENCES

1. *Gier, J., Kügler, E., and Seiffer, W.D.: "Fiber optic communications", Telcom Report, Vol. 1, pp. 51-56, 1978.*

2. *Keck, D.B., and Boullie, R.: "Measurements on high-bandwidth optical waveguides", Optics Communications, Vol. 25, No. 1, pp. 43-47, April 1978.*

3. *Hackmeister, D.: "IC push into data communications relieves prosessors, cuts software", Electronic Design 9, pp. 42-48, April 1978.*

4. *Keck, D.B.: "Transmission properties of optical fiber waveguides", SPIE, Vol. 63, Guided optical Communications, pp. 3-7, 1975.*

5. *Barnoski, M.K., and Personick, S.D.: "Measurements in fiber optics", Proc. IEEE, Vol. 66, No. 4, pp. 429-441, April 1978.*

6. *Gloge Detlef : "Optical waveguide Transmission", Proc. IEEE, Vol. 58, pp. 1513-1522, October 1970.*

DISTRIBUTED PROCESSING NETWORK SURVIVABILITY (1)

Gene Hilborn

Ford Aerospace & Communications Corporation

A new methodology is given for describing the effectiveness of communicating networks of processing subsystems which are subject to communication and processing failures or losses. The network attributes are described as a graph of nodes and edges. A class of performance measures called the "survival index" is defined axiomatically on a connectivity state for the network graph and its reductions. This definition is based on the idea of a team of members cooperating or sharing resources by use of communication links, and on the existence of synergism or a "team effect," where a group of connected members are more effective than isolated members. The present work is restricted to homogeneous teams - where members have equal capability.

INTRODUCTION

We seek to develop a new methodology for describing and comparing the survival effectiveness of connected networks of processing subsystems (termed "distributed processing networks") which are subject to deliberate or natural failures or losses of processing or communication components. Examples of distributed processing networks include:

- resource - sharing computer networks
- military force teams
- distributed architecture computers
- distributed sensor-processing systems
- distributed ballistic missile defense systems

(1) Research supported by Ballistic Missile Defense Command, Huntsville, ALA, while the author was at Systems Control, Inc.

D. G. Lainiotis and N. S. Tzannes (eds.), Advances in Communications, 313–320.

AN AXIOMATIC SURVIVAL INDEX

Formulation

The graph G = (N,E) of the distributed processing network before any element is removed (by attack or failure) is connected. Removal of some elements of G creates a reduced graph G´. The reduced graph contains ℓ isolated subgraphs, where each subgraph is connected. (These subgraphs are also called the component subgraphs.) More precisely, if G´ = (N´, E´) is a reduction of G, then

$$G' = \bigcup_1^\ell G_i' = (\bigcup_1^\ell N_i', \bigcup_1^\ell E_i') ,$$

where each component subgraph G_i' is connected and no path exists between any nodes, in distinct sets N_i' and N_j'.

We now define a connectivity state vector $\underline{x}$ on any reduction G of G as the nxl vector with components equal to the size of respective component subgraphs:

$$x_i = |N_i|, \ i \varepsilon I_n = \{1,2,\ldots,n\}.$$

However, any re-indexing of the state space components does not reflect any change in connectivity or function - all such permutations being equivalent. As a convenience, this equivalence can be formalized by adopting the convention that the subgraphs are indexed such that $|N_1'| > |N_2'|$, etc. Then,

$$n \geq x_1 \geq x_2 \geq \ldots \geq x_\ell > 0$$

Let Ω_n denote the space of possible connectivity states on reductions of a connected graph of n-nodes.

Let

$\underline{x}_c$ = (n, o,...,o) (connected graph)

$\underline{x}_I$ = (1,...,1) (isolated node graph)

$\underline{0}$ = (0,...,0) (empty graph)

We now define a survival index as a function $\sigma(\cdot)$ on Ω_n onto the unit interval [0,1] satisfying the following axioms:

A0) $\sigma : \Omega_n \rightarrow [0,1]$

A1) $\sigma\ (\underline{0}) = 0$

A2) $\sigma\ (\underline{x}_c) = 1$

A3) $\sigma\ (\underline{x}) < 1 \Longrightarrow \Delta_i\ \sigma(\underline{x}) > 0\ ,\ i\varepsilon\ I_{\ell+1}$

A4) $x_i > x_j \Longrightarrow \Delta_i\ \sigma(\underline{x}) > \Delta_j \sigma(\underline{x}) \quad i,j\varepsilon\ I_{\ell+1}$.

The notation "$\Delta_j \sigma(\underline{x})$" refers to a partial difference or increment:

$$\Delta_i \sigma(\underline{x}) = \sigma(\underline{x}+\underline{\delta}_i) - \sigma(\underline{x}) \quad , \text{ for } \underline{x},\ \underline{x} + \underline{\delta}_i\ \varepsilon\ \Omega_n$$

where $\underline{\delta}_i$ is an nxl unit vector with 1 in the ith position.

Axioms A0 - A2 are conventions. Axioms A3 and A4 have the following interpretations/motivations.

Axiom A3 means that adding a node to the graph always improves system performance.

Axiom A4 further orders the connectivity state space by asserting that performance is improved more by adding a node to the larger of the different connected subgraphs. This also includes the empty subgraph ($x_{\ell+1}=0$); thus, in particular, adding a node to an existing connected subgraph is better than adding it as an isolated node. Axiom A4 formalizes the notion of a "team effort" or synergy effect.

The defining of $\sigma(\cdot)$ as a function on the space of the connectivity state is equivalent to the assumption that all team members are of equal importance. This homogenity is the ideal in decentralized systems - no member is more vulnerable or important to the overall system than any other.

SHAPING

If we have any function $f(\cdot)$ on the state space which is a survival index (i.e., satisfies A0 - A4), we may wish to modify or <u>shape</u> it to model the behavior of a class of proposed systems. The following theorem allows us to do this and still preserve the axiomatic properties of our function.

<u>Theorem</u> (Shaping)

If $f(\cdot)$ is a survival index function and $g(\cdot)$: $[0,1]\to[0,1]$ is monotone increasing with $g(\underline{0}) = 0$ and $g(1) = 1$, then the composite function $g\circ f$ is also a survival index.

ORDERING AND BOUNDS

Let

$\tilde{\Omega}_m = \Omega_m - \Omega_{m-1}$. That is, $\tilde{\Omega}_m$ is the subset of Ω_n $(n \geq m)$ for which $\Sigma x_i = m$. Let $\ell(\underline{x})$ denote the number of connected subgraphs of $\underline{X}$. Thus, we have:

Theorem (Ordering)

$\tilde{\Omega}_m$ is ordered as follows: For any distinct $\underline{x}, \underline{y} \in \tilde{\Omega}_m$, with $\ell(\underline{x}) \leq \ell(\underline{y})$, then

(a) $\ell(\underline{x}) < \ell(\underline{y}) \Longrightarrow \sigma(\underline{x}) > \sigma(\underline{y})$

(b) $\ell(\underline{x}) = \ell(\underline{y})$, $\Longrightarrow$ for min i such that $x_i > y_i$, $\sigma(\underline{x}) > \sigma(\underline{y})$.

This ordering provides us with bounds on $\sigma(\underline{x})$, for any graph G with n nodes and its reduced graph G′ with m nodes. Let $\underline{x}_c$ (m) denote the connectivity state vector for a subgraph which is connected and of size m, and $\underline{x}_I(m)$ denote the connectivity state for a subgraph consisting of m isolated nodes. That is,

$\underline{x}_c(m) = (m)$ and

$\underline{x}_I(m) = (1,1,\ldots,1)$.

The bounding result is thus stated:

Theorem (Bounding)

For any connected graph G, and any reduced subgraph with connectivity state $\underline{x} \in \tilde{\Omega}m$,

$$\sigma(\underline{x}_I(m)) \leq (\underline{x}) \leq \sigma(\underline{x}_c(m))$$

The inequalities being strict for m>1 and $\underline{x}_I(m) \neq \underline{x} \neq \underline{x}_c(m)$.

ATTACKS AND FAILURES

If G = (N,E) is the original connected network graph, reduced graphs are generated in two ways or combinations:

- Node removals
- Edge removals

These "removals" could be from successful attacks or from failures.

NODE ATTACKS

A node attack of size k, denoted A_k, removes k nodes from N (and the associated incident edges). A_k is the set of k nodes in N. For any node attack A_k, there is a corresponding connectivity state vector $\underline{x}(A_k)$. Let Γ_k denote the set of state vectors resulting from all possible node attack of size k against G. Then an optimal node attack A_k^* is defined as any node attack of size k such that if $\underline{x}_k^*$ is the corresponding state in Γ_k, we have

$$\sigma(\underline{x}_k^*) \leq \sigma(\underline{x}) \text{ for any } \underline{x}\varepsilon\Gamma_k .$$

A performance function under optimal node attack is called the node survival function, μ_k,

$$\mu_k = \sigma(\underline{x}_k^*) = \min_{\underline{x}\varepsilon\Gamma_k} \sigma(\underline{x}) .$$

From the bounding theorem we have

$$\sigma(\underline{x}_I(n-k)) \leq \mu_k \leq \sigma(x_c(n-k)).$$

The smallest value of k for which μ_k is less than the upper bound is precisely the connectivity $\omega(G)$ of the graph, discussed in prior work (i, ii) and defined as the size of the smallest node disconnecting set.

An alternate measure to the node survival function is to use random node attacks or failures of size k. By selecting at random any set of k nodes A_k and the corresponding $\underline{x}$ (A_k) ε Γ_k. This may correspond more realistically to random attack leakage or natural failure. Thus, we define the mean node survival function as

$$\bar{\mu}_k = E\{\sigma(\underline{x})|\underline{x} \ \varepsilon \ \Gamma_k\}$$
$$= \int \sigma(\underline{x}(A_k)) \ d \ F(A_k) .$$

Since μ_k is minimal over the same Γ_k as $\bar{\mu}_k$ we have the bounds:

$$\mu_k \leq \bar{\mu}_k \leq \sigma(\underline{x}_c(n-k)) .$$

LINK ATTACKS

Analagous measures of survival against link attack/failure can be defined, with the point of departure from the upper bound being equal to the cohesiveness of the graph $\lambda(G)$, defined in prior work as the size of the smallest link disconnecting set, (i, ii).

16-NODE NETWORK EXAMPLES

Consider the problem of comparing survivability and cost of alternative topologies for a 16-node network, where the nodes are arrayed in a regular 4x4 grid. We suppose that the cost per unit distance for edges (communication cable) is one unit between nodes which are vertical or horizontal neighbors. We assume further that the basis of comparison is ability to function under optimal node attack.

Figure 1 illustrates seven connection topologies in order of increasing cost, and plots the node survival function (μ_k) for these topologies using the underlying survival index of

$$\sigma(\underline{x}) = \left[\sum (x_i/n)^2\right]^{1/2}$$

(As previously noted, this particular choice of σ creates a numerical scale, but any other choice of a survival index would not affect the vertical ranking of points on the graph.)

We can make several useful qualitative observations from study of this example:

a. Topologies of equal connectivity (or cost) are not otherwise of equal strength (string tree vs. star tree, and loop vs. grid).

b. Performance is not uniformly ordered by connectivity or cost (X-grid vs. 4-grid).

c. Large cost increases may be required to get small improvements in overall survivability performance.

A final example is illustrative of the relationship between node degree and survival performance, and shows that even the connectivity is not determined by the minimum degree of the nodes. We know (i)

$$\omega \leq \lambda \leq d_{min} ,$$

and regular graphs can be constructed to make $\omega = d_{min}$. But the inequality can exist. For example, it is possible to make the 16-node graph regular (equal degree of all nodes) and of degree 4 but have $\omega = 1$ (a single vulnerable node). Cost for this poor design is about 52 units, (Figure 2).

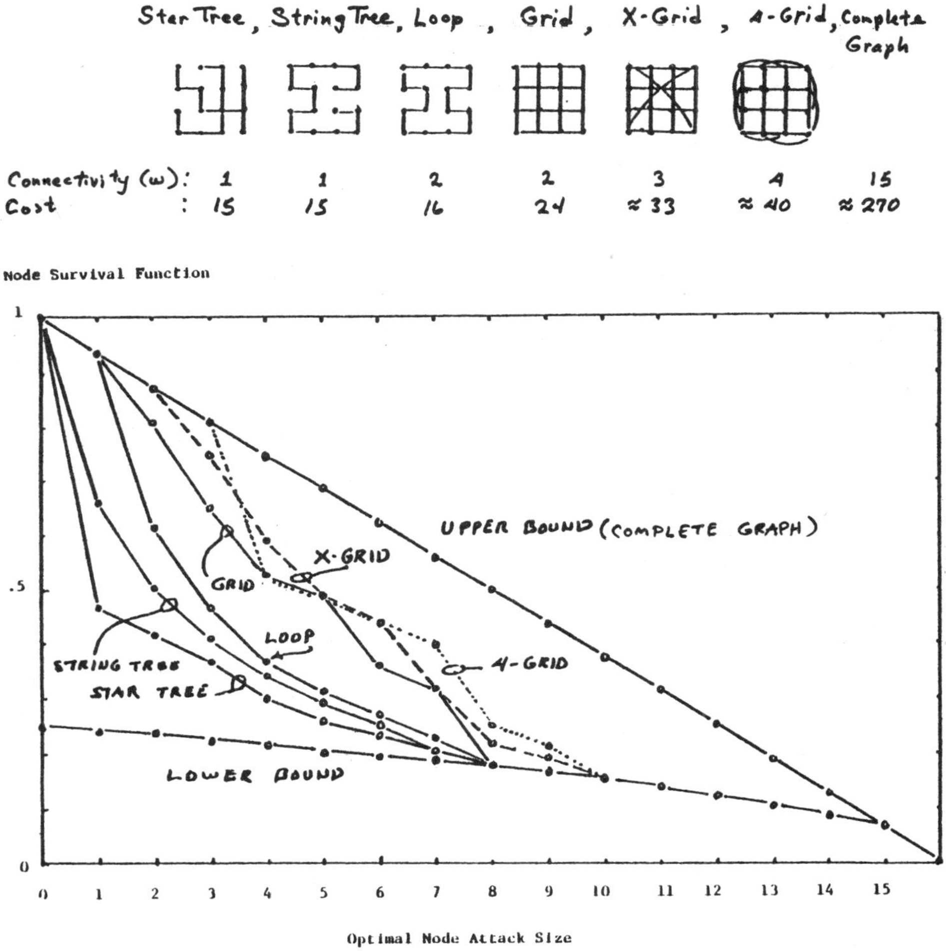

FIGURE 1 Node Survival Functions for 16-Node Examples

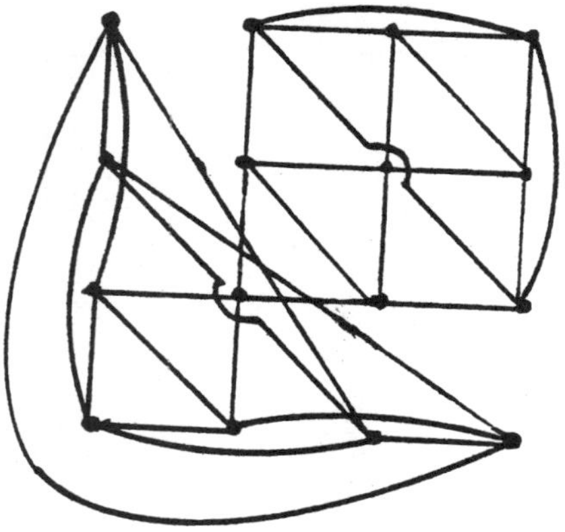

FIGURE 2 Degree-4 Graph with Single Vulnerable Node

CONCLUSIONS AND MORE QUESTIONS

This paper has summarized a new methodology for evaluating the survival performance of distributed processing networks which are subject to losses of nodes and/or links due to failure or attacks. The axiomatic definition of "survival index" accommodates a wide class of actual system performance characteristics, yet preserves performance rank. The present work is restricted to the homogeneous case - where member nodes have inherently equal processing capability.

The basic advance of the survival index approach over the prior notions of connectivity and cohesiveness are that performance degradation beyond the change from connected to disconnected is measured.

The ranking equivalence of all survival indices provides a convenience in computer searches for optimal attacks or evaluation of random attack effects and a freedom from concern with exact quantitative values of the model.

REFERENCES

(i) R. S. Wilkor, "Analysis and Design of Reliable Computer Networks," IEEE Trans. on Communications, June 1972.

(ii) H. Frank and I. T. Frisch, Communication, Transmission, and Transportation Networks, Reading, Maryland: Addison-Wesley, 1971.

INFORMATION THEORETIC APPROXIMATIONS FOR M/G/1 QUEUING SYSTEMS

John E. Shore

Naval Research Laboratory, Washington, D.C. 20375

ABSTRACT

The behavior of single server queuing systems is characterized by various "performance distributions", including distributions of queue length, waiting time, residence time, busy period, number served in a busy period, etc. In principle, if the arrival and service time distributions are known exactly, then these performance distributions can be computed using standard techniques. We consider the problem of estimating the distributions when only the first few moments of the service time distribution are known. Our approach uses standard relations to compute moments of the performance distributions from the known moments of the service time distribution and the principles of maximum entropy and minimum cross-entropy to estimate the performance distributions themselves. For M/G/1 systems with known average arrival rates, we derive analytic results for cases when one or two moments of the service time distribution are known, and we show how one can compute results using as many moments of the service time distribution as are available. For G/G/1 systems, our results are limited to the case in which only the average arrival and service rates are known. Among the results obtained with this information theoretic approach is a new light-load approximation for the M/M/1 busy period probability density. Throughout the paper, we illustrate our approach using M/M/1, $M/H_2/1$, and M/D/1 examples.

I. INTRODUCTION

For single server queuing systems, we consider the problem of

D. G. Lainiotis and N. S. Tzannes (eds.), Advances in Communications, 321–332.

estimating various interesting probability densities and distributions when only partial knowledge of the service time distribution is available. In particular, we consider M/G/1 systems: customers arrive with independent, exponentially distributed interarrival times from an infinite customer pool, wait in an infinite capacity queue, are served independently by a single server with a general service time distribution, and return to the customer pool. The performance of such systems depends on the details of the service time distribution and is characterized by various probability distributions and densities, including queue length, busy period length, number served during a busy period, waiting time, etc. We refer to these as the "performance distributions." In principle, given the service time probability density s(t), one can compute the performance distributions using standard techniques [1]-[3]. But suppose, instead of s(t), one knows only its first n moments

$$s_m = \int dt\ t^m s(t) \quad . \tag{1}$$

What is the best way to use this information in estimating the performance distributions?

Our approach exploits the fact that moments of the performance distributions are themselves determined by the service time moments (1) and the average arrival time (a sufficient statistic of the exponential interarrival time density). For example, the first n moments of the distribution of the number of customers served during a busy period can be expressed in terms of the average arrival time and the first n moments of s(t). Thus, the information (1) giving moments of s(t) is equivalent to information giving moments of the performance distributions. Given this information, we use the principle of maximum entropy to estimate the performance distributions themselves. Because entropy maximization has been shown to be a uniquely correct, self-consistent method of inference about probability distributions when new information is in the form of expected values [4], [5], we refer to the resulting estimates of the performance distributions as information theoretic approximations.

II. ENTROPY MAXIMIZATION AND CROSS-ENTROPY MINIMIZATION

A. The Maximum Entropy Principle and the Minimum Cross-entropy Principle

Suppose you know that a system has a set of possible states x_i with unknown probabilities $q^{\dagger}(x_i)$, and you then learn <u>constraints</u> on the distribution $q^{\dagger}$: either values of certain expectations $\sum_i q\ (x_i) f_k(x_i)$ or bounds on these values. Suppose you need to choose a distribution q that is in some sense the best estimate of $q^{\dagger}$ given what you know. Usually, there remains an infinite set of distributions that are not ruled out by the constraints. Which one should you choose?

The principle of maximum entropy states that, of all the distributions q that satisfy the constraints, you should choose the one with the largest entropy $-\sum_i q(x_i)\log(q(x_i))$. Entropy maximization was first proposed as a general inference procedure by Jaynes [6]. Since then, it has been applied successfully in a remarkable variety of fields, including traffic networks [7], and queuing theory [8]. For a lengthy list of applications and references, see [5].

The principle of minimum cross-entropy is a generalization that applies in cases when a prior distribution p that estimates $q^{\dagger}$ is known in addition to the constraints. The principle states that, of the distributions q that satisfy the constraints, you should choose the one with the least cross-entropy $\sum_i q(x_i)\log(q(x_i)/p(x_i))$. Minimizing cross-entropy is equivalent to maximizing entropy when the prior is a uniform distribution. Unlike entropy maximization, cross-entropy minimization generalizes correctly for continuous probability densities. One then minimizes the functional

$$\int dx\, q(x)\log(q(x)/p(x)) \quad . \tag{2}$$

The principle of minimum cross-entropy was first proposed by Kullback [9, p. 37], who called it minimum discrimination information.

B. Justifying the Principles as General Methods of Inference

Until recently, entropy maximization was justified best on the basis of entropy's unique properties as an uncertainty measure. That entropy has such properties is undisputed: one can prove, up to a constant factor, that entropy is the only function satisfying axioms that are accepted as requirements for an uncertainty measure [10]. Intuitively, the maximum entropy principle follows quite naturally from such axiomatic characterizations. For example, Jaynes states that the maximum entropy distribution "is uniquely determined as the one which is maximally noncommittal with regard to missing information" [6, p. 623]. Similar justifications can be advanced for cross-entropy minimization. Like entropy, cross-entropy has various properties that are desirable for an information measure [11], [12], and one can argue [13] that cross-entropy measures the amount of information necessary to change a prior p into a posterior q. The principle of minimum cross-entropy then follows intuitively much like entropy maximization.

To some, entropy's unique properties make it obvious that entropy maximization is the correct way to account for constraint information. To others, such an informal and intuitive justification yields plausibility but not proof --- why maximize entropy; why not some other function? As a result, entropy maximization has remained controversial despite its success.

Recently, we have obtained a new, formal justification for entropy maximization using a different approach [5]. This approach is based on the observation that previous justifications are weak, not only because they rely on informal, intuitive arguments, but also because they are indirect --- they are based on a formal description of what is required of an information measure rather than on a formal description of what is required of a method for taking new information into account.

Our approach in [5] was to formalize the requirements of inductive inference directly in terms of a set of consistency axioms that make no reference to information measures or properties of information measures. All of the axioms are based on a single fundamental principle: If a problem can be solved in more than one way, the results should be consistent. We were then able to prove that the principle of maximum entropy is correct in the following sense: Given information in the form of constraints on expected values, there is only one distribution satisfying these constraints that can be chosen in a manner that satisfies the axioms; this unique distribution can be obtained by maximizing entropy. This result for entropy maximization was obtained both directly and as a special case (uniform priors) of an analogous, more general result for the principle of minimum cross-entropy.

C. Mathematics of Entropy Maximization and Cross-entropy Minimization

We treat entropy maximization as a special case of cross-entropy minimization. Let $\underset{\sim}{x}$ denote a single state of some system that has a set $\underset{\sim}{D}$ of possible system states and a probability density $q^{\dagger}(\underset{\sim}{x})$ of states. We assume that the existence of $q^{\dagger}$ is known but that $q^{\dagger}$ itself is unknown. Given $p(\underset{\sim}{x})$, a <u>prior</u> density that is our current estimate of $q^{\dagger}$, we wish to select a <u>posterior</u> $q(\underset{\sim}{x})$ based on new information that $q^{\dagger}$ (and therefore q) must satisfy the expected value <u>constraints</u>

$$\int_{\underset{\sim}{D}} d\underset{\sim}{x}\; q(\underset{\sim}{x}) g_r(\underset{\sim}{x}) \equiv \langle g_r \rangle = \bar{g}_r \, , \tag{3}$$

for a known set of functions $g_r(\underset{\sim}{x})$ and numbers $\bar{g}_r$, $r = 1,\ldots,n$. The solution to this inference problem is obtained by varying $q(\underset{\sim}{x})$ so that the cross-entropy

$$H(q,p) = \int_{D} dx\, q(x)\log(q(x)/p(x)) \tag{4}$$

is minimized subject to the constraints (3) and the normalization constraint

$$\int_{D} dx\, q(x) = 1 \quad . \tag{5}$$

Mathematically, the solution is obtained using the method of Lagrangian multipliers and standard techniques from the calculus of variations. The minimization condition is

$$\log(q(x)/p(x)) + 1 + \alpha_0 + \sum_r \beta_r g_r(x) = 0 , \tag{6}$$

where the β_r and α_0 are Lagrangian multipliers corresponding to the constraints (3) and (5). The solution of (6) is

$$q(x) = p(x)\exp(-\alpha - \sum_r \beta_r g_r(x)), \tag{7}$$

where $\alpha = \alpha_0 + 1$. It is convenient to write (7) in the form

$$q(x) = Z^{-1} p(x)\exp(-\sum_r \beta_r g_r(x)) , \tag{8}$$

where Z is the "partition function,"

$$Z = \exp(\alpha) = \int_{D} dx\, p(x)\exp(-\sum_r \beta_r g_r(x)) \quad . \tag{9}$$

The values of the multipliers β_r are determined by the known expectation values $\bar{g}_r$ in (3). One can express the posterior q directly in terms of the values g_r by solving the equations

$$\begin{aligned} \bar{g}_r &= -Z^{-1} \frac{\partial Z}{\partial \beta_r} \\ &= -\frac{\partial}{\partial \beta_r} \log(Z) \end{aligned} \tag{10}$$

for the β_r, or by substituting (8) into the constraint equations (3) and solving for the β_r. Such solutions are often difficult or impossible to obtain analytically, but one can obtain them computationally in general [4, Appendix B], [14].

The principle of maximum entropy applies when one has constraint information (3) but no prior. In this case one selects the posterior by maximizing the posterior entropy

$$H(q) = \int_{D} dx\, q(x)\log(q(x))$$

subject to the constraints (3) and (5). The solution is the same as (6)-(9) with $p(x)$ deleted. In general, selecting a posterior by maximizing entropy is equivalent to assuming a uniform prior and minimizing cross-entropy [5].

III. M/G/1 QUEUE LENGTH DISTRIBUTION

We consider an M/G/1 queuing system with an average arrival rate λ and a service time probability density $s(t)$ with moments s_i (see (1)). Let $q_c(k)$ be the probability that k customers are in the system (queued or being served), and let c_m be the moments

$$c_m = \sum_{k=0}^{\infty} k^m q_c(k) .$$

The first moment c_1 is just the expected number of customers in the system. In this Section, we use the Pollaczek-Khinchin formula to express c_1 in terms of the first two service time moments s_1, s_2, and we derive a maximum entropy estimate of $q_c(k)$ given c_1. We then derive a formula expressing c_2 in terms of s_1, s_2, and s_3, and we compute maximum entropy estimates of q_c given c_1 and c_2. As examples, we consider M/M/1, M/H_2/1, and M/D/1 systems.

The result of maximizing the entropy of q_c subject to the single known constraint c_1 and the normalization constraint $\sum_k q_c(k) = 1$ is

$$q_c(k) = Z^{-1}e^{-\beta k} , \tag{11}$$

where

$$Z = \sum_{k=0}^{\infty} e^{-\beta k} = (1 - e^{-\beta})^{-1} \tag{12}$$

(see (8)-(9)). We apply (11) in order to express the multiplier β in terms of the constraint c_1,

$$c_1 = -\frac{\partial}{\partial\beta}\log(Z) = (e^{\beta} - 1)^{-1} .$$

This yields $e^{-\beta} = c_1/(1 + c_1)$, which we use (with (12)) to eliminate β from (11):

$$q_c(k) = \frac{1}{1 + c_1}\left(\frac{c_1}{1 + c_1}\right)^k \tag{13}$$

This expression gives the maximum entropy estimate of q_c directly in terms of the known information c_1.

Now, knowledge of s_1 and s_2 yields knowledge of c_1 by the Pollaczek-Khinchen formula [3, p. 187]

$$\begin{aligned} c_1 &= \rho + \rho^2 \frac{(1 + C^2)}{2(1 - \rho)} , \\ &= \rho + \frac{\lambda^2 s_2}{2(1 - \rho)} , \end{aligned} \tag{14}$$

where $\rho = \lambda s_1$, and C is the coefficient of variation $C = (s_2 - s_1^2)^{\frac{1}{2}}/s_1$. Thus, (13) and (14) provide an information theoretic approximation to q_c for an M/G/1 system given the average arrival rate and the first two moments of the service time density.

As an example application, we consider an $M/H_2/1$ system solved exactly by Kleinrock [3, pp. 195-96]. The service time distribution is

$$s(t) = \frac{1}{4}\lambda e^{-\lambda t} + \frac{3}{4}(2\lambda)e^{-2\lambda t} , \tag{15}$$

for which $\rho = \lambda s_1 = 5/8$ and $C^2 = 31/25$. Substituting these values into (14) yields $c_1 = 1.79166$. The information theoretic approximation (13) then becomes

$$q_c(k) = .358209(.641791)^k . \tag{16}$$

The exact solution for q_c is [3, p. 196]

$$q_c(k) = \frac{3}{32}\left(\frac{2}{5}\right)^k + \frac{9}{32}\left(\frac{2}{3}\right)^k . \tag{17}$$

We compare the one-moment approximation (16) with the exact solution (17) in the first three columns of Table I.

The extremely close agreement arises because the exact solution (17) is the sum of two similar geometric terms, which can be approximated closely by a single geometric term (16). In general, the single-moment result (13) can be thought of as providing the geometric distribution that is the best information theoretic approximation to q_c.

If the exact solution itself happens to be geometric, then the approximation (13) will be the same as the exact solution. For example, suppose that the service time distribution is exponential $s(t) = \mu e^{-\mu t}$. Then (14) reduces to $c_1 = \rho/(1 - \rho)$, with $\rho = \lambda s_1 = \lambda/\mu$, and the approximation (13) becomes $q_c(k) = (1 - \rho)\rho^k$, which is the well-known exact solution for the M/M/1 system [3, p. 96].

If other moments besides c_1 are known, the maximum entropy estimate of q_c will no longer in general be geometric. For example, c_2 is given by

$$c_2 = \rho + \frac{\lambda^2 s^2}{2(1-\rho)} + \frac{\lambda^4 s_2^2}{2(1-\rho)^2} + \frac{\lambda^3 s_3}{3(1-\rho)} + \frac{\lambda^3 s_1 s_2}{(1-\rho)} + \lambda^2 s_2 , \tag{18}$$

but the maximum entropy solution for $q_c(k)$ given c_1 and c_2 cannot be expressed analytically in terms of the moments s_m, so we resort to numerical techniques. We use an APL function written by Johnson [14] that computes maximum entropy distributions given arbitrary expected value constraints.

For the $M/H_2/1$ example, we have $c_1 = 1.79166$ from before. The moment s_3 is easily obtained from (15), and $c_2 = 8.68055$ follows from (18). Using the APL function to find the maximum entropy appproximations for q_c given c_1 and c_2, we obtain the results shown in the fourth column in Table I. This approximation, which was computed for 50 points, required 1.5 CPU seconds on a DEC PDP-10 KI processor. It is worth noting that single-moment results from the APL function agreed with the analytic expression (13) up to eight digits.

As an additional example, we consider a system with constant ("deterministic") service time $1/\mu$ --- i.e., M/D/1. The service time probability density is $s(t) = \delta(t - 1/\mu)$, with moments

$$s_m = 1/\mu^m \quad . \tag{19}$$

We use (19), (14), and (13) to obtain a single-moment approximation for q_c, and we use (19), (18), (14), and the APL function to obtain a two-moment approximation. For $\lambda = 1$ and $\mu = 2$, the results are shown in Table II together with simulation results. The simulation result $q_c(k)$ is the relative amount of time the system had k customers present during an overall period covering 5000 arrivals. The two-moment approximation in Table II required 1.6 CPU seconds.

Approximations involving more moments can be computed similarly since c_m can in general be expressed as a function of $s_1,\ldots,s_{m+1}$ --- one method is to differentiate the Pollaczek-Khinchin transform equation [3, p. 194]. But the accuracy of the two-moment approximation for the $M/H_2/1$ and M/D/1 examples, which have radically different service time densities, and the reduction of the one-moment approximation to the exact result in the M/M/1 case, together suggest that the two-moment approximation will in general be quite accurate for M/G/1 systems. This is only a conjecture, however, and more detailed studies are needed.

IV. OTHER DISTRIBUTIONS

The techniques used in Section III can also be used to obtain approximations for other distributions, including number served in a busy period, busy period length, residence time, and waiting time. For various results, see [15], which also includes some results for non-uniform priors and for G/G/1 approximations.

V. DISCUSSION

There are at least three possible uses for the results in this paper. First, the techniques presented could be used as a general method of computing the performance distributions in cases where all of the service density moments are available, i.e., when the density s(t) is known exactly. Second, the analytic approximations --- (14), and similar ones given in [15] --- could be useful in various studies whenever explicit forms for the performance distributions are required. Third, and probably best, the techniques provide a means of estimating the performance distributions when only the first few moments of s(t) are known and s(t) itself is not known.

How accurate are these information theoretic approximations? Unfortunately, about all that can be said in general is that the approximations are the least-biased choices given the information available. To use the language of statistics [11], the approximations are the hypotheses that are best supported by the information available. Depending on the actual performance distribution and the number of moments considered, an information theoretic approximation may or may not be a good approximation in the mean-squared-error sense, although it is true that the mean-squared-error can always be made sufficiently small by taking sufficiently many moments into account. On the other hand, it is not generally known what kind of error measure is best for judging the accuracy of performance distribution approximations. It may well be that measures such as mean-squared-error are less important than information measures such as cross-entropy.

Queuing models, particularly ones with Poisson arrivals and exponentially distributed service times, have been used with remarkable success in the performance modeling and analysis of computer systems. Because computer systems do not satisfy many assumptions made by the stochastic process models that are used, this success has been somewhat puzzling. The results presented in this paper and [15] show that the information theory viewpoint may be the best one from which to understand this success. For example, we show in [15] that various M/M/1 formulas are also information theoretic approximations for G/G/1 systems. That is, the M/M/1 formulas are the best hypotheses about G/G/1 behavior given only the mean arrival and service rates. This fact has nothing at all to do with the various assumptions that must be debated when considering the applicability of stochastic models.

ACKNOWLEDGMENTS

I thank R. W. Johnson, H. Vantilborgh, and A. Ephremides for helpful discussions. I also thank J. Stroup for her help in preparing the manuscript.

Table I

Comparison of exact and approximate solutions for $M/H_2/1$ queue length distribution

k	$q_c(k)$ (exact)	$q_c(k)$ (1 moment approx.)	$q_c(k)$ (2 moment approx.)
0	.375	.358	.367
1	.225	.230	.229
2	.140	.148	.144
3	.0893	.0947	.0914
4	.0580	.0608	.0583
5	.0380	.0390	.0375
6	.0251	.0250	.0243
7	.0166	.0161	.0158
8	.0110	.0103	.0104
9	.00734	.00662	.00688
10	.00489	.00425	.00458

Table II

Comparison of information theoretic approximations and simulation results for M/D/1 queue length distribution. (λ= 1 and μ = 2)

k	$q_c(k)$ (simulation)	$q_c(k)$ (1 moment approx.)	$q_c(k)$ (2 moment approx.)
0	.50	.57	.51
1	.33	.24	.30
2	.12	.10	.13
3	.038	.045	.044
4	.0093	.019	.011
5	.0025	.0083	.0022
6	.00047	.0035	.00033
7	.0000081	.0015	.000037
8	0	.00065	.0000032

REFERENCES

1. J. W. Cohen, The Single Server Queue, Amsterdam, North-Holland, 1969.

2. D. R. Cox and W. L. Smith, Queues, London, Chapman and Hall, 1961.

3. L. Kleinrock, Queueing Systems, Vol I: Theory, New York, John Wiley, 1975.

4. J. E. Shore and R. W. Johnson, "Axiomatic Derivation of the Principle of Maximum Entropy and the Principle of Minimum Cross-Entropy," NRL Memorandum Report 3898, Dec 1978, Naval Research Laboratory, Washington, D. C. 20375.

5. J. E. Shore and R. W. Johnson, "Axiomatic Derivation of the Principle of Maximum Entropy and the Principle of Minimum Cross-Entropy," IEEE Trans. Inf. Theory, IT-26, Jan., 1980.

6. E. T. Jaynes, "Information Theory and Statistical Mechanics I," Phys. Rev. 106, 1957, pp. 620-630.

7. V. E. Benes, Mathematical Theory of Connecting Networks and Telephone Traffic, Academic Press, New York, 1965.

8. J. E. Shore, "Derivation of Equilibrium and Time-Dependent Solutions to M/M/ //N and M/M/ Queueing Systems Using Entropy Maximization, Proceedings 1978 National Computer Conference, AFIPS, 1978, pp. 483-487.

9. S. Kullback, Information Theory and Statistics, Wiley, New York, 1959.

10. C. E. Shannon, "A Mathematical Theory of Communication," Bell System Tech. Jour., 27, 1948, pp. 379-423.

11. A. Hobson and B. Cheng, "A Comparison of the Shannon and Kullback Information Measures," J. Stat. Phys. 7, No. 4, 1973, pp. 301-310.

12. R. W. Johnson, "Axiomatic Characterization of the Directed Divergences and Their Linear Combinations," accepted by IEEE Trans. Inf. Theory.

13. A. Hobson, "A New Theorem of Information Theory," J. Stat. Phys. 1, No. 3, 1969, pp. 383-391.

14. R. W. Johnson, "APL Functions for Determining Probability Distributions by the Principles of Maximum Entropy and Minimum Cross-Entropy," Proceedings APL 79, May 1979.

15. J. E. Shore, "Information Theoretic Approximations for M/G/1 Queuing Systems," NRL Memorandum Report 4047, Naval Research Laboratory, Washington, DC 20375, July 1979.

QUEUING ANALYSIS OF RANDOM ACCESS PROTOCOLS

Anthony Ephremides

Electrical Engineering Department
University of Maryland

Abstract

Analysis of the Random Access channel with a finite number of buffered users is considered. The dependence between these users is taken into account. The average waiting, service and total delay times are obtained.

1. INTRODUCTION

In a slotted ALOHA system, all users transmit packets into channel time slots independently. If two or more transmit in the same time slot, collision occurs and they have to attempt retransmission again.

In this paper, we analyze the slotted ALOHA, with a finite number of M identical users and with each user having an infinite buffer capacity. We make the usual simplifying assumptions (like instantaneous sense of collision, geometric distribution of retransmission times, identical terminals, Poisson like arrivals, etc.)

The interaction among the terminals, due to the nature of the protocol, makes this problem a non-trivial one. Earlier attempts to analyze the system and obtain the average waiting time per packet [1-4] have failed either due to inaccurately modeling each

---

This work was supported by NSF Grant ENG-77-22752.

D. G. Lainiotis and N. S. Tzannes (eds.), Advances in Communications, 333–340.

user as a standard queuing system (as in [1-3]), ignoring thereby the interaction between the users, or due to seeking to obtain the entire distribution, rather than just the average value, of the waiting time (as in [4]), an approach that reduces to a rather complex version of a Riemann-Hilbert problem. Of course, other approaches that assume packet rejection rather than queueing in the buffer have succeeded in obtaining the values of average delay in a variety of schemes [5-10].

Our approach here is based on a combination of two Markov chain models, one that describes the status of the buffer contents of the typical user and one that describes the status of all the users of the channel. The first chain is quite similar to one used recently in [11] and the second one is quite similar to one used in [12]. Here the two are combined to yield the method of obtaining the average total delay per packet.

2. THE MODEL

The user terminal consists of a buffer, a transmitter and a retransmission unit, as in Fig. 1. The buffer has unlimited capacity. The packet at the head of line (HOL) in the buffer enters the transmitter when ready for transmission and stays there until a determination is made as to whether the transmission was successfully completed. We assume that sensing of collisions is instantaneous. If a collision occurs, the terminal is said to be in the <u>blocked</u> state, the packet enters the retransmission unit and sent to the transmitter with probability P. In case of success, the terminal stays in the <u>unblocked</u> state. The unblocked terminal can be in one of two substates; <u>idle</u> (when the buffer is empty) or <u>active</u> (when the buffer is non-empty). The active terminal sends the HOL packet into the transmitter with probability one. A packet that arrives to an idle terminal goes directly to the transmitter without incrementing the buffer. The arrival process is modeled as a Bernoulli process (time is considered slotted) with average rate of arrivals (or, equivalently, probability of an arrival in a given slot) equal to σ.

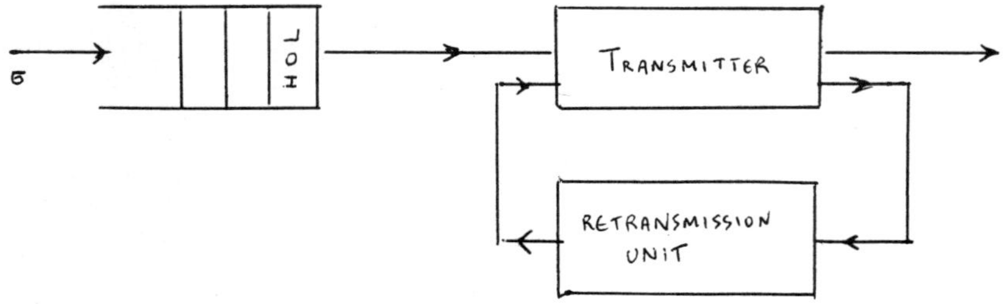

Fig. 1

The distinction of the three alternatives (blocked (B), idle (I), active (A)) is necessary as we shall see, because it affects the success probability. As usual no co-ordination among the M terminals is assumed.

We consider two Markov chains, one that models each terminal's status, and one that models the status of all M users.

a) The User Markov Chain

We consider the state of each terminal to be adequately characterized by the size of the buffer contents and by the indication whether it is blocked or unblocked. Thus the state is denoted by (i,j), where i is zero if the terminal is blocked and one if it is unblocked (ilde or active), and where j = 0, 1, 2, ... indicates the number of packets in the buffer.
Let $\pi_{ij} \triangleq$ steady state probability of the state (i,j)

$$G_i(z) \triangleq \sum_{n=0}^{\infty} \pi_{i,n} z^{-n} \qquad , \quad i = 0,1$$

In order to characterize the transition probability of the user Markov chain, we need to define the following additional quantities:

r = Pr [successful transmission/user is blocked]
q_1 = Pr [collision/user is idle]
q_2 = Pr [collision/user is active]

In the next section, we shall derive expressions for these quantities r, q_1, q_2. In terms of them we can already characterize the transition probability of the user Markov chain as it appears in Fig. 2. We leave it to the reader to verify their correctness since, so far, the model is very similar to the one considered in [11]. It can be shown that $\pi_{1,0}$, $G_0(1)$, $G_1(1)$, $\pi_{0,0}$, $\pi_{1,1}$ and Q, the average queue size in buffer, are as given by the following Eqs.

$$\pi_{1,0} = \frac{r(1-\sigma)-\sigma q_2}{r(1-\sigma)-\sigma(q_2-q_1)} \tag{1}$$

$$G_0(1) = \frac{(1-\sigma)\sigma q_1}{r(1-\sigma)-\sigma(q_2-q_1)} \tag{2}$$

$$G_1(1) = \sigma+(1-\sigma)\frac{r(1-\sigma)-\sigma q_2}{r(1-\sigma)-\sigma(q_2-q_1)} \tag{3}$$

$$\pi_{1,1} = \frac{\sigma}{1-\sigma}\ \pi_{0,0} \tag{4}$$

$$\pi_{0,0} = \frac{\sigma q_1}{\sigma(1-q_2)+r(1-\sigma)} \pi_{1,0} \tag{5}$$

$$Q = \frac{\sigma^2(1-\sigma)q_1}{[r(1-\sigma)-\sigma q_2][r(1-\sigma)-\sigma(q_2-q_1)]} \tag{6}$$

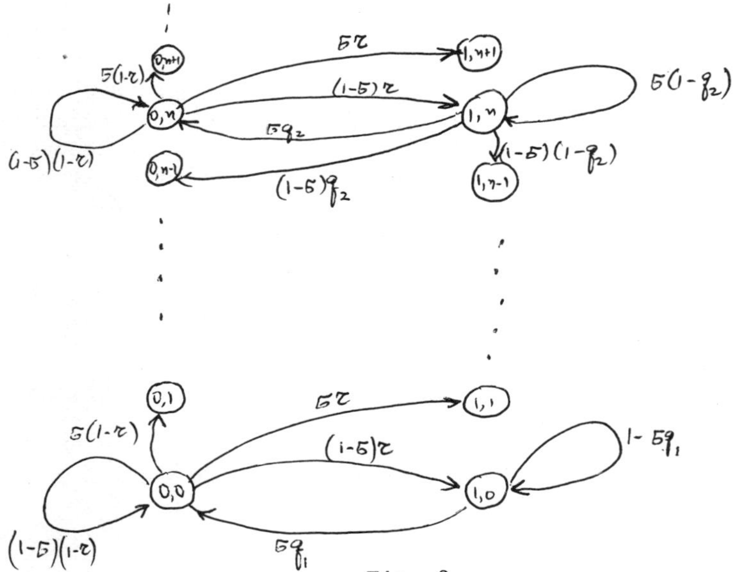

Fig. 2

b) The System Markov Chain

The state of the system is described by the number of terminals that are in each of the three states, blocked (B), active (A), and idle (I). Let

$n \triangleq$ number of blocked terminals

$n_1 \triangleq$ number of active terminals

then

$M-(n+n_1) \triangleq$ number of idle terminals

It is important to observe that n_1 can either be 0 or 1. The reason is the following: If there were two or more active terminals at the beginning of a slot, then there should have been two or more successful transmissions during the preceeding slot (an impossibility), since an active terminal attempts transmission with probability one and a terminal becomes active only if it transmits successfully a "new" or an "old" packet in the previous slot. Thus, the state of the system is described by the pair

(n_1, n), where $n = 0, 1, \ldots, M$ if $n_1 = 0$, and $n = 0, 1, \ldots M-1$ if $n_1 = 1$.

The Markov chain is illustrated in Fig. 3 where the transition probabilities are only marked by symbols which can be calculated.

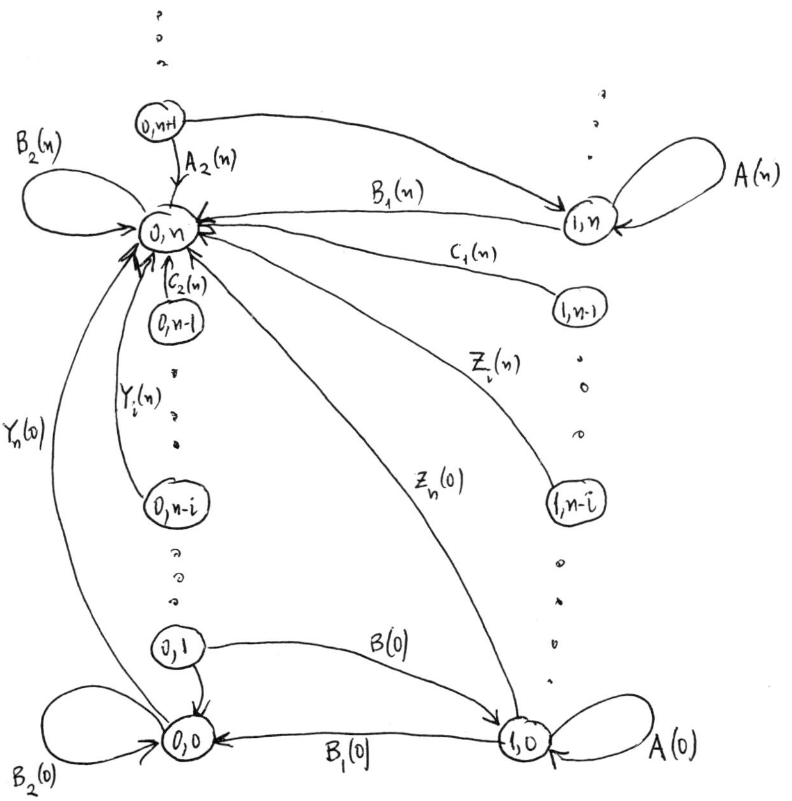

Fig. 3.

We denote by $P_{n_1, n}$ the steady state probability that the system is in state (n_1, n). The above model bears a similarity to the one considered in [12].

3. SUCCESS AND COLLISION PROBABILITIES

The calculation of the probabilities r, q_1 and q_2 as defined in Section 2, is straight forward based on the described operation. Note first that

$$G_0(1) = \Pr \quad [\text{a user is blocked}] = \sum_{j=0}^{\infty} \pi_{0j}$$

$$G_1(1) = \Pr \quad [\text{a user is unblocked}] = \sum_{j=0}^{\infty} \pi_{1j}$$

$$\pi_{10} = \Pr \quad [\text{a user is idle}]$$

$$G_1(1) - \pi_{10} = \Pr \quad [\text{a user is active}]$$

and $G_0(1) + G_1(1) = 1$.

By conditioning on the state of the system, we obtain;

a)
$$r = \frac{p\sum_{n=1}^{\infty} P_{0,n}(1-p)^{n-1}(1-\sigma)^{M-n}}{1 - P_{0,0} - P_{1,0}}$$

where the expression in the denominator represents the probability that there is at least one blocked user, p in the numerator represents the probability that the blocked user under consideration attempts transmission, $(1-p)^{n-1}$ the probability that no other blocked user transmits and $(1-\sigma)^{M-n}$ the probability that no idle user receives a packet arrival (and, therefore, transmits). Finally, $P_{0,n}$ is the probability of state (0,n).

b)
$$q_2 = 1 - \frac{\sum_{n=0}^{M-1} P_{1,n}(1-p)^n(1-\sigma)^{M-n-1}}{\sum_{n=0}^{M-1} P_{1,n}}$$

c)
$$q_1 = 1 - \frac{\sum_{n=0}^{M-1}(1-\sigma)^{M-n-1}(1-p)^n P_{0,n}}{1 - P_{0,M} - P_{1,M-1}}$$

Note that, if we don't assume an arrival to the idle terminal under consideration, the probability of successful transmission is given by

$$h = \sigma(1-q_1)$$

Eqs. (1) through (6) provide the values of the quantities that were needed in order to solve the system equations. However, these quantities are not expressed in terms of only the system parameters M,p,σ, but, in addition, in terms of the success and collision probabilities r, q_1, q_2. These last probabilities were calculated above in terms of the system parameters and the system probabilities p_{n_1}, n.

Thus, in total, we have a set of simultaneous coupled non-linear equations in r, q_1, q_2 and in $\{p_{n_1}, n\}$. Solving these equations requires computational methods. Assuming the solution is obtained, it is possible to then calculate the average waiting time.

REFERENCES

1. Ephremides, A. and Mowafi, O., "Multiple Access Schemes for Packet-Switched Channels with Queuing", John Hopkins Conference on Information Sciences and Systems, Baltimore, Md., March 1978.

2. Mowafi, O., A New Class of Multiple Access Schemes for Packet-Switched Broadcast Channels, Ph.D. Dissertation, University of Maryland, Electrical Engineering Department, August 1978.

3. Ephremides, A., "Two Adaptive Multiple Access Protocols", IEEE International Symposium on Information Theory, Grignano, Italy, June 1979.

4. Fayolle, G. and Iasnogorodski, R., "Two Coupled Processors- The Reduction to a Riemann-Hilbert Problem", Wahrscheinlichkeitstheorie, 1797, pp. 1-27.

5. Capetanakis, J. I., "Tree Algorithms for Packet Broadcast Channels", IEEE Trans. on IT, Vol. 25, No. 5, September 1797, pp. 505-515.

6. Capetanakis, J. I., "Capacity Allocation Techniques in a Multi-User Satellite System", Technical Note 1979-34, Lincoln Labs, MIT, July 1979.

7. Lam. S. S., and Kleinrock, L., "Dynamic Control Schemes for a Packet Switched Multi-access Broadcast Channel", AFIPS Conf. Proceedings, 1975 National Computer Conference, Vol. 44, pp. 143-153.

8. Kleinrock, L. and Tobagi, F., "Packet Switching in Radio Channels: Part I - Carrier Sense Multiple Access Modes and Their Throughput-Delay Characteristics", IEEE Trans. on Communications, Vol. 23, 1975, pp. 1400-1416.

9. Heitmeyer, C., Jullback, J., and Shore, J., "A Survey of Packet Switching Techniques for Broadcast Media", NRL Report 8035, October 1976.

10. Rubin, I., "Access-Control Disciplines for Multi-Access Communication Channels", Reservation and TDMA Schemes", IEEE Trans. on IT, Vol. 25, No. 5, pp. 516-536.

11. Kamal, S. S., and Mahmoud, S. A., "A Study of User's Buffer Variations in Random Access Satellite Channels", IEEE Trans. on Communications, Vol. 27, No. 6, June 1979, pp. 857-868.

12. Fayolle, G. Gelenbe, E. and Labetoulle, J., "Stability of Optimal Control of the Packet Switching Broadcast Channel", JACM, Vol. 24, No. 3, July 1977, pp. 375-386.

SHARING A PROCESSOR FOR MONITORING CHANGES IN QUALITY OF SEVERAL COMMUNICATION LINKS*

P. Papantoni-Kazakos
The University of Connecticut
Storrs, Connecticut 06268

and

Dimitri Kazakos
State University of New York at Buffalo
Amherst, New York 14260

The problem of sharing one processor for controlling the quality of more than one network links is studied. For the case of two links being assigned to one processor, a specific steering strategy is proposed, which adapts for each link the sampling rate of processed jobs, in a way depending on the values of the monitoring algorithms. The steering parameter is a threshold δ that is smaller than the decisive for the monitoring algorithms threshold Δ.

SUMMARY

The theme of this paper lies within the major objective of designing a auxiliary minicomputer network, which will perform the monitoring operations for a large scale communication network. In particular, the main algorithm proposed in [4] for the quality monitoring of direct links and extended in [5] for application to indirect links also, will be considered, as implemented by a single processor assigned to several links.

When one processor is shared by two links, each time the processor accepts a job, it identifies the link through which it is routed, it observes the job throughout its duration, it assigns it the binary number zero or one accordingly, and finally it updates the monitoring algorithm in [4] for that link. The two updated values W(1) and W(2) of the quality monitoring algorithms

*Research supported by the Air Force Grant AFOSR-78-3695 to the University of Connecticut, and NSF Grant ENG-76-20295 to the State University of New York at Buffalo.

D. G. Lainiotis and N. S. Tzannes (eds.), Advances in Communications, 341–346.

for the links 1 and 2 respectively are stored in the memory of the processor. The acceptance of a job is determined by the selection procedure enforced by the controller. This selection procedure controls directly the time intervals between consecutive processed jobs of types 1 and 2 (routed through links 1 and 2 with characteristic parameters (λ_1, μ_1) and $\lambda_2, \mu_2)$ respectively); thus it controls the time needed by the monitoring algorithms for deciding on the quality of the links. Also, the selection procedure steers the sampling rate for jobs type 1 and 2, that are processed by the processor. The arrival rates at the input of the processor are the natural rates λ_1 and λ_2, which satisfy the equation:

$$\sum_{j=1}^{2} \lambda_j \mu_j^{-1} = \alpha^{-1}(1-\varepsilon) = c>1 \tag{1}$$

; where α a known number between zero and one, and ε: $0<\varepsilon<<1$. The sampling rates for each job type are determined by the sharing procedure implemented by the controller.

We will consider two extreme sharing procedures first, and then a third procedure that is a combination of the first two.

In all procedures, the processor whenever idle, it initially accepts the first upcoming job. However, the processing of a job may be interrupted under certain sharing schemes. If this happens, the interrupted job is completely lost to the monitoring operation.

The three sharing procedures analyzed in this paper are the following:

a. <u>First come-first serve</u>. The processor, whenever idle, it processes without preemptiness (no interruption) the first upcoming job, independently of type.

b. <u>Priority for type 1 (2) jobs.</u> The processor interrupts the processing of a type 2 (1) job, in favor of an inbetween arrived type 1 (2) job. A type 1 (2) job is never interrupted.

c. <u>Threshold steering</u>. In this scheme, an intermediate threshold $\delta<\Delta$ is used to steer the selection procedure from a. to b. and vice-versa; where Δ the decisive threshold in (1).

In particular, whenever the values W(1) and W(2) of the monitoring algorithms for links 1 and 2 respectively, have both never exceeded the threshold δ or have both crossed this threshold at least once in the past but not Δ, the sharing procedure a. is applied by the controller. If W(1) has crossed δ at least once in the past but not W(2), and both W(1), W(2) have not crossed Δ yet, the sharing procedure b. is applied with priority to type 1 jobs. Similarly if W(1) and W(2) are reversed.

The statistical characteristics of procedures a., b., and c., with emphasis to operational regions of effective throughputs, are examined in detail in this section.

A. First Come-First Serve

The selection procedure is in this case independent of the values W(1) and W(2) of the monitoring algorithms. The expected length of the time interval between two consequtively processed jobs of the same type can be computed through a number of simple analytical steps.

Define:

$E\{t_i;A\}$; i=1,2 :The expected length of the time interval t_i between two consequtively processed jobs of type i, under the sharing procedure A.

Then, due to the memorless of the exponential distributions considered, we easily obtain the following expressions for the effective throughputs under the present scheme:

$$S_1(A) = \mu_1^{-1} \cdot E^{-1}\{t_1;A\} = \frac{G}{1+c}$$
$$S_2(A) = \mu_2^{-1} \cdot E^{-1}\{t_2;A\} = \frac{c-G}{1+c} \qquad (2)$$

where $G = \lambda_1\mu_1^{-1}$ is the traffic directed through link 1, and $c-G = \lambda_2\mu_2^{-1}$ is the traffic through link 2.

B. Priority for Type 1 Jobs

Under this procedure, the processor always preempties in favor of a type 1 arriving job. The procedure is again independent of the values W(1) and W(2) of the monitoring algorithms.

Adopting the same statistical characteristics as in the case of the first come-first serve procedure, we easily obtain:

$$S_2(B) = [c-(1+2c)S_1(B) + (1+c)S_1^2(B)] \cdot [1+c(1+\beta^{-1}) - (1+[c+1][1+\beta^{-1}])S_1(B)]^{-1} \quad (3)$$

; where $0 \leq S_1(B) \leq c[1+c]^{-1}$, $\beta = \lambda_2\lambda_1^{-1}$.

The parameter β involved is the ratio of the incoming arrival rates destined for the two links 1 and 2.

C. Priority for Type 2 Jobs

The characteristics of this procedure are similar to the characteristics of procedure B. The operational regions of the effective throughputs are derived by substituting the parameters S_1, β and G in (3) by S_2, β^{-1} and c-G respectively. If the traffic $G = \lambda_1\mu_1^{-1}$ through link 1 is smaller than $c\beta[1+\beta]^{-1}$, procedure C favors both S_1 and S_2 throughputs, compared to both procedures A and B. The benefit gained increases then with decreasing β.

D. Threshold Steering

A study of figures 3-5 in [6] (which exhibit the operational regions of effective throughputs realized by the sharing procedures A, B and C) clearly shows that for at least certain value regions of the statistical characteristics β, G, speed in the recognition of a faulty link can be gained, if the monitoring process that corresponds to the faulty link can be accelerated somehow. A threshold $\delta<\Delta$ can be used to steer the selection process towards such acceleration. The steering is realized in the following way:

1. When both the values W(1) and W(2) of the quality control algorithms for the two links have not yet exceeded δ at least once, the controller applies the sharing procedure A.

2. If one of the algorithms has crossed δ (but not Δ) at least once, but not the other algorithm, the controller dictates priority to jobs corresponding to the first algorithm (sharing procedure B or C).

3. If both the algorithms have crossed δ (but not Δ) at least once, the controller imposes the sharing procedure A again.

4. If one of the algorithm crosses Δ for the first time,

the controller reinitializes it, while it continues the processing of the other algorithm following the procedures stated by steps 1-3.

We will focus our attention to the speed achieved by the threshold steering sharing procedure, in the recognition of one failing link, when the other link operates normally. The operational characteristics of the threshold steering approach when both links are either failing or operating normally, are as in the sharing procedure A by close approximation.

Define:

$E_i\{p+\eta;\delta,\Delta\}; i=1,2$: The expected time length of the first Δ crossing by the algorithm W(i), under the condition that the threshold δ is used for steering, the link i is in condition $p+\eta$, and the other link is in condition p.

$P_{p+\eta}\{N;\Delta\}$:the probability that the algorithm in (1) crosses the threshold Δ for the first time at the Nth processed job, given that the monitored link is in condition $p+\eta$.

$P_{p+\eta}\left\{ n;\delta / N;\Delta \right\}$: the probability that the algorithm in (1) crosses the threshold δ for the first time at the nth processed job, given that the monitored link is in condition $p+\eta$ and the first Δ crossing occurs at the Nth processed job.

$S_1(\delta,\Delta)$, $S_2(\delta,\Delta)$: the effective throughputs when the thresholds δ,Δ are used.

$$\zeta(\delta,\Delta) = \frac{\sum_{N=o}^{\infty} P_{p+\eta}\{N;\Delta\} \sum_{n=o}^{N} n\, P_{p+\eta}\left\{ n;\delta / N;\Delta \right\}}{\sum_{N=o}^{\infty} N\, P_{p+\eta}\{N;\Delta\}}$$

Then if δ is high enough, and β, G such that:
For M large enough to satisfy:

$$\max\left(M\left(\frac{\beta}{1+\beta}\right)^M, \left(\frac{\beta}{c-G+\beta}\right)^M, \left(\frac{1}{G(G+\beta)}\right)^M\right) <<1$$

it is still true that:
$\alpha(p,p+\eta,\delta,M) <<1$
we find:

$$S_2(\delta,\Delta) = \frac{c-S_1(\delta,\Delta)[2+(c-1)\zeta(\delta,\Delta)]}{(1+c)-S_1(\delta,\Delta)[1-\zeta(\delta,\Delta)]\ [2+c+c\zeta(\delta,\Delta)} \quad (4)$$

From numerical results and the corresponding figures of effective throughputs included in [6], we observe that procedure D results always in improved throughput regions. The same approach can be used in the case where more than two links are assigned to the same processor. The effective throughputs will again improve then, while the corresponding operational regions can be easily plotted if the links are divided in two groups, and the collective throughputs for each group are considered.

References:

[1] E.S. Page (1954), "Continuous Inspection Schemes," Biometrika 41, 100-115.

[2] E. S. Page (1955), "A Test for a change in a Parameter Occuring at an Unknown Point," Biometrika 42, 523-527.

[3] G. Lorden (1971), "Procedures for Reacting to a Change in Distribution," Ann. Math. Stat. 42, 1897-1908.

[4] P. Papantoni-Kazakos (1979), "Algorithms for Monitoring Changes in Quality of Communication Links," IEEE Trans. Comm., April issue.

[5] P. Papantoni-Kazakos (1979), "The Potential of End-to-End Observations in Trouble Localization and Quality Control of Network Links," IEEE Trans. Comm., Vol. COM-27 No. 1, pp. 16-30.

[6] P. Papantoni-Kazakos, D. Kazakos, and A. Polydoros, (1979) "Sharing a Processor for Monitoring Changes in Quality of Several Communication Links," Technical Report TR-79-9, The University of Connecticut, June.

ADAPTIVE BIFURCATED ROUTING ALGORITHM WITH THRESHOLD*

W. Waung and A.N. Venetsanopoulos

University of Toronto
Department of Electrical Engineering
Toronto, Ontario, Canada M5S 1A4

ABSTRACT

With cost of computation decreasing, packet switched computer communication networks are becoming increasingly cost effective. To ensure robustness, mesh type topologies are often used. However, such network configurations require efficient and effective routing algorithms.

In this paper, a bifurcation with threshold routing technique is considered. It is an adaptive algorithm using the update information available on the ARPANET. Various traffic conditions were simulated to compare the proposed routing algorithm with others. Link failure effects were also examined and these observations are discussed. Observations are made about its possible performance on actual networks.

I. INTRODUCTION

Over the past two decades, the cost of communications has been steadily decreasing. However, the last few years have also witnessed a drop in cost of computations almost thirty times faster than that of communications [1]. This trend, expected to continue, has brought about a change in the concept of communication networks. No longer are simple switching nodes the most cost effective and efficient methods of using network resources. Intelligent terminals with the ability to make proper decisions on more efficient distribution of network capacities are now possible.

---

* This present work was supported by the Natural Sciences and Engineering Research Council of Canada, under grant No. A-7397.

D. G. Lainiotis and N. S. Tzannes (eds.), Advances in Communications, 347–356.

A major means of interconnecting computers separated geographically is by the use of terrestrial links. Many different topological configurations can be used for these networks. When a compromise between reliability and cost is needed however, mesh topologies with a connectively of two usually result. However, since not every node is connected to all other nodes directly routing control is required.

Routing algorithms can be divided into three main categories. They include static, dynamic and quasistatic or adaptive routing schemes.

Static algorithms do not allow routing parameters to vary with time and are thus susceptible to changing traffic conditions. Dynamic algorithms, on the other hand, adjust routing parameters instantenously, adjusting to network conditions. This flexibility is achieved at the expense of large and often impractical information transfers.

Adaptive algorithms adjust the routing parameters to accomodate small or slow changes in network conditions. To do so, some information about the network, but not the vast amount used in dynamic routing schemes, must be available. The ability to cope with changing network conditions, albeit slow changes, make these algorithms practical and useful in large systems. [2]-[4].

The routing computations required to react to changing network conditions can be achieved either centrally or through distributed methods. With centralized computation a network routing centre (NRC) receives global information about the system and routing decisions are made on these data. Distributed computation implies that routing calculations are performed at all nodes of the network. In most cases these calculations are performed using only local information available at the individual nodes. This local information may lead to some ambiguity about the network conditions and thus disallow optimal behaviour. However, this may be tolerable if the network conditions change slowly with respect to the routing update intervals. The most important advantage of distributed computation lies in the area of reliability. With decision making capabilities distributed, single node failures would not render the entire system inoperative. This reliability is the impetus behind computer communication networks, which is to provide robust communication channels.

In this paper, adaptive algorithms using the desirable characteristics of distributed computation are designed. The next section presents details and interpretations of some adaptive routing algorithms. The proposed algorithms are then explained and their expected performance analyzed.

compare different algorithms, along with their results. The paper then concludes with a discussion of the proposed algorithms.

2. ADAPTIVE ROUTING ALGORITHMS

The ARPANET routing algorithm is a very successful adaptive algorithm using simple updating methods. During updates, each node sends an update message to all its immediate neighbours containing a vector indicating its expected delay to all nodes. An illustration showing the message transfer between nodes is given in Figure 1 [7]. Upon reception of these vectors from all its neighbours, a node processes this information. First, the length of the queue to a particular neighbour is added to all entries of the vector received from that neighbour. Then a value of 4, chosen as a bias value, is added to all entries. Furthermore, the expected delay to itself is set to zero. The modified vectors from all neighbours are then compared and the neighbour with the lowest expected delay to a certain destination is chosen as the next node for all message with that destination. Two vectors are stored by the nodes after an update for future update purpose. The first is a vector of next nodes indicating the best neighbour to which traffic with a particular destination should be routed. The second vector indicates the expected delay if a message is sent to the chosen neighbour node. At the time of the next update, this expected delay vector is sent to all neighbours. Therefore, no information about the network is collected between updates. In the current implementation on the ARPANET, updates occur periodically three times every two seconds. During low traffic conditions, updates may occur more often.

The bifurcation routing algorithm [8] considers up to two output links for every message. At network initialization, the topology of the entire network is examined. First, the shortest delay path for each source-destination node pair is determined. Then, secondary paths are picked for some node pairs of the network. The secondary path is the shortest alternate path less than two hops longer than the primary path. If no alternate path satisfies the criterion, none is chosen.

During the operation of the network, a message is routed over the primary route if only one path is permitted. When two output paths are allowed for a message, a choice is made. This choice is achieved by using pseudo-random number generators and predetermined ratios. The ratios, set at the initialization of the network, control the proportion of the total traffic to be sent over the primary path.

'Delta routing' [9] is an adaptive routing algorithm, which uses a combination of centralized and distributed computation. Rudin proposed this algorithm with the intention of gaining advantages from both thpes of routing calculation methods. This mixture is achieved by conducting a centralized route calculation with less frequency than the distributed update. At the initialization of the network, up to two paths are considered for each source-destination node pair. These are recorded at a node acting as the network routing centre (NRC). During a centralized update, if two paths are allowed at a node, the expected delay on these two paths are examined. If the difference in expected delay of the two paths is above a preset threshold value, the lower delay path is chosen for all messages with that destination. This chosen path will then be the only allowed path until the next centralized update decision. On the other hand, if the difference in expected delay on the two paths is below the present threshold value, the path choice is left to the individual nodes. If the decision authority has been relegated to the node, the lower expected delay path is chosen. This choice is made with only local queue length information. Unlike bifurcation algorithms, only one output path is allowed at all times.

A logical evolution combining the advantages of these routing algorithms is proposed here. The algorithm uses bifurcation techniques to spread traffic and evenly use network resources.However, it does not do so by splitting traffic unnecessarily during low network utilization. This second criterion is met by the inclusion of a threshold level as in 'delta routing". Lastly, the algorithm incorporates an efficient computation method without the need for a network routing centre. This last condition avoids the reliability problems characteristic of centralized computation.

The distributed computation method decided on is a modification of the ARPANET routing algorithm. The new adaptive routing using bifurcation uses update information identical to that available on the ARPANET. The justification for this choice is that numerous current computer networks are using routing algorithms based on the ARPANET algorithm [1],[7]. This 'popularity' means an improved algorithm based on the ARPANET update information can benefit a large number of networks. For a bifurcation algorithm, the two best paths need to be chosen and the traffic divided between them. The secondary path should be the path with the lowest delay when the primary path is excluded and can be chosen using the ARPANET update information, with little increase in complexity.

To avoid bifurcation of traffic at low utilization, a threshold mechanism is used. With the information available, only a threshold involving the difference in expected delays of the primary and secondary path is practical. The ARPANET update information contains data about the number of hops to the destination and queue lengths along the path. During low traffic conditions, queue build-up is minimal, therefore the updating information contains numbers that are multiples of 4. A difference between expected delays of 4 thus indicates a path length difference of a single hop.

A choice of a threshold of 3 is expected to be a good compromise providing early bifurcation but reducing the possibility of unnecessary division of traffic. In the routing algorithm, if the difference in the expected delay between primary and secondary paths is less than or equal to the threshold value, then the traffic is bifurcated. Otherwise, the traffic is sent on the primary path only. An illustration of this process is presented in Figure 2.

The algorithm proposed in this work splits all traffic allowed to be bifurcated evenly. This even proportion of bifurcation is decided on because,1) with the threshold, traffic is allowed to be sent on a secondary path only when the difference in expected delays is small, 2) this alternate queue splitting technique produces a stream of traffic whose interarrival times have a lower variance than that of traffic subject to random splitting. A lower variance in interarrival times of customers will improve performance of a queueing system [10]. Thus also improving the performance in a network.

In the following sections, different algorithms' capabilities to handle a variety of topology and traffic conditions are examined by various simulations. The algorithms compared include a static routing scheme designed as a shortest path algorithm. This algorithm gives excellent performance at low or balanced traffic conditions. Then the ARPANET routing algorithm is simulated to provide an indication of the performance levels currently available using the ARPANET update information. The other algorithm used for comparison purposes is a bifurcation algorithm using ARPANET update information but without any threshold.

Those mentioned algorithms are all compared to the proposed bifurcation with threshold routing algorithms. The alternate splitting bifurcation algorithm is also simulated with a range of threshold values. These are compared among themselves to determine the best threshold value. Details about the simulation networks used are discussed in the next section.

3. SIMULATION PROGRAMS

In the simulations performed in this work, the delay encountered by each packet is recorded and their average is calculated. Related to the packet delay are two useful measures. These are the variance of the packet delays and the 95% delay level. Both these measures are useful from the user's point of view in determining the importance of the stated average delay. These measures are provided by all simulations.

Although the routing algorithms to be evaluated are designed for operation on large networks, the complexity of such networks make a good understanding of the routing effects difficult. To circumvent this complexity problem, small networks are examined first. Insight about the varios algorithms is gained from their respective performances in these small networks. Then simulation is performed on larger and more complex networks to insure that the observed differences are not limited to the small networks. Different traffic conditions on all these networks are simulated to observe each algorithm's ability to react to changes in network conditions.

After some knowledge is gained from the performance of the various routing algorithms in the small networks, they are examined on larger networks. The first large network topology is that shown in Figure 3. It is an eight node unbalanced topology. An unbalanced topology is one where all the links are not of similar lengths and capacity.

Various traffic conditions are simulated on this larger network. These include low balanced traffic, low unbalanced traffic and high unbalanced traffic.

Another aspect, the robustness of the routing algorithms' behaviour, is examined by simulation on another large network. The main concern here is a routing scheme's ability to provide acceptable performance levels during link failures.

Only a large network is used because link failures in small networks often result in trivial topologies. An unbalanced topology was considered because these topologies result in very uneven loading of network links. This uneven loading produces links, heavily utilized, whose failures can cause major degradation of the network performance.

4. NETWORK SIMULATION RESULTS

In this section, the various simulation results are discussed. Conclusions drawn from the simulations on the small networks are given first.

The simulation results showed that in all cases, the bifurcation with threshold algorithm results with a smaller variance in packet delays than the ARPANET adaptive routing algorithm. These smaller variances coupled with lower average values imply a lower maximum value. This expectation is supported by a lower 95% delay level of the bifurcation with threshold (T=3) algorithms in all cases.

The average link utilization of the static routing and bifurcation with threshold routing differed very little in the four cases examined. The bifurcation without threshold routing had average utilization values that are noticeably higher in all cases. This may indicate some inefficiency in the routing algorithm. Possibly some packets were sent along paths requiring more links unnecessarily and thereby raising the average link utilization. The variance of the link utilizations cannot be compared confidently due to the small number of links on these networks.

Simulations on larger networks also showed the bifurcation with threshold algorithm's ability to handle unbalanced and heavy traffic without sacrificing low traffic performance. The ARPANET algorithm performed very well but was consistently bettered by the proposed algorithms. This confirmed the expectations resulting from the small network simulations. Next to be considered is the algorithm's ability to handle link failures. This is again done by simulation. From these simulations it was found that the bifurcation algorithm provided slightly better performance than the ARPANET routing during normal conditions. But during link failure conditions, up to 20% improvement can be had with the bifurcation with threshold algorithm.

More detailed discussions of the simulations and results can be found in [13].

5. CONCLUSION

In this paper, routing algorithms used on computer communications networks were examined. Knowledge gained from the characteristics of other routing algorithms lead to the proposal of bifurcation with threshold algorithms. Various threshold values and two bifurcation techniques were examined and compared to other pertinent algorithms through simulations.

From the simulation results, it is expected that bifurcation without threshold routing algorithms would perform poorly in large networks. Static routing algorithms are ineffective in handling heavy or unbalanced traffic conditions. ARPANET adaptive routing performs very well under almost all traffic conditions but is not as efficient as the bifurcation with threshold algorithms proposed. The best compromise for thresholds appear to be a value of 3. This choice provides very good performance to the overall network traffic

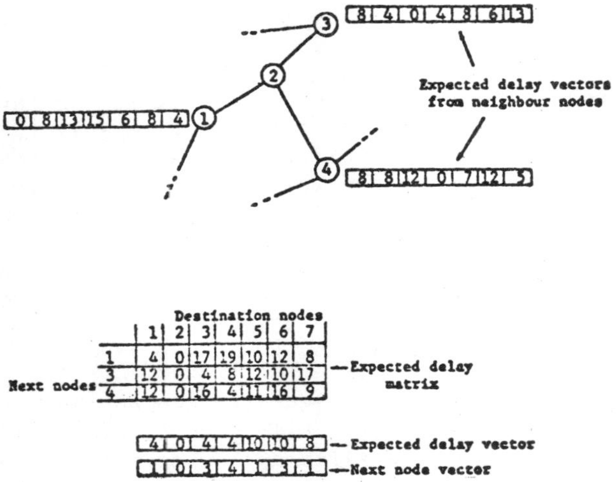

Figure 1 Update Information Transfer on the ARPANET.

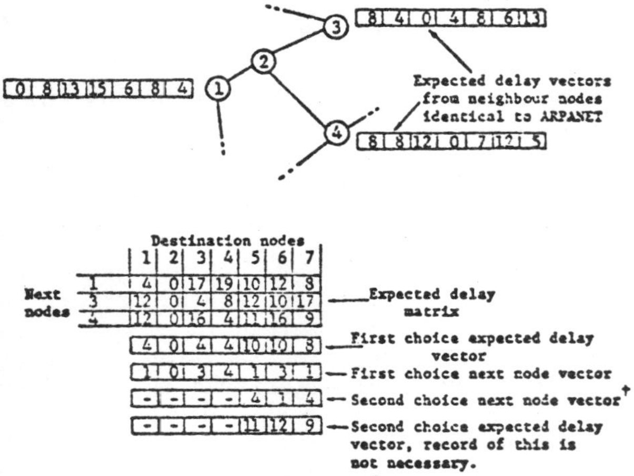

† Threshold used in this example is 3.

Figure 2 Second Best Neighbour Selection in Bifuraction with Threshold Routing.

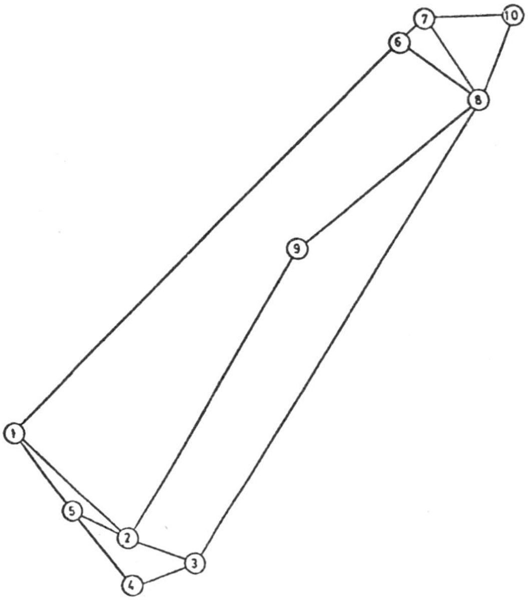

Figure 3 Ten Node Unbalanced Topology

while not neglecting portions of the network with heavy traffic. Of course this choice of 3 is related to the bias of 4 used in the routing updates.

The link failure simulations also showed the bifurcation with threshold routing algorithms ability to react to and handle topological changes. The proposed algorithm appeared to provide much smaller percentage degradation of service during failures and thus makes it a very useful routing technique.

The overall performance advantage of the proposed bifurcation with threshold routing algorithm, when compared with the ARPANET routing algorithm, makes it a very useful and effective extension.

REFERENCES

1. Roberts, L.G.: 1978, "The Evolution of Packet Switching", Proc. of the IEEE, vol. 66.
2. Gallager, R.G.: 1977, "A Minimum Delay Routing Algorithm Using Distributed Computation", IEEE Trans. on Communications, vol. COM-25.
3. Segall, A.: 1977, "The Modeling of Adaptive Routing in Data-Communication Networks", IEEE Trans. on Communications, vol. COM-25.
4. Stern, T.E.: 1977, "A Class of Decentralized Routing Algorithms Using Relaxation", IEEE Trans. on Communications, vol. COM-25.
5. Baran, P: 1977, "Some Perspective on Networks - Past, Present and Future", Information Processing'77, IFIP, North Holland Publishing Company.
6. Pouzin, L.: 1977, "Packet Networks - Issues and Choices", Information Processing'77,IFIP, North Holland Publishing Company.
7. McQuillan, J.M.: 1977, "Routing Algorithms for Computer Networks", Proc. 1977 National Telecommunications Conference.
8. Price, W.L.: "Adaptive Routing in Store-and-Forward Networks and the Importance of Load Splitting", Information Processing'77, IFIP, North Holland Publishing Company.
9. Rudin, H.: 1976, "On Routing and "Delta Routing": A Taxonomy and Performance Comparison of Techniques for Packet-Switched Networks", IEEE Trans. on Communications, vol. COM-24.
10. Kleinrock, L.: 1976, Queuing Systems, Volume II: Computer Applications, J. Wiley and Sons.
11. Foschini, G.J. and Salz, J.: 1978, "A Basic Dynamic Routing Problem and Diffusion", IEEE Trans. on Communications, vol. COM-26.
12. McQuillan, J.M., Falk, G. and Richer, I.: 1978, "A Review of the Development and Performance of the ARPANET Routing Algorithm", IEEE Trans. on Communications, vol.COM-26.
13. Waung, W.: 1979, "An Adaptive Bifurcation Routing Algorithm Using ARPANET Update Information", M.A.Sc. thesis, University of Toronto, Toronto, Canada.

CONGESTION AVOIDANCE IN HIERARCHICAL PACKET SWITCHED NETWORKS

W.L. Price & C.M. Solomonides

National Physical Laboratory, Teddington, UK

G.R.A. Gomberg

Logica Ltd., London, UK

ABSTRACT

The paper examines in general terms the requirements of a congestion avoidance system applied to packet switched networks. Two specific systems are then discussed, namely the system based on allocation of packet buffers in source nodes and, as an alternative, the isarithmic flow control scheme. It is seen that both systems are effective in single-level networks. The reasons for requiring a hierarchical topology are given, followed by an account of a simulation study of flow control, routing and congestion avoidance in such a context. It is shown that the behaviour of such a network in these respects is very different from that of a single-level network.

INTRODUCTION

The efficient control of commodity flow is an important consideration in any communication or transportation network. The network provider wishes to see the resources of his system used to the fullest possible extent, whilst the network user is concerned that his traffic shall be carried swiftly and that the network will always accept from him the level of traffic which he is entitled to offer under the terms of the network service specification. In order to achieve these joint aims it is vitally important that the network shall be prevented from

D. G. Lainiotis and N. S. Tzannes (eds.), Advances in Communications, 357–366.

entering any state which detracts from its performance. Perhaps the most damaging condition which can affect performance is that of congestion, when the network contains so much traffic that efficient flow is prevented; in the extreme case, congestion can become so serious that traffic flow ceases altogether.

Because, by definition, congestion in a packet switched network arises because of the presence within the system of an excessive number of packets, congestion avoidance must rely upon well designed techniques of controlling packet acceptance. The most important tool for achieving sensible packet acceptance is that of flow control, applied to the interface where packets enter the network as they are offered to the latter by the user. Unless one adopts the principle of deleting excess traffic within the network, as was done in the Cyclades network (1), or that of storing excess traffic temporarily in node backing store, as was done in the SITA network (2), then there is no alternative to preventing excess packet acceptance at the source.

Rigid control of traffic acceptance at the source interface may be sufficient to prevent global congestion of a network, but gives no guarantee that some degree of local congestion may not arise. However, provided that a complete lock-up is avoided by good system design (3), local congestion may be expected to disperse, though possibly slowly, into neighbouring uncongested areas.

Performance is improved by the provision of a fully integrated flow control system at all relevant physical and logical network levels. These include node-to-node flow control (inter-node links), end-to-end flow control (source node to destination node), source host-to-node flow control and node-to-destination host flow control. It is plainly vital that no network component should be allowed to overwhelm another component with which it communicates by imposing on the latter an excessive flow of traffic.

CONGESTION AVOIDANCE BY CONTROL ON PACKET ACCEPTANCE

Where a network is operated under a system of setting up calls between source and destination, total traffic may be limited by a combination of (a) placing a maximum on the number of packets that may be in transit for each call and (b) placing a maximum on the number of calls that may be active at any one time. The control on numbers of packets in each call is commonly exercised by operating a windowing number scheme coupled with positive acknowledgements. However, if the number of packets allowed to be in transit per call is set at such a level that global network congestion is avoided even if all possible calls are currently active, then, in a situation where only a few calls are active, each active call will carry traffic at a level less than that which could be carried safely; parts of the system will then be underused unnecessarily.

Another method of flow control of source traffic that has been suggested (4,5,6) is that of source node buffer scheduling. This system operates by allocating a limited number of source node buffers to new traffic seeking to gain acceptance; tandem traffic passing through the node is allowed access to additional buffers. In a simulation experiment at NPL a form of this protocol was tested under conditions of heavy offered overload.

This experiment assumed that each node output queue was limited in length; a few buffers allocated to each queue were reserved for tandem traffic. A new packet offered at the user interface was held in an input buffer whilst it was submitted to the routing algorithm to determine which output queue it should join. If the designated output queue was full (in respect of new packets), then the packet waited in the input buffer until space was available; note that packets of tandem traffic might still find space on the output queue which source packets were being prevented from joining. Whilst a packet was held in the input buffer, no further packets were accepted from the local source of packets, even though some of these later packets might wish to join a queue on which space was available.

In this experiment, with fixed routing, the flow control system, though crude, was sufficiently strong to prevent network congestion even in the face of a severe overload of traffic offered at all source nodes. If adaptive routing was introduced, then the strength of the flow control system was reduced; this was because greater choice of output queue was allowed for each incoming packet.

This type of flow control is not sufficient in itself to prevent congestion under all circumstances. The experiment did not simulate a destination unwilling to accept packets. If such should occur, then response of source buffer scheduling flow control would be sluggish, requiring a chain of full node queues extending from the destination to the source before remedial action could be initiated. This shows that, though this method of controlling traffic is cheap and effective in some circumstances, further additional measures, such as signalling packets from destination back to source, would be necessary in order to create a viable complete flow control system. Such signalling packets have been suggested in various contexts (7,8). It is worth noting the NPL experiments (9) in which sudden complete breakdown of single network components was simulated. Information packets were used to inform the network nodes of the nature and location of the failure. Source buffer scheduling flow control was in operation and congestion was prevented even in the presence of serious component failure.

A virtue of the method is that it achieves a degree of balance between the handling of traffic already in transit and new traffic seeking to enter at a source node. To this extent the control system is able to act in a distributed manner, discretion being given to local nodes and not reserved for some central controller.

Yet another suggestion for handling packet flow at the source node user interface was that for isarithmic (= constant number) control (10,11). This system requires provision of a limited number of permits, some to be allocated to each network node at switch-on time. When a new packet seeks acceptance at a source node, this is only allowed if the packet can acquire a free permit; if it cannot acquire a free permit, the packet is made to wait and acceptance is suspended. A permit acquired by a packet stays with that packet until the latter is accepted by the relevant destination and leaves the network; at that time the permit becomes free and can be used to allow acceptance of an incoming packet at the destination node. To prevent some nodes acquiring an excess of free permits at the expense of other nodes because of unbalanced traffic, only a small number of permits is allowed to be held at any one node; permits surplus to this number are sent off to other nodes which may require them. A balance must be struck between the three states in which a permit may find itself - free (waiting at a node), mobile (looking for a node at which to wait), in use (accompanying a packet in transit).

If the system parameters are well chosen, simulation results show that a network operating under isarithmic control can successfully avoid a congested state and deliver traffic efficiently, even when a highly excessive demand is made upon it by the users. Precautionary measures that must be taken include an audit check on the total number of permits; accidental creation or loss of permits from the network could defeat the aims of the control system.

HIERARCHICALLY ORGANISED PACKET SWITCHING NETWORKS

As packet switching networks become larger, so the problems associated with their efficient operation become more severe. It is already certain that in the future users of networks will wish to communicate on a wider and freer basis than ever.

Flow control, with the associated requirement of congestion avoidance, also presents new and difficult problems in the context of large networks. The chief difficulty is that of regulating a packet flow according to the effect it is having on the state of other parts of the network at various distances from the source of flow.

If a limit on the size of individual packet switched networks becomes general practice, then communication between such networks will assume greater importance. Several studies have been published which address the question of network interconnection (12,13); inter-network gateway nodes are usually postulated. However, where a large number of networks require intercommunication, it may prove insufficient simply to provide gateways between neighbouring networks. Carriage of long distance traffic through a chain of interconnected networks does not commend itself as an efficient system. The alternative is to create a hierarchy of networks.

In a hierarchical system each individual network may be allocated one or two nodes at a superior logical level ("super-nodes"). These nodes have the function of handling inter-network traffic and do not act as normal sources and destinations. The super-nodes are themselves connected by wider bandwidth links to form a network. Inter-network traffic is sent within a network to the local super-node; the latter sends the traffic to the appropriate distant super-node, which delivers it to the local network and so on to the destination node. In a really large configuration it might be necessary to group the super-nodes into groups each of which was served by a hyper-node; hyper-nodes would themselves be connected into a hyper-network.

The configuration need not be strictly hierarchical. One can still postulate gateway nodes at any level between geographically adjacent groups, thus cutting down the route length traversed by packets. Packets for adjacent nodes in neighbouring networks would go straight across without going to higher hierarchical levels. This additional, non-hierarchical, provision could result in complicating the operating system of the whole network.

SIMULATION EXPERIMENTS ON A HIERARCHICAL NETWORK

The network simulation study programme at the National Physical Laboratory (14) has covered a very wide range of networks and operating methods. For a study of hierarchical networks NPL placed a contract with Logica Ltd., the results of which have been published fully elsewhere (15,16).

In this study it seemed that the number of nodes in a simulated system could not be less than 50; fewer would have made the experiment unrealistic, more would have made it too expensive. The number of hierarchical levels selected was 2 and the number of individual networks was 5. Various configurations were simulated; they may be found in reference 17 . For reasons of efficiency in the simulation, the same topology was chosen for each individual local network. In each local area nine nodes were designated "low-level"; these served in the usual way as sources and destinations of traffic. The tenth node in each area was designated as "high-level" and was connected to two of the low-level nodes. The five high-level nodes were connected together by wide bandwidth links. Link speeds were - low-level, 250 Kbits/sec, - all others, 1.5 Mbits/sec. Inter-node link protocols were operated on a positive acknowledgement basis, with retransmission after timeout. Transmission errors were not simulated in this work.

The simulation experiments were designed in general terms to study aspects of network performance influenced by routing and flow control. In the present paper we shall, in the main, restrict ourselves to the means required to avoid congestion and congestive failure of the network.

No attempt was made to set up any end-to-end flow control in the simulation experiments, but both source buffer scheduling and isarithmic control schemes were tried at various stages. We have seen in the earlier part of this paper how both these schemes have been profitably applied in the context of single-level (non-hierarchical) networks.

In the first set of experiments source buffer scheduling was present in the low-level nodes and flow on all links was controlled by windowing protocols. No specific flow control provisions were built into the high-level nodes; source buffer scheduling had no effect at nodes which were not themselves connected directly to sources of traffic.

Satisfactory operation of the network was achieved for moderate loads, but the result of applying an overload was to produce a complete congestive lockup in the high-level network. This could be seen to be of the type of "indirect store-and-forward" lockup, as categorised by Kahn and Crowther (3). This lockup arose because the high-level network had admitted an excess of traffic, so that a ring of nodes with full output queues had resulted.

Evidently, it was necessary to devise some means of keeping the amount of traffic concurrently in the high-level network at a level below the critical number of packets at which a lockup was possible. The first expedient tried was to reduce the window-widths of the link protocols controlling traffic flowing from low-level nodes to the high level. It was expected that warning of congestion could be then propagated more rapidly back down these links. Experiments showed that this method was not effective in preventing excessive traffic acceptance and lockup failures.

It was then argued that the technique of source buffer scheduling could be extended to the high-level nodes, treating the latter as acceptors of source traffic from the low-level networks. This should allow traffic in the high-level network to keep moving, without lockup, and allow feedback of congestion signals to the low-level nodes, which in turn could prevent acceptance of excess traffic from the user interfaces.

The result of this step was to prevent the high-level lockups previously observed. An unexpected effect was the reduction of the incidence of full output queues on high-level links without a reduction in the total number of packets in the high-level network. The maximum carrying capacity of the network was slightly increased.

Introduction of source buffer scheduling to the high-level nodes gave a measure of priority to traffic already moving around the high-level network. If more than two network levels existed, then a logically sensible priority scheme at every node would give highest priority to packets moving down the hierarchy, second priority to packets moving on the one level across the hierarchy and least priority to packets seeking to move up the hierarchy, thus giving, at source nodes, least priority to traffic seeking to enter the net. This priority scheme can be expressed to a limited extent on the two level network, but it must be recorded that performance was not thus improved when this was tried.

To achieve greater certainty of lockup avoidance in the high-level network than that afforded by source buffer scheduling implemented alone, isarithmic control was introduced in addition in the high-level only. This meant that the number of packets which could co-exist in the high-level was prevented from exceeding the total number of isarithmic permits provided. Acceptance of packets from low-level nodes by high-level nodes was made to depend on possession of free permits by the latter as well as on the buffer allocation rules. Permits became free when packets were accepted by low-level nodes from high-level nodes, thus leaving the high-level network.

As congestive lockup had only been observed in the high-level network when more than 30 packets were present at that level, the initial number of permits allocated was 6 for each of the 5 high-level nodes. As expected, no lockups occurred when an extremely heavy offered load was applied. However, it was demonstrated that introduction of isarithmic control carried penalties in the form of reduced throughput and an increase in delays. Increased delay is to be expected, as packets are likely to spend more time in low-level nodes awaiting acceptance by high-level nodes.

To reduce this delay, the number of permits per node was doubled. This indeed gave a reduction in delay with a small increase in throughput. However, although no lockup occurred, it was judged that one nearly formed in one experiment, causing temporary reduced performance before it was dispersed.

Experiments showed that introduction of separate isarithmic control systems in each local low-level network as well as the high-level network produced a condition of permit lockup. This arises when a packet cannot cross the boundary between levels because no permit is available at the other level. No permit will become available until one is released by a packet crossing the boundary in the other direction, which will not be possible until a permit is freed by export of a packet. This situation would also arise if separate isarithmic control schemes were applied to two networks joined by a gateway; this is directly analogous to direct store-and-forward lockup (3). In theory the lockup can be dispersed by arranging to free a permit when a packet joins the appropriate output queue for the cross-boundary transfer before it has actually left the node. This implies a slight relaxation of the congestion avoidance rules and must be introduced with care.

Imbalance in traffic flows produces a severe strain on the isarithmic scheme; permits become free at points where they are not required for the admission of new traffic. The result is that new traffic is throttled more than is necessary and performance suffers accordingly. Therefore the design of the algorithm for redistributing free permits to the points where they are needed requires careful attention. In the simulation experiments described in this paper the worst of the problem was avoided by restricting the isarithmic scheme to the high-level network (where it was really needed), rather than introducing one universal isarithmic scheme covering both levels of the system.

CONCLUSIONS

The experimental results with the 50-node hierarchical network have demonstrated that the problems of controlling such networks have features quite unlike those encountered in studies of single-level networks. Our work has only touched the fringe of the problem. Both source node buffer scheduling and isarithmic control have been shown to be relevant and useful in preventing congestion.

If, as we believe, hierarchical organisation is the right approach to the inter-connection of a large number of individual networks, then further investigation must be carried out on the lines which we have initiated. The remedies proposed in this paper are somewhat ad hoc, but it is more than likely that the techniques investigated will prove to be valuable and important components of a future integrated hierarchical network control system.

REFERENCES

1. Irland, M. "Reflections on simulating Cigale at the University of Waterloo". Journees de Travail, Modelisation et Simulation de Reseaux d'Ordinateurs, IRIA, Rocquencourt, France, 1976, pp. 199 - 218.

2. Brandt, G.J. & Chretien, G.J. "Methods to control and operate a message-switching network". Proc. Computer-Communications Networks & Teletraffic Symposium, Brooklyn Polytechnic, 1972, pp. 263 - 276.

3. Kahn, R.E. & Crowther, W.R. "Flow control in a resource-sharing computer network". Trans. IEEE, Vol. COM-20, 1972, pp. 539 - 546.

4. Price, W.L. "Simulation studies of data communication networks operating in datagram mode". Computer Journal, Vol. 21, 3, 1978, pp. 219 - 223.

5. Price, W.L. & Haenle, J.O. "Some comments on simulated datagram store-and-forward networks". Computer Networks, Vol. 2, 1978, pp. 70 - 73.

6. Lam, S.S. & Reiser, M. "Congestion control of store-and-forward networks by input buffer limits". IBM Research Report RC6593, 1977 (Also presented at National Telecommunications Conference, Los Angeles, 1977).

7. Pouzin, L. "Flow control in data networks - methods and tools". Proc. ICCC, Toronto, 1976, pp. 467 - 474.

8. Pineda, L. "Le controle de flux dans le reseau Aramis". Journees de Travail, Modelisation et Simulation de Reseaux d'Ordinateurs, IRIA, Rocquencourt, France, 1976, pp. 95 - 169.

9. Price, W.L. & Cowin, G.W. "Simulation studies of the effect of link breakdown on data communication network performance". NPL Report COM77, 1975.

10. Davies, D.W. "The control of congestion in packet-switching networks". Trans. IEEE, Vol. COM-20, 1972, pp. 546 - 550.

11. Price, W.L. "Simulation studies of an isarithmically controlled store and forward network". Proc. IFIP Congress 74, Stockholm, 1974, pp. 151 - 154.

12. Cerf, V.G. & Kahn, R.E. "A protocol for packet network Interconnection". Trans. IEEE, Vol. COM-22, 1974, pp. 637 - 648.

13. Sunshine, C.A. "Interconnection of computer networks". Computer Networks, Vol. 1, 1977, pp. 175 - 195.

14. Price, W.L. "Data network simulation: experiments at the National Physical Laboratory, 1968-76". Computer Networks, Vol. 1, 1977, pp. 199 - 210.

15. Kerr, I.H., Gomberg, G.R.A., Price, W.L. & Solomonides, C.M. "A simulation study of routing and flow control in a hierarchically connected packet switching network". Proc. ICCC, Toronto, 1976, pp. 494 - 501.

16. Gomberg, G.R.A., Kerr, I.H., Richards, J., Price, W.L. & Solomonides, C.M. "A design study of a hierarchically connected packet-switching network using simulation techniques". Computer Networks, Vol. 3, 1979, pp. 114 - 135.

CONGESTION CONTROL IN PACKET NETWORKS

Jürgen O. Hänle

SESA Deutschland GmbH
Frankfurt, F.R. Germany

An overview of the congestion control mechanisms designed for the GMD-Network and investigated by simulation is presented. In more details a generalization of "congestion control by input buffer limits" is pointed out. These investigations were performed at the institute for teleprocessing of the Gesellschaft für Mathematik und Datenverarbeitung (GMD) in Darmstadt.

1. NETWORK DESIGN CHARACTERISTICS

The GMD-network design on which are based the following congestion control mechanisms, is characterized by:

- Virtual channel switching that is to say the route of a data stream is determined at connection set-up time and remains fixed until its clearing. Switching nodes have knowledge about traversing data streams (= virtual channels) and are able to handle virtual channels individually.

- The communication protocols are structured recursively. This means that the protocol which manages the data transport of a virtual channel between two adjacent switching nodes (node-node protocol) equals the end-to-end protocol being applied between source- and destination node of a virtual channel.
  Normally a communication protocol structure with two layers is applied, but in principal it can be extended by another protocol layer, e.g. in case of virtual channels, traversing several packet networks. A detailed description is given in [1].

D. G. Lainiotis and N. S. Tzannes (eds.), Advances in Communications, 367–376.

The buffer pool of each switching node is structured into a set of subsets of buffers, called buffer classes as outlined in figure 1. At its source node a virtual channel may only access to buffer class 0 (= input buffer class) at the next node, buffers from buffer class 1 may be allocated and so on. At destination node, buffer class i must be assigned whereas i is the number of hops the virtual channel has covered. As can be proven mathematically this mechanism prevents buffer allocation deadlocks [2].

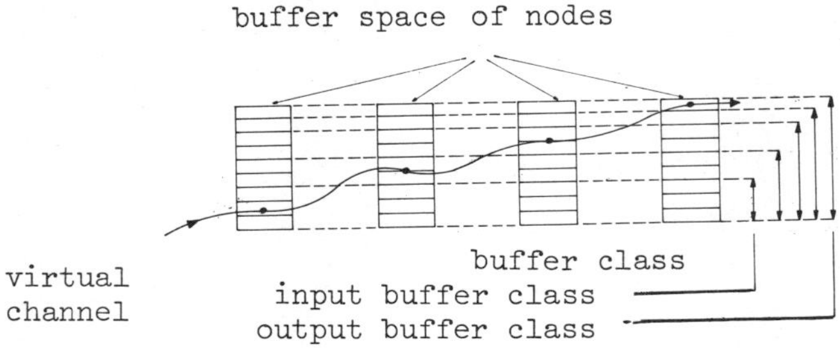

Figure 1: Access to buffer classes

Indeed for deadlock prevention it is sufficent if every buffer class is provided with at least one buffer, that does not belong to the next lower class. So the lowest class (= input buffer class) and the uppermost (= output buffer class) at each node may only differ by a few buffers. Besides all virtual channels can use at their destination nodes the uppermost buffer class without violating the buffer prevention principle.

2. INDIVIDUAL FLOW CONTROL

In the initial system design congestion control was restricted to the so called individual flow control of each virtual channel. This mechanism is described in detail in [1] and can be roughly summarized as follows:

> According to the recursive protocol layering, flow control of each virtual channel is performed twofold namely in a stepwise manner (i.e. between two adjacent nodes) and between source and destination node (end-to-end flow control). Both flow control principles are identical and are based on a dynamically controlled packet window. This means that the number of packets being within a pertinent transmission section is automatically adjusted to the packet acceptance rate of the next transmission section.

In the initial design step it was expected, that this flow control mechanism together with the buffer classes copes with congestions. However this is not true but a throughput degradation can occur when a network becomes saturated. This has been proven by simulation (figure 2, curve 1). The reason for throughput degration is, that although all virtual channels reduce their windows to the smallest size too many packets can enter the network in case of high applied load.

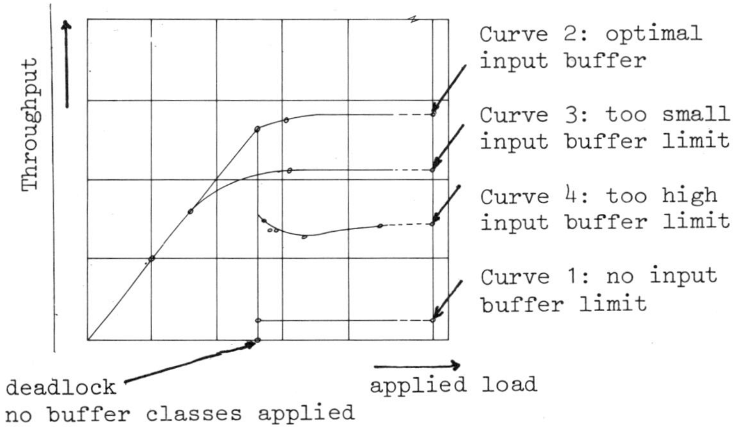

Figure 2: Throughput vs applied load for different input buffer limits

3. FLOW CONTROL BY FIXED INPUT BUFFER LIMITS

A detailed analysis of throughput degradation (figure 3) shows, that in case of critical applied load the input buffers attempt to monopolize the total buffer space and so create a lack of buffers for through traffic. This leads to similar congestion control mechanisms than already proposed by others (e.g. [4], [5]), namely to restrict the buffer space for network entering packets at each node in favour of through traffic by introducing limits for buffer class zero, called input buffer limit. As demonstrated by simulation the input buffer limitation is a very efficient congestion control mechanism (fig. 2). The only problem is to find the optimal input buffer limit. Too high or too low limits affect network performance (fig. 2, curve 2, 3 and 4).

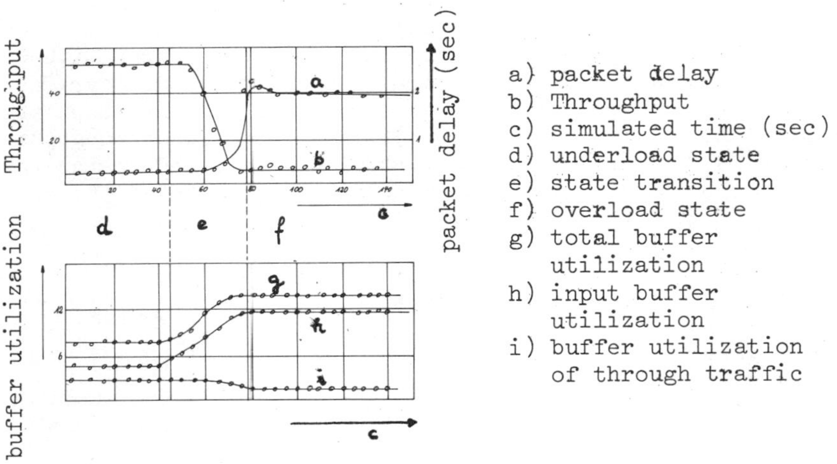

Figure 3: The process of performance degradation

However to only apply flow control by input buffer limits is not sufficient to permanently guarantee high network performance. For instance if some data streams is assigned a too low packet acceptance rate at their sink nodes and no individual flow control is applied then network performance is affected even with optimal input buffer limits (histogram 1) whereas by additionally applying individual flow control good network performance is obtained (histogram 2). Therefore individual flow control is always applied in addition to other control mechanisms.

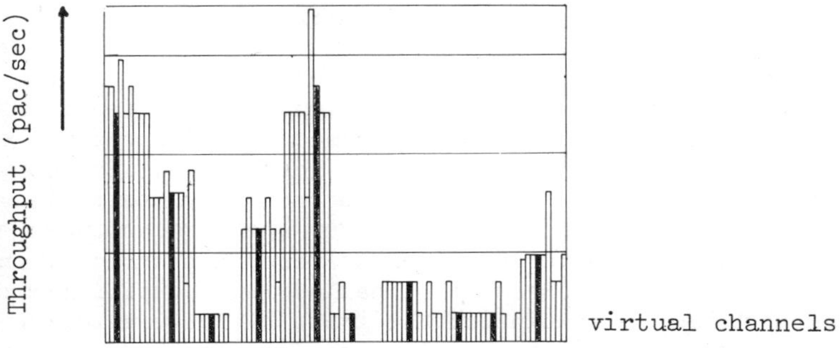

Histogram 1: Throughput of virtual channels when no individual flow control is applied and low sink rates are assigned to some virtual channels (hatched columns)

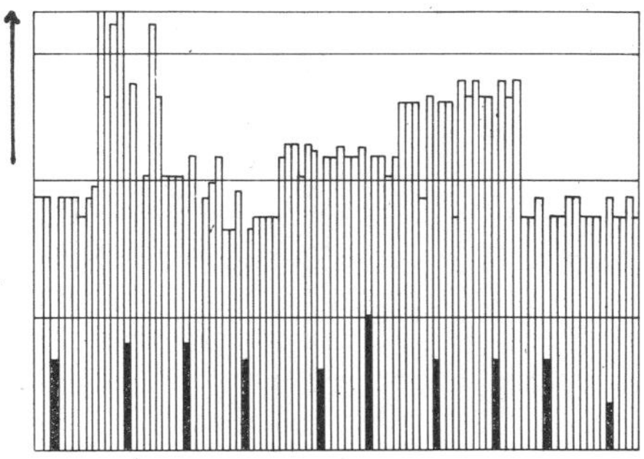

Histogram 2: Throughput of virtual channels when individual flow control is applied and low sink rate is assigned to some virtual channels (hatched columns)

4. FLOW CONTROL BY DYNAMICALLY CONTROLED BUFFER CLASS LIMITS

4.1. The need for buffer class limits

The optimum input buffer limit depends on the traffic distribution and may variate from node to node (for analytical approach see [6]). So instead of a fixed input buffer limit for all nodes it is desirable to have an algorithm that dynamically adapts at each node the input buffer limit to traffic distribution.

In addition when the packet acceptance rates of virtual channels are too low, packets are blocked at their destination nodes, i.e. the output buffer classes try to monopolize the total buffer spaces and so affect the network performance. Therefore it is necessary to limit the output buffer class.

Besides assume a network consisting of a set of special concentrator nodes being provided only for interfacing network subscribers and switching local traffic and of some other switching nodes which are only concerned with switching through traffic (figure 4). In this case the concentrator nodes are only provided with input and output buffer classes whereas the switching nodes are equipped with all buffer classes except input and output class. If there are too many concentrator nodes, congestions within the

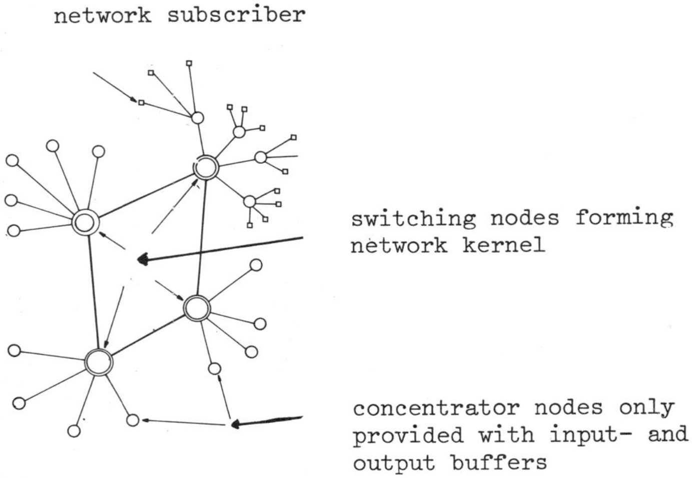

Figure 4: An example for a network configuration requesting limitation of all buffer classes.

network kernel may happen even when all concentrator nodes reduce their input buffer space to minimum. So in order to get an effective congestion control within the network kernel one has to limit not only input- and output class but also all other classes.

From these remarks results that:
- congestion control should include all buffer classes, i.e. to all buffer classes should be assigned limits

and
- these limits should be controlled automatically and dynamically by an internal driven congestion control mechanism.

4.2 An algorithm for dynamically controlled buffer class limits

The algorithm distinguishes between actions performed at the 'congested' node and actions performed at that node, to which a congestion is reported by sending a negative acknowledgement as response to an arriving but not acceptable packet.

Actions at the congested node:
Assume an arriving packet referring to buffer class i cannot be assigned to a buffer. This may be caused by two different reasons:

Case 1: There is no free buffer in class i even if the class limit would be unrestricted.
Case 2: Free buffers are available.
However the buffer class limit is too small.

Reaction in case 1:
In this case the congestion could only occur because some other buffer classes are provided with too many buffers. So a reduction of buffer class limits must be performed.
The problem is to find out the most appropriate buffer class for reduction. Several criteria for selection were investigated, e.g. most buffer consuming class, buffer class with the smallest index and combinations of both criteria. Bearing in mind that for reduction only those classes may be considered whose limits have not yet reached their lower bounds it turned out to be best to select among all those classes that one with the smallest index.

This means that in case of congestions first
- the local incoming traffic is throttled as far as possible. Then if congestions continue to occur
- the traffic generated at adjacent nodes is throttled and then
- traffic coming from the next but one nodes

and so on.

After reduction of the selected buffer class the packet is rejected and a congestion indicating response is transmitted to that node from which the packet came.

Reaction in case 2:
Normally the buffer class limit is increased and the packet is assigned to a free buffer.
However flow control might become ineffective if buffer class limits are increased too frequently without waiting until formerly performed modifications have taken effect.
Therefore the buffer limit is only increased after at least a predefined number of buffers has been released from that class since the last limit modification. Otherwise the class limit remains unchanged. The required number of buffers that must be released (so-called release counter) before the limit can be modified should be proportional to the number of utilized buffers within the class.
This has the following effect:
- The smaller the buffer utilization of a class the greater the chance for increasing its limit.
- The greater the rate by which buffers are released within a class the faster the limit may be increased.

Actions at the congestion causing node:
Actions at the congestion causing node are only performed if congestion indicating responses are received.
On receipt of such a response the release counter of that buffer class to which the rejected packet belongs is reset. So is avoided that the buffer class limit is increased in the near future.

Remark:
It is recommended to delimit the buffer limits to a reasonable upper bound. Otherwise the algorithm does not avoid in case of high applied load that each node accepts packets as long as buffers are available. As there is a size of buffer space per node which should not be exceeded because it does not result in much more throughput, but only increases packet deleays, buffer utilization should be limited.

4.3 Simulation results

Figure 5 shows simulated throughput for a given network versus buffer pool size assuming unlimited applied load. Curve 1 represents flow control by fixed input buffer limits, curve 2 depicts flow control by dynamically controlled buffer class limits.

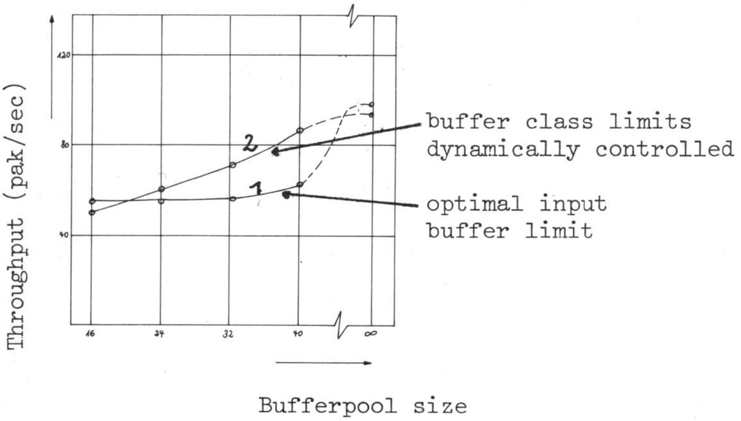

Figure 5: Throughput vs bufferpool size for unlimited applied load

As one has not only to maximize throughput but also to minimize packet delay a convenient measure for flow control effectiveness is the term throughput/packet delay, called "power". Figure 6 represents the power for both flow control mechanisms. These results demonstrate that at a whole both flow control mechanisms provide good network performance.

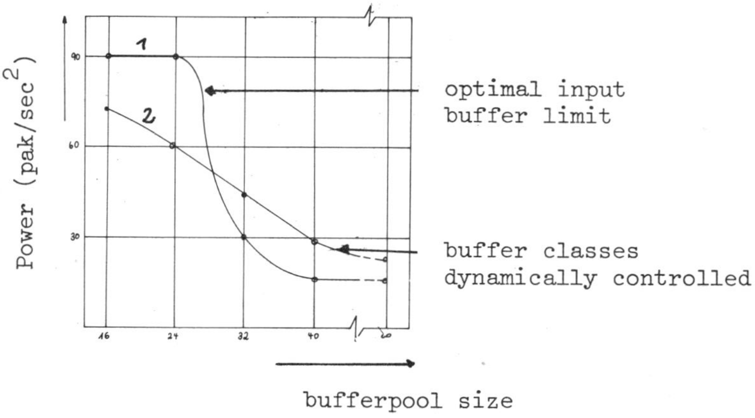

Figure 6: Power vs bufferpool size for unlimited applied load

5. CONCLUSION

These investigations may be summarized by the following statements:

- Individual flow control is necessary but not sufficient
- Buffer classes avoid deadlock but do not provide congestion control
- Input buffer limits provide good congestion control. However it is desirable
  - .. to have a mechanism for selfadapting input buffer limits
  - .. to impose these limits on all other buffer classes.
- There is a mechanism for automatically adapting buffer class limits which provides good network performance.

Acknowledgement
The author thanks Mr. Giessler, Mrs. Mäser and Mrs. Jägemann for their assistance in these investigations.

References

[1] E. Raubold, 'Flow Control in the GMD-Networks', Lecture given at the Computer-Network Workshop, Dublin 1977

[2] K.D. Günther, 'Prevention of buffer deadlocks in packet-switching networks', report presented at the IFIP-IIASA workshop on data communication, Luxemburg, Austria 1975

[3] A. Gießler, J. Hänle, A. König, E. Pade, 'Free buffer allocation - an investigation by simulation', 1978 Computer Networks, pp. 191 - 208

[4] W.L. Price, 'Data Network Simulation Experiments at the National Physical Laboratory 1968 - 1976', 1977 Computer Networks

[5] W. Chou, M. Gerla, 'A Unified Flow and Congestion Control Model for Packet Networks', Proc. Third International Conf. on Computer Communication, Toronto, Aug. 1976

[6] Simon S. Lam, M. Reiser, 'Congestion Control of Store- and-Forward Networks by Input Buffer Limits I : Analysis', IBM T.J. Watson Research Center, Yorktown Heights, N.Y. 10598

CONGESTION CONTROL TECHNIQUES FOR PACKET NETWORKS

Simon S. Lam

Department of Computer Sciences
University of Texas at Austin

Abstract

Congestion control mechanisms are needed for store-and-forward packet networks to maintain a high level of throughput. The basic requirements of network congestion control and some specific control techniques are examined. The use of input buffer (IB) limits for network congestion control is introduced. The rationale for the effectiveness of IB limits is discussed. Strategies for the design of IB limits have been investigated using both queueing analysis and simulation experiments. Some of our preliminary findings are presented.

INTRODUCTION

Consider a store-and-forward packet communication network of N nodes. Each node is connected to a multiplicity of packet sources and sinks. A source generates data packets (at some rate) which may be destined for any sink in the network. Routes followed by packets may be fixed or adaptive. If a packet arrives at the node connected to its destination sink, the packet will be consumed in a finite amount of time. Each node must hold on to a packet that it has accepted until the packet has been successfully consumed or forwarded to the next node on its route, at which time its buffer can then be freed. Each node has a finite number of packet buffers. When all these buffers in a node are in use, packets offered by its sources or forwarded to it by its neighbors are discarded; packets being forwarded by its neighbors will, of course, be retransmitted later. The network throughput is defined to be the aggregate rate at which packets are consumed under the assumption of statistical equilibrium conditions.

D. G. Lainiotis and N. S. Tzannes (eds.), Advances in Communications, 377–385.

Store-and-forward packet networks without flow control mechanisms have been shown to exhibit the throughput-load relationship illustrated in Figure 1 [KAHN 72, DAVI 72, PRIC 77, GIES 76, KLEI 78]. A characteristic, typical of many contention systems, is that as the offered load is increased from zero, the network throughput increases to a maximum and then turns down and decreases sharply to a low value (possibly zero). Realistically, when a store-and-forward network is congested, some "processes" may be blocked, and data packets may be lost or delayed due to a lack of resources [OPDE 74]. In either case, work is not conserved; hence, the degradation in network throughput.

Degradation in network throughput is often caused by deadlocks [KAHN 72, DAVI 72, OPDE 74]. However, networks which are deadlock-free may still be degraded in the sense that the throughput, though nonzero, is relatively low [GIES 76]. Hence, flow control mechanisms are needed to prevent throughput degradation whether or not a network can be formally proved to be deadlock-free.

From now on, a network is said to be congested when it operates in the region of negative slope in Fig. 1.

NETWORK VERSUS END-TO-END CONTROL

By network congestion control we mean any mechanism with the primary objective of preventing the network from operating in the congested region for any significant period of time. Typically, most networks are based upon the concept of virtual channels (or connections, sessions etc.) and are end-to-end flow-controlled between pairs of sources and sinks. Examples of end-to-end controls are SNA pacing [IBM 75], RFNM in the ARPANET [OPDE 74], and various window mechanisms [POUZ 73, CERF 74]. An important function of such end-to-end controls is synchronization of the source input rate to the sink acceptance rate. All of them work by limiting the number of packets permitted in a virtual channel. Suppose L_i is the maximum number of packets in virtual channel i and the network has a total of K virtual channels. The maximum number of packets permitted to enter the network is thus

$$N_{max} = L_1 + L_2 + \ldots + L_K.$$

The fact than N_{max} is bounded does not imply that network congestion control is not necessary. In fact, one of the motivations for a store-and-forward network in the first place is that data traffic sources are typically bursty [LAM 78]. In other words, virtual channels require actual transmission capacity only intermittently with a small duty cycle. If, for example, a network is operated such that N_{max} is at point B in Figure 1, it is obvious that network congestion control is not necessary. However, due to the bursty traffic, the average utilization of the network will be very low (such as at point A). It is therefore desirable for store-and-forward networks to operate on the principle of overcommitment such that N_{max} is far to the right (such as at point C) in Figure 1 and through averaging, the network utilization is at point B with a correspondingly high throughput. An immediate consequence is that network congestion control is now necessary to prevent the network operating point from going over the peak of the curve as a result of statistical fluctuations.

NETWORK CONGESTION CONTROL TECHNIQUES

A network congestion control mechanism must be capable of: (1) detection of network congestion, and (2) shutting off input into the network according to some rule.

The isarithmic technique proposed by DAVIES [DAVI 72] and studied by Price [PRIC 77] does the above functions by limiting the number of packets permitted to enter the network. This is accomplished by circulating a fixed number of "containers" in the network. A newly generated packet will be accepted by a network node only if it can get hold of an empty container. A second technique is to control the window size of each virtual channel as a function of network load [KERM 77]. Both techniques are attractive in theory but are quite difficult to implement in practice because individual network nodes do not have fresh and accurate information about the rest of the network.

A third technique that we have studied, assumes some knowledge about the pattern of traffic flows in the network. It attempts to control network inputs by differentiating between input and transit traffic at each node and imposing a limit on the fraction of buffers in a node's buffer pool that input traffic can occupy. This fraction will be referred to as the input buffer limit. Note that transit traffic can occupy all buffers in the buffer pool. In times of extreme congestion, input traffic may be shut out by transit traffic but not vice-versa; a desirable property. The advantage of discriminating against input traffic was first noticed by Price [PRIC 77]. He observed that if one or two buffers are dedicated to transit traffic, the network throughput can be much improved. A similar idea was also suggested by Chou and Gerla [CHOU 76]. This idea, however, is most clearly demonstrated and investigated in the GMD simulation studies [GIES 76]. In addition, they have also shown that if the buffer pool is structured into nested subsets of buffers and packets are assigned to these subsets according to the number of hops they have covered, then it can be proved that store-and-forward deadlocks of the type described in [KAHN 72, OPDE 74] can be avoided.

DESIGN STRATEGIES FOR IB LIMITS

Although the GMD studies explored the use of input buffer (IB) limits for congestion control, it was not known how to design such limits. We have studied design strategies for IB limits using both queueing analysis and simulation experiments. Our objective is to achieve the maximum network throughput as well as to provide some safety margin for errors and uncertainties in our traffic flow assumptions.

We employed an analytic model based upon our earlier work on the modeling of store-and-forward networks [LAM 76] and an extended class of queueing networks [LAM 77]. Using the analytic model, we were able to study the tradeoffs among network offered load, nodal buffer capacity and IB limit. From these results, we found that the key to designing effective IB limits hinges upon a very intuitive "capacity law" [LAM 79].

In Fig. 2, we have shown analysis results for a homogeneous network of identical nodes. Fixed routing is used. Let σ_{max} be the theoretical maximum network throughput rate assuming infinitely many nodal buffers; σ_{max} can be easily calculated from the channel capacities. λ is the rate of offered load to the network (i.e. aggregate packet generation rate of all sources). N_T is the number of buffers in a node. An important observation from the analysis results in Fig. 2 is that if $\lambda < \sigma_{max}$, IB limits are not necessary. If $\lambda > \sigma_{max}$, then IB limits (<1 in value) are necessary for congestion control. In fact, under such a heavy loading condition, IB limits are what determines the amount of traffic that can be admitted into the network. The results in Fig. 2 clearly show that there is a critical value of IB limit beyond which the traffic carrying capability of the network is seriously impaired. We shall refer to this critical value as the IB capacity.

The explanation for this critical behavior turns out to be rather intuitive. For each new packet that the network admits into an input buffer, additional buffers are needed elsewhere for its subsequent journey to the destination. Therefore, there is a natural ratio, say α_0, of the number of input buffers to the number of total buffers in the network that serves as an upper bound for IB limits.

For a homogeneous network with fixed routing and fixed input traffic pattern, the IB capacity is identical for each node and the ratio α_0 can be easily determined [LAM 79]. (α_0 is equal to 0.344 for our example in Fig. 2.) If IB limits exceeding α_0 are used, it will then become possible for input buffers to become completely filled such that the network will not have enough buffers to accommodate the resulting transit traffic! This explains the critical behavior observed in Fig. 2.

Another important observation from Fig. 2 is that if the IB limits are properly designed, the offered load λ can go to infinity without degrading the network throughput! Note that the throughput curves in Fig. 2 all have a wide "plateau" around the maximum point, meaning that the IB limit can be made much smaller than the IB capacity without sacrificing much throughput. Thus a large safety margin can be provided for errors in our evaluation of the IB capacity. Errors may arise due to time and statistical flucuations in user traffic as well as uncertainties in our estimate.

The above observations from analysis results are supported by simulation results of a four-node homogeneous networks. In Fig. 3, results from our analysis and two different simulators are shown. The first simulator assumes fixed-length packets. The throughput results from this simulator (consider the curve λ= 2.2, N_T=30) are actually better than the analytic throughput results which assume exponentially distributed packet lengths. The second simulator uses exponentially distributed packet lengths (but no Independence Assumption [KLEI 64]) and give throughput values which are slightly less than the analytic results.

Note that the part of an analytic throughput curve with negative slope is not realizable in simulation. This is because the assumption of equilibrium used in the analysis breaks down under such congested conditions. In practice, the network should not operate in this region.

Another very interesting observation which we have not yet been able to provide a satisfactory explanation is: if retransmissions are given priority over the transmission of other messages, the network behaves much better in terms of a higher throughput and a larger IB capacity.

For a nonhomogeneous network, the network nodes will have different IB capacities. Furthermore, for networks with adaptive routing, the IB capacities will not be fixed. These problems are currently being studied. Some heuristic methods for desiging IB limits have been proposed.

A promising approach is under investigation for networks based upon virtual channels. (We note that the new generation of packet networks are mostly virtual channel networks [ROBE 78]. In fact, the CCITT X.25 international standard for network interface is designed for virtual channel networks.) Virtual channels are individually end-to-end flow controlled via a window mechanism. Knowing the set of virtual channels and their window sizes, we use a queueing network model to obtain estimates of (peak) traffic flows in the network. From these estimates, we have developed various heuristic assignment strategies for designing IB limits.

Simulation results are shown in Fig. 4 for a 7 node (nonhomogeneous) network with 84 virtual channels. Each network node has 30 packet buffers. Each virtual channel uses a fixed route. The window size of each virtual channel is set equal to 3 times the number of hops (links) in its route. The network throughput is plotted versus the ratio N_I/N_T, where N_T is the total number of buffers and N_I is the total number of input buffers in the whole network. Two assignment strategies are considered: a uniform assignment strategy which uses the same IB limit at each node and a heuristic assignment strategy that uses IB limits proportional to our estimates of traffic flows.

We found that both assignment strategies achieve approximately the same maximum network throughput. However, the IB capacity of the heuristic assignment strategy is better than that of the uniform assignment strategy. The heuristic assignment strategy also gives rise to a lower value of mean network delay for packets at the same level of network throughput.

However, at small values of N_I/N_T, say 0.1 to 0.2 in Fig. 4, we found that the heuristic assignment strategy does not perform well because of the small number of input buffers available in the network; some IB limits become excessively small according to the heuristic assignment strategy thus restricting the network throughput.

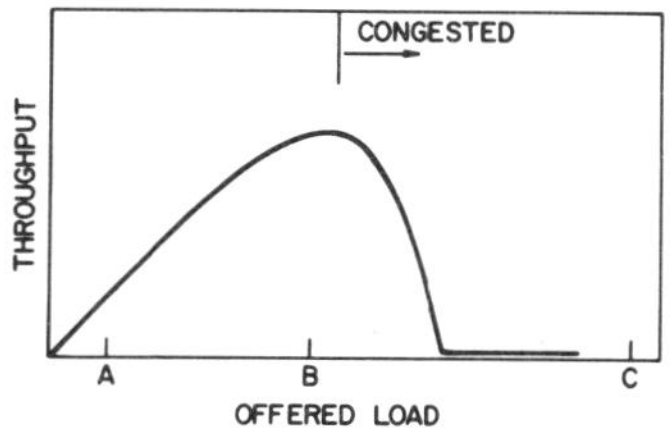

Fig. 1. Throughput versus offered load.

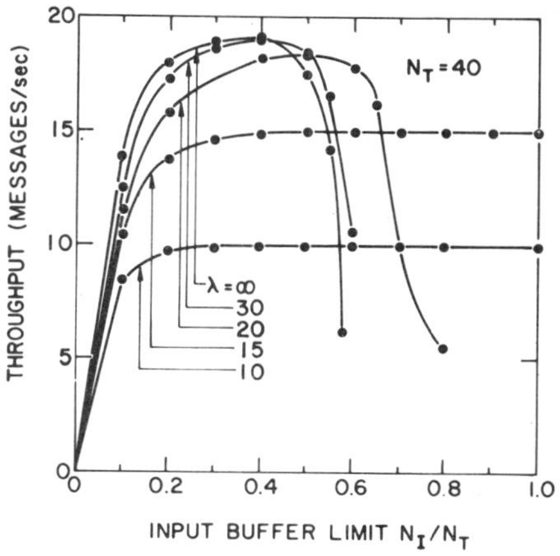

Fig. 2. Throughput versus IB limit.

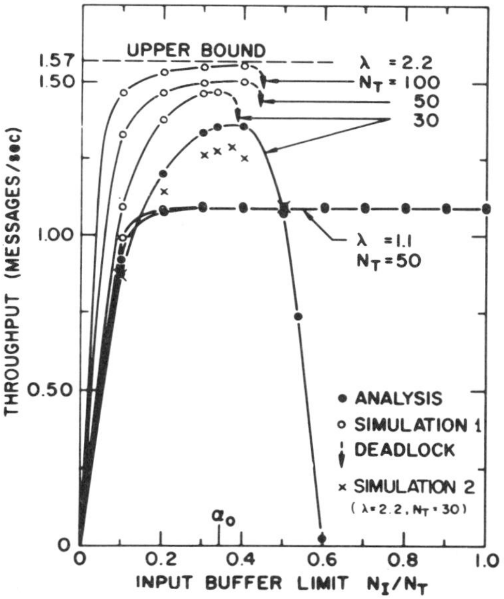

Fig. 3. Analysis and simulation results (four-node homogeneous network).

The following strategy appears to be a reasonable compromise. Each IB limit consists of two components: (1) the first component is fixed and guarantees a minimum acceptance rate at the node, (2) the second component is designed proportional to network traffic flow estimates.

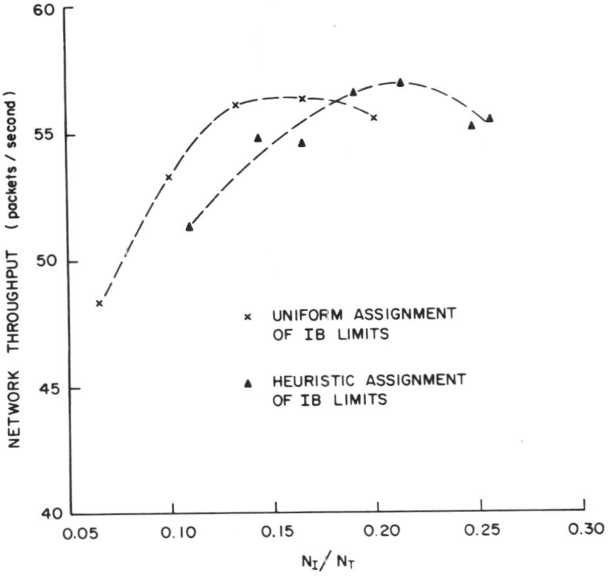

Fig. 4. Simulation results (seven-node nonhomogeneous network).

CONCLUSION

We discussed the need for congestion control mechanisms in store-and-forward packet communication networks. The rationale for the effectiveness of IB limits as a congestion control mechanism was examined. Design strategies for IB limits have been investigated using both queueing analysis and simulation experiments. Some of our preliminary findings are presented.

ACKNOWLEDGEMENT

This work was supported by National Science Foundation under Grant No. ENG78-01803.

REFERENCES

CERF 74 Cerf, V. and R. Kahn, "A protocol for Packet Network Intercommunication," IEEE Trans. on Commun., Vol. COM-22, May 1974.

CHOU 76 Chou, W. and M. Gerla, "A Unified Flow and Congestion Control Model for Packet Networks," Proc. Third International Conf. on Computer Communication, Toronto, August 1976.

DAVI 72 Davis, D. W., "The Control of Congestion in Packet Switching Networks," IEEE Trans. on Commun., Vol. COM-20, June 1972.

GIES 76 Giessler, A., J. Haenle, A. Koenig and E. Pade, "Packet Networks with Deadlock-Free Buffer Allocation--An Investigation by Simulation," GMD Report, Darmstadt, Germany, 1976.

IBM 75 IBM Corp., Systems Network Architecture General Information, GA27-3102-0, January 1975.

KAHN 72 Kahn, R. and W. Crowther, "Flow Control in a Resource-Sharing Computer Network," IEEE Trans. on Commun., Vol. COM-20, June 1972.

KERM 77 Kermani, P., "Switching and Flow Control Techniques in Computer Communication Networks," Ph.D. Dissertation, Computer Science Department, UCLA, 1977.

KLEI 64 Kleinrock, L., Communication Nets: Stochastic Message Flow and Delay, McGraw Hill, New York, 1964.

KLEI 78 Kleinrock, L., "On Flow Control in Computer Networks," Conf. Rec. ICC'78, Toronto, June 1978.

LAM 76 Lam, S. S., "Store-and-Forward Buffer Requirements in a Packet Switching Network," IEEE Trans. on Commun., Vol. COM-24, April 1976.

LAM 77 Lam, S. S., "Queueing Networks with Population Size Constraints," IBM Journal of Res. and Develop., Vol. 21, July 1977.

LAM 78 Lam, S. S., "A New Measure for Characterizing Data Traffic," IEEE Trans. on Commun., Vol. COM-26, Jan. 1978.

LAM 79 Lam, S. S. and M. Reiser, "Congestion Control of Store-and-Forward Networks by Input Buffer Limits -- An Analysis," IEEE Trans. on Commun., Jan. 1979.

OPDE 74 Opderbeck, H. and L. Kleinrock, "The Influence of Control Procedures on the Performance of Packet-Switched Networks," Proc. National Telecommunications Conference, San Diego, California, Dec. 1974.

POUZ 73 Pouzin, L., "Presentation and Major Design Aspects of the CYCLADES Computer Network," Data Networks: Analysis and Design, Third Data Communications Symposium, St. Petersburg, Florida, Nov. 1973.

PRIC 77 Price, W. L., "Data Network Simulation Experiments at the National Physical Laboratory 1968-1976," Computer Networks, Vol. 1, 1977.

ROBE 78 Roberts, L. G., "The Evolution of Packet Switching," Proc. IEEE, Vol. 66, Nov. 1978.

ROUTING AND FLOW CONTROL IN VIRTUAL CIRCUIT NETWORKS*

Mario Gerla
University of California
Los Angeles, USA

The concept of available bandwidth to each destination is used to establish routes and flow-control data in a virtual circuit network in which traffic requirements are relatively steady with time. A distributed algorithm to evaluate available bandwidth between any two nodes in the network is presented.

1. VIRTUAL CIRCUIT SWITCHING

Virtual circuit switching is a variant of the packet switching scheme, and consists of establishing a "virtual" circuit for each user connected to the network. The virtual circuit switching mechanism is similar to the traditional circuit switching mechanism (implemented, for example, in telephone switching networks) in that a path is dedicated to a user at call set up time for the entire duration of the session. All the packets transmitted within the session follow the preestablished path. Bandwith, however, is not dedicated to a user; rather, it is shared among all users in a statistical fashion. Thus, a virtual circuit can be viewed as a "virtual pipe" of variable (statistical) bandwidth that is used to transfer a sequential stream of packets from a given source to a given destination. (1), (2), (3),

Packets travel along the virtual pipe in a store-and-forward mode, being reassembled and checked for errors and, if necessary, retransmitted at each intermediate node. Several pipes are

*This research was supported in part by the Advanced Research Projects Agency of Dept. of Defense, Contract MDA 903-77-C-0272, and in part by Office of Naval Research, Contract N00014-79-C-0866.

D. G. Lainiotis and N. S. Tzannes (eds.), Advances in Communications, 386–394.

generally multiplexed on the same trunk; thus, each pipe must be distinguished from the others by a unique identifier, called virtual circuit number, which is stamped in the header of each packet travelling on that pipe. At each intermediate node a map translates incoming virtual circuit numbers into outgoing trunks. The map is updated each time a connection is set up or is cleared.

In comparing the virtual circuit switching scheme with the conventional packet switching scheme (also known as "pure packet" or "datagram") we immediately recognize the following advantages of virtual circuit switching:

- the packet header overhead is reduced since the destination ID (which in datagram consists of a string with destination node ID, destination Host ID and destination process ID) is replaced by the virtual circuit number.
- the routing overhead is reduced since routing decisions are made only at call set up time, and routing messages do not need to be frequently exchanged between nodes as in the datagram case.
- no looping can occur since paths are fixed.
- the network can selectively flow control traffic on a virtual circuit basis, since packets carry unique virtual circuit numbers.
- no packet reordering is required at the destination, before delivery to the Host, since packet sequencing is maintained withing the virtual circuit.

The virtual circuit scheme has also some drawbacks. The weakest point is probably the requirement of setting up a path for each connection. In transaction type applications, for example, where the actual data transfer consists of only one inquiry and one response packet, the line overhead and delay introduced by the call set up procedure would be prohibitive. For connections of longer duration (such as, for example, a file transfer, or an interactive terminal-computer session) the overhead of the call set up phase is negligible with respect to the volume of data transferred during the active phase.

These observations lead to the generally accepted conclusion that virtual circuit switching is better for long sessions and datagram is better for extremely short sessions. The best of all worlds is of course the integration of the two services. Indeed, standards for the integration of virtual circuit and datagram are now being prepared by CCITT for inclusion in the X.25 recommendation (4).

The purpose of this paper is to describe a flow control scheme for virtual circuit networks which combines selective flow control and routing. In Section 2, we introduce the concept of selective flow control, we describe the conventional technique used for its implementation, namely, backpressure, and we propose a new technique called bandwidth control. In Section 3, we further elaborate on bandwidth control and propose a distributed algorithm implementation. Section 4 presents some concluding remarks.

2. FLOW CONTROL IN VIRTUAL CIRCUIT NETWORKS

It is well known that packet networks are prone to congestion. Namely, if input rates are increased beyond a safe threshold, buffers at intermediate nodes become full and start to overflow; the overflow causes packet retransmissions; and the retransmissions cause wastage of bandwidth resources. The end result is that useful bandwidth drastically decreases because of retransmission overhead; therefore, useful throughput actually decreases when input rates increase beyond a safe value. To prevent congestion, a number of flow monitoring and regulation techniques are generally implemented and are commonly referred to as flow control. Examples of flow control are the permit scheme, which allows only up to a fixed number of packets to be simultaneously present in the network (5) and the input buffer limit scheme, which sets a limit on the number of buffers in a node that can be used to store packets from the Hosts (as opposed to "transit" packets) (6). These schemes may be regarded as global flow control schemes since all packets are treated equally, independently of the user session they belong to.

In addition to global flow control, virtual circuit networks permit to exercise selective flow control, i.e. they permit to individually regulate the flow on each user session. Selective flow control is made possible by the fact that packets are distinguishable (by ID numbers) on a virtual circuit basis, and virtual circuit routes are fixed for the entire duration of a session. As we shall see, selective flow control becomes particularly effective when the following conditions are met:

- average traffic flow on each connection can be predicted at call set up time (e.g., in a packet voice connection, the flow is determined by the vocoder rate).

- network traffic pattern is relatively stable.

Indeed, these conditions are often verified in VC networks, due to the relatively long duration of each user connection.

The selective flow control method most commonly applied in virtual circuit networks is the backpressure method (1). This method sets an upper limit on the number of packets that can be stored at each intermediate node along a virtual circuit path. Upon reaching this limit, an intermediate node temporarily shuts down the virtual circuit, i.e. it rejects packets coming from the preceding node in the path until the congestion is cleared. If congestion persists, the preceding node also reaches its packet limit and becomes blocked (backpressure). Eventually, by virtue of backpressure, the congestion propagates all the way to the traffic source, thus temporarily blocking further packets from that source.

Backpressure can selectively stop input flow at the origin. However, a drawback of this method is the fact that the "pipe" must be full before the flow control action is taken. This is not very desirable since in the event of simultaneous backpressure on several virtual circuits, the network may become flooded with packets and may fall into a global congestion state, with severe consequences on overall user performance.

To overcome the global congestion tendency of the backpressure method, we propose here the <u>bandwidth control</u> scheme. The basic principle of this scheme consists of maintaining at each node fresh information on the average bandwidth available to each destination. More precisely, each node has the following information:

(1) next leg on the shortest path, to each destination.
(2) available bandwidth on the shortest path.
(3) next leg on the maximum bandwidth path to each destination (the maximum bandwidth path is the path with maximum available bandwidth; in some cases, it may coincide with the shortest path).
(4) available bandwidth on the maximum bandwidth path.

The Bandwidth Control procedure assumes that the average bandwidth requirement for each session is either declared by the user, or is estimated by the network. The entry node, upon receiving a call request from the user, will compare the requested bandwidth with the bandwidth available on the shortest path, and, if feasible, it will forward the call request packet on the shortest path. If the bandwidth request exceeds the shortest path bandwidth, the maximum bandwidth path is inspected. If the bandwidth request exceeds also the maximum bandwidth, the user is given two options: either he drops the call, or he accepts a reduction in his original bandwidth request.

The advantage of this method over backpressure is that acceptance and routing decisions are made at the entry node, so that user demands which exceed network capacity are immediately blocked at network entrance, rather than being accepted and later blocked internally in the network. It is precisely this internal blocking that causes global congestion in the backpressure scheme.

Bandwidth control can be viewed as a combined routing and flow control scheme. The routing function consists of maintaining two routes for each destination (the shortest path and the maximum bandwidth path), which are used to establish the virtual circuit path. The flow control function consists of maintaining the knowledge of available bandwidth, which is used to decide acceptance or rejection of virtual calls at the network access points.

The feasibility of the bandwidth control scheme rests on the existence of an efficient algorithm to produce and update the shortest route and available bandwidth information. Such an algorithm is described in the following section.

3. BANDWIDTH CONTROL: A COMBINED ROUTING AND FLOW CONTROL SCHEME

Our goal is to define a distributed algorithm (i.e., an algorithm performed in parallel by all network nodes) for the computation of the following tables: shortest route; bandwidth available on the shortest route; maximum bandwidth route; and, maximum bandwidth. There is one set of tables for each node; and there are N entries in each table, where N = number of possible destinations (i.e., number of nodes in the network). For example, entry j in the maximum bandwidth table residing at node i represents the value of the maximum bandwidth from node i to node j.

The algorithm that we are going to introduce assumes that each node monitors the average utilization on each outgoing trunk. Several methods can be proposed for this type of measurement. The most straightforward method consists of counting the characters transmitted during the past T seconds, and defining utilization U as follows:

$$U = \frac{\#\text{ of characters}}{T \bullet C}$$

where C is the trunk capacity (chrs/sec). The average available bandwidth B on a trunk is then given by:

$$B = C(1-U)$$

The available bandwidth b(q), on path q is defined as the smallest value of B over all the trunks on path q, namely:

$$b(q) = \min_{i \in q} B_i$$

The algorithm is based on the periodical exchange of information between neighboring nodes, and on local table updating based on the information received from the neighbors. The procedure is essentially a dynamic programming procedure which closely resembles the ARPA routing scheme (7).

The algorithm is implemented with two independent modules - the shortest path module and the maximum bandwidth module - described below.

3.1 Shortest path module

The data structure of the shortest path module consists of the following set of tables for each node i (i=1...N):

<u>routing vector</u> $RV_i(j)$, where j is the destination (j=1...N) and $RV_i(j)$ is the next leg on the shortest path (in terms of number of intermediate links) to j.
<u>bandwidth vector</u> $BV_i(j)$, where j is the destination (j=1...N) and $VB_i(j)$ is the available bandwidth on the shortest path to j.
<u>hop vector</u> $HV_i(j)$, where j is the destination (j=1...N), and $HV_i(j)$ is the number of hops on the shortest path to j.

Periodically, node i receives from all its neighbors, say k, ℓ,...p, the bandwidth vectors BV_k, BV_ℓ,...BV_p and the hop vectors HV_k, HV_ℓ...HV_p. When all the vectors have been received, the table updating at node i proceeds as follows:

$$HV_i(j) \leftarrow \min_{t\varepsilon\{k,\ell...p\}} [HV_t(j)+1]$$

Let t_j^* be the neighbor that minimizes the hop count to destination j.

Then: $BV_i(j) \leftarrow \min[BV_{t_j^*}(j), B_{i,t_j^*}]$

where B_{i,t_j^*} is the available bandwidth on the trunk from i to t_j^*.

Finally, the routing vector is updated as follows:

$$RV_i(j) \leftarrow t_j^*$$

When node i is first brought up in the network, its tables are initialized as follows:

$$RV_i(j) \leftarrow j \quad , \quad j=1,...N$$

$$\begin{cases} BV_i(j) \leftarrow 0 & \forall j \neq i \\ BV_i(j) \leftarrow \infty & j=i \end{cases}$$

$$\begin{cases} HV_i(j) \leftarrow \infty & \forall j \neq i \\ HV_i(j) \leftarrow 0 & j=i \end{cases}$$

At network start up, all the tables are reinitialized. The repeated application of the update procedure will drive the tables to their steady state values in a finite number of steps. (The typical number of steps to convergence is on the order of the network diameter).

3.2 Maximum Bandwidth Module

The data structure of the Maximum Bandwidth Module consists of the following set of tables (for node i):

routing vector $MRV_i(j)$, indicating the next leg on the maximum bandwidth path to node j.
bandwidth vector $MBV_i(j)$, indicating the maximum available bandwidth from i to j
hop vector $MHV_i(j)$, indicating the hop number of the maximum bandwidth path.

The update procedure is very similar to the one described for the shortest path computation. Periodically, node i receives vectors MBV and MHV from all the neighbors, and updates its vectors as follows:

$$MBV_i(j) \leftarrow \max_{t \in \{k,\ell \ldots p\}} [\min\{MBV_t(j), B_{i,t}\}]$$

Let t_j^* be the neighbor of i that maximizes the bandwidth $MBV_i(j)$ from node i to destination j.

Then: $MHV_i(j) \leftarrow MHV_{t_j^*}(j)+1$

If two neighbors tie in the value of maximum bandwidth provided, the neighbor with smaller hop number MHV(j) is chosen. If both maximum bandwidth and hop numbers are tied, any tiebreaker can be used.

Finally, the routing vector is updated as follows:

$$MRV_i(j) \leftarrow t_j^*$$

The initialization procedure and the convergence behavior of the maximum bandwidth algorithm are identical to those of the shortest path algorithm.

3.3 Overhead

The line and processing overhead introduced by the update procedure is proportional to the number of nodes N and the update frequency f. In particular, assuming each table entry is 1 byte, line overhead (in characters/seconds) on each trunk is given by:

$$\text{line overhead} = \frac{4 \times N}{f}$$

As an example, for a 10 node network with update frequency = 1 update/sec, the trunk overhead is 40 characters/sec.

3.4 An Example

To illustrate the concepts of routing, mini hop and maximum bandwidth tables, Figure 1 shows the final vector values generated by the algorithm for a 4 node sample network.

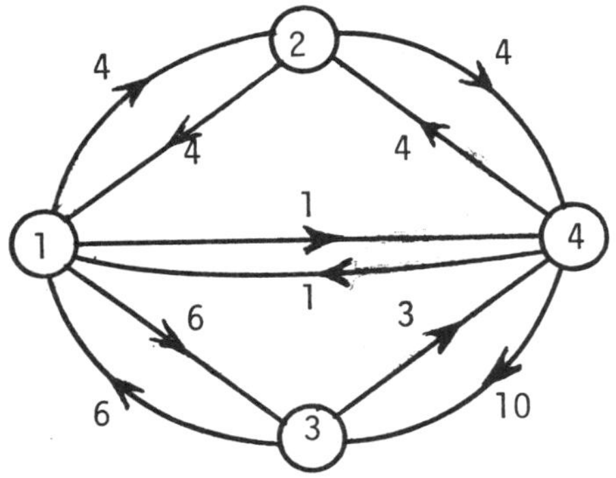

Labels indicate available bandwidth on trunks.

$BV_1(4)=1; RV_1(4)=4; HV_1(4)=1$

$MBV_1(4)=4; MRV_1(4)=2; MHV_1(4)=2$

Fig. 1 - Final vector values for a 4 node sample network

4. CONCLUSION

In this paper we have introduced a novel scheme, Bandwidth Control, for the prevention of congestion in virtual circuit networks. This scheme differs from previous schemes in that it controls and, if necessary, blocks virtual call requests at the network entry points, rather than letting all traffic indiscriminately enter the network until congestion occurs. Routing and flow control are intimately combined since a call is accepted only after a feasible route has been found. The method is based on the existence of bandwidth tables at each node indicating the available bandwidth on the shortest path as well as the maximum bandwidth over all possible paths. Bandwidth tables are maintained and updated using a very efficient and reliable distributed algorithm.

Work is now under way to apply the Bandwidth Control Scheme also to integrated datagram and virtual circuit networks. It appears that minimum hop and maximum bandwidth routes can be efficiently used to create alternate routes for datagram traffic. At the same time, maximum bandwidth availability seems to be a promising criterion also for the input flow control of datagrams.

Other areas related to Bandwidth Control which are currently under investigation are: the automatic path recovery after link and node failures; the use of multiple virtual circuits for load leveling and for bandwidth expansion; and the recovery from deadlocks.

REFERENCES

Rinde, J., "Tymnet I: An Alternative to Packet Technology", Proc. ICCC 76, Toronto, Canada, August 1976.

Roberts, L.G., "Telenet: Principles and Practice", Proc. Eur. Computing Conference on Comms Networks, London, England, 1975.

Poncet, F. and P. Allaire, "Message Exchange Protocol for Transpac and Euronet Network Management", Computer Network Protocol Symposium, Liege, Belgium, February 1978.

Folts, C. "International Standards in Computer Communications", NTC Proceedings, November 1979.

Davies, D.W., "The Control of Congestion in Packet Switching Networks", IEEE Transactions on Communications, COM-23, January 1972.

Lam, S. and M. Reiser, "Congestion Control of Store-and-Forward Networks by Input Buffer Limits", NTC 77 Conference Record, December 1977.

McQuillan, J. et al, "Improvement in the Design and Performance of the ARPA Network", AFIPS Conf. Proc. FJCC, 1972.

TOWARDS A UNIFIED DEFINITION OF TRAFFIC CONTROL IN COMMUNICATIONS NETWORKS

A. CHATTERJEE
BELL CANADA
160 ELGIN STREET, 11TH FLOOR
OTTAWA, ONTARIO, CANADA
K1G 3J4

ABSTRACT

This paper presents a unified concept of flow and congestion control applicable to both circuit switched and packet switched networks. The congestion control mechanism and techniques to realize the above are examined in detail. In particular, the relative influence of 'link' and 'external customers' on the source - destination link-congestion in a packet switched network having End-to-End congestion control are mathematically evaluated. It is deduced that if the network is subdivided by the set of such intersecting links which could be controlled, the network would be free of congestion.

1. INTRODUCTION

Any system where finite resources are shared by a number of contending, independent, and unrelated users requires the adoption of a control scheme for its proper functioning. Controls may also be needed to optimize performance under varying load conditions (1), (2), (3), (10). In our view, the object of traffic control in a telecommunications network (packet or circuit) is to optimize throughput and avoid excessive delay. This goal is achieved by the application of flow and congestion control.

This paper will discuss Flow and Congestion Control, their respective functions and means of realization. The general problems of overall network design incorporating features of error control, routing or end-to-end acknowledgement will not be examined except in so far as they relate to Flow and Congestion control.

D. G. Lainiotis and N. S. Tzannes (eds.), Advances in Communications, 395–408.

1.1 Flow Control

In general terms, flow can be defined as a physical displacement of goods, commodities, or information of significance or value. A flow network is composed of nodes, which act as sources, sinks or tandem points, and transmission facilities.

In any flow network, a scheme to allocate resources amongst users is required. Further, if there is a finite possibility that the total instantaneous resource demands exceed those available, a scheme to resolve contention must also be used. A flow control scheme is an algorithm adopted by the network to settle how resources will be allocated between (independent and unrelated) users and to settle how contention will be resolved between the bidders when resources are exceeded by demand.

Thus by the definition of Flow Control we mean:

1) Allocation of Resources by either,

   (i) First Come First Served basis or (ii) Last Come First Served basis or (iii) Establishment of Priority Classes basis.

2) Resolution of Contention by either,

   (i) Blocked and Lost basis or (ii) Blocked and Delayed basis.

1.2 Congestion Control

Congestion is a well experienced, if imprecisely defined phenomenon. For the purpose of a communication network, it is convenient to define congestion as a situtation where the network throughput starts decreasing in response to increasing traffic (14), (15). The reduction in throughput can be readily observed and can be computed for simple networks (2). By definition a network consisting of a single server never congests, but only degrades in performance. The purpose of adopting congestion control is to prevent or minimize loss of throughput as the load on the network increases.

There are two basic mechanisms in any congestion control:
a) Detection of Congestion and
b) Activation of a suitable Control procedure.

A simple case illustrating the occurrence of congestion is shown in Fig. 1. Identification of performance corresponding to point A on the curve would coincide with the occurrence of congestion (14), (15). In actual practice, measurement of congestion in a real network may not be an easy task. Measurement of round trip delays on 'pilot' packets or the measurement of buffer fill at a node may provide identification of congestion to an acceptable degree of accuracy.

The effectiveness of a particular congestion control mechanism could best be measured by a comparison of throughput obtained without any control (curve a) and throughput obtained when control is being exercised (curve b). A single measurement would generally be sufficient. Such a measurement could be defined as the percentage reduction in throughput corresponding to a given percentage of traffic increase compared to the maximum throughput without congestion control. More efficient congestion control techniques may improve the throughput as shown in curves c and d of Figure 1.

2. FLOW AND CONGESTION CONTROL IN THE TELEPHONE NETWORK

The congestion control schemes adopted in the telephone network are based on properties of the hierarchieal switching system employed for call routing, and have been formulated on the basis of network simulation and actual experience. In the call routing hierarchy used in North America calls between one end office and another are completed through as low a hierarchical chain as possible.

The following three methods of congestion control are adopted in the telephone network (1) - (4):

1. Reduction of Alternate Routing
2. Directionalization of Traffic Flow
3. Prevention of Feedback - Amplification of sender delays and reduction of intra-office switching delays.

The cancellation or reduction of alternate routing may appear at first to reduce the call carrying capacity of the network. This is indeed true under normal conditions, but under overload conditions as the offered load increases, the proportion of messages that are alternately routed also increases. This results in more trunk usage per message so that the number of messages that can be carried is less than it would be without alternate routing.

Directionalization of traffic flow gives preference to traffic moving from an office of higher rank to one of lower rank at the expense of traffic moving from lower rank to higher rank. In other words, use of this scheme results in a nearly completed call being given further preference to enhance the probability of its completion. This results in a higher overall completion rate. The net effect, with regard to the whole network, is to maximize the completion rate under varying offered load, and make the network behave like the single server (which never congests but only degrades in performance).

Senders are devices which receive, store and forward the address digits of an attempted call. Sender congestion tends to propagate from one office to another under overload conditions and amplify in the process. A reduction of 'time-out' interval during a period of sender congestion results in reduction of congestion feed-back and improves the overall performance.

Several conclusions of a general nature could be drawn from the above. The first and foremost of these is that several practices which tend to increase the call carrying capacity of a network during normal operating conditions tend to congest the network during overload conditions. Chief factors amongst these are the procedures which 'delay' the progress of a call with a view to enhancing its completion probability. While perfectly valid under normal conditions, such delays actually lower the total number of completed calls during an overload condition, thus degrading the network performance. The emphasis during normal conditions is to introduce delay, if necessary, by providing longer time outs etc., while the emphasis during overload conditions is to reduce delays as much as possible in call completion thereby giving a preferential treatment to all calls which can be completed quickly.

The fundamental mechanisms which cause congestion are similar in both the telephone network and packet switched computer network. There are two important differences, however. The telephone network is a circuit switched network traffic engineered by and large on a blocking and loss basis whereas a packet-switched computer network is traffic engineered on a blocking and delay basis. Further, in a circuit switched network all the effects of delay or congestion are experienced only during call set-up.

On the other hand, a computer network, which is generally a store and forward network, will provide a degradation in terms of reduced throughput or higher delay to calls which are already in progress (critical call based networks). For this reason, congestion in a packet switched network can be more serious and a greater design effort is required to reduce its effects under overload conditions. In a circuit switched network, a call in progress (during the establishment phase) can be abandoned at any stage of its progress, messages in progress should not be purged in a computer network under overload conditions. A message should be throttled in such a case right at source. In a packet switched network it is apparent that congestion control should be activated before the onset of congestion.

3. CONGESTION CONTROL IN A PACKET-SWITCHED COMPUTER COMMUNICATION NETWORK

There are three ways of controlling congestion in a packet-switched network as reported in the literature. These are: 1) End-to-End Control, (6); 2) Local Control, (6); and 3) Isarithmic Control, (8)

Similar techniques can be applied to circuit switching in the call establishment phase while trying to establish a physical circuit through pilot messages (which could be treated as packets). None of the above three techniques can prevent congestion completely on its own.

In the following analysis, we concentrate on the End-to-End congestion control (6), (7) and we show later how local control in conjunction with End-to-End control and fixed routing (13) may prevent congestion in a link and hence in the network.

We define links as a logical connection between a source-destination pair. In the link as defined above, there may be service demands by many packets (customers). As we cannot satisfy all customers because of limited resources, the first phase of congestion control is to choose N, the maximum number of customers allowable in that link. N is dependent on the buffer-space available in each node of the link and the network transit delay, and is generally called the Window Size. As the nodes of that particular link and the buffer-space in each node are shared by other source-destination link customers; the proper management of local buffers is very important in avoiding congestion. For our analysis, we will partition the network link by link.

Figure 3(a) shows a link of M stages with fixed routing, where λo is the average link arrival rate at stage 1, λi is the average external arrival rate at stage i which belongs to other source-destination pairs and μi is the average service rate at each stage for both the link and the external customers. This link can be modelled under End-to-End congestion control as shown in figure 3(b) to evaluate the sum of the expected queue lengths of the link and the local customers (6). For a uniform system where all λi's and μi's are equal and given by λ and μ respectively, we define the following parameters as in (6).

E_ι = Expected queue length of link customers at each stage.
E_ϵ = Expected queue length of external customers at each stage.

It is known as in (6) that:

$$E_\epsilon = \frac{\lambda}{\mu-\lambda}(1 + E_\iota)$$

We further define,

$$Z_M^N = \sum_{\substack{\text{all partitions} \\ \text{of } n \leqslant N \text{ in } M}} \prod_{i=1}^{M} \left(\frac{\lambda}{\mu-\lambda}\right)^{n_i} \qquad \text{.........................(3.1)}$$

Let T = Sum of the expected queue lengths of the link and external customers at each stage

$$= E_\iota + E_\epsilon$$

T can be shown to be given by (6) as:

$$T = \frac{X}{Z_M^N} \frac{dZ_M^N}{dx} + \frac{\lambda}{\mu - \lambda} \left(1 + \frac{X}{Z_M^N} \frac{dZ_M^N}{dx}\right) \qquad (3.2)$$

$$= E_\ell + \frac{\lambda}{\mu - \lambda} (1 + E_\ell) \qquad (3.3)$$

where, $X = \frac{\lambda o}{\mu - \lambda}$ (3.4)

Now we would like to see the influence of λo and λ on T.

Differentiating (3.3) with respect to λo, we get:

$$\frac{dT}{d\lambda o} = \frac{\mu}{(\mu - \lambda)^2} \cdot \frac{dE_\ell}{dx} \qquad (3.5)$$

Further, $\frac{dT}{d\lambda} = \frac{\mu \lambda o}{(\mu - \lambda)^2 (\mu - \lambda)} \cdot \frac{dE_\ell}{dx} + \frac{\mu (E_\ell + 1)}{(\mu - \lambda)^2}$ (3.6)

E_ℓ , the expected queue length of link customers saturates as X increases, because, $\frac{dE_\ell}{dx} = 0$ for large X.

At a high value of X, $dT/d\lambda >> dT/d\lambda o$ which is an indication of higher buffer requirement with respect to external customers compared to the link customers assuming there is common buffer space for both. At this phase of link operation Local Control of external customers at each node of the link is necessary to achieve equitable usage of buffer-space. The second term in equation (3.6) indicates how this buffer requirement varies with respect to X.

As $dT/d\lambda$ is always increasing with X contrary to $dT/d\lambda o$ which is monotomically decreasing to zero, we can define the limiting value of X for which both $dT/d\lambda$ and $dT/d\lambda o$, are equal. Let

$$\frac{\mu}{(\mu - \lambda)^2} = K$$

then, from (3.3) and (3.6) we get:

$$K \frac{dE_\ell}{dx} \left(\frac{\lambda o}{\mu - \lambda} - 1\right) + K (1 + E_\ell') = 0 \qquad (3.7)$$

where, $E_\ell = \sum_{n=1}^{N} \frac{X^n (n + M - 1)!}{(n - 1)! M!}$ (3.8)

Solution of the equation (3.7) will give equal buffer requirement for the limiting value of X. Using equation (3.4), equation (3.7) can be rewritten as,

$$\frac{dE_\ell}{dx}(x - 1) + 1 + E_\ell = 0 \qquad \ldots\ldots\ldots\ldots\ldots\ldots(3.9)$$

Equation (3.9) can be readily solved to yield,

$$(1 + E_\ell)(1 - X) = 1 \qquad \ldots\ldots\ldots\ldots\ldots\ldots\ldots(3.10)$$

Equation (3.10) can be solved for X and a numerical value of λ (called λ^*) obtained, provided λo is assumed. λ^* will determine the limit up to which local traffic can rise without producing a relatively higher common buffer requirement) than that of the link traffic. Equation (3.10) can also be interpreted as presenting the relative influence of λo and λ on the average buffer requirement. A similar analysis will also hold for a non-uniform system (6) with random routing and so could be applied to the telephone network during call establishment phase.

4. UNIFIED DEFINITIONS

As a unified approach we have defined flow control as a means of allocating resources and resolving contention and congestion control as a means to maintain steady throughput for both the circuit switching and the packet switching. So far in the literature the definition of flow and congestion control has been used intermixed. We feel that isolation of the two definitions will help attack the problem for better understanding and analysis. It is evident from all the discussions that implementation of some flow control techniques could ease the implementation of congestion control and vice versa because there will always be some cross-relation between the two.

5. A SPECIFIC EXAMPLE

In this section we give a numerical example to illustrate the influence of X on congestion. For this, let us take N = 3, M = 3. Where N = window size and M = number of stages in the link. With these values equation (3.10) becomes:

$$20X^3 - 6X^2 - 3X - 1 = 0$$

There are three solutions for X, the acceptable one being X = 0.6499. Table 1 below shows the relationship of E_ℓ , E_ϵ and T for the same X, but different combination of λo and λ.

Table 1 Influence of and on T (Constant X)

X	μ	λ_o	λ	E_l	E_ϵ	T
0.6499	2.0	0.3	1.53	1.853	9.28	11.14
		0.5	1.23		4.56	6.4
		0.7	.923		2.44	4.29
		0.9	.615		1.26	3.12
		1.105	0.3		0.503	2.356
		0.975	0.5		0.951	2.804
		0.845	0.7		1.536	3.389
		0.715	0.9		2.334	4.187

In Table 1. λo and λ have been varied between 0.3 and 0.9 in the upper and lower halves respectively while keeping X constant. It can be seen that for the same ratio of variations of $\lambda / \lambda o$, higher values of λ necessitate much higher requirements for buffers. The example indicates that variations in λ are more demanding on the network resources and thus the external customers requires a greater control compared to the link customers at the time of congestion. From this, we can conclude that End-to-End control by itself is not sufficient and so end-to-end control in combination with local control is more effective for the control of congestion problem.

To qualify our statements further we plot the following curves in figures 4 and 5:

1) Throughput and delay vs λo or λ
2) Expected queue length vs λo or λ

Our aim is to maximize throughput through the link under consideration. From curve (a) of figure 5, we see that the requirement of total expected buffer size grows at a much faster rate with than with λo. Examining the combined expected delays (including both link and external customers), we see that delay also increases at a faster rate with external customers as compared to the link customers.

So at the start of congestion illustrated by a rectangular block of figures 4 and 5 we should start accepting link customers only and blocking external customers as far as possible because of rapid consumption of buffer-space by external customers. This can be done first by accepting customers from the node which is the part of the link and thus serving those belonging to the link we are concerned with. If this fails to decrease the buffer occupancy we should start blocking link customers i.e., modify window size (the maximum number of outstanding customer N).

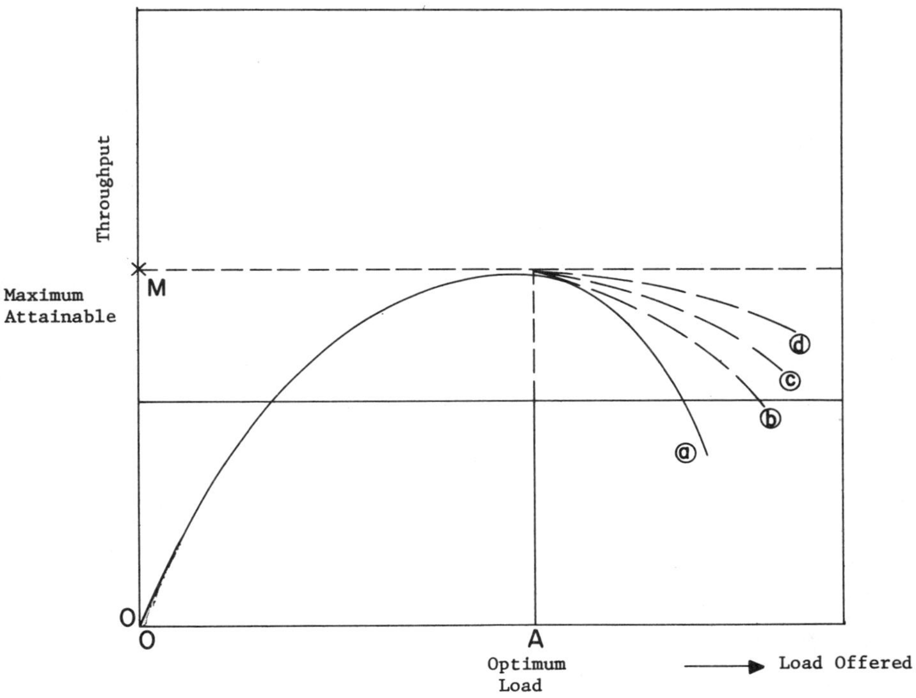

Figure 1: Throughput and offered load relation in a telecommunication network.

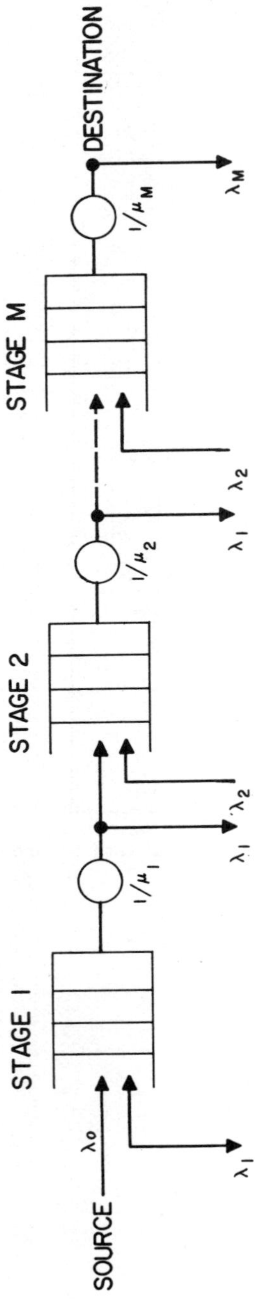

FIGURE 3 (a) A QUEUEING MODEL OF A LOGICAL LINK

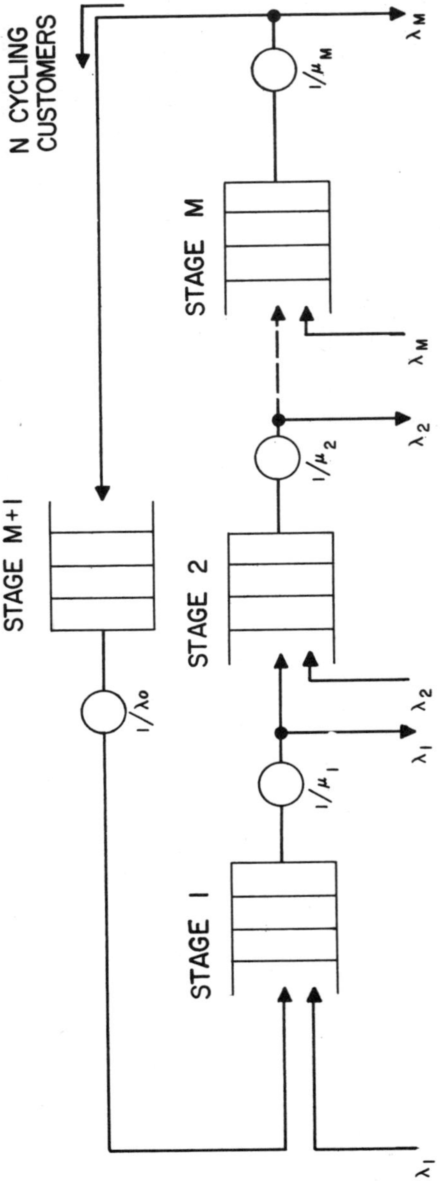

FIGURE 3 (b) AN EQUIVALENT CLOSED SYSTEM FOR END TO END CONTROL

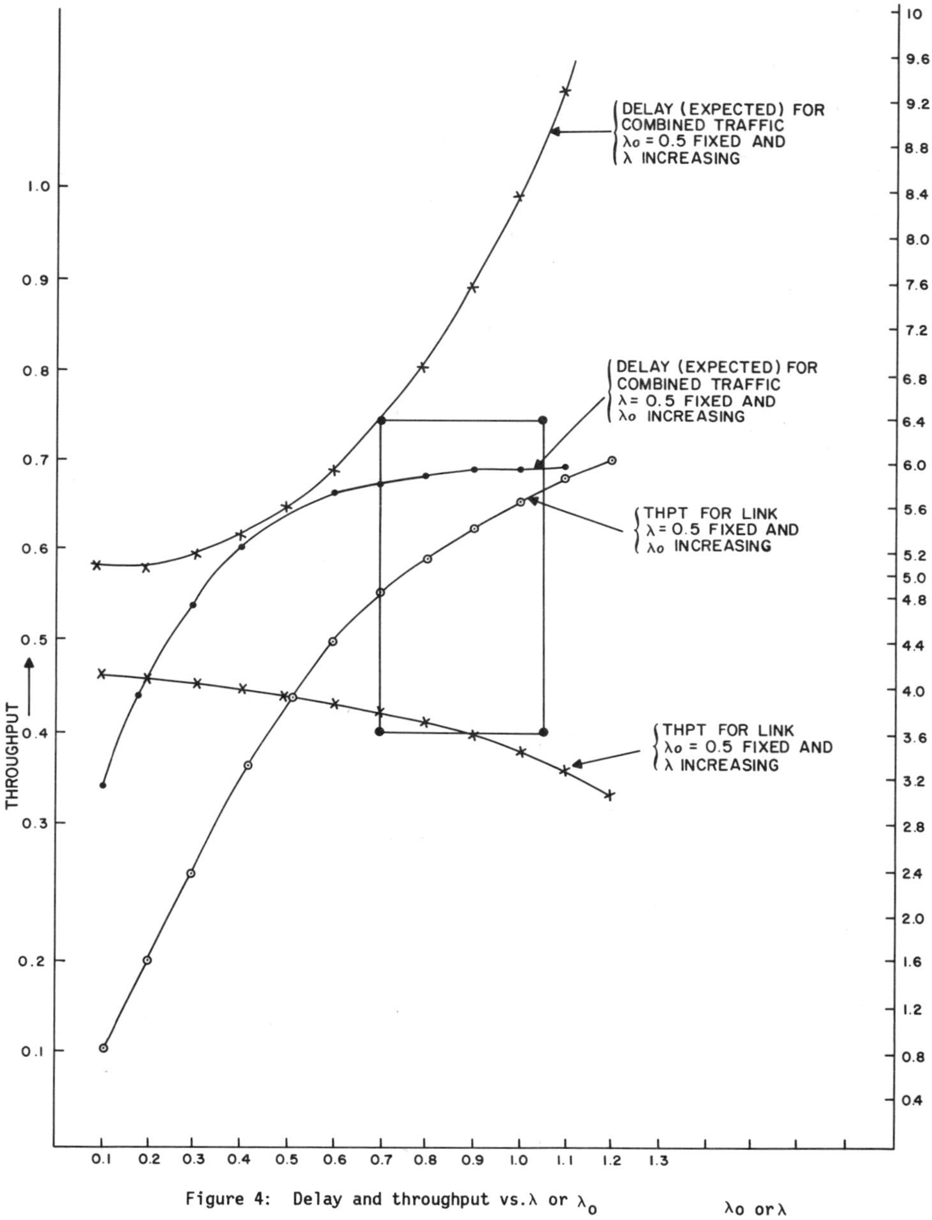

Figure 4: Delay and throughput vs. λ or λ_0

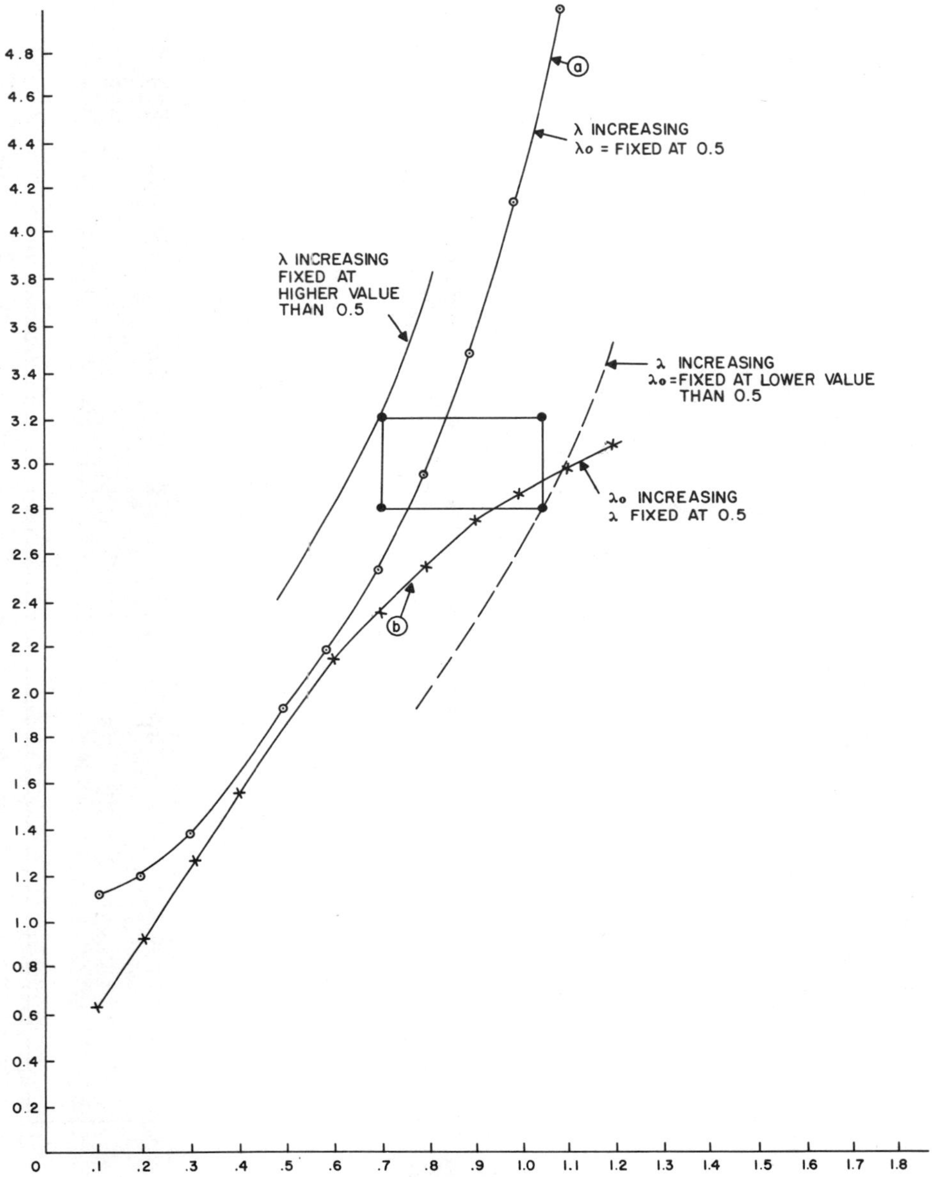

Figure 5: Queue Length vs. λ or λ_0 for $\mu = 2$

Summarizing the various discussions we see that five techniques can be employed for controlling congestion in the link after it has been observed in any node:

(1) Prioritising to terminating traffic
(2) Blocking external customer to that link
(3) Reducing the window size
(4) Any combination of the above
(5) Purging packets

The last technique can be used if congestion has not been lessened and all the other four techniques (1-4) prove ineffective. A criterion to purge certain types of packets can also be established depending on the durations of such packets in the network.

6. CONCLUSION

This paper has attempted to present a unified concept of flow and congestion controls for the management of traffic in any communication network. A distinction between Flow and Congestion controls has been made and examples of such controls are cited for the telephone network and a packet-switched network.

It has been experienced from the telephone network that multiple routing technique results in throughput enhancements at the low and medium load levels, but at high load levels fixed routing is desirable. We have shown that such an approach will have comparable effects on both circuit and packet-switched networks.

As a means of controlling congestion in a packet-switched network, it has been substantiated that if external and link customers can be differentiated, control of external customers for a particular link would yield better performance than control of link customers. An End-to-End Congestion Control technique together with a Local Control technique has been proposed as a solution to the congestion problem of packet-switched networks.

REFERENCES

1. Burke, P.J. "Automatic Overload Controls in a Circuit-Switched Communication Network", Proceedings of the National Electronic Conference, Vol. XXIV, 1968.

2. Rudin, H., "Chairman's Resources, Session 16", ICCC'76, Toronto

3. Laude, J.L., "Local Network Management Overload Administration of Metropolitan Switched Networks", Proceedings of the National Electronics Conference, Vol. XXIV, 1968.

4. Franks, R.L. and Rishel, R.W., "Optimum Network Call- Carrying Capacity", The Bell System Technical Journal, Vol. 52, No. 7, 1973.

5. Weber J.H., "A Simulation Study of Routing and Control in Communications Networks", The Bell System Technical Journal Vol. XLIII, No. 6, 1964.

6. Agnew, C.E., "Dynamic Modelling and Control of Congestion - Prone Systems", Operation Research, Vol. 24, No. 3, 1976.

7. Penotti, M.C. and Schwartz, M., "Congestion Control in Store and Forward Tandem Links, IEEE Transactions on Communications, Vol. Com-23, No. 12, Dec., 1976 and also Penotti, M.C., "The Control of Congestion in Message- Switched Networks", PH.D. Dissertation, Ploytechnic Inst., New York, Brooklyn, N.Y., 1974.

8. Chatterjee, A., Georganes, N.D. and Verma, P.K., "Analysis of a Packet-Switched Network with End-to-End Congestion Control and Random Routing", ICCC'76, Toronto.

9. Davies, D.W., "The Control of Congestion in Packet-Switching Networks", IEEE Transactions on Communications, COM-20, No. 3, January 1972.

10. Talley, D. "Basic Telephone Switching Systems", Hayden Book Company, Inc. New York, 1969.

11. Kahn, R.E. and Crowther, W.R., "Flow Control in a Resource Sharing Computer Network", ACM-IEEE, Ind Symp. on problems in the optimization of data comm. systems (Oct 1971), Palo Alo.

12. Standard Network Access Protocol - The Computer Communications Group, Trans-Canada Telephone System, Ottawa, March 31, 1976.

13. Naylor, W.E. and Kleinroek, L., "On the Effect of Periodic Routing Updates in Packet-Switched Networks", NTC, Dallas, Texas, Nov, 1976.

14. S.S. Lam and M. Reiser, "Congestion Control of Store and forward Networks by Input Buffer Limits - An Analysis, IEEE transactions on Communications, January, 1979, vol. Com-27, No. 1.

15. C. Lemieux, 'Flow Control in Computer Networks,' Participants Edition, North Holland, February 1979.

AN INFORMATION THEORETIC APPROXIMATION FOR MODELLING COMPUTER NETWORK NODAL QUEUING DELAY: A CONGESTION MEASURE

C. A. Niznik†

SUNY at Buffalo
Department of Electrical Engineering
Amherst, New York 14260

A method for modelling the congestion measure of computer node delay due to the queuing representative of general (GI/G/1) and specific queuing systems is developed. The basic technique is the formulation of a computer node queuing mean waiting time from a discrete marginal customer per time slot solution of a continuous marginal time per customer pdf. The information theory idea of quantization is utilized to numerically determine the areas of sections (quantiles) of the pdf of the marginal customers per time slot. These quantiles represent the entries of the probability transition matrix of the change of the number of customers in the queue for the irreducible Markov Chains of uniform quantization upper and lower bound steady state solutions for the number of customers in the queue at time slot j. The marginal pdf area considered is normalized to one resulting in the computation of normalized quantile areas for the probability transition matrix required for the steady state upper and lower bound buffer occupancy solutions. The mean waiting time is then the summation of the L+1(L=length of the computer node queue) steady state buffer occupancy solutions multiplied by the corresponding number of customers in the computer node queue.

The numerical procedure of increasing the number of quantiles or samples within a specific area of the service minus inter-arrival pdf shows the closest comparison with the classical solution for quantile width ranges characteristic of traffic intensity greater than one and less than one. This finding was observed in numerical classical solution comparison experiments

†This research was supported in part by the National Science Foundation under Grant No. ENG 76-20295.

D. G. Lainiotis and N. S. Tzannes (eds.), Advances in Communications, 409–416.

for M/M/1 and E_2/M/1 queuing systems. Currently, these experiments are being formulated for the less general queuing systems of G/M/1, M/G/1 and GI/G/1. The solution is particularly relevant for computing the mean waiting time when only a few points of the service minus interarrival probability density function are available for the computer node.

I. INTRODUCTION

The present topic of major theoretical interest in the area of computer communication networking is the measure and control of congestion. The interpretation of the measure of congestion adhered to is the increase in waiting time to the customer as a result of hardware and software originating delays in the various component sections of a computer network. The steady state probabilities $g_j(m)$ describing the number (m) of customers in a finite size buffer during time interval j are therefore of interest in computer network modelling for the purpose of congestion control. The explicit closed form steady state solution of $g_j(m)$ is required for computing the mean waiting time due to the change of the number of messages in the buffer of the computer node. A quantization approximation (QA) method for an exact upper and lower bound solution for the steady state mean waiting time of a GI/G/1 queue with finite capacity is derived. The (QA) solution realizes two finite state Markov Chains which are the upper and lower bounds of this mean waiting time. The research by Wyner (6), derives an asymptotic expression for the buffer overflow probability of a finite size buffer. In the results derived here, Wyner's initial formulation for the buffer probability, the properties of Markov Processes,and a quantization approach determine the closed form solution for the steady state probability distribution of buffer state for the finite buffer solution.

II. BASIC MODELS

Two models, a buffer model by Wyner (6) and a queueing model by Calo and Schwartz (1) are imbedded within the Quantization solution to realize the mean waiting time bound solution. Two Markov Chains represent the lower and upper bound steady state buffer occupancy distribution solutions. The basic model is the irreducible Markov Chain equations of the buffer loss model (4) reflecting the marginal overflow Z_j of customers per time slot j (for a finite length buffer) which a channel cannot accept from a customer source (computer node).

Buffer Model

The formulation of the basic buffer model follows. Let X_j be a sequence of independent, identically distributed, integer random variables where $Pr\{X =k\}=P(k)$, $k=0,1,\ldots,S$. The random variables, $Z_j = X_j$ (arriving customers to be transmitted on the channel), $- Y_j$ (customers channel will accept), take the values $(-\alpha_1, \ldots. 0, \ldots, \alpha_2)$ with $\alpha_1 = C$, and $\alpha_2 = S-C > 0$, and indicate the increase or decrease in buffer storage at the jth time slot, measured in data units. The following definitions are also stated:

$$E\{Z_j\} = \sum_{n=-\alpha_1}^{\alpha_2} nP(n) < 0 \text{ for stability } (\rho < 1), \text{ and}$$

$P(-\alpha_1)$, $P(\alpha_2) > 0$. Then, W_j, $j \geq 0$, is the state (or queue length) of the buffer at the jth time slot, $L =$ buffer size, $W_o = 0$.

$$W_j = \begin{cases} W_{j-1} + Z_j, & 0 \leq W_{j-1} + Z_j \leq L \\ 0 & , \quad W_{j-1} + Z_j \leq 0 \\ L & , \quad W_{j-1} + Z_j \leq L \end{cases}$$

$g_j(n) = Pr[W_n = n] =$ steady state occupancy distribution at time slot j for a message entering the queue. $P(k) =$ probability of a change of k messages in the buffer.

$$W_L = \frac{1}{\mu} \sum_{m=0}^{L} g(m)m = \text{average probabilistically weighted waiting time for a message in a buffer of length L, the mean wait (sec/time slot j) (11)} \tag{1}$$

$$\frac{1}{\mu}[g(m) \cdot m] = \text{probabilistically weighted waiting time for a message at position m in the buffer.} \tag{2}$$

Realizing that one of the L+1 equations in the Wyner solution is a linear combination of the others, a nonsingular set of the L+1 equations is formed by substituting

$$\sum_{m=0}^{L} g(m) = 1$$

for the $g(0)$ equation. The accuracy of $g(0)$ has a negligible effect on W_L since $g(0) \cdot 0 = 0$ in the summation in equation (1). It is observed that this substitution will somewhat affect the numerical accuracy of all $g(i)$, $i=0,\ldots,L$ solutions because the equations are coupled. The new set of equations, substituted into the $G = GP$ formulation is:

$$\sum_{m=0}^{L} g(m) = 1 \tag{3a}$$

$$g(n) - \sum_{m=-\alpha_1}^{\alpha_2} g(n-m)P(m) = 0 \tag{3b}$$

$$g(L) - \sum_{m=0}^{\alpha_2} g(L-m) \sum_{k=m}^{\alpha_2} P(k) = 0 . \tag{3c}$$

In equations (3a-c) it is implied that $P(n)=0$ for $n \ni (-\alpha_1,\alpha_2)$. In matrix form, these equations become $[I-P]G = E$ or $G=GP$. The following vector partitioning is then required for the matrix solution:

$$\begin{bmatrix} 1 & V_1 & 1 \\ \hline M_2 & R & M_1 \\ \hline 0 & V_2 & b \end{bmatrix} \begin{bmatrix} g(0) \\ \hline g(m) \\ \hline g(L) \end{bmatrix} = \begin{bmatrix} 1 \\ \hline 0 \\ \vdots \\ 0 \\ \hline 0 \end{bmatrix} \tag{4}$$

where $m=1,\ldots,L-1$, and R is a banded Toeplitz matrix. If, $A = 1-V_1R^{-1}M_2$, $B = -V_2R^{-1}M_2$, $C = 1-V_1R^{-1}M_1$, $D = b-V_2R^{-1}M_1$. Then,

$$g(0) = D/[AD-BC] \tag{5a}$$

$$g(L) = B/[CB-DA] \tag{5b}$$

$$g(m) = -R^{-1}M_1g(L)-R^{-1}M_2g(0). \tag{5c}$$

The Average Fractional Overflow for the Wyner solution with the inclusion of the service rate from Wald's equation (5) is:

$$Q_1(L) = Q(L)/E[X_j] \tag{6}$$

where $Q(L) = \frac{1}{\mu} \sum_{m=0}^{\alpha_2} g(L-m) \sum_{k=0}^{\alpha_2-m} kP(m+k)$, and $E[X_j] = 1/\lambda$.

Queueing Model

To obtain the numerical values for the $P_{j,j+k}(k)$, the probability of a change of k customers in the buffer (j are the number of customers presently in the buffer), a queuing model is appended to the buffer formulation. The assumption of a FCFS queuing discipline ensures that W_j equals the waiting time for the $(n+1)^{th}$ customer in a comparable queueing system. The queuing model of the marginal overflow rate per customer n is then stated as: $u_n = x_n - t_{n+1}$, $E[u_n] < 0$. Where, x_n = service time for customer n, t_{n+1} = interarrival time between customer n and n+1, and W_n = waiting time in queue for customer n. The pdf of u_n for this queuing model is quantized to approximate its amplitude continuous characteristic by two finite state Markov Chains with steady state probability vectors given by $\bar{G} = \bar{G}\bar{P}$ and $\underline{G} = \underline{G}\underline{P}$ representing upper and lower bound buffer steady state probability distributions. Refer to Figure 1 for the graphical representation of the transition probability space for the W_n solutions. Note that the queuing model marginal overflow in time per

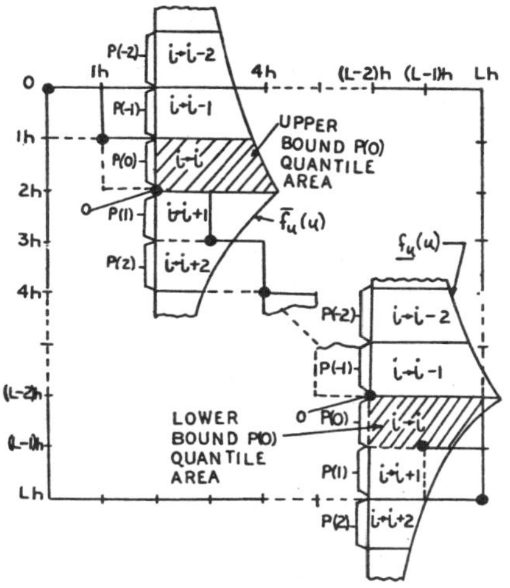

Figure 1. P(0) Area Computation ($\alpha_1=\alpha_2=2$).

customer n is analogized to the buffer model marginal overflow in customers per time slot (T.S.) j.

III. QUANTIZATION APPROXIMATION MODEL

The quantization model approximation is derived realizing the previous buffer model steady state solution for GI/G/1 queue buffer occupancy and the queuing model. Quantization is defined for this application as the procedure of taking a finite set of intervals comprising a finite area of the probability density function of the service time minus interarrival time to replace the entire probability density function area. Each quantile represents the probability of a specific number of changes to the buffer occupancy ($-\alpha_1$ to α_2 changes). A normalization is performed on the quantile areas to include the remaining pdf area in the quantile range of interest for the quantile areas, with the following expression, since

$$\sum_{j=-\alpha_1}^{\alpha_2} P(j) = 1 \text{ is true, } P(i)\Big|_{\text{normalized}} = P(i) \Big/ \sum_{j=-\alpha_1}^{\alpha_2} P(j) \qquad (7)$$

This quantization of the queuing model therefore approximates an amplitude continuous system by a finite state Markov Chain representing the buffer states.

The areas of the quantiles of $\bar{f}_u(u)$ and $\underline{f}_u(u)$ correlated to the entries of the probability transition matrix P are graphically illustrated by placing the pdf's in Figure 1 on the space of the probability transition matrix. To compute the probability P(i) for a position i in the buffer, for both lower and upper bound steady state solutions, the zero amplitudes of $\bar{f}_u(u)$ and $\underline{f}_u(u)$ are placed at the edge of staircase step i, i.e. at the respective right or left of the $i \to i$ quantile. This results in the right side of the quantile area for $\underline{P}(0)$ occurring on the dotted lower bound staircase function and the right side of the quantile area for $\bar{P}(0)$ occurring on the solid upper bound staircase function. Here h is defined as the width of the quantile, i.e. a percentage of (1) the standard deviation σ or (2) the length along the u axis from u=0 to the mean (XM) for various values of the ρ and μ. The width of the entire quantization area considered is $W=(\alpha_1+\alpha_2+1)h$. Each step represents the state of the system (the number of messages in the buffer). The optimal quantile width h is determined by the smallest allowable absolute error comparison with the classical solution for the queuing system considered. The quantile area expression within the area of the quantization interval $((-\alpha_1 h$ to $\alpha_2 h)$ about zero), allows computation of $P_{ij}(i)$ for an optimal comparison solution with the classical mean waiting time and steady state solution.

$$P_{ij}(i) = \int_{ih}^{ih+h} f_u(u)du \tag{8}$$

The value of $\bar{P}(i)$ and $\underline{P}(i)$ at the probability transition matrix boundary is the summation of the values at and outside the boundary.

IV. EXPERIMENTAL RESULTS

The following observations are made in reference to the steady state distribution, mean waiting time and average fractional overflow QA solutions for the M/M/1 (refer to Fig. 2) and E_2.M/1 systems. The Finch classical solution (2) for finite buffer M/M/1 systems and the Lindley Integral Equation solution for infinite buffer E_2/M/1 systems are the comparison solutions.

Error Explanation. Boundary error is observed between the steady state distribution QA solution and the classical solution as a result of the following factors: (a) The substitution of Eq. (3a) for the Wyner coupled equation to achieve a nonsingular set of solution equations, results in boundary error at m=0. (b) Quantization error at the boundaries (m=0 and L) is due to the summation required for quantile area computation at these boundaries. (c) The E_2/M/1 QA is a finite buffer solution whereas the Lindley solution is an infinite buffer solution.

Mean Waiting Time QA Solution. Also, for both queuing system's QA solutions, the following characteristics of the mean waiting time (W_L) are observed: (a) decreases with increasing service rate (ρ varying for M/M/1 (refer to Fig. 2)), and ρ constant for $E_2/M/1$. (b) increases with increasing traffic intensity. (c) increases when the length of the buffer is increased with constant allowable change of messages in the buffer (α_1 and α_2). Referring to the M/M/1 mean waiting time solution (Fig. 2, $\alpha_1=\alpha_2=10$, L=21),

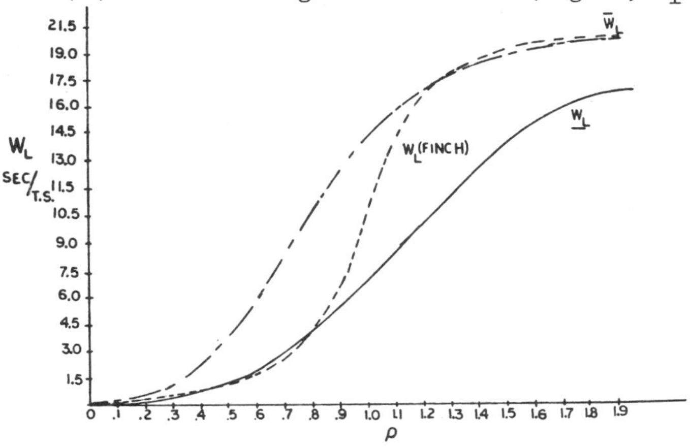

Figure 2: Mean Waiting Time vs ρ(M/M/1, L=21,$\alpha_1=\alpha_2=10$, h=$\sigma/4$)

there is an interval of closest comparison to the classical solution characteristic of each bound, i.e. $\bar{W}_L$ for $\rho>1.2$, $\underline{W}_L$ for $\rho<.8$.

Increasing QA Mean Waiting Time Solution Accuracy. Reducing the individual quantile width, thereby increasing the number of quantiles while holding constant the area of consideration, i.e. the area under $f_u(u)$ from -10σ to 10σ over a width of 21σ, results in an increased accuracy comparison with the classical solutions for mean waiting time. A decrease in the Mean Square Error (MSE) of the steady state solution with the Finch solution for the M/M/1 system is observed when the number of quantiles within a 21σ interval width along u is increased from 21 to 33 . For both systems the convergence characteristics of the lower and upper bound QA mean waiting time solutions are characterized in the following manner with respect to the classical solutions: below, for large quantiles, surround closely for optimal quantile range, above for very small quantiles.

Average Fractional Overflow (AFO) Solution. The AFO for both M/M/1 and $E_2/M/1$ decreases with increasing buffer length and decreases when the number of customers allowed to enter and leave the buffer (α_1,α_2) is decreased for the same width of quantile area.

V. NUMERICAL ANALYSIS THEORY AND RESULTS

QA Algorithm Complexity. The main factor in the complexity of the Quantization Mean Waiting Time Algorithm is the operational complexity of inverting the R matrix. Using the IBM MPS 360 linear programming sparse matrix inversion complexity of $O(n^3/2)$, the total operational complexity for this algorithm is $O((n^3/2) + 2(n-2)^3)$, for $n = L+1$.

QA Algorithm Robustness. The robustness of this algorithm is evaluated for the following factors: (a) The QA method has been shown here to be applicable to the less general M/M/1 and E_2/M/1 queuing systems with close comparison of the mean waiting time solution to the classical solution and optimal quantization interval. (b) It is easier to characterize the service minus interarrival distribution by the few available data points required for a specific decimal accuracy, than to compute the Lindley integral equation solution requiring the continuous pdf.

QA Algorithm Measurement Accuracy vs Measurment Complexity. The results observed by Hughes of a threshold measurement complexity (number of samples required for the optimal measurement accuracy of a specific number of finite data points) is consistently observed for the QA solution for various finite areas of $f_u(u)$. For an M/M/1 system ($L=61, \rho=.5, \alpha_1=\alpha_2=10, W=\sigma, 5\sigma, 10\sigma$) plots of Measurement Accuracy(MA) = (1-MSE) vs Measurement Complexity (MC) = the number of quantiles ($\alpha_1+\alpha_2+1$) in the quantization area of $f_u(u)$ considered, indicate the following results: (a) As the length of sampling interval W is increased, the increased MA occurs at a larger value of Peak (threshold) MC. (b) Also, finer quantization of a specific $f_u(u)$ area yields increased MA up to a threshold value of MC, beyond which the MA decreases.

REFERENCES

1. Calo, S., Schwartz, S.: 1975, Proceedings of the 13th Annual Allerton Conference on Circuit and System Theory, October.
2. Finch, P.D.: 1958, Journal of the Royal Statistical Society, Series B, 20, pp. 182-186.
3. Hughes, G.F.: 1968, IEEE Transactions on Information Theory, IT-14, No.1, pp. 55-63, January.
4. Niznik, C.A.: 1978, Ph.D. dissertation, Department of Electrical Engineering, SUNY at Buffalo, April.
5. Saaty, T.L.: 1959, Mathematical Models of Operations Research, McGraw-Hill Book Co., Inc.
6. Wyner, A.D.: 1974, SIAM Journal of Applied Mathematics, 27, 4, pp. 540-570.

IN-ORBIT MEASUREMENTS OF THE RF SUBSYSTEM OF COMMUNICATIONS SATELLITES

V. Riginos, C. Mahle, and I. Atohoun

COMSAT Laboratories
22300 COMSAT Drive,
Clarksburg, Maryland 20734, U.S.A.

ABSTRACT

This paper describes the methods and equipment developed by the Communications Satellite Corporation to perform measurements on the RF subsystem of communications satellites in orbit. As communications satellites have increased in capacity and complexity, in-orbit measurement techniques have evolved from simple evaluation type manual measurements to sophisticated computer driver measurements.

The equipment described was used to perform in-orbit testing of the INTELSAT and COMSTAR spacecraft and to measure flux density for saturation, spacecraft e.i.r.p., antenna patterns, beam isolation, gain transfer characteristics, gain-to-noise temperature ratio, receiver local oscillator frequencies, traveling wave tube turn-on transients, in-band and out-of-band frequency response, and cross polarization. In addition, a new system presently being developed for the INTELSAT V in-orbit tests will be briefly discussed.

INTRODUCTION

After the launch of a communications satellite, it is necessary to conduct in-orbit tests of the communications subsystem. A comparison of the data obtained from in-orbit tests with the pre-launch data determines whether all the communications subsystems of a spacecraft have successfully survived the launch and placement into geosynchronous orbit about 42,000 km away from the earth station. In addition, during in-orbit testing, highly accurate

D. G. Lainiotis and N. S. Tzannes (eds.), Advances in Communications, 417–430.

far-field antenna patterns of the satellite are obtained for the first time, since prelaunch antenna patterns are generally measured on nonideal ranges.

The technical objective of an in-orbit test has been to measure, as completely and accurately as possible, the performance of the satellite. In addition, there are two other very important considerations:

a. The test should be performed in the shortest possible time to make the satellite available for commercial operation as soon as possible.

b. The tests should be performed with the least expenditure of onboard fuel since the useful lifetime of the spacecraft depends on the amount of available fuel.

BACKGROUND

The basic elements forming the link from an earth station to a spacecraft under test and back to an earth station are outlined in Figure 1. A test signal is amplified by a 6-GHz high-power amplifier (HPA) to levels ranging from a few watts to several kilowatts and beamed through the earth station antenna toward the spacecraft under test. After encountering a "space loss" of the order of 200 dB, this signal is picked up by the spacecraft antenna, then it is is amplified, and translated by the spacecraft transponder. This transponder typically consists of a 6-GHz low-noise amplifier (LNA) and a down-converter to 4 GHz followed by driver and transmitter amplifiers providing an RF power level of several watts. The test signal is then beamed to the earth station by the spacecraft transmit antenna. After another space loss of about 196 dB, it reaches the earth station, where it is collected by the large-aperture antenna, amplified by the LNA, and processed in the receiver.

To achieve an acceptable measurement accuracy in an in-orbit test requires careful calibration of the microwave power meters at 6 and 4 GHz, the passive microwave components such as couplers in up-and down-link setups, and the earth station antenna gain in both transmit and receive bands. Measurement of the earth station gain is a vital part of in-orbit testing because sufficient errors can be a introduced. Use of an earth station antenna dedicated to in-orbit testing has greatly enhanced accuracy and repeatability of the measurements. During the early years of satellite testing, a variety of earth station antennas with apertures ranging from 13 m to 30 m were used; today's level of precision was achieved through the use of a dedicated antenna which yielded an excellent correlation between the data obtained from star tracking, boresight measurements, and near-field techniques after several years of effort.

For INTELSAT IV-A and COMSTAR, the latest generation of COMSAT-tested satellites, measurement methods were developed (1) to evaluate polarization and beam isolation performance, since both satellites provide frequency reuse via beam or polarization isolation. The simplified block diagram shown in Figure 2 of the RF section of INTELSAT IV-A spacecraft reveals a considerable system complexity. The transponder includes several redundant elements to achieve reliable operation. The receivers are connected via redundant switching networks to the antennas. In COMSTAR, diplexers enable the same antenna to be used for both receiving and transmitting at different frequencies and polarizations; separate receive and transmit antennas are used in INTELSAT IV-A. For each transponder all redundant paths must be evaluated during the in-orbit test. This complexity and the higher sensitivity necessary for beam isolation measurements in conjunction with stringent limitations on the allowable test time have led to development of the semiautomatic test system.

IN-ORBIT TEST SETUP

The semiautomated test setup allows rapid measurements at a number of discrete frequencies (the transponder band centers) without any major manual adjustments. The down-link carrier from the spacecraft is compared with an injected reference carrier. Both of these signals are synchronized as the measurement frequency is shifted from one transponder to another, presenting a fixed spectrum display for evaluation by the test team. An AM nulling technique (2) is used to accurately determine the spacecraft TWT amplifier (TWTA) saturation level. The setup can be operated in either a computer-controlled or completely manual mode.

The setup consists of three racks of equipment which are controlled by a computer. The system's basic functions are as follows:

a. To generate a phase-locked signal of known power level at the up-link frequency. This up-link signal, which is fed to the HPA at the earth station, is transmitted to the spacecraft.

b. To generate a similar signal at the down-link frequency which is injected in front of the station's LNA.

c. To display and measure the LNA's output.

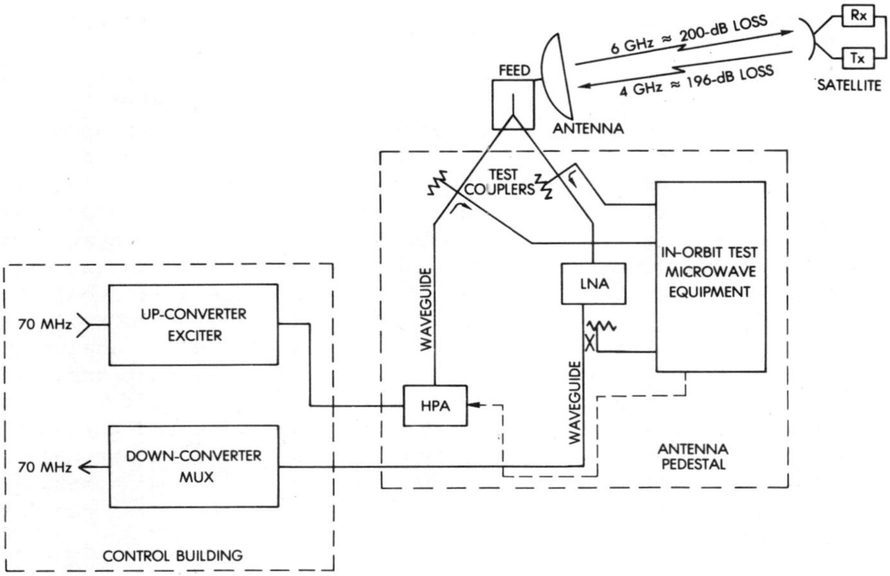

Figure 1. Link Under Test

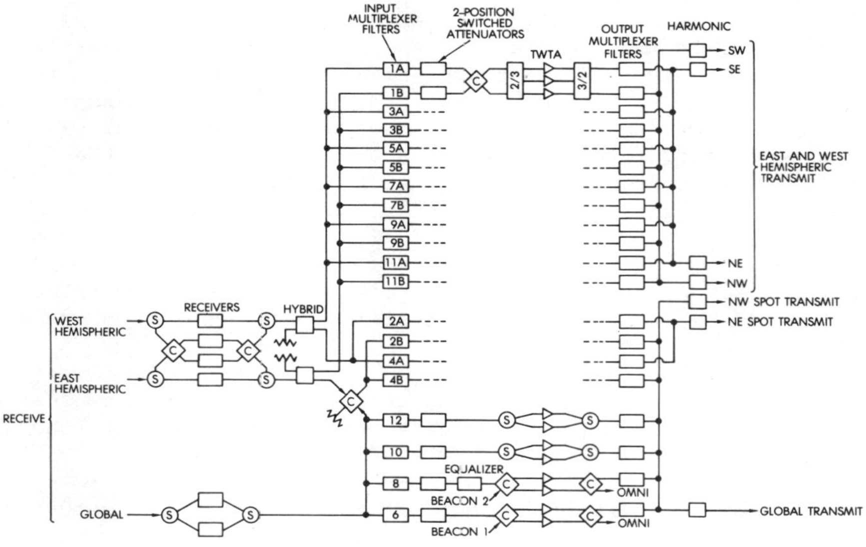

Figure 2. Simplified INTELSAT IV-A Block Diagram

Figure 3 is a schematic diagram of the RF part of the system. The up-link signal is generated by a sweep oscillator which is phase-locked to synthesized signal generator #1 mixer. The HP9830A controller tunes the sweeper to the desired up-link frequency. The controller also adjusts the synthesizer frequency so that the fourth harmonic of this signal mixed with that of the sweeper is 20 MHz. The frequency synchronizer, which contains a 20-MHz reference oscillator, then locks the sweeper. The frequency of this phase-locked signal is monitored by frequency counter #1.

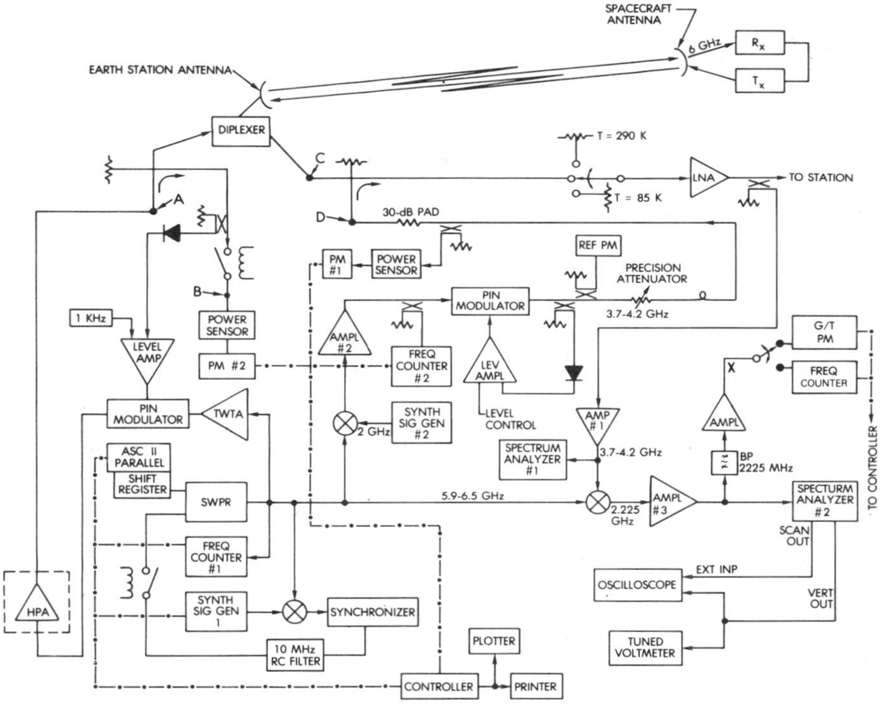

Figure 3. RF System Schematic Diagram

The up-link signal is channeled to the earth station's HPA, fed to the antenna, and monitored by a crystal detector mounted on a cross-guide coupler. The output of the detector is amplified by a DC amplifier which drives a PIN modulator connected at the output of the TWTA, thus providing a leveling loop. The up-link power level can be set by adjusting the DC offset of the leveling amplifier; the signal's power level is monitored by a power head and meter at the cross-guide coupler. In addition, the up-link signal can be amplitude modulated by a 1-kHz sinusoidal wave through the leveling amplifier.

The down-link reference signal is generated by down-converting the up-link signal through a mixer. The local oscillator (LO) frequency for this down-conversion is provided by synthesized signal generator #2, which is set at 2,225.1 MHz across the 3.7- to 4.2-GHz band. This signal, whose frequency can be monitored by counter #2, is fed through a precision attenuator, which is used to adjust its power level, and is injected into the station's LNA through a cross-guide coupler. The level at the input of the precision attenuator is made frequency independent by using a leveling loop and can be manually adjusted if necessary. It is monitored by the reference power meter to ensure proper stability.

After amplification, the output signal from the station's LNA is displayed in spectrum analyzer #1. It is also mixed with the up-link signal to yield a product which is fixed in frequency and corresponds to the spacecraft's LO frequency. After being amplified, this signal is displayed in spectrum analyzer #2, which is used for comparing the received down-link signal with the injected reference signal. The display of the oscilloscope on rack 1 is slaved to the display of spectrum analyzer #2 so that it can be monitored by a second operator.

To determine the power required to saturate the spacecraft output TWTAs, the following procedure is used. The up-link signal is amplitude modulated by a 1-kHz sine waveform when the satellite is illuminated in the linear operating region. As the operator increases the up-link power, the amplitude of the sine wave decreases, goes to zero at saturation, and increases again past the saturation point. The exact power required for saturation is found when the tuned voltmeter, which is connected to the vertical output of spectrum analyzer #2, shows a minimum.

The input signal to spectrum analyzer #2 is monitored through a coupler. After amplification, this signal can also be fed to a power meter for G/T measurements or to counter #2 for spacecraft LO frequency measurements.

The computer controller, in addition to setting the phase-locked up-link frequencies, provides automatic instrument calibration and power meter zeroing, and performs data acquisition from the power meters, frequency counters, and digital clock. These data are then properly formatted, printed, and recorded for plotting, if desired, at the completion of the measurements.

MEASUREMENTS

Antenna Patterns

Antenna patterns are taken at fixed elevation and variable azimuth angles (3). The spacecraft is oriented for the desired elevation cut, and the despun platform is oriented so that the antenna boresight points in the direction of the earth station. The up-link power required to saturate the spacecraft TWTA and the down-link power are measured as described in subsequent subsections on flux density and e.i.r.p. These values are recorded and stored, and the process is repeated until both the up-and down-link levels at saturation for all the pertinent channels in this antenna orientation have been measured. The antenna pattern for each channel is plotted as it is measured to help determine the next orientation position for the antenna, and to provide a check on the consistency of the measured values. The antenna platform is then rotated to the new azimuth position relative to the earth station, and the sequence is repeated.

When sufficient azimuth points have been obtained, the stored values are plotted in their final form. For a given channel, the plot of the up-link power for saturation versus azimuth angle represents the receive antenna pattern, while that of the down-link power or the down-link attenuator represents the transmit antenna pattern. This technique is successful because these antenna patterns can be obtained relative to the output TWTA saturation point, which is, of course, independent of the antenna orientation. Typical transmit and receive antenna patterns for the COMSTAR satellite are shown in Figure 4. The figure includes both the in-orbit measured points and the values predicted from prelaunch measurements.

Flux Density for Saturation and e.i.r.p.

An up-link signal is transmitted to determine the flux density necessary to saturate a spacecraft output TWTA. This signal is amplitude modulated by a 1-kHz sine wave. The received down-link signal is displayed on the down-link spectrum analyzer (#2) which is set for a zero scan; the transmit power is then adjusted until a minimum is found on the voltmeter. The power flux density corresponding to this level is then computed by using the power measured by power meter #2.

The power flux density required to saturate every output TWTA is measured at the center frequency corresponding to the appropriate channel after commanding the spacecraft to provide an RF path through the tube that is being measured. The down-link spectrum analyzer is also set for a linear display with a scan width that allows simultaneous viewing of both the down-link receiver carrier and the injected reference signal. The power level of the injected signal is then adjusted by the down-link attenuator so that it is equal to the received signal level. From this measurement the spacecraft e.i.r.p. is computed. Figure 5 shows typical e.i.r.p. data.

Gain Transfer Characteristic

The gain transfer characteristic of a given spacecraft channel corresponding to a particular output TWTA is obtained by measuring and plotting the down-link level versus the up-link level. The results of such a measurement, together with prelaunch data for the COMSTAR D-1 satellite, are shown in Figure 6.

Spacecraft Receiver LO Frequency

The spacecraft receiver's LO frequency is determined by connecting a frequency counter instead of a power meter at point X (Figure 3) and measuring the frequency while an up-link signal is being transmitted. Since the frequency measured by the counter results from mixing the signal transmitted by the spacecraft and the up-link signal, it equals the spacecraft LO frequency which was used to down-convert the up-link signal. The accuracy of this measurement is generally limited by the resolution and stability of the counter and the additional error introduced by a possible Doppler shift.

Spacecraft G/T

Figure 7 illustrates the received noise power as a function of frequency with

a. the earth station (E/S) antenna pointed at the sky,

b. spacecraft,

c. the input of the LNA connected to a hot elevated temperature termination, and

d. a cryogenic termination.

From these data and the gain measurements, the spacecraft gain temperature ratio can be computed as previously reported (2).

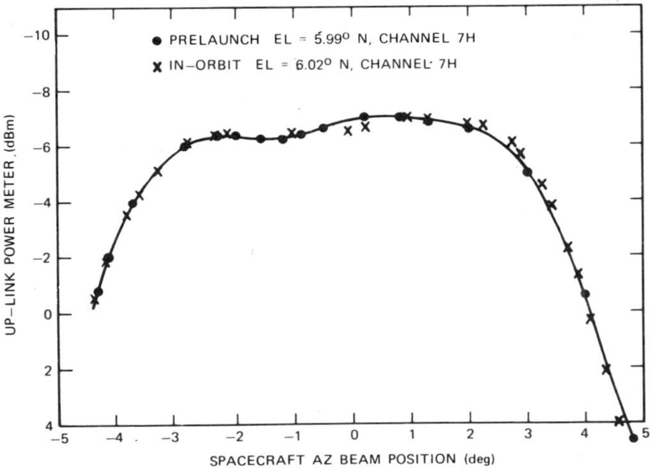

Figure 4. Receive Antenna Pattern for COMSTAR D-1

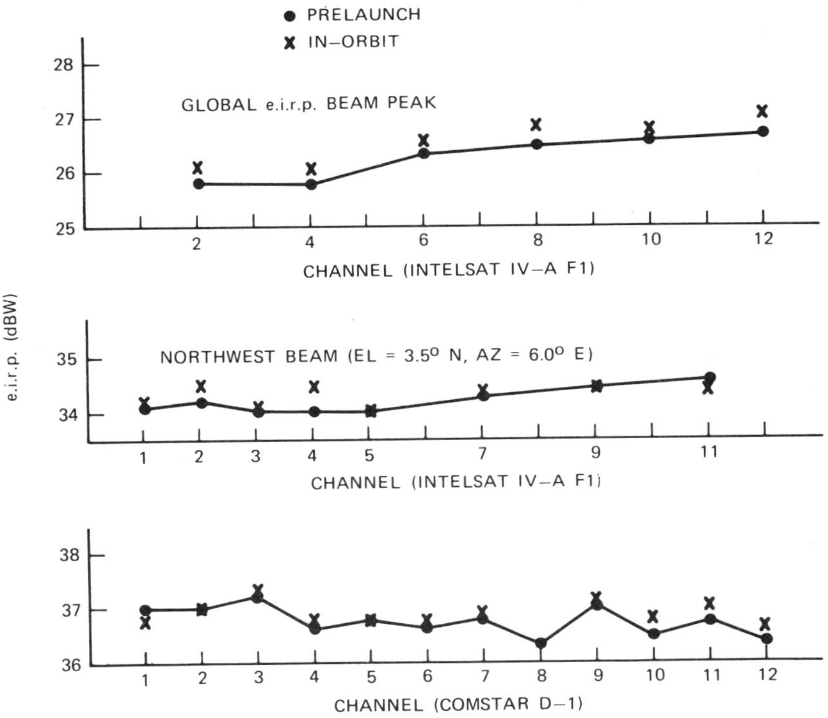

Figure 5. INTELSAT IV-A F-1 and COMSTAR D-1 e.i.r.p.

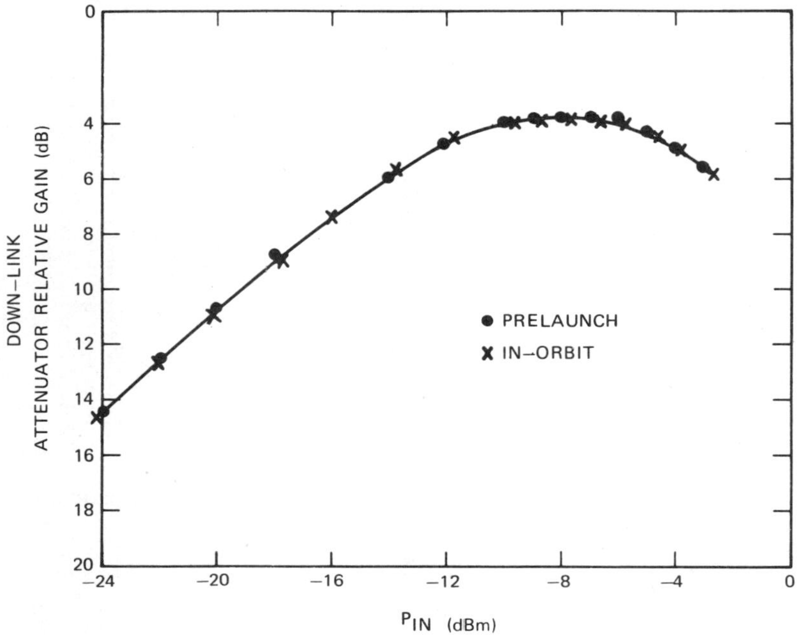

Figure 6. Channel 9 TWTA Transfer Characteristics

In-Band and Out-Band Frequency Response

The setup (Figure 3) is configured for nondigital operation, and the phase-lock loop is disabled. To provide a swept up-link signal, the swept frequency generator is set for an analog sweep of 70 MHz centered at the central frequency of the channel to be measured. On the down-link side, spectrum analyzer #2 is set for linear display, and its vertical output is connected to one of the vertical channels of the X-Y recorder. Spectrum analyzer #1 is fed the signal from the G/T port of the setup and set for a logarithmic display; its vertical output is connected to the second vertical channel of the X-Y recorder. The horizontal channel of the recorder is driven by the sweep signal from the sweeper. The trace from the linear display will show at maximum resolution the in-band frequency response of the spacecraft filters, corresponding to the channel under test, while the logarithmic trace will provide the out-of-band rejection information for this channel. Typical results for INTELSAT IV are shown in Figure 8.

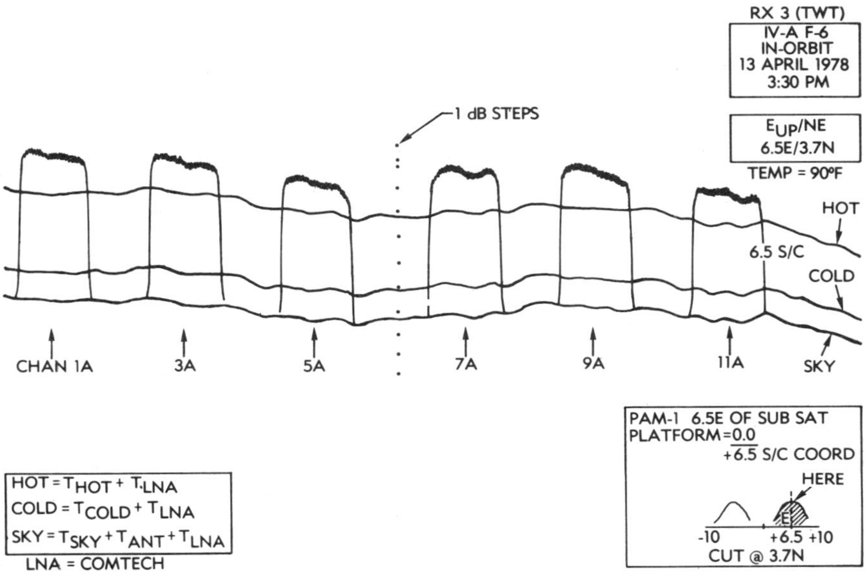

Figure 7. Receiver 3 INTELSAT IV-A F-6 G/T Measurements

INTELSAT V IN-ORBIT TEST SYSTEM

The next generation of communications satellites is INTELSAT V (I-V). These satellites will be capable of both left- and right-hand circularly polarized signals at 6 GHz and a linearly polarized signal at 14 GHz. They will then transmit circularly polarized signals at 4 GHz and a linearly polarized signal at 11 GHz. Any received signal will be retransmitted at the desired polarization. There are 588 possible RF interconnections.

A new system is now being built for the in-orbit testing of these satellites. Figure 9 depicts a simplified functional diagram of this system, which has evolved from the one described in this paper. It will perform all the necessary measurements with significantly increased accuracy and speed.

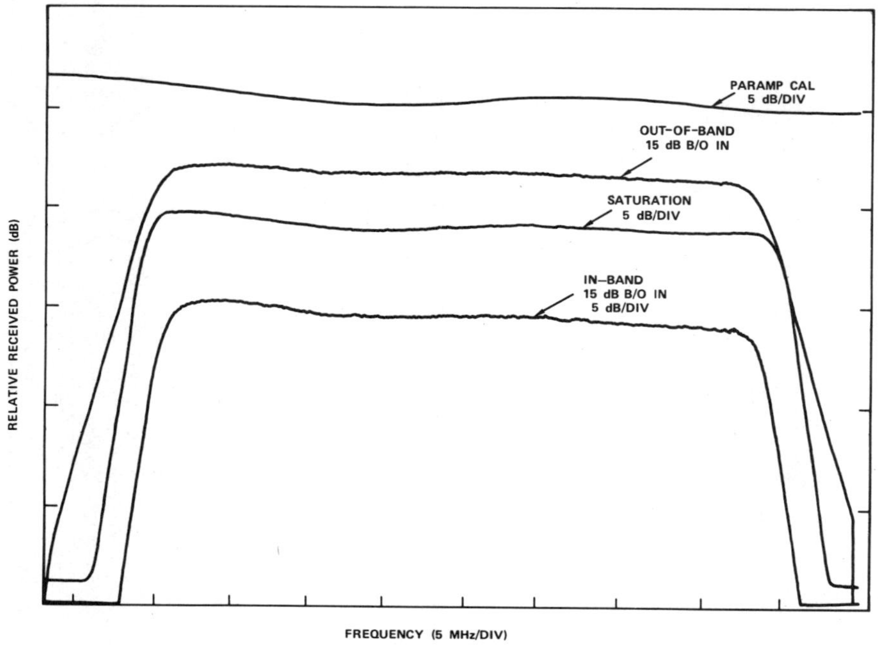

Figure 8. Spacecraft Filter Frequency Response

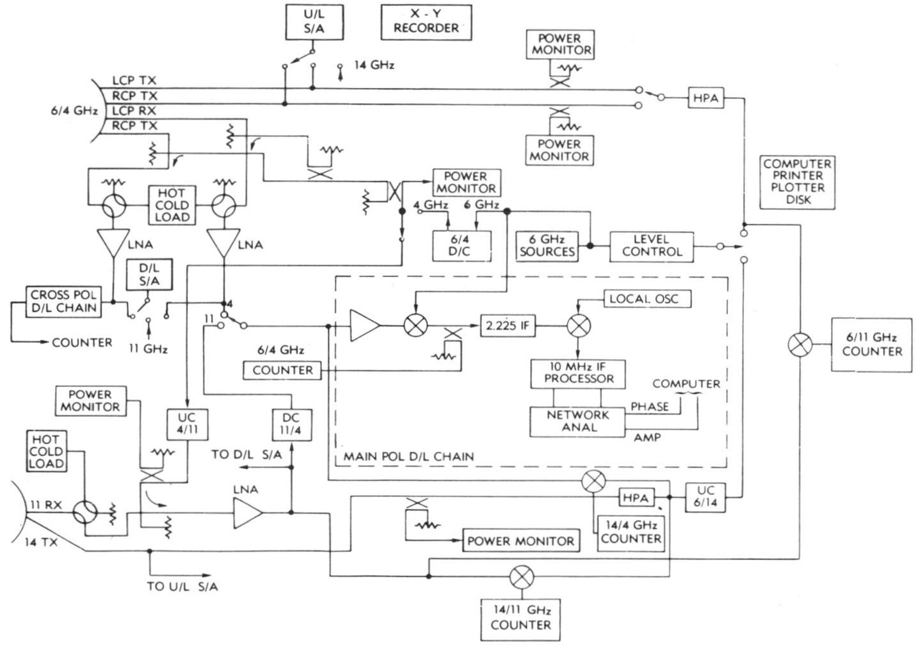

Figure 9. INTELSAT V In-orbit Test System Functional Block Diagram

ACKNOWLEDGMENT

This paper is based upon work performed in COMSAT Laboratories under the sponsorship of the Communications Satellite Corporation.

References

1. I. Dostis, C. E. Mahle, V. E. Riginos, and I. G. Atohoun, "In-Orbit Testing of Communications Satellites," COMSAT Technical Review, Volume 7, Spring 1977, pp. 197-226.

2. A. F. Standing, "A Null Technique for Accurately Determining the Drive Level Required to Saturate a TWT," COMSAT Technical Review, Volume 4, Fall 1974, pp. 493-497.

3. E. T. Jilg, editor, "The INTELSAT IV Spacecraft," COMSAT Technical Review, Volume 2, Fall 1972, pp. 277-294.

MULTIPLE ACCESS AND LOCATION PERFORMANCE OF THE TIROS-N DATA COLLECTION SYSTEM

Dr. James C. Morakis
NASA/Goddard Space Flight Center
Greenbelt, MD 20771

Abstract

The TIROS-N satellite series consists of 8 environmental spacecraft that will be operational in pairs during the time period 1978 to 1985 or beyond.

In designing the data collection system for TIROS-N, trade-off studies were conducted in order to optimize the multiple access capacity (number of transmitters in view of the spacecraft antenna), the location accuracy, the probability of acquisition and processing and storage equipment.

Parametric solutions have been found for the transmission repetition rate and duration, acquisition time, time-frequency threshold strategy for acceptance or rejection of signals and data rate. These solutions result in a repetition period of 50-60 seconds, duration of 360 msec, acquisition time of 160 msec and data rate of 400 bps for a multiple access capacity of 300 transmitters in view, location accuracy less than 1.5 nautical miles, probability of acquisition 0.99 and bit error probability better than 10^{-5}.

D. G. Lainiotis and N. S. Tzannes (eds.), Advances in Communications, 431–436.

INTRODUCTION

Due to the growing utility of satellite based data collection systems, the quantity of data to be handled by these systems is rapidly increasing; therefore, in designing these systems an attempt is made to maximize the capacity without degrading the performance.

The data collection system (DCS) consists of 3 subsystems. The DCS platforms (DCP), the onboard equipment, and the data processing centers. Since the satellite is polar orbiting, it is possible to determine the location and average speed of the DCP by measuring the doppler at the satellite.

LOCATION

While the satellite is moving over the earth, as shown in Figure 1, it receives transmissions from a DCP with a doppler frequency f_d which is given by the equation.

$$fd= \frac{v}{c} \ f_c$$

where
f = transmission frequency
v = satellite velocity
c = the speed of light
θ = the angle between v and the DCP line of sight

for f, c and v constant O can be determined as a function of the measured doppler (f_d). In other words each doppler measurement defines a cone of half angle θ. The surface of which is the locus of the DCP. More doppler measurements in other satelite positions can furnish enough data to solve for the intersection of the cones which represents the DCP location. Due to uncertainties such as satellite location, and speed, DCP transmission frequency variation and others, many transmissions are necessary in order to perform a least squares estimation. In general, one needs at least 4 to 5 transmissions to make a rough estimate and about 10 for a good estimate.

RANDOM ACCESS

The DCP's are transmitting with a repetition period T. When the satellite is in view of N such DCP's. The incoming transmission can be considered random, with the time of arrival having a poisson distribution. If the transmission duration is τ the probability of interference, (2 or more simultaneous transmissions during τ) is:

$$P_1 = 1 - e^{-\frac{N\tau}{T}\frac{2f_1}{F_T}}$$

This expression would be true if all incoming frequencies were equal. However, due to doppler effects the DCP oscillator tolerance, the frequency distribution of such a system is as shown in Figure 2. Approximating this to a uniform distribution one can combine the time and frequency statistics to obtain the final expression for P_1 as

$$P_1 \cong \frac{2N\tau f_1}{T\,F_T}$$

where f_T is the total range of the frequency distribution and f_1 is the smallest frequency separation before two signals can interfere. To optimize the system, we can hold N (and consequently the amount of data to be transmitted) fixed and decrease P_1; to reduce P_1 one should reduce N, τ and f_1 and increase T and f_T. If τ is decreased then the data rate must be increased and this increases f_1 an undesirable change. Similarly if f_1 is decreased then τ must be increased also an undesirable change. If T is increased then we may not allow for an adequate number of transmissions to get a good estimate of the location. Finally, increasing F_T will decrease the sensitivity and tracking capability of the system because of the decrease of the total signal to noise ratio.

Clearly, there exists interdependence between the variables, and varying one parameter may have opposing effects on this system. A parametric analysis was conducted, where different values were assigned and the variables and consideration was given to different modulation techniques. Based upon the above, the interference statistics of a platform were estimated in terms of the probability of acquiring a transmission and performing the demodulation function with an interfering power level from other platforms.

The above analysis resulted in the values listed below:

N = the number of platforms within communication view of the satellite equals to 200

τ = the duration of the platform's transmission (360 ms)

R = the platform's data rate (400 bps)

T = the time interval between platform transmissions (about 60 sec)

The above analysis assumes that all non-interfering data that is received in the spacecraft can be processed and successfully put into storage. In other words the spacecraft is capable of processing all signals that may arrive simultaneously. If the spacecraft has a finite number of data recovery units (DRU) then some of the transmissions will be lost.

In Figure 3, a parametric presentation of the anticipated interference statistics for the TIROS-N DCS is shown. The ordinate is the probability that a randomly selected transmission from the specified platform will be acquired and demodulated by the satellite. This probability is presented as a function of the overall system data rate and parametrically for 3,4,6, and 8 data recovery units available.

From Figure 3 it can be seen that using more than 4 DRU's does not significantly improve the performance. Thus, 4 DRU's were used for the TIROS-N DCS.

DESIGN GOALS

The design goals for position location are in Table 1.

TABLE 1

error	buoy	balloon
position	3 Km	5 Km
velocity	0.5 m/sec	1.5 m/sec

these goals were given under the following assumptions:
- altitude error less than 50m for buoys
- altitude error less than 500m for balloons
-initial frequency duration less than 500 Hz
- stability (mean) 10^{-8} in 20 min
- speed less than 100 m/sec
- acceleration less than 1 m/sec
- probability of correctly processing one transmission 0.99
- bit error probability $\leq 2 \times 10^{-5}$,
- false alarm probability; better than 10^{-5}.

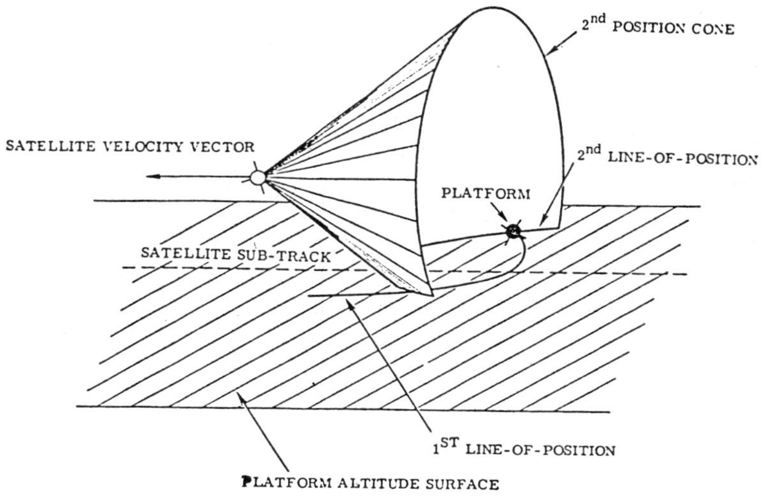

FIGURE 1. DOPPLER POSITION LOCATION

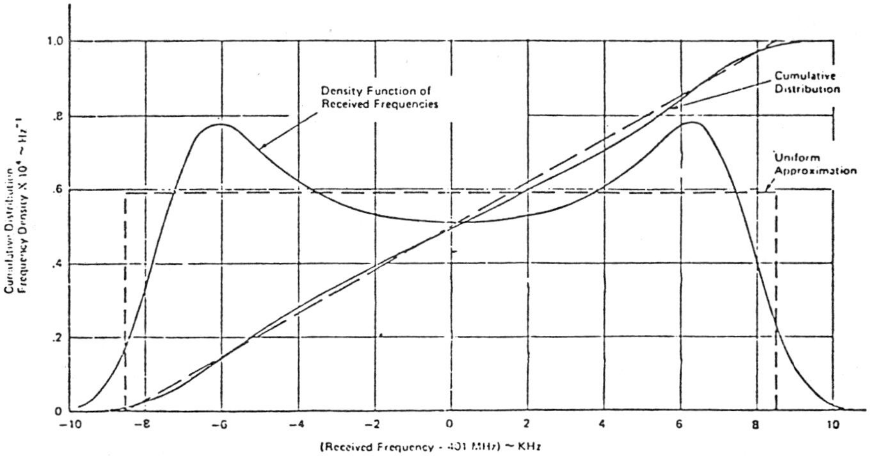

FIGURE 2 FREQUENCY DISTRIBUTION

2 Hour Orbit Period
5° Horizon Angle Limit
Mean Transmission Frequency ~ 401 MHz
Standard Deviation of Transmission Frequencies ~ 2000 Hz

PROBABILITY OF SIGNAL ACQUISITION

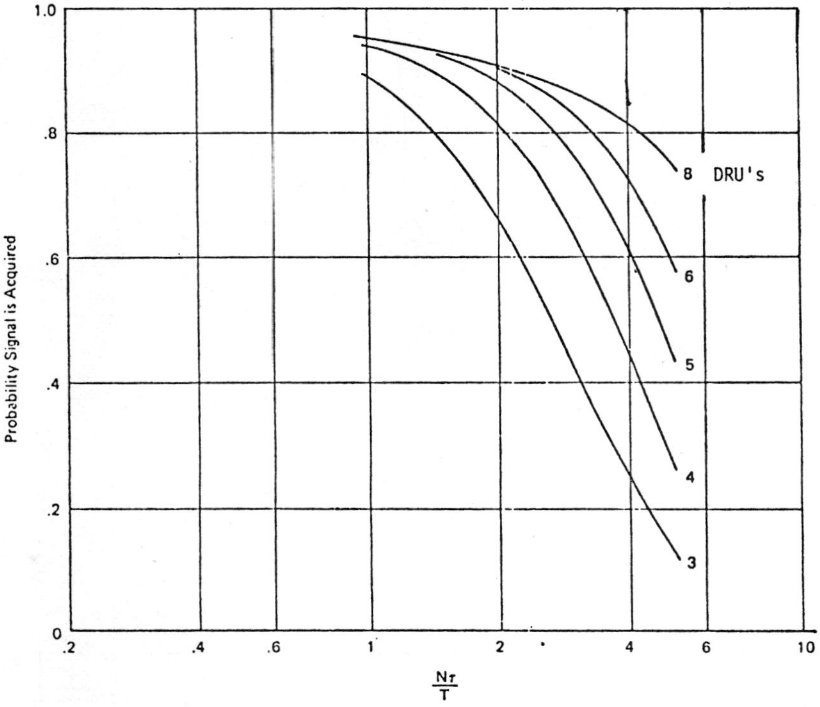

FIGURE 3 PROBABILITY OF SIGNAL ACQUISITION

RESULTS

The location accuracy is about 200 to 400 meters based on some post launch tests that lasted for 3-5 weeks for a fixed platform with a good oscillator.

The mobile platforms had a location accuracy within specifications.

It is anticipated on the basis of preliminary results that the system can accomodate 300 platforms in view.

SYNCHRONIZATION AND SYNCHRONOUS OSCILLATORS

V. Uzunoglu

COMSAT Laboratories
Clarksburg, Maryland 20734

ABSTRACT

Synchronization is an important phenomenon which has applications for reliable and efficient communications systems and other fields. The concept of synchronization originated in the 17th century with Huygens, who reported that two clocks which were slightly out of step with each other became synchronized when hung on a thin wooden board. In electronics, when an external signal of frequency f is applied properly to an oscillator circuit that is oscillating with a frequency f_o, the beats of two frequencies can be observed. If the frequency f approaches f_o, the beats disappear suddenly and only the frequency f remains. Synchronizations of this nature can be used in various communications systems.

INTRODUCTION

Outside the field of communications and power generating systems, the phase-lock principle, or synchronization, has been referred to as the principle of entrainment. Apparently, Huygens (1629-1695) was first to observe the phenomenon when two clocks which were slightly out of step with each other became synchronized when hung on a thin wooden board. Also, Lord Raleigh noticed that two organ pipes of slightly different frequencies vibrate at the same frequency when placed near each other. However, if their frequencies are sufficiently far apart, one hears the frequency of beats. Synchronization is not a unique property of electronic elements or systems, and it appears in any process with the proper nonlinear features, regardless of its physical

D. G. Lainiotis and N. S. Tzannes (eds.), Advances in Communications, 437–448.

nature. One means of achieving synchronization is the synchronous oscillator, which finds many uses in space and digital communications.

SYNCHRONIZATION IN NATURE

Studies of circadian* rhythms have established light as a major engineering agent for most living organisms (1). People also become entrained to the time of day in demonstrating their peak abilities. For crucial tasks, such as those performed by a fighter pilot or an athlete, peak performances can be achieved by gearing activities to the optimum time of day.

Entrainment is also useful for explaining certain epileptic seizures observed via electroencephalography. When a lamp is flickered into the eye at 10 Hz, the rhythm in the brain will become phase locked to this signal. An interesting phenomenon which may link epilepsy to entrainment is the photo-flicker effect. Pilots of propeller driven aircraft have noticed that occasionally the sunlight shining through propeller blades caused them to experience a peculiar form of vertigo, and in extreme cases, loss of consciousness. Effects of a similar kind were also considered in the design of the proposed tunnel between England and France.

The electrocardiogram (EKG) clearly indicates a complex electromechanical as well as biological synchronization phenomenon. It appears that such waveforms are due to the simultaneous phase locking of many random nonlinear oscillators, which occur in parallel operation of electrical generators and in a communications network of mutually synchronized oscillators.

A similar situation occurs in oscillating electrical networks when the damping factor, μ, of the circuit is increased. In this case, the oscillations are not sinusoidal; they become relaxed and resemble distorted square waves.

To synchronize the waveforms in the brain, drugs containing caffeine are recommended for humans. The effective increase of μ in the brain is associated with mental or physical exhaustion. Physiologically, the exhaustion is accompanied by a decrease of oxygen availability which causes metabolism in the brain. One of the most predominant characteristics of sleep is low (delta) frequency waves. If the oscillations in the brain are nonrelaxed, introducing a low-frequency stimulation to certain subcortical structures (which connect to the cortex) induces sleep.

---

*Pertaining to rhythmic biological cycles.

OSCILLATOR SYNCHRONIZATION

If the frequencies of two oscillators are close to each other, mutual coupling results in the frequencies to become synchronized (2). If the coupling is unidirectional, one of the oscillators will be the controlling oscillator whose frequency will be unaffected, and the other will be forced to assume the frequency of the controlling oscillator. The process of synchronization is nonlinear. To produce the synchronization effectively, the amplitude, waveform, and phase of the controlling oscillator must be adjusted properly. The controlled oscillator can not only be synchronized to the source frequency, but also to its subharmonics and superharmonics (3,4). The system can also be built to contain analog memory, so that even if some pulses of the controlling oscillator are missing the oscillator will continue to be locked to the frequency of the main oscillator. Such networks, termed synchronous oscillators, can find great applications in present communications systems (1).

ANALYSIS OF A SYNCHRONIZED OSCILLATOR

Figure 1 is a block diagram of a synchronized oscillator (SO). The feedback network comprises all positive feedback networks. The injected voltage to the SO is e_1, and the resultant voltage to the input of the active network is e_{in}. The equivalent circuit of Figure 1 can be drawn as shown in Figure 2, and its vector diagram as shown in Figure 3. It is assumed that e_1 is constant and fixed in phase; it is used as a reference voltage. Also, β is the phase difference at a given time between the feedback and the input voltages, and α is the phase difference between the feedback and the output voltages, Ae_{in}.

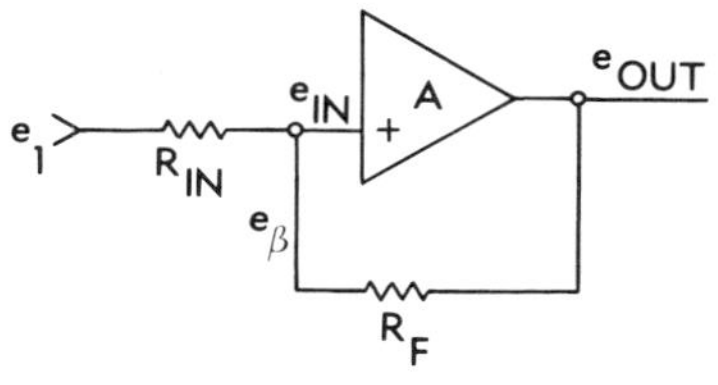

Figure 1. Block Diagram of Synchronized Oscillator

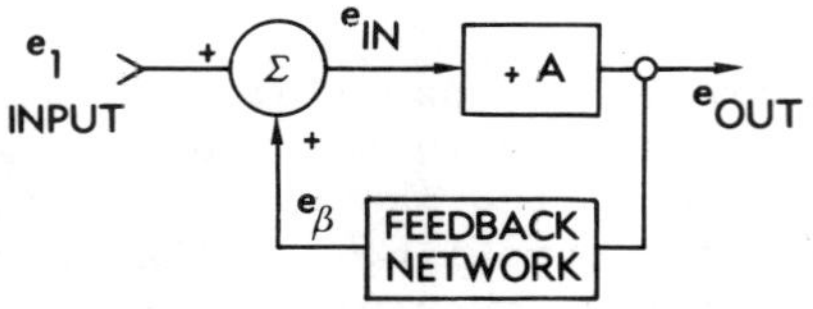

Figure 2. Equivalent Circuit of Figure 1

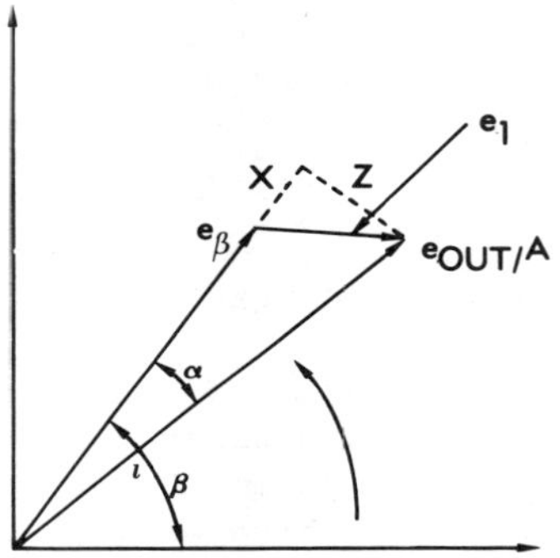

Figure 3. Vector Diagram of Synchronized Oscillator

From Figures 1 and 2,

$$\tan \alpha = - \frac{Z}{e_\beta + X} \tag{1}$$

$$\sin \beta = \frac{Z}{e_1} \tag{2}$$

$$\cos \beta = \frac{X}{e_1} \tag{3}$$

Substituting equations (2) and (3) into equation (1) results in

$$\tan \alpha = \frac{e_1 \sin \beta}{e_\beta + e_1 \cos \beta} \tag{4}$$

For small α,

$$\tan \alpha \to \alpha$$

$$e_1 \cos \beta << e_\beta$$

as e_1 becomes very small. Thus, equation (4) becomes

$$\alpha = - \frac{e_1 \sin \beta}{e_\beta} \tag{5}$$

For a single tuned circuit, the phase versus frequency variation is given in Figure 4. For operation around the center in a linear region,

$$K = \frac{d\alpha}{d\omega} \tag{6}$$

where K is a constant. The integration of equation (6), yields

$$\int_{\omega_0}^{\omega} K d\omega = \int_0^{\alpha} d\alpha \tag{7}$$

$$K(\omega - \omega_0) = \alpha \tag{8}$$

where ω is the instantaneous frequency, and ω_0 the natural frequency. Substituting equation (8) into equation (5) results in

$$K(\omega - \omega_0) = - \frac{e_1 \sin \beta}{e_\beta} \tag{9}$$

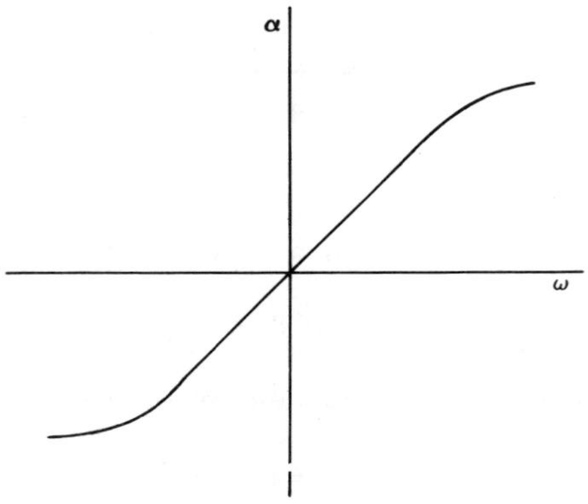

Figure 4. Phase vs Frequency Variation

Adding and subtracting input frequency ω_1 from equation (9) yields

$$K[(\omega - \omega_1) - (\omega_0 - \omega_1)] = - \frac{e_1 \sin \beta}{e_\beta} \tag{10}$$

or

$$K(\Delta\omega - \Delta\omega_0) = - \frac{e_1 \sin \beta}{e_\beta} \tag{11}$$

Since

$$\Delta\omega = \frac{d\beta}{dt}$$

expression (11) can be written as

$$K\left(\frac{d\beta}{dt} - \Delta\omega_0\right) = - \frac{e_1}{e_\beta} \sin \beta \tag{12}$$

When

$$B = \frac{1}{K} \frac{e_1}{e_\beta} \tag{13}$$

equation (12) can be written as

$$\frac{d\beta}{dt} = -B \sin \beta + \Delta\omega_0 \tag{14a}$$

or

$$\Delta\omega = -B \sin \beta + \Delta\omega_0 \tag{14b}$$

which is a nonlinear differential equation. Adding the input frequency on both sides of expression (14b) gives

$$\Delta\omega + \omega_1 = -B \sin \beta + \Delta\omega_0 + \omega_1 \tag{15}$$

which results in

$$\omega = -B \sin \beta + \omega_0 \tag{16}$$

or

$$\omega - \omega_0 = -B \sin \beta \tag{17}$$

For a single tuned circuit,

$$\tan \alpha = 2Q \frac{\omega - \omega_0}{0} \tag{18}$$

For a small α, equation (18) becomes

$$\alpha = 2Q \frac{\omega - \omega_0}{\omega_0} \tag{19}$$

From expression (8),

$$2Q \frac{\omega - \omega_0}{0} = K(\omega - \omega_0) \tag{20}$$

or

$$\frac{2Q}{\omega_0} = K \tag{21}$$

since

$$B = \frac{1}{K}\frac{e_1}{e_\beta}$$

Equation (21) can be written as (5)

$$\frac{2Q}{\omega_0} = \frac{1}{B}\frac{e_1}{e_\beta} \tag{22}$$

or

$$B = \frac{E_1}{e_\beta}\frac{\omega_0}{2Q} \tag{23}$$

Equation (14a) can be written as

$$\frac{d\beta}{dt} = -\frac{\omega_0}{2Q}\frac{e_1}{e_\beta}\sin\beta + \Delta\omega_0 \tag{24}$$

For the steady state,

$$\frac{d\beta}{dt} = 0 \tag{25}$$

Thus,

$$\sin\beta = -\frac{2Qe_\beta}{e_1}\frac{\Delta\omega_0}{\omega_0} \tag{26}$$

which gives the stationary phase angle between the feedback voltage of the oscillator and the input signal. Since $\sin\beta$ can only assume values between +1 and -1, no steady state is possible if the right side of equation (26) is outside this range. This gives the condition for synchronization (6)

$$2Q \frac{e_\beta}{e_1} \frac{\Delta\omega_0}{\omega_0} < 1 \tag{27}$$

or

$$\frac{e_1}{e_\beta} > 2Q \frac{\Delta\omega_0}{\omega_0} \tag{28}$$

PULL-IN CONDITION

The pull-in and lock-in are important considerations in understanding the operation of an SO. In equation (24), $\Delta\omega_0 = 0$. This means that the input frequency is equal to the oscillator frequency, and that locking will occur eventually. The expression

$$\frac{d\beta}{dt} \simeq B \sin \beta \tag{29}$$

shows what happens when the input signal e_1 is suddenly switched on with an initial lag, β_1, behind the free running oscillator. Equation (29) is almost equal to

$$\frac{d\beta}{dt} = B\beta \tag{30}$$

when β is small.

Expression (30) has the familiar solution

$$\beta = \beta_1 e^{-BT} \tag{31}$$

In Laplace transform notation, equation (31) becomes

$$\beta(s) = \frac{\beta_1}{s + B} \tag{32}$$

Physically, this means that the oscillator phase approaches that of the input signal. The speed of this process, according to equation (23), is proportional to the ratio of the input voltage to the oscillator feedback voltage and to the bandwidth of the tuned circuit. If the free running frequency is not equal to that of the input frequency, but close enough to permit locking according to expression (28), the manner in which the steady state is reached must still follow the same trend. If the input signal is suddenly removed, the return of the oscillator to its free running frequency will require some time since $e_\beta = e_{in}$ because of the high feedback, and the system remains stable for some time. External effects and noise will eventually force the oscillator to return to its free running mode. This effect can be considered as a hysteresis in the oscillator which will continue to oscillate in the synchronous mode even if the external synchronization signal is turned off.

TRANSFER FUNCTIONS

In Laplace transform notation, equation (24) can be written as

$$s\beta(s) = -\frac{\omega_0}{2Q}\frac{e_1}{e_\beta}\beta(s) + \frac{\Delta\omega_0}{s} \qquad (33)$$

where for small β

$$\sin\beta \to \beta$$

Rearranging equation (33) results in

$$\beta(s)\left[s + \frac{\omega_0}{2Q}\frac{e_1}{e}\right] = \frac{\Delta\omega_0}{s} \qquad (34a)$$

$$\beta(s) = \frac{\Delta\omega_0}{s[s + (\omega_0/2Q)(e_1/e_\beta)]} \qquad (34b)$$

Output phase versus input frequency variations can be written as

$$\frac{\beta(s)}{\Delta\omega_0} = \frac{1}{s[s + (\omega_0/2Q)(e_1/e_\beta)]} \qquad (35)$$

Expression (35) is a second-order servo loop with one pole at the origin. If the output frequency variations are of interest, the following equation applies:

$$\Delta\omega = \frac{dQ}{dt} = \frac{d\beta(s)}{dt} \qquad (36)$$

$$\frac{\beta(s)}{dt} = s\beta(s) \tag{37}$$

Therefore,

$$\frac{d\beta(s)}{dt} = s\beta(s) = \frac{\Delta\omega_0}{s + (\omega_0/2Q)(e_1/e_\beta)} \tag{38}$$

Expression (24) is a nonlinear differential equation. Expressions (34), (35), and (38) are solutions of expression (24) for a linearized case.

SUMMARY OF EQUATIONS

The steady-state phase angle between the feedback voltage and the input voltage is

$$\sin\beta = -\frac{2Qe_\beta}{e_1}\,\frac{\Delta\omega_0}{\omega_0} \tag{26}$$

which is the condition for synchronization. Also, the tracking range equals $\Delta\omega_0$, which is proportional to input e_1, and inversely proportional to the gain of the oscillator:

$$\frac{e_1}{e_\beta} > 2Q\,\frac{\Delta\omega_0}{\omega_0} \tag{28}$$

Phase variation when e_1 is suddenly applied to the SO is expressed as

$$\beta = \beta_1\, e^{-BT} \tag{31}$$

Phase variation with respect to change in frequency is

$$\frac{\beta(s)}{\Delta\omega_0} = \frac{1}{\{s[s + (\omega_0/2Q)(e_1/e\beta)]\}} \tag{35}$$

ACKNOWLEDGMENT

This paper is based upon work performed at COMSAT Laboratories under the sponsorship of the Communications Satellite Corporation.

REFERENCES

1. Willian C. Lindsay, Synchronization Systems in Communications and Control, Prentice Hall, 1972, Chapter 5.

2. Vasil Uzunoglu, Semiconductor Network Analysis and Design, McGraw-Hill Book Company, 1964, Chapter 11.

3. J. W. Edwards and N. Minorsky, Introduction to Nonlinear Mechanisms, Ann Arbor, Michigan, 1947.

4. Vasil Uzunoglu, Analysis and Design of Digital Systems, Gordon and Breach, New York, 1973, Chapter 8.

5. L. J. Paclorek, "Injection Locking of Oscillators," Proc. of the IEE, Vol. 53, No. 11, November 1965, pp. 1723-1727.

6. R. Adler, "A Study of Locking Phenomena in Oscillators, Proc. of the IRE, June 1946, p. 351.

SECTION III

PATTERN RECOGNITION

RELEVANCE OF A POSTERIORI PROBABILITY ERRORS IN BAYESIAN CLASSIFICATION

Gustavo BELFORTE - Basilio BONA

CENS - Torino, Istituto di Elettrotecnica Politecnico di Torino, Istituto Elettrotecnico Nazionale "G. Ferraris" Torino

ABSTRACT

When a bayesian classification algorithm has to be performed on a set of continuous variables, their probability density functions (pdfs) have to be known for each one of the pattern classes.

Different kinds of pdfs have been proposed in literature; in most cases these functions are parametrized through a suitable vector that, in general, has to be estimated on a training sample. Thus the computed a posteriori probabilities become stochastic variables and are affected by uncertainty.

A method to compute these uncertainties was still proposed. In this paper attention is devoted to some examples; they show the relevance that a posteriori probability errors may have in a bayesian classification.

1. PROBLEM STATEMENT

The problem of classifying subjects using a set of r continuous variables has been widely studied since Fisher's first paper in 1936 [1-5].

One of the most familiar techniques to perform the classification is the use of Bayes' theorem. To this extent the probability density functions (pdfs) of the different groups, into which the classification has to be performed, must be available.

D. G. Lainiotis and N. S. Tzannes (eds.), Advances in Communications, 451–456.

Two families of pdfs have been proposed in literature beside the case, not relevant in practice, of "a priori" perfectly known pdfs.

The first family is composed by pdfs that are functions both of the vector $\underline{x} = (x_1, x_2, \ldots x_r)$ of the variables and of a suitable vector $\underline{\lambda} = (\lambda_1, \lambda_2, \ldots \lambda_n)$ of n parameters. This last must be estimated on the base of the variables measured on a set T of "a priori" correctly classified subjects, called the training sample.

Thus let p_i be the pdf of class i $i = 1, M$ where M is the total number of allocation classes; it results that:

$$p_i = p_i(\underline{x}, \underline{\lambda}_i(T)). \tag{1}$$

As an example, to this family belongs the multinormal distribution with unknown mean vector and covariance matrix. Their elements must be estimated from the training sample and constitute the λ_i vector.

The second family is the one in which the pdfs are functions of the t_j elements of the set T, $j = 1, N_T$ (N_T is the number of elements in the training sample) both through a parameter vector $\underline{\lambda}$ and directly. In this case it results that:

$$p_i = p_i(\underline{x}, \underline{\lambda}_i(T), t_1, t_2, \ldots t_{N_T}) \tag{2}$$

As an example, to this family belong the pdfs estimated by means of Parzen estimators [6].

The pdfs of both families are parametrized through the vector $\underline{\lambda}$, that, being estimated on a finite size sample, results to be a stochastic vector. Its variability is related to the assumptions on the statistic of the group and on the estimator's properties.

Bayes theorem states that:

$$F_i = F_i(\omega_i | \underline{x}) = \frac{p_i \ P(\omega_i)}{\sum_{m=1}^{M} p_m \cdot P(\omega_m)} \tag{3}$$

where p_i is either (1) or (2), $P(\omega_i)$ is the a priori probability of occurrence of the i-th class ω_i, $F_i(\omega_i | \underline{x})$ is the "a posteriori" probability of membership in class ω_i given the variables' vector $\underline{x}$.

Since the p_i are, as shown, stochastic variables it follows from (3) that F_i is itself a stochastic variable.

The F_i are commonly used to perform classification employing some selected decision rule. This rule must take into account both the predominant groups attribution and its reliability. If perfectly known pdfs would be available, then the Bayes rule would give both the required informations. As it has been shown, when the pdfs are evaluated using a finite training sample, the F_i in (3) are stochastic variables; therefore they alone cannot anymore give the required information on group attribution and its reliability.

In this case it seems to be necessary some information on the reliability of the F_i themselves.

To this extent a proper tool in a stochastic framework seems to be the computation of confidence intervals for the F_i.

This problem can be numerically solved if the estimator of the $\underline{\lambda}$ is such that some information on the distribution of the estimate vector is available.

As an example if a maximum likelihood estimate is performed, the distribution of the estimate is at least asimptotically normal with covariance matrix given by the Cramer-Rao limit.

An algorithm has been proposed in [7-8] to perform this computation in the case of pdfs that are either multinormal or mixtures of multinormal ones.

In the following some applications will be shown. They are concerned with a problem of phisiopathological discrimination among subjects on the base of two measures of a laboratory analysis: the rate K of Bromsulphalein (BSP) plasma disappearance and % RC 45, the quantity of conjugated BSP in blood at 45 minutes expressed as a percentage of the injected dose [9-10]. 350 subjects were available with their classification into 16 groups which were further grouped into 5 classes.

Probability of membership in these last 5 groups were computed for ranges of K and % RC 45 of clinical interest. The results are reported in figure 1 and figure 2.

Figure 1 is the sketch of the curves of equal F_i for the different groups when the p_i are estimated as mixtures of normal pdfs whose parameters are estimated from the 16 subgroups.

Figure 2 is the case in which the 5 classes are considered normally distributed; the parameters being estimated from the sample directly divided into the 5 considered classes.

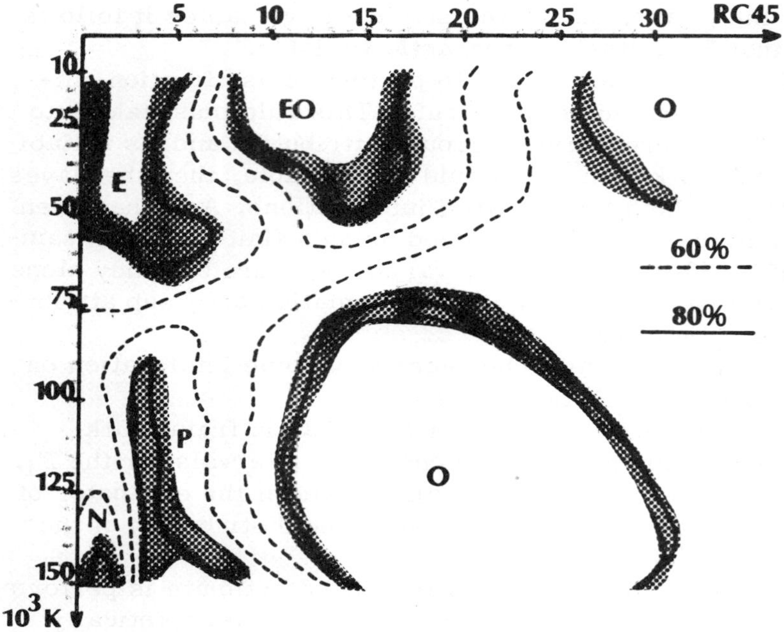
5
10
15
20
25
30
RC45
10
25
50
75
100
125
150
10^3 K
E
EO
O
P
N
O
60%
80%

Fig. 1

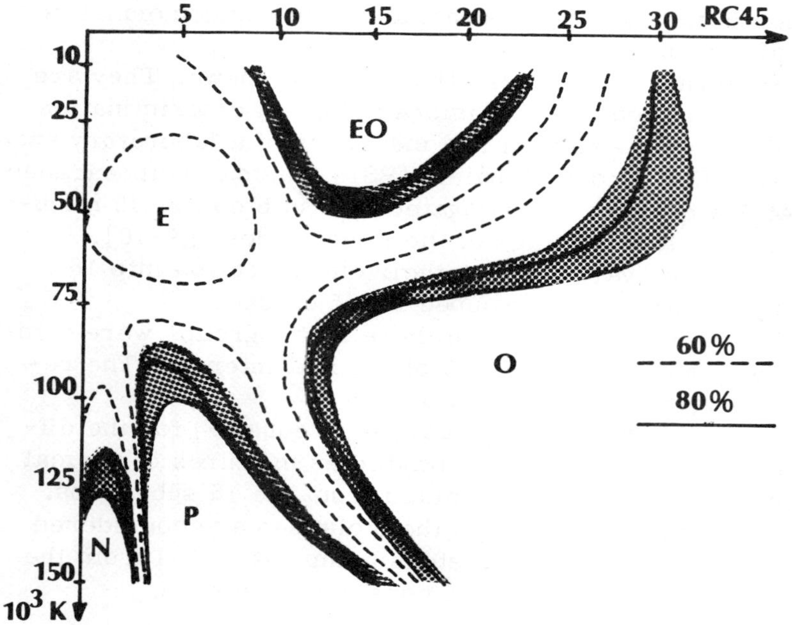
5
10
15
20
25
30
RC45
10
25
50
75
100
125
150
10^3 K
EO
E
O
P
N
60%
80%

Fig. 2

Table 1

K	%RC45	% of attribution	Q=0.95	Q=0.68	Q=0.38	Q=0.24
0.082	2.8	N = 6	+ 4- 2	+ 2- 1	+ 1- 1	+ 1- 0
		P = 65	+13-12	+ 7- 7	+ 3- 3	+ 2- 2
		O = 7	+ 9- 6	+ 5- 4	+ 2- 2	+ 1- 1
		E = 21	+ 8- 7	+ 4- 4	+ 2- 2	+ 1- 1
		EO = 2	+ 1- 1	+ 1- 1	+ 0- 0	+ 0- 0
0.060	11.	N = 0	+ 0- 0	+ 0- 0	+ 0- 0	+ 0- 0
		P = 7	+ 7- 3	+ 3- 2	+ 1- 1	+ 1- 1
		O = 18	+20-15	+10- 9	+ 5- 5	+ 3- 3
		E = 32	+22-15	+11- 9	+ 5- 5	+ 3- 3
		EO = 43	+22-17	+11-10	+ 5- 5	+ 3- 3
0.054	0.6	N = 2	+ 4- 2	+ 2- 1	+ 1- 1	+ 0- 0
		P = 11	+ 6- 4	+ 3- 2	+ 1- 1	+ 1- 1
		O = 0	+ 5- 0	+ 1- 0	+ 0- 0	+ 0- 0
		E = 64	+10-14	+ 5- 6	+ 3- 3	+ 2- 2
		EO = 23	+ 9- 7	+ 4- 4	+ 2- 2	+ 1- 1
0.046	12.6	N = 0	+ 0- 0	+ 0- 0	+ 0- 0	+ 0- 0
		P = 1	+ 1- 1	+ 0- 0	+ 0- 0	+ 0- 0
		O = 8	+21- 8	+10- 6	+ 5- 3	+ 3- 2
		E = 10	+12- 7	+ 6- 4	+ 3- 2	+ 2- 1
		EO = 81	+14-24	+ 9-12	+ 5- 6	+ 3- 3

N = Normal subject
P = Paranormal subject
O = Obstructive subject
E = Hepatocellular subject
EO = Both Obstructive and Hepatocellular subject

In both figures is reported a shaded zone for each curve of 80% probability of membership in one of the classes. These zones show the range of variability of the 80% curves for a confidence coefficient Q = 0.24.

In table 1 are reported the probabilities of membership of some representative subjects and their confidence intervals for different confidence coefficients Q.

Table 1 is computed in the case of normal pdfs therefore it may be compared with figure 2.

2. DISCUSSION

As can be seen from table 1, when dealing with small Q the confidence intervals of the F_i are not relevant, however for a reasonable value of Q, Q = 0.95 the confidence intervals become quite relevant and must be taken into account by the decision rule.

In fact in some cases no classification can be reasonably performed since there may be an overlapping of the confidence intervals of two (or more) probability of membership.

REFERENCES

(1) J. Roy. Coll. Physics : London: 1975, Special Issue on statistical methods for clinical decision making, 9, pp. 191-269.
(2) Mendel, J.M., Fu, K.S.: 1970, Adaptive, learning and pattern recognition systems, Academic Press, New York.
(3) Fukunaga, K.: 1972, Introduction to statistical pattern recognition, Academic Press, New York.
(4) Meisel, W.S.: 1972, Computer-oriented approaches to pattern recognition, Academic Press, New York.
(5) Anderson, T.W.: 1962, An introduction to multivariate stàtistical analysis, John Wiley & Sons, New York.
(6) Parzen, E.: 1962, Ann. Math. Stat., 33, pp. 1065-1076.
(7) Belforte, G.: to appear in Lecture Notes in Medical Informatics, Springer Verlag.
(8) Belforte, G.: 1979, Internal Report CENS-CNR, Torino.
(9) Molino, G., Milanese, M., Villa, A.: 1978, J. Lab. Clin. Med., 91, p. 3.
(10) Belforte, G., Molino, G., Rossi, G.: 1977, 1st Mediterranean Conf. on Med. and Biol. Eng., Sorrento, Italy.

A SPATIAL FILTERING TECHNIQUE FOR MSS UNSUPERVISED CLASSIFICATION PROBLEMS

H. DO-TU, M. INSTALLE, J. WILMET

Louvain University - B 1348 Louvain-la-Neuve-Belgium

ABSTRACT

The authors first examine several techniques of unsupervised classification of multispectral scanner (MSS) data using spectral-spatial information and spectral information only. The shorcomings of these techniques are pointed out and a spatial filtering technique based on the contrast property is proposed in order to improve the classification using spectral information only. In addition to the greater accuracy, the advantages of the classification using this filtering technique are the preservation of realy small and filiform objects, the great consideration of local contour details of the image and the results are in a more desirable form for interpretation. Two urban-area images of Landsat 1 are used to illustrate the results of the proposed technique.

I. INTRODUCTION

One important problem in remote sensing of the earth's resources is to find a meaningful description of a set of remotely sensed data of an earth's scene provided by aircrafts or satellites. A frequently used description of a scene is the representation of all "homogeneous" objects, each of which is formed by an aggregation of spectrally similar adjacent picture elements. Without prior information about these objects, such a description implies first the recognition of homogeneous objects by computer, with or without human assistance. The recognized homogeneous objects are then interpreted (for insistance water, forest, urban area, ...) and mapped for specific uses.

D. G. Lainiotis and N. S. Tzannes (eds.), Advances in Communications, 457–464.

A picture element called pixel is defined in a geographical space (image). The information from each pixel is provided by spectral power or intensity measurements in k separate spectral bands. Let a pixel P be represented by a k-dimensional vector X called pattern in the spectral measurement space (mathematical space), the value of its ith component ; i = 1, ..., k, is the spectral intensity value measured in the ith spectral band. The recognition of homogeneous objects previously mentioned and dealt with in this paper consists in classifying all pattern X, one by one into p non-overlapping classes without any prior knowledge about these classes. Each class is formed by an aggregation of patterns. The pixels represented by the patterns of a class have then similar spectral values. If these pixels are connected one to another in the geographical space, they form a homogeneous object.

The classification method which is used is the unsupervised classification method or clustering method.
The remotely sensed information we can use to distinguish two objects on the earth's surface is their spectral measurements (visible or invisible radiation emitted and reflected by the objects), their spatial (*) characteristics(form, size, contrast, texture, structure,...) and their temporal variations. The multispectral data corresponding to one earth's image does not inform us on the temporal variation (**) and the spatial characteristics may only be obtained by processing the spectral data.

Using spectral information and spatial information, we may distinguish two kinds of methods used for processing and mapping multispectral data. In the first methods, each pixel is assigned individually to an object on the basis of its spectral measurement alone. The classification is realized in the spectral measurement space. In the second methods, the information from spectral measurements and from spatial dependance of pixels are combined for the classification. These methods assume that spatially adjacent pixels tend to belong to the same object except at boundaries.

Table 1 summarizes several methods used in the literature. It is worth noticing that : 1) The unsupervised classification of MSS data based only on spectral information may usually give unsatisfactory results and the resulting image has a noisy aspect, complex for interpretation. In order to obtain better results, it is preferable to complete them by a complementary processing which exploits spatial information. [Braconne et al., 74], [Goldberg et al., 75].....

(*) the word "spatial" refers to the pixel position in the geographical space, not that of pattern position in the mathematical space.

(**) the multitemporal classification uses more than one image and is more expensive, we are not interested in its use in this paper.

Type	Methods	Comments
1) Methods based on spectral information	1.1 migrating-means methods (ISODATA, dynamic clusters, K-means clustering,...) 1.2 histogram methods (Lallemand and Thomas, Goldberg and Schlien, ...) 1.3 histogram migrating-means methods (Dotu and Installe)	Clustering by use of spectral information and may be followed or preceded by use of spatial information for improving the results
2) Methods based on spectral and spatial information	2.1 boundary extraction and classification methods (Nagy Haralick and Dinstein,...) 2.2 object extractions and classification methods (Kettig and Langrebe,....)	Extraction of objects by use of spatial information followed by use of spectral information for clustering

2) The object extraction and classification methods are very interesting, it is realized by i) grouping first pixels into lxm pixel cells, ii) annexing homogeneous cells to form homogeneous objects, iii) then classifying homogeneous ogjects and iv) finally classifying pixels of unhomogeneous cells by the classification technique using spectral information. This method largely exploits the statistical decision theory and hypothesis testing.
From a theoretical point of view, this method answers to the texture notion in photointerpretation where a textural element is the smallest homogeneous set of pixels discernable in a photo.
From a practical point of view, the classification is based on the recognition of homogeneous objects whose dimension is larger than or equal to the basic lxm pixel cell, some small and lengthy forms (filiforms) such as road, canal, river, ..., may be mis-recognized (*).

The combination of spectral and spatial information is necessary in MSS unsupervised classification. The spatial information used in the previous techniques is simple. One may think about a more sophisticated spatial information to be used, for ins-

* In order to realize this fact, the reader may refer to the work of Swain [Swain, 76] and compare the classification results of Clermont and Bloomington (classified with spectral data only and with the ECHO tecnique, pp. 40,48,54,55).

tance the notion of texture. Until now, such an information is difficult to be used in MSS unsupervised classification, it may be combined with spectral information but does not yet give better results [Swain, 76], [Wiesma and Landgrebe, 76]. The difficulty seems to occur in the fact that the unsupervised classification is typically a pointwise classification, hence the textural information is only consistently defined in considering a sufficiently large and "homogeneous" cell around each pixel and such a cell may not be found at the boundaries of objects.

In this paper, we shall use a simple form of spectral information, that is the contrast property of adjacent pixels, an important notion in remote sensing, in order to construct a new spatial filtering technique which may be implemented for improving the unsupervised classification with spectral information only. The proposed spatial filter must precede the classification. The forms of homogeneous objects will be preserved by this technique. The proposed spatial filetring may also be used to improve the supervised classification with spectral information only.

II. SPATIAL FILTERING BASED ON IMAGE CONTRAST

Our main goal in the MSS classification problem is the recognition of homogeneous ogjects with minimal error. An object is assumed homogeneous when its constituent pixels are similar one to another ; in other words the spectral difference between two pixels of a same object is less significant than the spectral difference between two pixels belonging to two different objects. The boundary separating two objects may be discerned by their spectral contrast, it is located on the abrupt change of spectral values of two adjacent pixels. We may then test whether two adjacent pixels belong to a same object or not

On the other hand, what we are interested in, is not the spectral value of individual pixels but the spectral value which represents objects in order to recognize them. The classification result may not be changed if we replace the spectral value of two adjacent pixels of the same object by their mean. A spatial processing algorithm may then be realized as follows :
Let decompose the image into several rectangular lxm pixel subimages called windows (*). The pixels in the window are examined by the growing square technique as schematized in figure 1.
The processing within the window is realized in two phases : examination of pixel contrast and modification of spectral value.
The main objective of the processing in a window is to merge similar pixels into homogeneous elementary units called cells (**)

* this technique is better than the column by column technique used in our previous paper [Bozet et al., 79]

** An object may be formed by one cell or several connected cells located in several windows and aggregated by clustering method

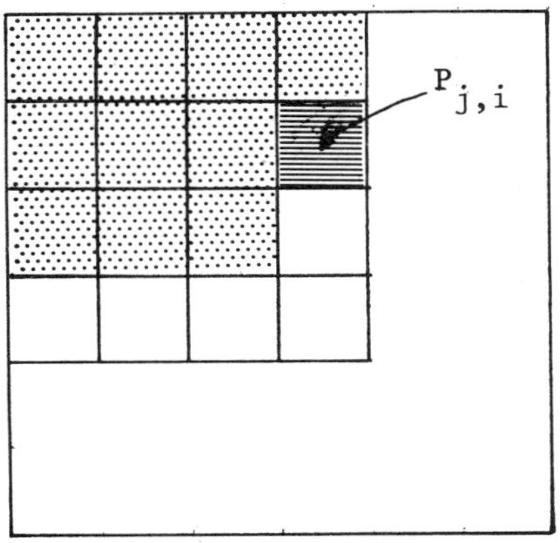

fig 1. pixels examined by the growing square technique in a 6 x 6 pixel window

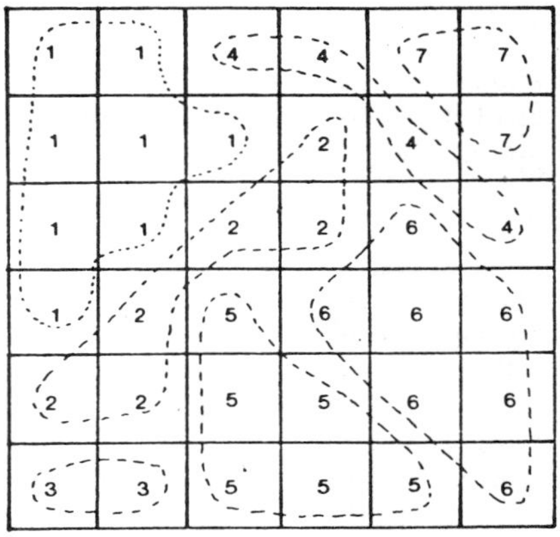

fig 2. Exemple of cells in a 6 x 6 pixel window

Phase 1 : Examination of pixel contrast

The spectral value X_{ji} of pixel $P_{j,i}$ is compared to that of its four (or less) already examined neighbors (see figure 1), corresponding to four possible directions in the image grid.

Let d denote the smallest distance of $P_{j,i}$ to its four neighbors (in spectral space). Pixel $P_{j,i}$ will be associated to the cell which contains the nearest neighbor of $P_{j,i}$ if the distance d is small, that is

$$d < \delta_o \qquad (1)$$

where δ_o is a preselected threshold.

Otherwise $P_{j,i}$ is considered to belong to a new cell. An example of a result of this phase may be schematized in fig. 2

Phase 2 - Modification of spectral value

The spectral value of every cell in the window is computed as the average of spectral values of its constituent pixels. The spectral value of each pixel is then replaced by that of the corresponding cell.

Many modifications to this basic test are possible and prefered. Indeed the test(1) does not prevent a poor contrast of several successive pixels and may allow for growing cells which become anormally large. The spectral value of a modified cell may be recursively computed after annexing a new pixel. A modified test may be implemented. In addition to condition (1), pixel $P_{j,i}$ is associated to the cell which contains its nearest neighbor if

$$\| X_{j,i} - M_h \| < \delta \qquad (2)$$

where M_h is the spectral value of cell h. If $P_{j,i}$ is annexed to cell h, the actual spectral value of this cell is recursively computed by

$$\frac{M_h \cdot n_h + X_{j,i}}{n_h + 1} \qquad (3)$$

where n_h is the number of pixels already associated to cell h prior to the observation of pixel $P_{j,i}$.

The proposed technique allows to amplify the homogeneity inside a cell and the contrast between two adjacent ones. Local contour details in the image ignored in the unsupervised classification technique with spectral information only are then well emphasized by the use of this spectral information . The essential role of the image window is to limit the size of a possible continuously growing cell and to delimit a geographical area for enhancing local contour details.... The enhancement of local contour details considers only pixels close to the examined pixel, then it implies the choice of a small size window. But, if the window size is too small, the proportion of pixels located inside but on the window boundary is high. This fact is not wanted since these pixels are only partially spatially filtered with two neighbors.

Moreover the spectral average inside a cell acts as a lowpass filter which removes additive noise due to various sources like measurement noise, spectrally non-homogeneous pixels, ... It may be expected that if the thresholdes δ_o and δ are approprietely chosen, the pixels of the filterd image will aggregate more closely to their center. Such a contraction of the cluster forms may result in an improvement of the classification.

III. A CLASSIFICATION STUDY

In order to illustrate the spatial filtering technique proposed in this paper, we shall study the unsupervised classification of two urban-area Landsat 1 images : Zaventem and Uccle (suburbans of Brussels). There are many unsupervised classification techniques using spectral information. In this paper, we shall use the technique recently proposed by [Dotu and Installe, 78] . The procedure used in this study may be briefly descried as follows :

Step 1 : Principal component analysis of the raw image. The two most important components are saved.

Step 2 : Application of the proposed spatial filtering (if desired) on the resulting two-component image.

Step 3 : Computation of the two-dimensional histogram from the two-component image resulting from step 2 (or step 1 if no filtering has been made). Unsupervised classification based on a modification of the ISODATA method.

Step 4 : Classification of pixels for displaying the image.

The previous classification technique was chosen because of its advantages :
1) The spectral values in different spectral bands being largely correlated, it is sufficient to use a principal component analysis and to retain the first two components
2) The two-dimensional histogram which contains all the information on the image may be stored in core memory
3) The total number of points to be clustered by the ISODATA method is not too large and the necessary computation time is low
4) The classification of pixels for display in step 4 may be done rapidly by use of a table look-up
5) It is preferable to have a large number of classes (15 or 20) since it is easier for the interpreter to merge several classes than to split a class. The class merging is facilitated by use of a plot of the histogram, the grouping in the histogram (which is the table look-up) on the listing and the display of the image on a TV screen.

For each of two studied images (128x128 pixel size), we carried out the classification with and without spatial filtering, using a square window. We have studied the effect of the window size by using a 4x4, 6x6, 10x10 and 20x20 square window and the effect of the threshold δ in using the values δ^2 equal to 4, 8 and 18 ($\delta_o^2 = 32$). It is observed that :
1) By using a spatial filtering technique, the filtered pixels are more concentrated to their center, that is the clusters are taller and more separated one from another. This fact is observed by comparing the histograms of the image and the filtered image. A great deal of classification noise (random errors) are removed by the filtering.
2) The filtering emphasizes the sharpness of contours of various objects. However, the filiform objects appear well when the contrast between them and their neighbor objects is important (highways, landing sites, highway access to town, ...), they tend to disappear if the contrast is poor(for example the small canal of Willebroek or some parts of highways located in the suburb.
3) The quality of the filtering depends on the choice of the window size and the threshold, namely δ. For an urban image, a window size of 10 x 10 or 6 x 6 and a threshold δ^2 equal to 18 or 8 give a better result.

IV. CONCLUSION

A spatial filtering technique is proposed in order to improve the unsupervised classification of MSS images. The filtering precedes the classification and is realized by comparing the Euclidean distance of the spectral values of the observed pixel to its nearest neighbors.

It seems that this spatial filtering may still be applied selectively in each spectral band, in this case, it appears as an enhancement of the MSS image before classification. It is the object our future study.

Acknowledgements :
The authors wish to thank M. Bozet for her discussions and comments.

REFERENCES

Bozet M., H. Dotu, M. Installé and J. Wilmet - Filtrage spatial des données Landsat en vue de la cartographie urbaine, Proc. 3rd Int. GDTA Conf., Toulouse 19-22 june 79.
Braconne S., A. Fontanel, M. Guy and C. Lallemand - Cartographie thématique automatique - Principes et réalisation partielle d'un système. Proc. Symp. on Remote Sensing and Photo-Interprétation, Banff, Alberta, Oct. 1974
Dotu H. and Installé M.- A fast clustering algorithm based on ISODATA with application on remote Sensing - Proc. 4 th Int. Joint Conf. on Pattern Recognition, Kyoto, nov. 1978
Goldberg M., D. Goodenough and S. Schlien - Classification methods and error estimation for multispectral scanner data-Proc. 3rd Canadian Symp. on Remote Sensing, Edmonton, Sept. 1975
Goldberg M. and Schlien S.- A clustering scheme for multispectral Images, IEEE Trans. on Syst. Man. Cybern., Feb. 1978
Haralick R.M. and I. Dinstein - A spatial clustering Procedure for Multi-Image Data - IEEE Trans on CAS, May 1975
Kettig R.L. and Landgrebe D.A. - Classification of Multispectral Image Data by Extraction and Classification of Homogeneous objects - IEEE Trans. Syst. Man. Cybern., Janv. 1976
Lallemand C. and G. Thomas - Traitement numérique des images multispectrales , 3e réunion de la Commission III du CFC, 9 mars 77
Nagy G. and Tolaba J.- Non supervised crop classification through airborn multispectral observations, IBM Res.Develop. vol. 16, n°2, March 1972
Swain P.H. - Land -Use Classification and Mapping by Machine Assisted Analysis of Landsat Multispectral Scanner Data, Final Rep., LARS inform. note 111276, Purdue Univ., Sept. 76
Wiersma D.J. and D. Landgrebe (1976) - The use of spatial characteristics for the improvement of multispectral classification of remotely sensed data, Proc. Symp. Machine Processing of Remotely Sensed Data, june 29 -july 1, 1976

FACTOR ANALYSIS AND CLUSTERING **

G. Della Riccia[o*] - F. De Santis[*] - A.Shapiro[o]

[o] Department of Mathematics, Ben Gurion University of the Negev, Beer Sheva, Israel
[*] Institute of Information Sciences, University of Salerno, Salerno, Italy

ABSTRACT

Two problems related to the factor analysis model are discussed in the sphere of pattern recognition applications: the determination of the communality values based on the minimization of the trace of a reduced correlation matrix and the relationship between factor analysis and Fisher discriminant analysis.

1. INTRODUCTION

Pattern recognition applications generally involve several major operations, including sensing, feature extraction and decision tasks. This paper discusses feature extraction in connection with the question of dimensionality reduction. The basic mathematical tool used is the factor analysis model. Emphasis, in our discussion, is put on two aspects: first, on the computational difficulties arising in the estimation of communality values which appear in any factor analysis treatment and, second, on an interesting connection between the factor analysis model and the Fisher's discriminant analysis.

** Work supported in part by the Group of Mathematical Information Sciences (GNIM) of the National Research Council (Italy).

D. G. Lainiotis and N. S. Tzannes (eds.), Advances in Communications, 465–469.

2. FACTOR ANALYSIS AND FEATURE EXTRACTION

Feature extraction is an essential process in a pattern recognition algorithm, since the identification and classification of patterns become a much simpler task when a "good" set of discriminatory features for each class in a collection S of objects can be singled out from a set of raw measurements.

Our approach to feature extraction is by means of the factor analysis model, in which we reduce the dimensionality of the measurement vector via a linear transformation such that correlations among input variables are preserved.

As usual in any factor analysis process (1), the starting point is the input data, represented by the outcome of $X = (X_1 \dots X_n)$ features measured on the objects of a certain population. Assuming that, for sake of simplicity, variables are normalized in the standard form (mean value 0 and standard deviation 1), the classical factor analysis model is based on the following transformation:

$$X = X' + D U \tag{1.1}$$

or equivalently,

$$X = A F + D U \tag{2.1}$$

where $F = (F_1 \dots F_m)$ and $U = (U_1 \dots U_n)$ are respectively the m-dimensional vector of common factors accounting for the correlated information in X (with $m < n$), and the n-dimensional vector of unique factors accounting for the uncorrelated information in X. Matrices A and D are respectively the $n \times m$ matrix of factor loading coefficients and the $n \times n$ diagonal matrix of unique factor coefficients. The main assumption in (2.1) is that the $n+m$ variables $F = (F_1 \dots F_m)$ and $U = (U_1 \dots U_n)$ are mutually uncorrelated. Advantages of using the vectors F and U as extracted features in pattern recognition applications were discussed in (2). Here, we only intend to recall briefly the major problems arising from model (1.1) and their possible solution.

According to the preceding remarks, let us assume that the vectors F and U are mutually uncorrelated and normalized in standard form, so that the correlation matrix R of the input data X can be written as:

$$R = R' + D^2 = A A^t + D^2 \tag{3.1}$$

where R', the so-called reduced matrix with rank $m < n$, is the covariance matrix of the new variables X'. Matrices R and R' are equal except for the diagonal terms, since D^2 is a diagonal matrix. Diagonal terms $\alpha_1^2, \alpha_2^2 \dots \alpha_n^2$, of R' are the so-called communalities of the problem. In classical factor analysis the goal is to find communality values which minimize the rank m of R' with the constraint that R' remains semipositive definite. The difficulties arising when such a problem is tried to be solved numerically are well-known and fully discussed in the current literature. In a previous work (2), it was suggested to determine the communality values which minimize the trace of R' instead of its rank. Also, it was shown in (3) that the domain of $\alpha_1^2, \alpha_2^2 \ldots \alpha_n^2$, such that R' is semipositive definite, is a convex set and that the set of communality values, such that

$$\sum_{i=1}^{n} \alpha_i^2 = tr(R')$$

is minimum, is unique. Consequently, the communality estimation problem can be easily solved using convex programming algorithms.

It is to be noticed that, in general, the minimum trace point does not coincide with the minimum rank point. Some sufficient conditions which guarantee that minimum of $tr(R')$ and minimum of rank (R') correspond to the same set of communality values have been indicated in (3). It is worthy noticing that one of such conditions requires that $m < n/2$, which implies a strong dimensionality reduction, that is a situation in which the factor analysis is particularly meeningful.

The last step in a factor analysis treatment is the computation of the factor scores that is the values assumed by $F = (F_1, F_2 \ldots F_n)$ on each object of the collection S. It is not easy to obtain the factor scores in a unique way from the decomposition (2.1), since there are $n+m$ unknowns, the m common factors $F_1 \ldots F_m$ and the n unique factors $U_1 \ldots U_n$, and only n given variables, the experimental features $X_1 \ldots X_n$. However, for practical purposes it is enough to use an estimation of the factor scores by the best approximation in the least-squares sense, namely

$$\tilde{F} = A^t R^{-1} X \qquad (4.1)$$

$$\tilde{U} = D^t R^{-1} X \qquad (5.1)$$

This is the representation of the input data either in the m-dimensional reduced space of the common factors F (correlated information) or in the n-dimensional space of the unique factors U (uncorrelated information), considering $\tilde{F}$ and $\tilde{U}$ as linear predictions derived from the observables X.

3. FACTOR ANALYSIS AND FISHER DISCRIMINANT ANALYSIS

In the preceding paragraph we have seen that a basic step in the factor analysis procedure is the decomposition of the input correlation matrix R into the sum (3.1).

It is natural, then, to examine what connection exists, if any, between this decomposition and the decomposition used in Fisher's discriminant analysis. According to the classical literature (5), a given correlation (or covariance) matrix R corresponding to a collection of objects divided into a certain number of classes, can be written as:

$$R = R_w + R_B \tag{1.2}$$

where R_w is the correlation -(covariance) matrix within classes and R_B the correlation -(covariance) matrix between classes. Thus equations (3.1) and (1.2) are two different ways of decomposing the same matrix.

According to Fisher discriminant method, the optimal directions on which the X data should be projected correspond to the eigenvectors of the matrix $R_w^{-1} R_B$.

On the other hand relation (4.1) used to determine the common factor scores, namely $\hat{F} = A^t R^{-1} X$ can be interpreted as a projection of the X data on the space spanned by the columns of the matrix $R^{-1}A$. It has been shown in (4) that, under certain rather general conditions, including the case in which the minimum of $tr(R')$ and the minimum of rank (R') are achieved for the same set of communality values, the subspace spanned by the columns of $R^{-1}A$ include the optimal Fisher's directions determined by $R_w^{-1}R_B$.

Since a factor analysis procedure does not require an a priori decomposition of the training set into given classes, it can also be used in clustering problems. What we have shown is that in clustering problems, factor analysis can have the same efficiency as Fisher discriminant analysis in classification problems, in the sense that Fisher analysis, when applicable, leads to certain optimal directions which are usually included in the subspace determined by the more general factor analysis procedure.

4. CONCLUSION

The minimum trace criterion has been experimentally tested by means of a certain number of examples and compared with other classification methods with satisfactory results, (6). Simple algorithms for the determination of the communality values which minimize $tr(R')$ have been written. However, the connection we have found between factor analysis and Fishers' discriminant analysis is a theoretical result, that we have not used yet in practice. It would be desirable in future reseach to develop a clustering algorithm taking advantage of this interesting property of factor analysis.

REFERENCES

(1) Harman, H.H.: 1967, "Modern Factor Analysis", Chicago, University Chicago Press.

(2) Della Riccia, G.: 1980, "Feature Processing by Optimal Factor Analysis Techniques in Statistical Pattern Recognition", IEEE Trans.on Pattern Analysis and Machine Intelligence, (to appear).

(3) Della Riccia,G., Shapiro,A.: "Minimum Trace and Minimum Rank of Correlation Matrices", submitted to Psychometrika.

(4) Della Riccia,G., Shapiro,A.: "Fisher Discriminant Analysis and Factor Analysis", submitted to Psychometrika.

(5) Duda, R.D., Hart,P.E.: 1973, "Pattern Classification and Scene Analysis", John Wiley and Sons, New York.

(6) Della Riccia,G., de Santis,F., Sessa, M.I.: 1978, "Optimal Factor Analysis and Pattern Recognition", Proc.of the 1978 Intern.Conf.on Cybernetics and Society, Tokyo-Kyoto (Japan), pp.1567-1583.

A METHODOLOGY FOR GRAPHICAL IDENTIFICATION

Jean-Claude COMMERÇON and ROBERT BADARD

Laboratoire d'Informatique Appliquée 502
I.N.S.A., 20, avenue Einstein
69621 - VILLEURBANNE CEDEX FRANCE

We give a graphical identification methodology which is based on a pattern recognition technique. The sampled responses of a parametered family are used as data for a factorial analysis of correspondences method. By a judicious sampling, the principal factorial plane is a good picture for the parameters estimation, in the sense that the isovalue lines are quasi- orthogonal. The experimental response is projected on the factorial axes from which the parameters are easily picked out.

In this paper we are going to deal with the parameter identification problem for elementary responses through pattern recognition techniques. Furthermore, we wish to make this identification with few primitive measurements and little calculation in order to obtain the result fast and easily. Elementary responses (impulse or initial step for example) are frequently met in practical work, for example : - Impulse response obtained by the use of covariance techniques with binary pseudo-random sequences (1). - Initial step (or impulse) response used in industrial environment or in specific work (calorimetry). - Initial step response with laser beam excited fluorescence in physical work, etc... In these situations, there are two solutions for identification : - the first uses "sophisticated programs" on top computer with batch or conversational terminal. This procedure is incompatible with the rapidity wished by the user. - The second uses the graphic method to be found in numerous papers (2), (3), ... However, these methods resolve very particular problems which are not generally the user's ones.
To solve this problem, we study methods which use few samples and work fast. In pattern recognition, we must make primitive measurements on the waveforms (impulse or initial step for exam-

D. G. Lainiotis and N. S. Tzannes (eds.), Advances in Communications, 470–477.

ple) from which we extract the features by means of a given method. We then look at two classifiers :

- the first one automatically chooses a pattern among the features, so giving the parameters tied to the waveform (for example, the two times constants for second order systems). We will principally use this method to show that the user must choose the primitive measurements with great care.
- The second shows the features on a plane representation, the abacus. The user himself chooses the "best pattern". Futhermore, in this case, the methodology is sufficiently general to enable graphic identification to be made in numerous cases. We have discussed in (6) many cases, but we cannot insert them in this paper for space consideration.

I. THE DIRECT METHOD (4)

In the direct method, we use the additional a priori hypotheses of the linear systems. For greater simplicity we first work with impulse response s (t) of second order linear invariant system with time constant τ_1, τ_2 .
The primitive measurements are sampled at time T,2.T,3.T,4.T,5.T
For feature selection, we built he array :

$$X = \begin{pmatrix} s(T) & s(2.T) & s(3.T) \\ s(2.T) & s(3.T) & s(4.T) \\ s(3.T) & s(4.T) & s(5.T) \end{pmatrix} \quad \text{Array 1 (N = 3 lignes)}$$

In three-dimensional space, these three points define a plane. The vector θ , orthogonal to this plane is the feature (the linear relation) from which we can easily extract the patterns : τ_1, τ_2. If we take additional measurements, (N lines in the array 1) we cannot generally define a plane. The classifier must then choose a plane for which the orthogonal quadratic error : λ_3, is minimum, for example. The feature, that is to say, the vector θ is easily found by solving the eigenalue problem (4) :

$${}^tX.X.\theta = \lambda\,\theta \qquad \lambda_1 \geqslant \lambda_2 \geqslant \lambda_3$$

where ${}^tX$ denotes the transpose array. However, there is much freedom for the choice of the sampling time : T, and the building of the array 1. We now have to study these "optimal choices", because with few measurement we must only lose the minimum of information for a good result, if the primitive measurements are noisy. Figure 1 shows the cluster of points corresponding to each line of the array 1 for $N = \infty$, $T/\tau_1 = 0.25$, $\tau_2/\tau_1 = 2$. However the ratio $\lambda_1/\lambda_2 = 85$ means that the second axis is not well defined in a noisy process ; the cluster is then rather like a thread and we must put on a plane on it ! This property once again explains the poor excitation of the system by an impulse input. To contend with this poor conditioning we must choose the optimal T to disperse the points, that is to say, to make the little eigenvalue λ_2 greater. Figure 2 shows the dependance of thos optimal sampling T upon N, τ_2/τ_1.

Furthermore, few primitive measurements do not give a very considerable loss of information since, for example, the ratio of the two optimal little eigen-values for $N = \infty$ or $N = 3$ is $\lambda_2^{\infty}/\lambda_2^{3} \simeq 1.3$ ($\tau_2/\tau_1 = 2$).

We shall now study the different methods of building array 1 for feature selection. Its general form may be :

$$X = \begin{pmatrix} s(nT_1) & s(n.T_1+T_2) & s(n.T_1+2.T_2) \\ s((n+1).T_1+T_2) & s((n+1).T_1+2.T_2) & s((n+1).T_1+3.T_2) \\ \ldots\ldots & \ldots\ldots & \ldots\ldots \end{pmatrix}$$

with for example $T_2=k.T_1$, $k \in 1, 2, 3\ldots$

In this way, for a given sampling rate T_1/τ_1 and a defined τ_2/τ_1, there is an optimal choice of $T_2=k.T_1$ to make the little eigen-value greater, that is to say, in order to obtain a good conditioning. For $N = \infty$, T_2 has a constant asymptotic value, then $T_1 \to 0$, that is to say, the waveform pattern is well caracterised by samples taken from T_2 to T_2. The little eigenvalue λ_2 is made greater if we reduce T_1 with optimal choice of $k = T_2/T_1$. For example, for $N = \infty$ and $\tau_2/\tau_1 = 2$ we have : $\lambda_2(k=10)/\lambda_1(k=1) = 5$.
We shall now briefly study the initial step response. The array for feature selection becomes :

$$X = \begin{pmatrix} s(n.T_1) & s(n.T_1+T_2) & s(n.T_1+2.T_2) & 1 \\ s((n+1).T_1+T_2) & s((n+1).T_1+2.T_2) & s((n+1).T_1+3.T_2) & 1 \\ \ldots\ldots & \ldots\ldots & \ldots\ldots & \end{pmatrix}$$

To obtain a free response, we can substract from each line element the means of the corresponding column ; or if we know the gain, we can substract the asymptotic line : $(s(\infty), s(\infty), s(\infty), 1)$. We thus obtain a X' array which can be used as the impulse array. The cluster in this case is still finer than that of the impulse response. It proves, once more, that the step response is worse than the impulse response. Table 1 compares this property for optimal sampling and $N = \infty$ and $\tau_2/\tau_1 = 2$.

TABLE 1

λ_1/λ_2 \ T_2/T_1	1	2	5	10
Initial step	280	425	590	680
Impulse	6.5	17	35	40

This identification method of elementary response viewed here as a pattern recognition technique is an easy and fast method. It can be solved on a programmable pocket calculator with an execution time of less than a minute. It has been shown that that we musr take great care with the choice and the use of the primitive measurements. This fact may be useful with higher sophisticated identification methods. We have also shown that the few numbers of initial measurements not very important for elementary response.

II. THE GRAPHIC METHOD

We start with some initial measurements on the unknown waveform from which we extract some features by a given method. The user must make a comparison of this unknown features with the corresponding features of known patterns. To build this known features, the user makes a learning set. It contains some representative elements of the whole user pattern specific range. The unknown pattern will be identified by searching for a feature of the learning set in its neighbourhood. As we are limiting our work to the graphic method, the pattern will be parametered with a little number of parameters, for example : 2. The method must reduce the primitive measurements to two features closely tied to the patterns. As in the direct method the choice of the primitive measurements is of the great importance. The method we have used to transform the primitive measurements into features is a factorial analysis where the principal factorial plane will be the abacus. We have used two factorial analysis : one, with the usual euclidian distance, the other with the χ^2 distance (factorial analysis of correspondances (5)). We slightly prefer the second method because it appears to give better features in our application. However, we are convinced that the fundamental hypotheses of this method aretrangressed ; for this reason, we would point out that we shall use a factorial analysis with a distance which has the same mathematical expression as that of the χ^2, with one constraint : The marginal must be positive ! We would also point out that we do not use the "correspondance idea" of this method.

Example 1 : Second order impulse response (real and complex poles). We will obtain a general abacus which covers the range $a=\tau_2/\tau_1=16$ to $a=1$ (real poles) and the range $\xi=1$ to $\xi=0.4$ for the damping factor (complex poles) in the models :

$K/((1+\tau_1 p).(1+\tau_2 p))$ (real poles) and $K/(p^2+2.\xi.\omega_n.p+\omega_n^2)$ (complex poles). These ranges hold, for the first parameter. The second parameter (second order process) is the time $\hat{t}_m$ of the first maximum of the waveform. However, we can make an error in its determination. Its range will be therefore: $t_m=\hat{t}_m.(1+\rho)$ with $\rho=\pm 10\ \%$. We can obtain : $\tau_2=\hat{t}_m(\tau_2/\tau_1-1)/\ln(\tau_2/\tau_1)$ (distinct real poles) ; $\tau_1=\tau_2=\hat{t}_m$ (double poles) ; $\omega_n=\mathrm{Arctg}(\sqrt{1-\xi^2}/\xi)/\sqrt{1-\xi^2}/\hat{t}_m$ (complex poles) : thus, the learning set has, for example, 30 different patterns with the combination of the two ranges : $a=16$; 5 ; 1 ; ξ = 0.8 ; 0.6 ; 0.4 ;

ρ =+ 10% ; + 5% ; 0% ; -5% ; -10%. The primitive measurements are obtained, for example, by sampling at : $t_m/2$; t_m ; $4.t_m$.

The abacus built is shown in figure 3. The two axes are closely correlated with each pattern. From the primitive measurements : ${}^tX = (x_1 ; x_2 ; x_3)$ of the unknown waveform, we obtain the entries : a_1 and a_2 of the abacus by : $a_1 = {}^tX.V_1/P$

$a_2 = tX.V_2/P$ with $P = \sum_i x_i$ and where V_1 and V_2 are the eigenvectors of the principal factorial plane. The patterns are obtained by interpoling between the iso-criterium curves around this point. If we are not on the 0% iso-criterium curve, we must correct the position of the maximum. For verification, we can sample the unknown waveform once again and obtain new entries on the abacus. If it is still not on the 0% iso-criterium curve, this will come from noise or non-second order process.

Example 2 : Second order initial step responses (real and complex poles). We will obtain a general abacus (figure 4), with the same range of the first pattern as for impulse responses. The second pattern is the time of the inflexion point with a range of ± 10%. This abacus has two advantages over the HOERNER (2) method : - It covers any range of the parameters since the range is given by the learning set in which the user gives his own waveforms. - The iso-value curves are more orthogonalised.

CONCLUSIONS

The methodology based on the pattern recognition techniques is an interesting manner of dealing with the identification of parameters tied to elementary responses. The direct method, a very usual one in parameters identification, shows that there is an "optimal manner" of using the primitive measurements. This fact is generally forgotten when we use higher sophisticated method. We would point out that before using optimal input, we must take more care with the existing data.

The graphic method also uses a pattern recognition technique. It uses a learning set which exactly covers the range of the user parameters. It will be seen as a model method where the classifier work is left to the user. Its principal feature is that the user may build an abacus very quickly. This abacus can be oriented towards his own applications by his choosing his own waveforms in the learning set.

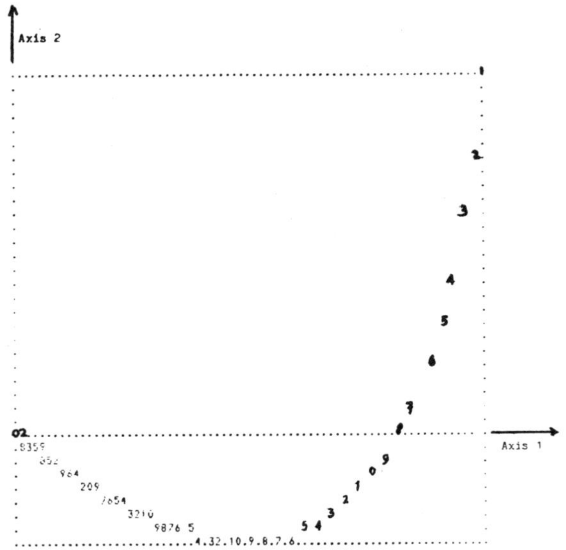

Figure 1 : $N = \infty$, $T_2/T_1=5$, $T_1/\tau_1=0.12$, $\tau_2/\tau_1=2$

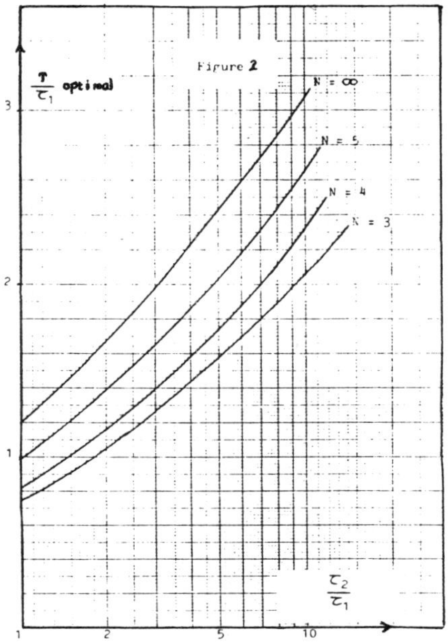

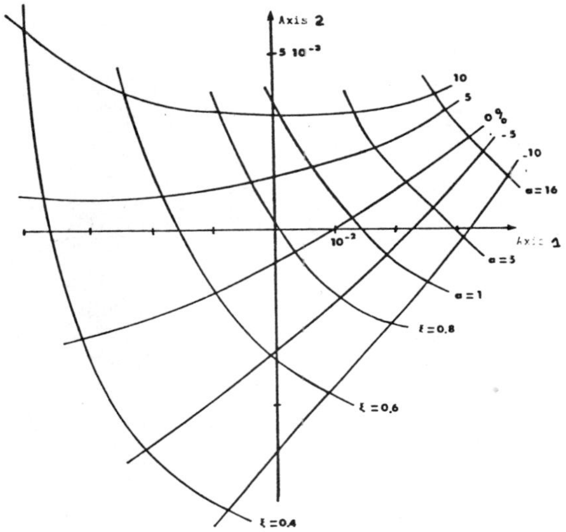

	Axe 1	Axe 2
v	$-1.41\ 10^{-1}$	$2.69\ 10^{-1}$
	$1.22\ 10^{-1}$	$-1.30\ 10^{-1}$
	$1.39\ 10^{-1}$	$6.79\ 10^{-3}$
λ	$\lambda_1 = 2.5\ 10^{-2}$	$\lambda_2 = 2\ 10^{-4}$

Figure 4

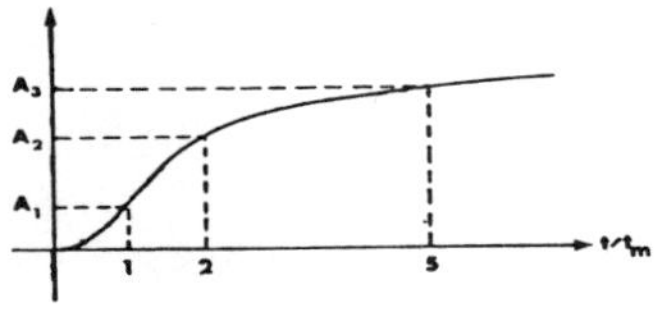

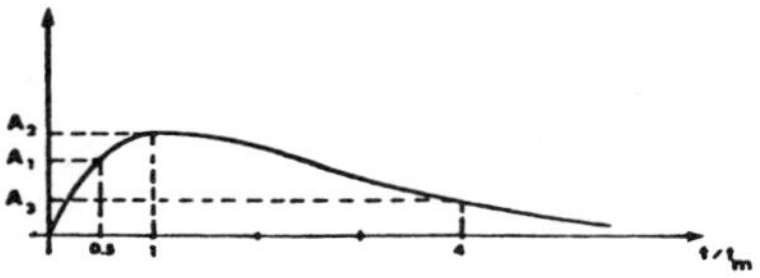

	Axe 1	Axe 2	
v	$3.27\ 10^{-2}$	$1.51\ 10^{-1}$	x: A_1
	$4.35\ 10^{-2}$	$1.21\ 10^{-1}$	A_2
	$-4.36\ 10^{-1}$	$1.73\ 10^{-2}$	A_3
λ	$\lambda_1 = 2.3\ 10^{-3}$	$\lambda_2 = 3.8\ 10^{-4}$	

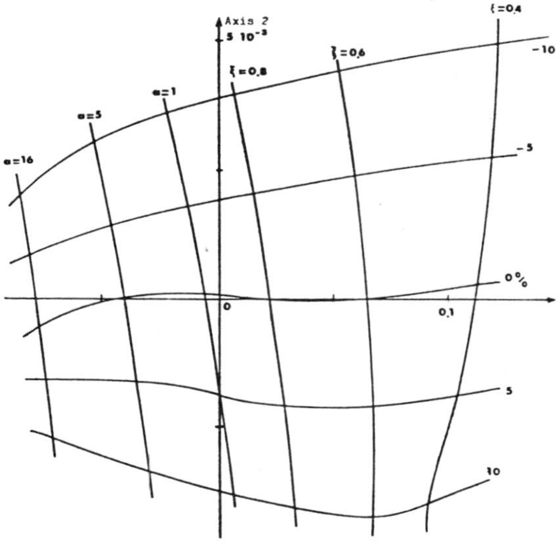

Figure 3

BIBLIOGRAPHY :

1 - J.C. COMMERÇON, R. BADARD
"Deconvolution et corrélation dans la détermination des reponses impulsionnelles" AUTOMATISME, n° 8, 1979, to appear.

2 - G.H. HOERNER
"Second order systems characteristics from initial step responses" CONTR. ENGNG., Vol. 9, n° 12, 1962, P. 93-94.

3 - V. STREJC
"Détermination approchée des caractéristiques d'un processus à réponse apériodique" AUTOMATISME, tome V, n° 3, 1960 p. 109-111.

4 - R. BADARD, J.C. COMMERÇON
"Building of robust "orthogonal estimators"" Second International Conference on Information Sciences and Systems. PATRAS, 1979.

5 - J.P. BENZECRI
"L'analyse des données", tome 2 : "L'analyse des correspondances", DUNOD, 1973

6 - J. C. COMMERÇON, R. BADARD
"A methodology for graphical identification" Report n° 20 05 79 Département Informatique Appliquée - INSA LYON.

TRANSMISSION LINE PROTECTION USING PATTERN CLASSIFICATION TECHNIQUES

Evangelos LEKATSAS & John SAKELLARIOU

Public Power Corporation Greece

ABSTRACT

In this paper the possibility of using a digital computer relay that will correctly recognize the state of a transmission line by taking a vector of measured quantities (voltages and currents) and classifying it appropriately using pattern classification methods is investigated.

The functioning of two such relays at the two ends of the line operating in conjunction via a communication link is examined with encouraging results. Each relay receives complete information of the state variables (voltages and currents) from both ends of the line, forms an overall pattern which is finally used for the classification process.

1. INTRODUCTION

In the literature of the past 15 years there have been a number of papers proposing the application of digital computers to various functions within a power system. Some papers propose the use of digital computers for various data acquisition functions and some limited monitoring and control functions. Some other papers present various algorithms for digital computer protective relaying functions.

To date there have been very few computers installed in power systems for protection purposes, the main reason being economics. However, the use of higher and higher transmission voltages increased the importance as well as the total cost of a transmission line and its associated equipment and thus augmented the need of faster and more accurate protection means and justified - from the economical point of view - the

D. G. Lainiotis and N. S. Tzannes (eds.), Advances in Communications, 478–484.

use of digital computers for this purpose. It must be emphasized here that it is the advent of microprocessors that enhanced the trend towards this direction.

The task of a relay protecting a transmission line is to continuously check the existing operating conditions of the line, sense the presence of an abnormality, classify it according to its type (e.g. three phase fault, single phase fault, etc.) and finally decide, when necessary to give the tripping signal to the circuit breaker.

To this end electromechanical and electronic relays have been succesfully used in the past. But, as explained above, once the decision is made in favor of a computerized protection scheme one has the possibility to abandon classical methods and use completely new concepts, thus further improving the performance of the system.

The possibility of using a pattern classification scheme performing all the functions of a single distance relay was successfully investigated in a previous paper [1] . In that paper a cross-correlation technique is described with which the fundamental harmonic of each input signal (voltage or line current) to the relay is recovered. The resultant filtered inputs are then transformed into an n-dimensional feature vector and fault classification is accomplished using a pattern recognition scheme with linear discriminant functions first described by Duda and Fossum [2] . Once the type of fault is isolated a distance calculation is performed via an impedance calculation. The proposed relay senses the presence of a fault within 6 to 15 milliseconds from the time of its occurence.

It is the purpose of this paper to extend these basic concepts of the previous paper [1] by examining the functioning of two relays at the two ends of the transmission line operating in conjunction via a communication line.

2. FUNCTIONAL DESCRIPTION OF THE PROTECTION SYSTEM

We shall assume that there are two idendical relays, one at each end of the transmission line to be protected. Each relay receives complete information of the state variables (i.e. voltages and currents) from both ends of the line and forms an overall pattern which is finally classified to the proper fault category with the aid of linear classifiers. No distance calculations are needed in this case and a much faster response is obtained. However, it must be pointed out here that the transmission of the necessary information from the one end of the line to the other and vice versa increases significantly the cost of the protection, but we hope that for the Ultra High Voltage (UHV) transmission lines, for which such a fast protection scheme is proposed, such a cost increase is negligible compared to the overall costs of the power line itself and the associated UHV equipment.
A functional diagram of the proposed system is shown in Fig. 1.

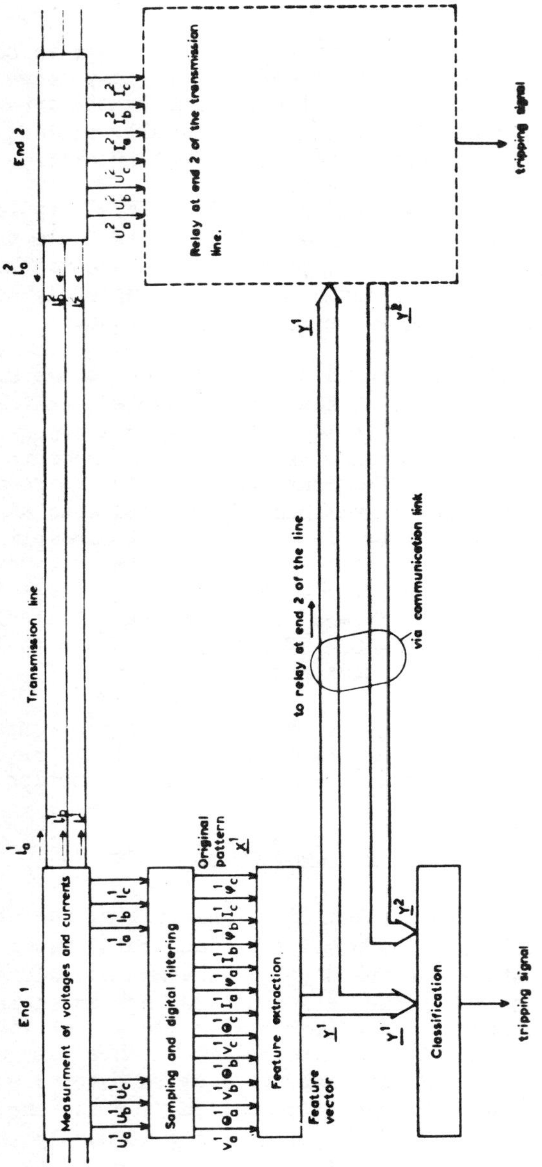

Fig. 1. Functional description of the computerized relaying system proposed.

At each end of the line all phase to ground voltages U_a^i, U_b^i, U_c^i and all line currents i_a^i, i_b^i, i_c^i (i = 1 or 2) are measured, sampled and filtered in the same way as was explained in the previous paper [1]. For the resulting original Pattern

$$\underline{X}^i = (V_a^i, \theta_a^i, V_b^i, \theta_b^i, V_c^i, \theta_c^i, I_a^i, \varphi_a^i, I_b^i, \varphi_b^i, I_c^i, \varphi_c^i) \qquad (1)$$

a suitable feature vector $\underline{Y}^i$ is formed as explained subsequently.

As is shown in Fig. 1 it is this vector $\underline{Y}^i$ that it is transmitted to the relay at the other end of the line . Of course one may choose a different solution e.g. transmit only the measured instant values of voltages and currents at end 2 $(U_a^2, U_b^2, U_c^2, i_a^2, i_b^2, i_c^2)$ and form from it the feature vector $\underline{Y}^2$ locally at end 1. The solution chosen in Fig. 1 is simpler from the point of view of the computations required at each end of line but imposes heavy requirements on the communication channel especially when vector $\underline{Y}^i$ is of high dimensionality (higher than 6). However, for the purposes of the present preliminary stage of the research carried out we may assume that at each end of the line we have at hand both feature vectors $\underline{Y}^1$ and $\underline{Y}^2$ no matter how the information is exchanged between the relays at the two ends of the line. Thus we assume that the overall feature vector is

$$\underline{Y} = (\underline{Y}^1, \underline{Y}^2) \qquad (2)$$

This vector $\underline{Y}$ is finally used in the classification process. The classifier classifies the pattern $\underline{Y}$ into one of the following 5 categories: (1) Normal operation (i.e. without any kind of fault present).(2) Three phase fault or double phase fault. (3) Single phase fault on phase a.(4) Single phase fault on phase b. (5) Single phase fault on phase c.
This is accomplished by computing five linear discriminant functions of the form

$$g_p(\underline{Y}) = \underline{C}_p^t\, \underline{Y} + T_p \quad (p=1,2,3,4,5) \qquad (3)$$

and then by classifying $\underline{Y}$ in category q if

$$g_q(\underline{Y}) = g_p(\underline{Y}) + \varepsilon \quad , \forall p \neq q \qquad (4)$$

In Eq.s (3) and (4) $\underline{C}_p$ and T_p are the vectors of weights and the thresholds for the p-th category and ε is a separation margin. The classifier's performance depends on the values of the weight vectors $\underline{C}_p$ and the thresholds T_p. These are found during a "training" phase by some iterative algorithm as was explained in [1] .

3. THE FEATURE EXTRACTION AND CLASSIFICATION PROBLEM

The extraction of the important characteristics (features) of a pattern in order to enhance the performance of the classifier is a difficult problem with no general formulation and solution, simply because it depends on the type of the pattern recognition problem itself. Having this fact in mind we tried intuitively to transform the vector $\underline{X}^i$ obtained from the digital filter, into an n-dimensional feature vector $\underline{Y}^i$ which we then used as input to the classifier.

Thus using the voltages and currents involved in e.g. (1) in complex form the following impedances are calculated:

$$Z_a^{oi} = \frac{V_a^{oi}}{I_a^{oi} + G^o I_R^{oi}} , \quad Z_b^{oi} = \frac{V_b^{oi}}{I_b^{oi} + G^o I_R^{oi}} , \quad Z_c^{oi} = \frac{V_c^{oi}}{I_c^{oi} + G^o I_R^{oi}} \qquad (5)$$

$$Z_{ab}^{oi} = \frac{V_a^{oi} - V_b^{oi}}{I_a^{oi} - I_b^{oi}} , \quad Z_{bc}^{oi} = \frac{V_b^{oi} - V_c^{oi}}{I_b^{oi} - I_c^{oi}} , \quad Z_{ca}^{oi} = \frac{V_c^{oi} - V_a^{oi}}{I_c^{oi} - I_a^{oi}}$$

where the factor G^o depends on the characteristics (per unit length positive and zero sequence impedance) of the line ($G^o = \frac{Z_0^{oi} - Z_1^{oi}}{3Z_1^{oi}}$) and $I_R^o = I_a^o + I_b^o + I_c^o$ is the residual current.

Because of their complex nature these six impedances give a 12-dimensional feature vector.

$$\underline{Y}^i = (Z_a^{oi}, Z_b^{oi}, Z_c^{oi}, Z_{ab}^{oi}, Z_{bc}^{oi}, Z_{ca}^{oi}) \qquad (6)$$

The above described classification procedure using the feature vector $\underline{Y}$ unfortunately did not give any substantially better results than those described in Ref. [1]. Thus it was decided to change the method of extracting the features of the patterns as explained in the following.

In order to visualize how the vector $\underline{Y}^i$ behaves during normal conditions as well as during abnormal conditions we can plot each one of the above six impedances as a point on the Z = R+jX complex plane. During normal conditions all six impedances are equal i.e.

$$Z_a^o = Z_b^o = Z_c^o = Z_{ab}^o = Z_{bc}^o = Z_{ca}^o \qquad (7)$$

because of symmetry. This means that during normal conditions all six impedances are mapped on the same point N on the complex plane as shown in Fig. 2 a,b and c. When a fault occurs on the line the points representing the above impedances on the R-X plane move in manner depending on the type of fault, as shown in Fig. 2., until, after a transient period of approximately 1 cycle, they settle down to new positions. In case of a three phase fault all six impedances settle down to the same final point P on the line characteristic as shown in Fig. 2.a. It is this difference in the way these points move on the complex plane during the transient period that is used by the

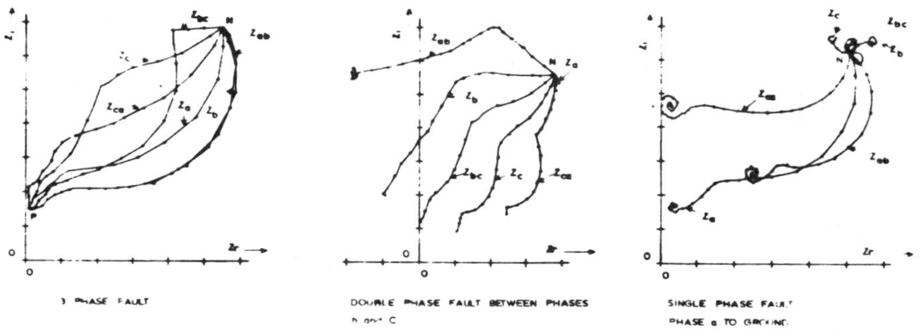

Fig. 2. Mapping of impendances on complex plane

classifier to correctly classify the patterns the soonest possible. Research is still going on towards this direction with encouraging results. A scheme that seems to be promissing involves 3 circles centered at point N (normal condition) as shown in Fig. 3. In this same figure the successive positions of the points representing the line impedances for the next 3 msec after a fault are shown. The following criteria are used: (a) A point is considered to have moved only if within this time interval (3 msec) it comes out of circle R_1. Other wise it is considered that it has not essentially moved. (b) A point is considered to move fast if within this time interval the point comes out of circle R_3 e.t.c. From Fig. 2 it must be observed that all six impedances move fast away from point N in case of a 3-phase fault; for a double phase fault between phase b and c Z_a is considered as essentially immovable; for a single phase fault on phase a all impedances involving phase a (i.e. Z^{o}_{a}, Z^{o}_{ca}, Z^{o}_{ba}) move fast away from point N.

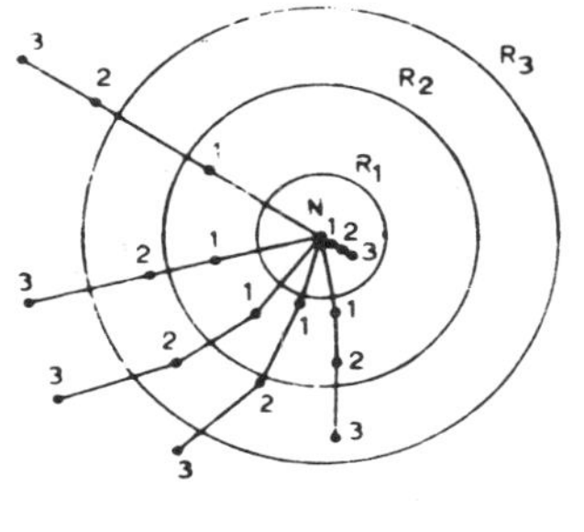

Fig. 3. Initial movement of impedances on complex plane after a fault

Using this facts and the criteria described above we deviced an algorithm with which, by proper adjustment of the radii R_1, R_2 and R_3, all types of faults can be correctly, classified.

It must be emphasized however that it will take time before all cases of faults at all positions along the line or outside it with all possible line loads and for various line lengths, will be examined and thoroughly tested.

REFERENCES

1. E. LEKATSAS; J. SAKELLARIOU: "Transmission Line Distance Protection Using Pattern Recognition" Proc. of the 1st International Conference on Information Sciences and Systems, Patras, Greece 1976 p.p. 770 ÷ 775.

2. R.O. DUDA; H. FOSSUM : "Pattern Classification by Iteratively Determined Linear and Piecewise Linear Discriminant Functions" IEEE Trans. Elec. Computers, EC-15, No.2, April 1966,p.p.220-232.

FIELD BOUNDARY LOCATION BY PATTERN RECOGNITION METHODS

Thomas M. Simundich

Department of System Science
University of California
Los Angeles, California 90024

ABSTRACT

Classical domain identification for Distributed Parameter Systems requires that the governing PDE be known, otherwise the problem is ill-posed. For many field problems such as those arising in Engineering the system parameters as well as the complete system structure is unknown. Hence, strong assumptions are made on the structure of the PDE as well as the parameters to facilitate domain identification. This often leads to unsatisfactory models. In this paper a methodology utilizing pattern recognition techniques is suggested as an alternative to classical domain identification. The patterns are cartesian products of the domain and boundary conditions. Features are extracted from the field observations allowing a classification of the boundary location.

INTRODUCTION

The boundary location problem, or as it is often referred to in the literature the domain identification problem (1,2), can be briefly stated as follows. Let Ω_0 and Ω_L be two bounded sets in R^m. Consider that there are given system observations

$$z(x,t) = H\phi(x,t) + \eta$$

where H is the restriction of ϕ on Ω_0, $x = (x_1,x_2,\ldots x_m)$, η is appropriate noise and the dependent variable, ϕ, is a solution to a partial differential equation (PDE), for example

D. G. Lainiotis and N. S. Tzannes (eds.), Advances in Communications, 485–489.

$$\sum_{k=1}^{m} \frac{\partial}{\partial x_k} \left[\sum_{i=1}^{m} a_k(x,t,\phi) \frac{\partial \phi}{\partial x_i} \right] + b_k(x,t,\phi) \frac{\partial \phi}{\partial x_i} \Bigg\} + c(x,t,\phi)$$

$$= \frac{\partial \phi}{\partial t} + u_\Omega(x,t) \qquad (x,t)\ \varepsilon \Omega x(0,t)$$

subject to the conditions

I.C. $\phi(x,0) = \theta(x)$ $\qquad x \varepsilon \Omega$

B.C. $d(\Gamma,t,\phi) \frac{\partial \phi}{\partial \Gamma} + e(\Gamma,t,\phi) = 0$

u_Ω is a distributed input and a_k, b_k, c, d and e are functions. Then, find $\Omega(\Omega_0 c \Omega c \Omega_L)$. That is locate the boundary Γ. (See figure)

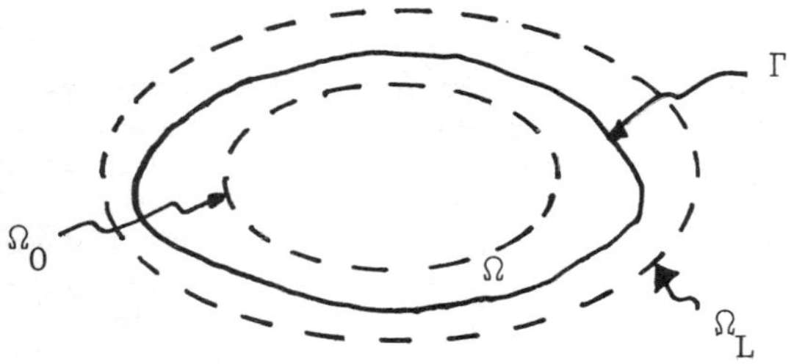

The classical solution methodology is to determine Γ by minimizing a functional

$$J(\Omega) = \| \phi(x,t) - z(x,t) \|_{\Omega_0}$$

with respect to Ω over the set Ω_0.

There are many problems in applying the classical methodology. First the problem is ill-posed since the following must be known:

(1) The structure of the PDE
(2) The numerical values of the parameters in the PDE
(3) The initial conditions
(4) The input u_Ω

Second there are numerical difficulties in minimizing J(Ω). Perturbating the boundary being one of the chief problems.

Hence a solution methodology is sought which assumes given only the system observations and the structure of the boundary conditions. The later is needed in that the mathematical location of a boundary varies with the type of boundary condition assumed. Such a methodology is next described which utilizes pattern recognition techniques.

PATTERN RECOGNITION

Pattern recognition (3,4) is the classification of input data $\{z_1,z_2,\ldots,z_n\}$ into identifiable patterns $\{P_1,P_2,\ldots,P_J\}$ by means of extracting features which are characteristic of a pattern. In order to employ pattern recognition techniques to the boundary location problems the following items must be determined: the patterns, feature extraction and classification.

SOLUTION METHODOLOGY

The first step of a pattern recognition methodology is the pattern designation. This is accomplished in two steps. First the class of boundary conditions is subclassed according to structure. For example

$$(B.C.)_1: \quad \phi(\Gamma) = 0 \qquad \text{Dirichlet Condition}$$

$$(B.C.)_2: \quad \left.\frac{\partial\phi}{\partial\Gamma}\right|_{\Gamma} = 0 \qquad \text{Neumann Condition}$$

In the second step distinct open connected sets $(\Omega_b)_i$ are designated which are likely to contain the boundary Γ. The $(\Omega_b)_i$ are to be designated such that

$$\bigcup_{i=1}^{m} (\Omega_b)_i = \Omega_L - \Omega_0$$

$$(\Omega_b)_i \cap (\Omega_b)_j = \emptyset \qquad i \neq j.$$

For example if in two dimensions $\Omega_0 = (x_1,x_2 | x_1^2+x_2^2 < r_0^2)$ and $\Omega_L = (x_1,x_2 | x_1^2+x_2^2 < r_L^2)$ where $r_0 < r_L$, then the $(\Omega_b)_i$ can be designated as concentric rings $(\Omega_b)_i = (x_1,x_2 | r_{i-1}^2 < x_1^2+x_2^2 < r_i^2)$.

Finally the patterns are designated by taking pairs of $(B.C.)_i$ and $(\Omega_b)_j$, for example

P_1: $(B.C.)_1 \times (\Omega_b)_1$

P_2: $(B.C.)_2 \times (\Omega_b)_1$

P_3: $(B.C.)_1 \times (\Omega_b)_2$

P_4: None of the above

Feature extraction may be effected as follows;

(1) Subset the system observations $\{z\}_\ell$, for example utilizing time as an index, that is the set of observations at time t_ℓ.

(2) Fit the data with a smooth function $F_\ell(x)$.

(3) Apply the boundary operator described by $(B.C.)_j$ to $F_\ell(x)$ and solve for the zeroes of the resulting equation.

(4) If a zero lies in $(\Omega_b)_i$ set

$$b_i \leftarrow b_i + 1$$

(5) Repeat steps 3 and 4 for all boundary operators described by $(B.C.)_j$. If step 4 does not result in incrementing a b_i then increment b_f that is

$$b_f \leftarrow b_f + 1$$

(6) Repeat steps 2, 3, 4 and 5 with another subset of observations.

Classification can be effected by implementing the decision rule.

Decide: P_i if

$$b_i > b_j \qquad i \neq j$$

and

$$b_i > b_f$$

otherwise make no decision.

FINAL REMARKS

This paper is a preliminary description of applying pattern recognition techniques to the boundary location problem. Some research areas being investigated by the author are:

(1) The influence of boundaries on observations. The dynamic behavior of a system of observed far away from a boundary may not be sensitive at all to the boundary location.

(2) The affect of the smooth function $f_\ell(x)$ on the pattern recognition methodology.

(3) The creation of confidence intervals for any classification.

REFERENCES

(1) Chenais, D. "On Existence of a Solution in a Domain Identification Problem": 1975, Jour. Math. Anal. and Appl. 52, pp. 189-219.

(2) Chen, W. and Seinfeld, J. "Estimation of the Location of the Boundary of a Petroleum Reservoir": 1975, Soc. Petr. Engr. Jour., pp. 19-38.

(3) Meisel, W. S. Computer-Oriented Approaches to Pattern Recognition: 1972, Academic Press, New York.

(4) Duda, R. D. and Hart, P. E. Pattern Classification and Scene Analysis: 1973, Wiley, New York.

A CASE OF SPEAKER IDENTIFICATION

M.Koukias and G.Kokkinakis

Laboratory of Wire Communications, School of Engineering, University of Patras, Patras, Greece.

ABSTRACT

The possibility of identifying a speaker is examined, when strong differences exist, both in the type of speech and in the voice-recording conditions.

The spontaneous speech of a person during several conversations was compared with the voice of the same person when reading a text. The conversations were telephone calls made in a period of one month and transmitted by a minitransmitter to a recorder several tens of meters away, without knowledge of the speaker. The read text was recorded in a single session, with the speaker speaking directly before a microphone.

Comparisons were performed by visual inspection and measurements of spectrograms and by listening, using identical words and phrases. Comparisons were also made on identical utterances of a small number of other speakers.

The results of listening tests, spectral envelope measurements, and matching tests of spectrograms are discussed.

I. INTRODUCTION

The work described in this paper has been performed some time ago, as a response to a court decision asking a judgment on the identity of a female speaker from a voice-allegedly belonging to her-which was recorded in a magnetic tape. The tape contained several telephone conversations, recorded-as we were informed-on

D. G. Lainiotis and N. S. Tzannes (eds.), Advances in Communications, 490–497.

a common tape deck via a wireless link of some tens of meters, in a period of one month.

In order to compare the "known" voice of the accused person with the unknown of the magnetic tape, this person was asked to read a text with several phrases contained in the tape. The text was recorded in our laboratory directly on a tape deck recorder, during a single session.

Before describing the methods used for the voice comparison and the results obtained, it is necessary to summarize the differences which existed from the beginning between the two voices:

a. The unknown voice came from spontaneous, conversational speech, while the known voice from ordinary reading.
b. The known voice was recorded in a few minutes on a single session and could therefore contain no major variations. The unknown voice was recorded in several occasions in a longer period and many factors have played a role during its formation (topic of the conversation, time of telephone call, physical and emotional condition of the speaker etc.).
c. The recording in the laboratory had probably caused, under the circumstances, an emotional stress to the speaker, affecting the voice, not existent in the telephone conversations.
d. In case of an identity of the two speakers it is probable that the known speaker had tried to change her voice speaking for example quickly or monotonically etc.
e. The recording conditions were quite different in both cases. The recording of the unknown voice contained several distortions and noise, due to the wireless transmission. The direct recording in the laboratory was in contrast of a good quality.

Because we did not then possessed the necessary instrumentation as to apply automatic procedures for spectrum analysis and processing of voice patterns, an audio sound spectrograph was used to this purpose. (A Sona-Graph 6061 B of KAY Electric Company).

Spectrograms of two types have been obtained from this spectrograph: Three dimensional time-frequency-intensity patterns for the display of words and phrases, and two dimensional frequency-intensity patterns, the so called "sections", taken in definite time points, for the frequency analysis of vowels.

The spectrograms of both speakers (known and unknown) were compared with each other and with the spectrograms of the same utterances obtained from five other "known" female speakers. These speakers were chosen by chance with the only criterion to be some how of the same age with the accused person. In addition, the voices of all speakers were compared by listening tests.

The comparisons have shown a rather high similarity between the accused and the unknown speaker as it is described below. But similarities or differences have little value concerning the validity of the identification method, in so far the probability of an error can not be estimated. As the unknown voice does not belong necessarily to a closed set of known size, but may belong to a person among million unidentified persons, it is impossible to estimate such a probability, until the comparison has included a sufficient large portion of the population. Such extensive studies have not yet been made to our knowledge with any method, and the available results are inadequate to establish the reliability of voice identification from an open set by spectrograms, listening, or other method (1,2).

So, concerning the legal aspect of the examined case we were obliged to underline in our report the limited value of the observed similarity between the two voices.

From a scientific point of view on the other hand the tests we have performed obtain a validity in the case that the unknown speaker was a member of the group which was examined.

In fact, the known and the unknown voice belonged to the same person, as it came out later. So the case reported in this paper can be interpreted as the identification of a speaker from a closed set of 7 persons given the peculiarities of the unknown voice described in points a to e.

II. IDENTIFICATION METHODS

The methods we have used in comparing the 7 voices comprised: a. Listening tests, b. Matching and sorting tests of spectrograms, and c. Spectral envelope measurements. The methods a and b are subjective, while method c is objective (1,2).

a. Listening tests

A set of 50 phrases of the unknown speaker containing some 250 words, was submitted to 10 listeners without special training in listening tests. After that the same phrases of the 6 known speakers were successively presented. The listeners were asked to judge, to which voice the unknown voice is similar. Each listener was permitted to listen the phrases many times before making a decision. The test was repeated three times with different phrases. An average of 56% of the listeners identified correctly the unknown voice. The remaining listeners attributed the unknown voice to a single other person of the set.

Listeners who gave correct answers stated after that, that they made their judgement on the basis of some characteristic features they discovered in the unknown voice, as for example the pronunciation of /s/ and /u/, the lower pitch, an abrupt termination of words and phrases, etc.

b. Spectrogram tests

The spectrogram tests we performed consisted in visual inspections of amplitude-time-frequency spectrograms taken with a broadband filter of 300 Hz in a frequency range from 80 to 8000 Hz (3). The tests were of two types: Sorting tests and matching tests. All tests were performed by untrained persons.

Sorting tests: In sorting tests the observer sorts a set of spectrograms containing a test word or phrase uttered a given number of times from each speaker, into individual speaker categories.

First, a set of 12 spectrograms was submitted to each of 16 observers. The set contained twice a phrase with 6 words from each of 6 speakers. The unknown speaker was not in the set. The task was to sort the spectrograms in 6 categories corresponding to the individual talkers. The test was repeated two more times with other phrases. An average of 73.44% of all couples was sorted correctly.

In three subsequent tests with 14 spectrograms each and different phrases the unknown voice was included. This time an average of only 32.2% was sorted correctly. The spectrograms of the unknown speaker were sorted correctly 20%. The high error rate in this case came out, because 66% of the observers interchanged the spectrograms of the unknown and the accused person, or 66% of the spectrogram couples of these persons were interchanged. As the interchange of spectrograms between other persons was scattered and was by far lower than this rate, this was an indication for the identity of the two persons.

Matching tests: In matching tests the observer identifies spectrograms of unknown talkers by matching them against spectrograms of known talkers, containing the same word or phrase.

We have performed three matching tests with 7 spectrograms each and different phrases. Each phrase contained a different number of words. The correct results had an average of 58.9%.

Incorrect matching was scattered to other talkers, the next higher identification percent of a talker being in each test considerably lower than the correct percent.

c. Spectrum envelope measurements

In order to get an objective measure of the similarity between the spectrograms, 5 sections of the vowels /a/ and /e/ were taken from different utterances of all speakers, i.g. from different consonantal environment of /a/ and /e/ (Fig.1).

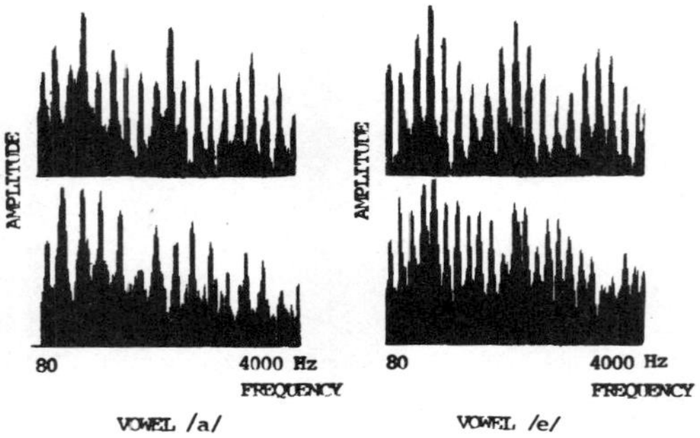

Fig.1. Two sections of /a/ and /e/ of a speaker

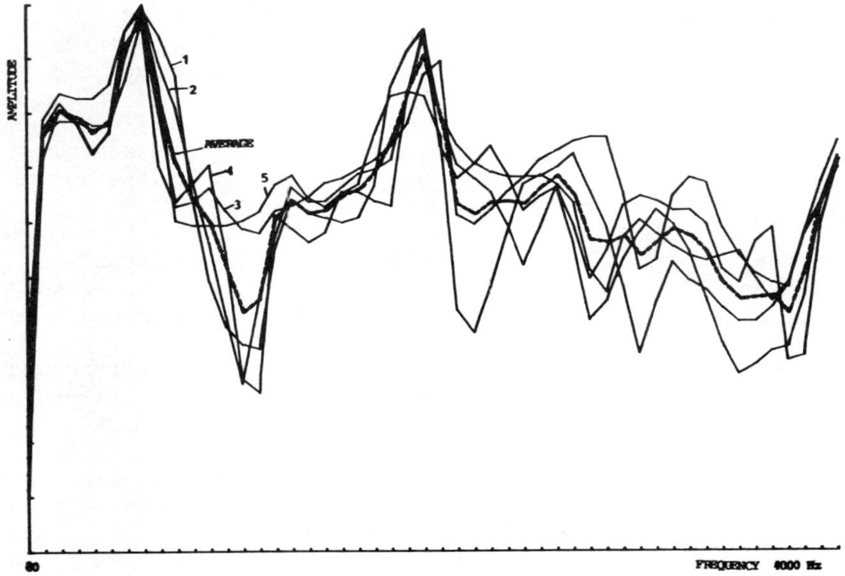

Fig.2. Average envelope of a speaker

The envelope of each section was digitized by means of a HP digitizer (HP 9874 A) in a range of 4000 Hz and supplied in to a computer (HP System 25). Fifty samples in a distance of 80 Hz from each other were taken in each section.
From the envelopes of the same talker an average envelope was computed (Fig.2). The average envelope of the unknown speaker was then correlated to each average envelope of the known speakers by means of the computer. The results are shown in Fig.3.

SPEAKER	$\bar{A}$ (accused)	$\bar{B}$	$\bar{C}$	$\bar{D}$	$\bar{E}$	$\bar{F}$	$\bar{X}$	VOWEL
$\bar{X}$	0.82827	0.82192	0.51047	0.28920	0.11826	0.46269	1.0	α
$\bar{X}$	0.82658	0.74347	0.72361	0.67891	0.44884	0.64202	1.0	ε

Fig.3. Correlation coefficients between the average envelopes of the speakers

For both vowels the correlation coefficient between the unknown and the accused person was high (82,82% and 82,65% correspondingly) and higher than the coefficients with the other speakers. That means, that the identification was 100% correct in this case.

III. EVALUATION OF THE RESULTS

From all the tests the unknown speaker has been identified correctly as the accused person, among the other speakers of the closed set. Considering the major differences mentioned at the beginning, which existed between the compared voices and the fact, that the tests were performed by quite untrained persons, this was a surprising result. It means that the voice of the identified speaker has preserved its distinctive characteristics when speach was changed from spontaneous-conversational to reading, when it was recorded in different sessions under different emotional conditions of the speaker, even when it was distorted to a high degree from bad transmission and recording.

Comparing the three methods we have used, we come to the following conclusions:

a. Sorting tests with spectrograms not including the unknown voice had a relatively low error rate. In contrast the same tests with the unknown spectrograms included, had a poor success. This is due, up to a point, to the interchange of spectrograms between the unknown and the accused person. A high and distinctive interchange of spectrograms in a sorting test can be evaluated as an indication of an identity between the interchanged voices.

On the other hand, a higher error in sorting the spectrograms of the other known speakers has been also observed in this case. It seems that including in the test the unknown spectrograms, which contained major variations has confused the observers. Perhaps, sorting tests can not be used very successfully for the identification of a speaker, when the compared talks have differences similar to that accounted in the introduction.

b. The scores in matching tests were considerably better than that of the sorting tests. Probably it is easier for an observer to find out to which spectrogram a given spectrogram has more similarities, despite of the differences which may exist between the spectrograms of the same speaker.

c. Listening tests were less successfull than matching tests, but the difference in error between them is small. As mentioned before, listeners who gave wrong judgement associated the unknown voice only to a single of the other voices. The spectrograms of that voice however had very little similarity with the spectrograms of the unknown and the accused person and were not related to those spectrograms in the visual tests. Perhaps the visual inspection of spectrograms gives better results than listening, under the described circumstances, at least with untrained listeners and observers.

d. The correlation of the envelopes of vowel sections has given the best results. Probably usefull speaker characteristic information is contained in all the frequencies of the vowels. Perhaps the envelope of vowels is a characteristic feature of a speaker. This hypothesis is strainghtened by the high correlation coefficients we have found between the average vowel envelope of a speaker and envelopes of the same vowel and speaker taken by chance.

In a set of measurements we have performed with all the four phonetically distinctive vowels of the Greek language and 3 male speakers independantly from the case described here, we obtained also very encouraging results. Fig.4 shows the correlation coefficients between the average vowel section of each speaker and sections of the same and other speakers. The distinction between the coefficients of the same speaker and the other speakers is obvious in 87.5% of the examined vowels.

As sections of vowels can be taken automatically by real time frequency analysers and processed immediately by a computer, the computation of the envelope correlation coefficients could be done easily. So in the case that this method would be proven acceptable, its application in identification and verification of speakers, would have no practical problems.

SPEAKER	A		B		C	
$\bar{A}$	/a/	0.791520573	/a/	0.743232164	/a/	0.784858308
	/i/	0.936370099	/i/	0.269514313	/i/	0.797083228
	/e/	0.870254068	/e/	0.917134867	/e/	0.667176092
	/o/	0.884342966	/o/	0.963923506	/o/	0.704963334
$\bar{B}$	/a/	0.744408197	/a/	0.909419734	/a/	0.790089316
	/i/	0.508750162	/i/	0.889267861	/i/	0.512557099
	/e/	0.850363987	/e/	0.846319347	/e/	0.727105714
	/o/	0.924738409	/o/	0.931943251	/o/	0.712760741
$\bar{C}$	/a/	0.689875054	/a/	0.814715666	/a/	0.883394259
	/i/	0.777763414	/i/	0.625957296	/i/	0.811325143
	/e/	0.677985458	/e/	0.791020381	/e/	0.876703620
	/o/	0.811506112	/o/	0.774505999	/o/	0.904950555

Fig.4. Correlation coefficients between the average envelopes of each speaker and the envelopes of the same and other speakers.

IV. CONCLUSION

The identification of a speaker from his voice, which was recorded under strongly different conditions, has been achieved in a closed set of speakers by untrained persons. Classical methods as listening tests and visual inspection of spectrograms have been used, together with a new method which correlates the vowel envelopes of the speaker's voice. This last method promises to give acceptable results in the case of the identification or verification of talkers.

REFERENCES

1. Lass,J.N., "Contemporary issues in Experimental phonetics", Academic Press, New York, pp.295-326, 1976.
2. David,E.E. and Denes P.B., "Human Communication-A Unified View", McGraw-Hill Book Co., New York, pp.369-397, 1972.
3. Flanagan J.L., "Speech Analysis, Synthesis and Perception", Springer-Verlag, Berlin, pp.141-147, 1972.

PREDICTION AND VERIFICATION OF HYPOTHESES IN A SPEECH RECOGNITION SYSTEM

Silvano RIVOIRA and Piero TORASSO

C.E.N.S. - Istituto di Scienze dell'Informazione
University of Turin - Turin - Italy

ABSTRACT

This work describes the mechanism of prediction and verification of lexical hypotheses in a speech recognition system which accepts meaningful sentences spoken in the Italian language. The knowledge source at the syntactic-semantic level is modelled by a transition network grammar.

Word hypotheses can be emitted in a bottom-up way on the basis of the phonemic description of the input words or can be predicted in a top-down way by the transition network. A verification process assigns a score to new hypotheses (both emitted and predicted) by evaluating a distance measure between the phonemic description of the input word and the prototype description of the hypothesized word. An island-driven parsing strategy based on the prediction and verification mechanisms is introduced and briefly discussed.

INTRODUCTION

A central problem for speech recognition systems consists of efficiently exploiting information from different sources of knowledge, because of the inherently errorful nature of speech processing.

Top-down strategies may suffer combinatorial explosion of the number of hypotheses to be verified in task domains characterized by large vocabularies and weak syntactic and semantic constraints. On the other hand, bottom-up strategies are effective tools only

D. G. Lainiotis and N. S. Tzannes (eds.), Advances in Communications, 498–505.

when the low-level classification processes are so accurate to emit very few hypotheses besides of the correct one.

Combined stategies of bottom-up and top-down methods have been proposed in the literature /1,2/. According to the hypothesize-and-test paradigm, these strategies use cooperating knowledge sources communicating with each other through a global data structure.

This paper describes the application of the hypothesize-and-test paradigm to a recognition system which accepts meaningful sentences spoken in the Italian language /3/.
The control strategy of the system produces an interpretation of the input sentence by activating separate processes for the emission, verification and prediction of lexical hypotheses, after the classification at the acoustic-phonemic levels has been performed.

The words detected in the input sentence are represented by linguistic descriptions whose symbols are phonemic labels. Such phonemic descriptions are the result of a sequence of classification processes which firstly subdivide the input words into segments corresponding to broad classes of phonemes and then assign to each segment a set of weighted phonemic labels. The classification processes are controlled by the rules of appropriate grammars and are organized in a hierarchical structure /4/.

Since the linguistic description of an input word is not, in general, a unique string but a finite language, it may be represented by a weighted finite state automaton. Similarly the prototype description of each word in the lexicon is modelled by a finite state automaton in order to contain not only the correct phonemic transcriptions but also their most frequent distorted versions.

In the following the syntactic-semantic knowledge representation, the mechanisms of emission, verification and prediction of lexical hypotheses and their control strategies are described in some detail.

SYNTACTIC - SEMANTIC KNOWLEDGE REPRESENTATION

The protocol chosen in the preliminary experiments is a robot command language consisting of orders, descriptions and questions related to a geometric-block world /3/. The vocabulary size is one hundred words with average branching factor equal to ten.

Even if the protocol used in the experiments could be represented by a finite state automaton, our system models the syntactic semantic knowledge by means of a transition network grammar /5/.

The advantages of this model consist of the possibility of capturing most of the regularities in the language (representing them as subnetworks) and by saving parsing efforts since the result obtained in the parsing of an input substring by means of a subnetwork may be used in different invocations of the same subnetwork.

In our system non-terminal symbols and the related subnetworks represent syntactic-semantic constructions which may be present in different contexts, while the terminal symbols are the words in the lexicon. The categories introduced in the model refer to classes of words with similar syntactic-semantic constraints (like numbers, colours, blocks,...) according to the concept of pragmatic grammar introduced in the HWIM system /6/.

The integration of the syntactic and semantic rules into a unique representation allows a strong reduction of the parsing time since:

- no attempt is made in order to expand interpretations which are syntactically correct but semantically inconsistent
- the number of words which can be hypothesized in a given context is substantially reduced.

EMISSION, VERIFICATION AND PREDICTION OF LEXICAL HYPOTHESES

Word hypotheses emission is performed on the basis of a simplified description of the incoming words, obtained from the complete phonemic description by considering only the most robust phonemic labels like those related to fricative consonants and vowels. The set of all the simplified phonemic descriptions derived from the prototypes of the words in the lexicon is represented by means of a decision tree where each branch is labelled with a vowel or fricative symbol. Each leaf of the tree is connected with the set of lexical hypotheses related to the words in the lexicon whose simplified description corresponds to the path from the root to the leaf.

The set of lexical hypotheses emitted for an input word is obtained as the union of the hypotheses associated to all the leaves in the decision tree which can be reached with strings in the simplified description of the input word. The emission mechanism provides the possibility of quickly selecting the set of words in the lexicon with a strong resemblance to an input word, since only few hypotheses are, in general, emitted for each word detected in the input sentence. It is also possible that no hypothesis can be emitted for some largely distorted descriptions of input words.

Each hypothesis is then verified by evaluating the dissimilarity between the complete phonemic description of the input word and the prototype description of the hypothesized word . The dissimilarity criterion between symbolic descriptions is based on the modified weighted Levenshtein distance /7/ and allows one to assign different cost values to each type of edit operations (insertions, deletions, substitutions of symbols) and to take into account the symbols every operation applies to.

The lexical verifier is based on an algorithm /8/ adopting the branch and bound technique in order to efficiently evaluate the distance between the weighted finite state automaton which represents the phonemic description of an input word and the finite state automaton which represents the prototype description of the hypothesized word.

Any new bottom-up hypothesis is then inserted in an ordered list, containing all the lexical hypotheses related to the input word under consideration; the lists are ordered on the basis of the dissimilarity score evaluated by the lexical verification process and are used during the syntactic-semantic analysis phase in order to speed-up the recognition process.

The interpretation of an input sentence is controlled by the parsing strategy which uses the information contained in the Transition Network to predict the existence of a specific word in a particular context. The prediction of a word (i.e. the activation of a lexical hypothesis in a top-down way) is performed every time the parsing strategy tries to expand the most promising interpretation of the sentence; the lexical verifier is invoked only if the strategy controller realizes that the word predicted by the Transition Network is not included in the list of lexical hypotheses previously emitted for the input word under examination. These lists represent, therefore, the global data structure shared by the processes operating at the syntactic-semantic level and the processes operating at the phonemic level. The mechanism of word prediction allows the consideration of all the word hypotheses which are acceptable in a given context, including those which were not emitted in the bottom-up way because of distorsion in their phonemic representations.

The system tries to construct a syntactic-semantic interpretation even for sentences which exhibit distorted representations, according to the following criteria:

- the possibility of inserting one input word is examined by the strategy controller everytime it is possible to follow a CAT arc starting from the current state in the network. In this case the parser can advance on the input string without following any arc in the network;
- the possibility of loosing an input word is considered only when the context allows a unique semantic interpretation of the sentence. The deletion of a word is performed following a CAT arc in the transition network without advancing in the input sentence. An appropriate mark distinguishes the CAT arcs related to words which can be deleted without introducing semantic ambiguity;
- the possibility of misclassifying an input word is taken into account everytime a CAT C arc can be followed. Two cases are then to be considered:
  a) the list of hypotheses previously emitted for the input word under examination contains some words belonging to the category C. In this case the lexical hypothesis in C with the best score is selected as the word to be used in the sentence interpretation;
  b) no hypothesis referring to words in C has been previously emitted. In this case all the words in C are predicted and verified by means of the mechanism described above. The hypothesis with the best score (the minimum distance from the description of the input word) will be used to expand the current interpretation of the sentence.

PARSING STRATEGIES

The main task of a parsing strategy, in this system, is the activation of the processes that perform the prediction and the verification of lexical hypotheses for different portions of the input sentence.

A left-to-right parser which searches the path in the network with the best acoustic evidence according to the branch-and-bound technique has been developed by the same authors and described in /9/. This algorithm has the capability of recovering errors due to word misclassifications, missing and extra words, according to the criteria previously described. The search strategy is essentially model driven and it develops the most promising interpretations in parallel. A good parsing efficiency has been achieved by ordering the parsing configurations on the basis of the values of an heuristic function and by choosing the configuration for which this function is minimum as the most promising interpretation to be expanded.

The same prediction and verification mechanisms are used by a data driven parsing strategy which develops interpretations of the input sentence starting from the lexical hypotheses with good acoustic evidence. This strategy is based on the concept of island introduced by the BBN systems /6/ and it seems to be particularly effective when the initial portion of the sentence is strongly affected by misclassification errors.

In our system an island may correspond to a lexical hypothesis for an input word or to the best interpretation developed by a subnetwork for a given portion of the input sentence.
Each island is represented as the quadruple ISL = (N,LD,RD,W) ,
where N is the name of the subnetwork (or the lexical hypothesis),
LD and RD are, respectively, the left and right delimiters of the island over the input sentence,
W is the score associated to the best interpretation of the input words in the positions LD,LD+1,...,RD-1,RD according to N.

Since the system recognizes protocol languages which can be modelled by transition networks without recursions, it is possible to establish a hierarchy between subnetworks according to the following rules: the highest level (equal to zero) is assigned to the main network; the level i-1 is assigned to a subnetwork S if it can be invoked by a subnetwork S' of level i.

The parsing algorithm, schematically described in the following, uses the ISLOPEN list to store the islands to be expanded and the ISLCLOSE list to store the islands corresponding to already expanded interpretations.

step 1 - Put in the ISLOPEN list the input words with good acoustic evidence.

step 2 - If the ISLOPEN list is not empty go to the next step, otherwise reject the input sentence.

step 3 - Find the islands in the ISLOPEN list that are at the lowest hierarchical level and select the island ISL with the best score.

step 4 - If the island ISL is a complete covering of the input sentence and it corresponds to a path in the network at the highest hierarchical level, then accept the input sentence, otherwise go to the next step.

step 5 - Remove ISL from ISLOPEN and insert it in the ISLCLOSE list.

step 6 - Expand the left side of ISL (following in the reverse direction all the arcs leading to the initial state of the island ISL) until the initial state of the subnetwork invoking ISL is reached. Prune the paths with bad acoustic evidence.

step 7 - If no path survives return to step 2, otherwise go to the next step.

step 8 - Expand the right side of ISL (following all the arcs leaving from the final state of the island ISL) until all the final states of the subnetwork invoking ISL are reached. Prune the paths with bad acoustic evidence.

step 9 - If no path survives return to step 2, othewise go to the next step.

step 10 - Add in the ISLOPEN list all the islands that correspond to complete paths from the initial state to a final state of the subnetwork.

step 11 - Return to step 2.

It is woth noticing that the selection (at the 3-rd step) of the island at the lowest hierarchical level as the first element to be expanded makes it impossible to expand an island until all the islands at a lower hierarchical level have been considered for expansion.

In this way an island may grow up during the expansion phase (step 6 and step 8) joining together islands that were previously completed. In fact it is possible to use an island ISL'= (N',LD',RD',W') during the expansion of the current island ISL everytime a PUSH N' arc may be followed and the pointer to the input sentence is equal to LD' (during a right expansion) or to RD' (during a left expansion); if these conditions do not hold, the expansion is performed by means of the mechanisms of prediction and verification previously described.

Different partial interpretations may be developed for the same input substring, but only the best complete interpretation (corresponding to a path from the initial to a final state of the current subnetwork N") is retained. The ISLOPEN list is then updated (step 10) inserting the new island ISL"= (N",LD",RD",W") where LD" and RD" are the delimiters of the portion of input sentence corresponding to the interpretation with cost W".

Notice that different islands with the same delimiters always correspond to different subnetworks, while a given subnetwork may correspond to different portions of the input sentence.

REFERENCES

(1) Lesser,V.R., Fennel,R.D., Erman,L.D., Reddy,D.R.: Organization of the HEARSAY II Speech Understanding System. IEEE Trans. on Acoustics, Speech and Signal Processing, ASSP-23,1975,pp.11-23

(2) Lesser,V.R., Erman,L.D.: A Retrospective View of the HEARSAY II Architecture. Proc. Int. Joint Conf. on Artificial Intelligence, Cambridge MA,1977,pp.790-800.

(3) Rivoira,S., Torasso,P;: Syntax and Semantics in a Word Sequence Recognition System. Proc. IEEE Int. Conf on Acoustics, Speech and Signal Processing, Washington,1979,pp.586-590.

(4) Rivoira,S., Torasso,P.: An Isolated-Word Recognizer based on Grammar Controlled Classification Processes. Pattern Recognition,10,1978,pp.73-84.

(5) Woods,W.A.: Transition Network Grammar for Natural Language Analysis. Communications ACM,13,1970,pp.591-606.

(6) Woods,W.A., Wolf,J.: The HWIM Speech Understanding System. Proc. IEEE Int. Conf. on Acoustics, Speech and Signal Processing, Hatford,1977,pp.784-787.

(7) Tanaka,E., Fu,K.S.: Error Correcting Parsers for Formal Languages. TR-EE 76-7, Purdue University,1976.

(8) Rivoira,S., Torasso,P.: A Verifier of Lexical Hypotheses in a Speech Recognition System. Proc. IEEE Int. Conference on Cybernetics and Society,Denver,1979,pp.679-683.

(9) Rivoira,S., Torasso,P.: Lexical,Syntactic and Semantic Processing for a Speech Recognition System. CENS Internal Report, November 1979.

SYNTHESIS OF GREEK SPEECH WITH ELEMENTARY SPEECH SEGMENTS

P.Stathopoulou, G.Kokkinakis

Laboratory of Wire Communications, University of Patras, Patras, Greece.

A.G.Mian

Institute of Electrical and Electronics Engineering, University of Padua, Italy.

ABSTRACT

Computer-synthesis of Greek speech has been performed by combining elementary speech segments, extracted from the utterances of a Greek speaker and stored in a digital computer. A text put into the computer via a teletypewriter is segmented into units according to the stored speech segments. Speech is then synthesized by a suitable program, with segments called successively from the memory and forwarded to a D/A converter combined with a loudspeaker. A list of 134 elementary speech segments similar to that used for the synthesis of the Italian speech has been adopted as a basis for a first synthesis approach. The results achieved from this study are discussed and the followed method in order to improve the intelligibility and the naturalness of the synthetic speech is described.

I. INTRODUCTION

There are two basic approaches to the automatic synthesis of speech: Synthesis by rule and synthesis from prerecorded speech elements.

In synthesis by rule a dictionary containing a phonemic transcription of each word, or each segment of a word, is stored in a computer along with grammatical information, information about stress, etc.

The text to be converted into speech, is supplied in printed form to the computer. Each word or each segment is located in the

D. G. Lainiotis and N. S. Tzannes (eds.), Advances in Communications, 506–515.

dictionary and the corresponding information is passed to a program which incorporates linguistic and syntactic rules of the language.

The result is an output in discrete symbols representing the sequence of phonemes and their individual pitch and duration. From these symbols, control signals are produced which command a speech synthesizer.

In synthesis with prerecorded elements, speech segments extracted from human utterances or artificially generated, are stored in a computer. A text put into the computer is segmented according to the stored speech elements. Speech is then synthesized by a suitable program that concatenates the appropriate segments.

In speech synthesis by rule the problem consists in determining the linguistic rules for going from the written text to a specific utterance and in devising synthesizers which adequatly represent the speech producing mechanism. This is a very difficult task. Nevertheless, there is considerable work being done in this direction as reported in the literature (1).

In synthesis with segments, on the other hand the segments contain the appropriate dynamic properties of the vocal tract. But the segmentation can lead to discontinuites in pitch and formant frequencies at the junctions, so there arises the problem of how the segmentation may be done.

As applied to English this method has had little success, and therefore it has been abandoned. The reason lies perhaps in the fact that in English each word has its own relevant pitch and the number of allophones is very large. So the necessary inventory of prerecorded elements becomes very high. For example it has been estimated that 8000 segments containing parts of two phonemes are required to synthesize an idiolect of American speech (2). It is, however, interesting to note that in recent years there has been seen a renewed interest for synthesis of Engish by dyads or demisyllables (4,5).

In contrast to English, synthesis with segments, has been accomplished for the Italian language (3) with a considerably high degree of intelligibility (an average of 98% sentence intelligibility, with a minimum value of 96% and a maximum value of 99%).

An application of the synthetic speech has been introduced for several years at the University of Padua. Each student here has the possibility of being connected with a computer by dialing a code in pushbutton telephones installed in the University and get from the computer in synthetic speech a report with his notes and other information (6),(9),(10).

This encouraging result, the similarity between the Greek and Italian languages, and the simplicity of Greek pronunciation, led us to try to synthesize Greek in the same way. To quote Divry's Greek-English Dictionary:"Greek pronunciation is very simple. As a rule each letter has the same phonetic sound. The exceptions are few and follow fixed rules. There are no variations in the pronunçiation of the Greek letters as in English, no silent characters, no long or short vowel sounds. The accented syllable is marked in each word".

At the preliminary stage described in this paper, we have applied the procedure used for the Italian speech synthesis. Our first approach to speech synthesis in Greek will be improved in the future, according to the conclusions which were drawn from the results.

II. DEFINITION OF THE SPEECH SEGMENTS

In general, there is no a priori method of determining how the speech segments must be chosen in order to produce readily intelligible standardized speech. A suitable answer can only be obtained by experiments which involve listening tests.

We have established the following three requirements for the definition of the segments:

a. Segments must allow reproduction of speech with a given degree of intelligibility, for example with 99% intelligibility of sentences, as it is accepted in telephony.
b. The number of segments has to be small.
c. It must be possible to get the segments from a teletyped text with the usual Greek orthography.

According to the Italian approach we have adopted the notion of "Standard Speech Reproducing Units (SSRU)", that is "elements extracted from the utterances of a speaker, which have the same pitch and a limited number of stated lengths and consent the reconstruction of speech with a given degree of intelligibility"(3).

Certainly, the limitations on pitch and length of the standard units result in a loss of naturalness, but this affects little the intelligibility in Greek (as in Italian).

As SSRU's we have chosen 134 segments, roughly corresponding to one or two phonemes. Fig.1 shows letters and clusters of letters corresponding to the 134 segments with phonetic transcription and their frequency of occurence in modern written Greek.

group C		group CV				group CV or CVV							
Π=/p/	= 0.749	BA=/va/	≤ 0.1945	NA=/na/	≤ 0.8223	BI=/vi/	≤ 0.0433	KY=/ci/	= 0.0070	[illegible]	≤ 0.1206	ΨH=/psi/	= 0.0406
P=/r/	= 0.5141	BO=/vo/	≤ 0.0690	NO=/no/	≤ 0.5892	BH=/vi/	= 0.0164	KOI=/ci/	≤ 0.0804	ZI=/si/	≤ 0.5285	ΨY=/psi/	
Σ=/s/	= 0.9476	BΩ=/vo/	≤ 0.0048	NΩ=/no/	≤ 0.2318	BY=/vi/	= 0.0013	KOE=/ki/	≤ 0.0804	ΣH=/si/	= 0.3511	ΨOI=/psi/	–
T=/t/	= 0.2523	ΓA=/γa/	≤ 0.1635	ΠA=/pa/	≤ 0.4532	BOI=/vi/		ΛI=/li/	≤ 0.2777	ΣY=/si/	= 0.3863	ΨEI=/psi/	≤ 0.0080
Φ=/f/	= 0.0442	ΓO=/γo/	≤ 0.1555	ΠO=/po/	≤ 1.0768	BEI=/vi/	≤ 0.0161	ΛH=/li/	= 0.3208	ΣOI=/si/	≤ 0.0161	BE=/ve/	≤ 0.0421
M=/m/	= 0.1304	ΓΩ=/γo/	≤ 0.1311	ΠΩ=/po/	≤ 0.0706	ΓI=/ji/	≤ 0.2675	ΛY=/li/	= 0.0878	ΣEI=/si/	≤ 0.1528	BAI=/ve/	≤ 0.1945
N=/n/	= 0.1265	ΔA=/ða/	≤ 0.0791	PA=/ra/	≤ 0.6265	ΓH=/ji/	= 0.0943	ΛOI=/li/	≤ 0.0563	TI=/ti/	≤ 0.8188	ΓE=/je/	≤ 0.1578
Δ=/ð/	= 0.0774	ΔO=/ðo/	≤ 0.2024	PO=/ro/	≤ 0.8253	ΓY=/ji/	= 0.0010	ΛEI=/li/	≤ 0.1206	TH=/ti/	= 1.3509	ΓAI=/je/	≤ 0.1635
Λ=/l/	= 0.0054	ΔΩ=/ðo/	≤ 0.0406	PΩ=/ro/	≤ 0.2471	ΓOI=/ji/	≤ 0.0080	MI=/mi/	≤ 0.2887	TY=/ti/	= 0.0749	ΔE=/ðe/	≤ 0.2700
Γ=/γ/	= 0.2296	ZA=/za/	≤ 0.0070	ΣA=/sa/	≤ 0.1594	ΓEI=/ji/	≤ 0.0483	MH=/mi/	= 0.2004	TOI=/ti/	≤ 0.0724	ΔAI=/ðe/	≤ 0.0791
B=/v/	= 0.1071	ZO=/zo/	≤ 0.0575	ΣO=/so/	≤ 0.1681	ΔI=/ði/	≤ 0.6210	MY=/mi/	= 0.0184	TEI=/ti/	≤ 0.0161	ZE=/ze/	≤ 0.0582
Θ=/θ/	= 0.1465	ZΩ=/zo/	≤ 0.0231	ΣΩ=/so/	≤ 0.0890	ΔH=/ði/	= 0.1885	MOI=/mi/	≤ 0.0161	ΦI=/fi/	≤ 0.3500	ZAI=/ze/	≤ 0.0070
K=/k/	= 0.4547	ΘA=/θa/	≤ 0.1403	TA=/ta/	≤ 1.3819	ΔY=/ði/	= 0.1209	MEI=/mi/	≤ 0.0563	ΦH=/fi/	= 0.0358	ΘE=/θe/	≤ 0.3043

Fig.1. Letters and clusters of letters corresponding to the 134 segments.

The segmentation of the text is performed after each vowel, resulting in clusters of the form V, CV or CCV where CC are similar consonants. Combinations of the form CCV, where CC are different consonants, are separated after the first consonant giving the form C/CV except a few cases where an independent segment CCV is formed as in NTV, ΓΓV, ΓKV, MΠV etc. Rising diphthongs as IA, IO, II etc., are not separated, i.e., they form independent segments. Falling diphthongs as OI, and AI are distinguished from a punctuation mark, which is placed after the second vowel, i.e. OI:, AI:, and are formed from two separate vowel segments.
The above segmentation is based on the following assumptions:

a. The allophones of the Greek vowels are little influenced by the presence of following consonants or vowels. Thus, every word can be cut after each vowel. This corresponds to the syllabication of the majority of Greek words.
b. The allophones of consonants are little influenced by the presence of a following consonant.
c. The redundancy of Greek speech is so high that the lack of stress does not affect seriously the intelligibility.
d. The prosodic (or suprasegmental) features of the Greek speech, intonation, stress and tempo are not important intelligibility factors. Fig.2 shows the segmentation of a text according to the above definitions.

Fig.2 Example of segmentation of the sentence:"συνθετική ὁμιλία στήν Ἑλληνική γλῶσσα" (synthetic speech in Greek language).

III. SEGMENTATION TECHNIQUES

The acquisition of the above defined 134 segments from the utterances of a speaker is accomplished experimentally with the aid of a computer. A set of words carefully selected, in order to contain each segment at least twice, was uttered by a speaker with good pronunciation (a speaker of the Hellenic Broadcasting Service-ERT-was used). After a few trials so that the speaker could avoid stress and maintain a nearly constant pitch, the words were recorded on magnetic tape. Afterwards they were low-pass filtered at 4.3 kHz, digitized on 12 bits at the sampling frequency of 10 kHz and stored on the magnetic disc of a computer (7) (Fig.3). From the disc, word segments, of variable length,

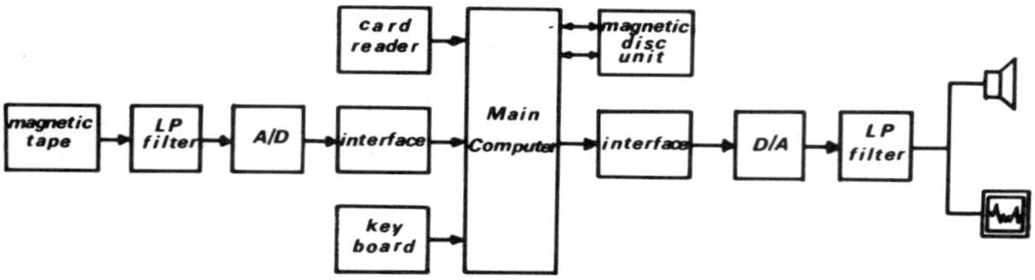

Fig.3. Technique of segmentation

could be addressed and extracted by a suitable program. The corresponding signal was presented in an oscilloscope and could also be heard simultaneously in a loudspeaker. Starting with a signal which roughly represented the segment C,V or CV to be examined, one could extract pieces of signals located before or after the initial signal. In this way the desired segment of a given vowel, consonant, cluster CV etc., was separated from the speaker's utterance; this signal was then standardized in length and level. In each segment we can distinguish three or four parts in general, depending on the type of the segment (Fig.4). In a vowel segment

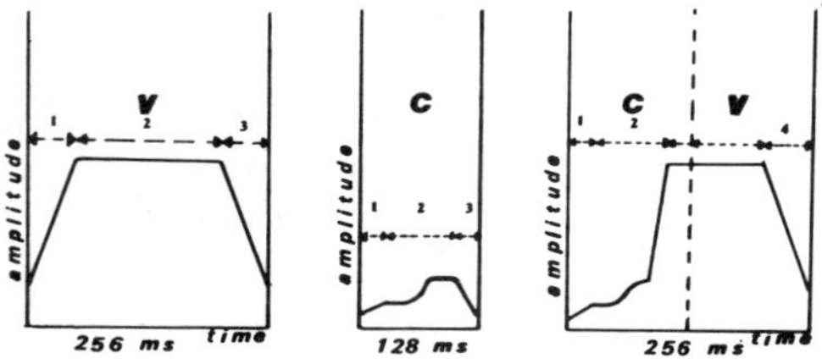

Fig.4. Idealized amplitude of vowels (V),consonants(C) and vowel-consonant clusters (CV).

for example we distinguish: the transition at the beginning, the quasi-steady state and the transition at the end. The duration of these parts relative to each other and the transitions from one to the other, define the intelligibility of the phonemes contained in the segment.

On the basis of the experiments carried out for the Italian language, two different standard lengths were adopted: 128 ms for each single consonant C and 256 ms for each single vowel V and each cluster CV or CCV. Within this length, the durations of the above mentioned three-segment-parts were adjusted in order to achieve the optimum possible intelligibility. The level of the segment was then adjusted stepwise to a nearly constant value for all segments. This procedure was accomplished for at least two segments per vowel, cluster CV etc., taken from different utterances of the speaker. Finally the acoustically best segment was adopted and stored on a second magnetic disc.

Vowels V were obtained from the first portion of clusters VCV or VVCV (VV not being a diphthong) found after pauses (Fig.5).

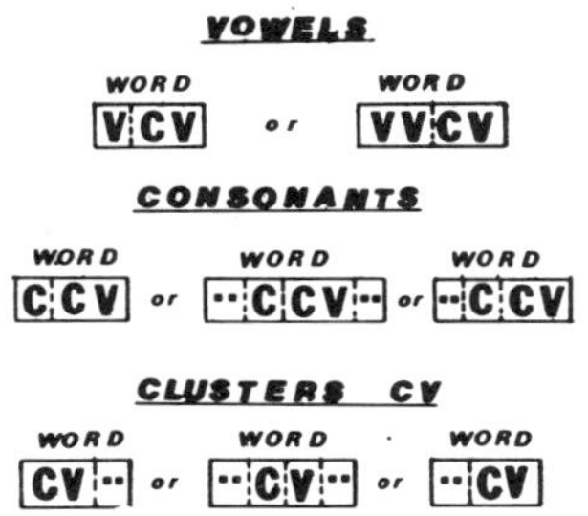

Fig.5. Extraction of vowels, consonants and CV-clusters

Consonants C were separated from the first part of clusters CCV found preferably after pauses, while clusters CV were separated from all parts of a word. Rising diphthongs, were treated as clusters CV. Fig.6 shows the amplitude -vs.- time of some segments yielded by this procedure. The procedure adopted differs from the one used for Italian in that CV clusters are formed in Italian from two separated segments of 128 ms length each. One segment is for each C to V transition and the second for the vowel V, which is used for all consonants. The result is that less memory space is required for storing the segments. With the present segmentation of Greek a memory of about 312 kWords is needed, while with the Italian segmentation the

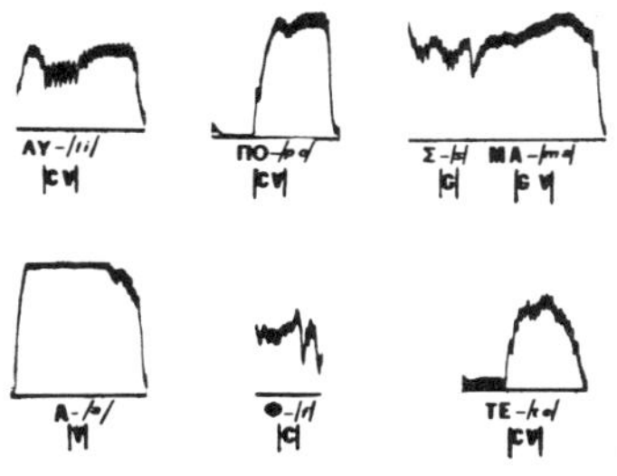

Fig.6. Amplitude vs.time of some segments

memory is only 180 kWords. On the other hand, independently from the considerable difficulty which the extraction of segments for vowels adaptable to all the phoneme consonants presents, the quality of combined segments is lower than that of unseparated segments (the synthesized speech sounds less monotonous). As in the whole problem of segmentation, a compromise is unavoidable here between memory size and synthetic speech quality. It should be noted, however, that recent significant reductions in memory cost, minimizes the effect of this factor. Actually, going from a wave form coding of speech segments to a parametric coding, as given by linear prediction techniques, the overall memory requirements of the segments used for Greek synthesis can be reduced to about 16 kWords.

IV. SYNTHESIS OF SPEECH FROM WRITTEN TEXT

The synthesis of speech from written text is accomplished in four stages shown in Fig.7. Fig.8 gives the block diagram of the

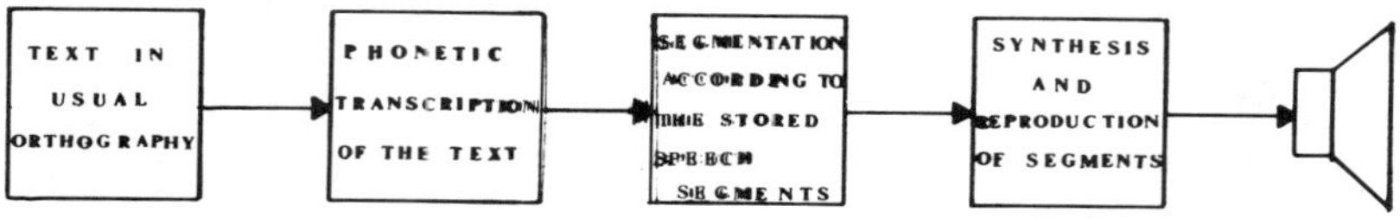

Fig.7. Procedure of speech synthesis from written text

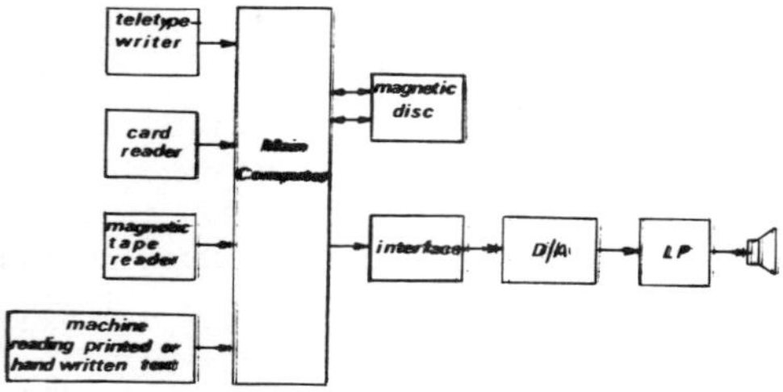

Fig.8. Block diagram of the system for the synthesis of speech from Greek text.

instrumentation employed for this purpose:

a. The text is transferredto the computer in usual orthography. This is done via a teletypewriter, a punched card reader, or a magnetic tape. But it could also be performed by any machine able to read printed or hand written text.
b. The text is transformed into its corresponding phonetic string.
c. The phonetic text is segmented according to the described rules in clusters, corresponding to the stored speech segments.
d. The segments representing the text are called successively from

the disk memory, forwarded to the D/A converter, low-pass filtered and finally sent to a loudspeaker.

For each of the above stages a suitable program has been written in FORTRAN language. Nevertheless, the work of the stages a. to c. can also be performed manually. The operator makes the segmentation and supplies the computer with the successive segments. He then initiates stage d. for the synthesis of the given text. This procedure is used during the experimental work of synthesis. Fig.9 shows spectrograms of words synthesized from the prerecorded elements.

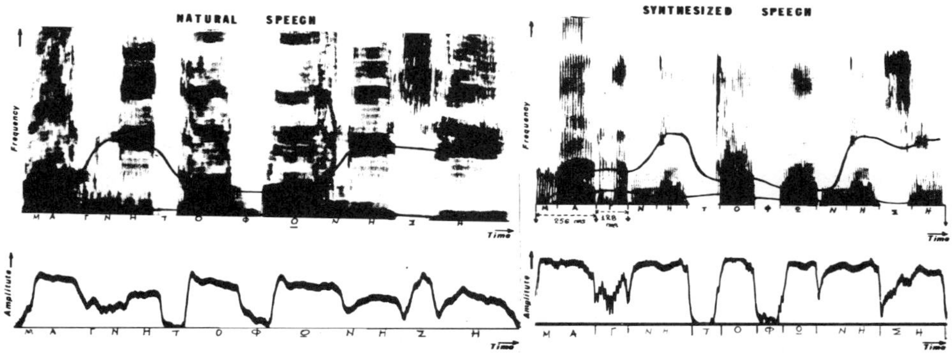

Fig.9. Spectrograms of natural and synthesized speech.

V. EVALUATION OF THE RESULTS

The synthetic speech produced is rather slow, monotonous, and lacks the naturalness of human speech. In order to evaluate its intelligibility a set of 40 synthetic words was submitted to 39 listeners. The test was repeated three times with the words presented in a different order each time. The maximum, minimum and average intelligibility achieved is presented in Fig.10. The results show clearly the influence of the listeners adaptation to the mechanical speech. An increase of 20% in average intelligibility has been noted between the first and third test. These results are encouraging taking into account that this synthetic speech was only the first attempt. Nevertheless, since synthetic speech must be highly intelligible to

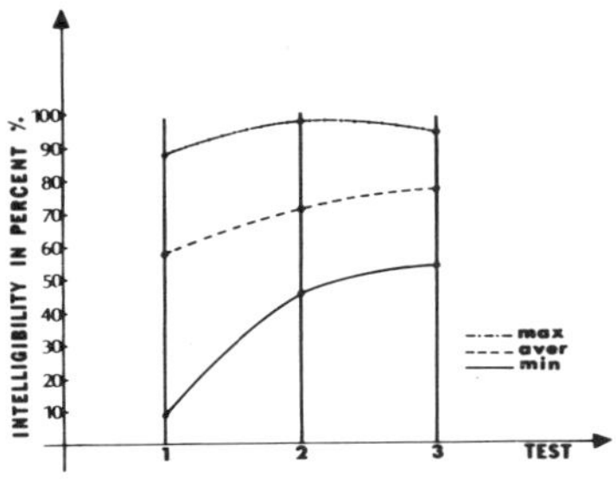

Fig.10. Max.,min. and average intelligibility of the synthetic speech.

people listening to it for the first time it is evident that the present speech has to be improved to reach an average sentence intelligibility of nearly 99%, which is also accèptable in telephony. Therefore, work will be done in the following areas:

a. Quality of segments:From the listening tests several parts of words with a poor intelligibility have been located. We will try to replace the corresponding segments with better ones. Pitch will also be controlled more carefully during recordings for the sake of less variation.
b. Length of segments:We will examine different lengths of segments in order to determine their influence on intelligibility and decide whether it is reasonable to introduce other lengths and/or more lengths than two.
c. Number of segments:We will examine the necessity of introducing new clusters together with the possibility of a further segmentation of existing CV clusters. This will lead perhaps to a new total number of segments.
d. Parametric coding:We will try to use the parametric representation of speech production according to linear prediction. This would drastically reduce the memory requirements and improve the quality of synthetic speech by reintroducing supresegmentals, as experienced for Italian (8).

VI. CONCLUSIONS

Computer speech synthesis from a written Greek text has been performed with 134 segments of natural speech, having a length of 128 or 256 ms. The method applied is the one which has been used for the synthesis of the Italian language. The preliminary results are encouraging and further work in improving the speech is being continued.

ACKNOWLEDGMENTS

We are much indebted to the Institute of Electrical and Electronics Engineering, University of Padua, Italy, where a major part of the experimental work described above was done. We acknowledge Prof. G.B.Debiasi and Ing. C.Offelli of the Institute of Electrical and Electronics Engineering, and Prof. A.Mioni of the Institute of Phonetics and Glottology, University of Padua, for the many useful discussions and advice given during this work.

REFERENCES

1. Flanagan J.L., "Computers that talk and listen: man-machine communication by voice", Proc. IEEE, 64, pp.405-415, 1976.
2. Harris C.M., "A study of the building blocks in speech"

J. Acoust. Soc. Am.25, 962-969, 1953.
3. Francini G., Debiasi G.B., Spinabelli R. "Study of a system of minimal speech reproducing units for Italian", J.A.S.A., 43, pp.1282-1286, 1968.
4. Olive J.P., "Rule synthesis of speech from dyadic units", Proc. 1977 IEEE Int. Conference Acoust., Speech, Signal Processing, pp.568-570, Hartford, May 9-11, 1977.
5. Lorius J.B., Macchi M.J., Fujimura O., "A demisyllable inventory for speech synthesis", J.A.S.A., 65, Suppl.1, S130, 1979.
6. Debiasi G.B., De Poli G., Mian G.A., Mildonian G., Offelli C., "Italian speech synthesis from unrestricted text for an automatic answerback system", Proc. 8th Inf. Congress on Acoustics, London, 1, 296, July 1974.
7. Debiasi G.B., "Sintesi elettronica della lingua Italiana: Premesse attuazione prospettive", Alta Frequenza, Febraio 1971.
8. Mian A., Morgantini F., Offelli C., "An application of the linear prediction technique to efficient coding of speech segments", Proc. 1976 IEEE Int. Conf. Acoustic, Speech, Signal Processing, pp.722, Philadelphia, April 12-14, 1976.
9. Mian G.A., Mildonian G., Offelli C., "Un programma per la Sintesi elettronica del parlato mediante unita normalizzante", Alta Frequenza, Vol. XLII, N.3, 1973.
10. Baccolini G., Offelli C., "Interfaccia e procedure per la gertione remota di un elaboratore tramite telefono a tastiera" Alta Frequenza, Vol.XLVI, N.7, 1977.
11. Tomic T., Trainito G., "Synthesis by speech segments of Serbocroatian", Speech Symposium of Szeged, 20-23 Aug. 1971.
12. Peterson G., William S.Y., "Segmentation Techniques in Speech Synthesis", J.Acoust. Soc. Amer. 30, 739-742, 1958.

AUTHOR INDEX

SUBJECT INDEX